Environmental
Sociology

Environmental Sociology

From Analysis to Action

Edited by Leslie King
and Deborah McCarthy

ROWMAN & LITTLEFIELD PUBLISHERS, INC.
Lanham • Boulder • New York • Toronto • Oxford

ROWMAN & LITTLEFIELD PUBLISHERS, INC.

Published in the United States of America
by Rowman & Littlefield Publishers, Inc.
A wholly owned subsidiary of The Rowman & Littlefield Publishing Group, Inc.
4501 Forbes Boulevard, Suite 200, Lanham, Maryland 20706
www.rowmanlittlefield.com

PO Box 317
Oxford
OX2 9RU, UK

British Library Cataloguing in Publication Information Available

Library of Congress Cataloging-in-Publication Data

Environmental sociology : from analysis to action / edited by Leslie King
and Deborah McCarthy.
p . cm.
Includes bibliographical references and index.
ISBN 0-7425-3507-X (cloth : alk. paper) —
ISBN 0-7425-3508-8 (pbk. : alk. paper)
1. Environmentalism—Social aspects. 2. Environmental justice.
3. Environmentalism—North America. I. King, Leslie, 1959-
II. McCarthy, Deborah, 1966- GE195.E588 2005
333.72—dc22   2004025145

Printed in the United States of America

♾™ The paper used in this publication meets the minimum requirements of
American National Standard for Information Sciences—Permanence of Paper
for Printed Library Materials, ANSI/NISO Z39.48-1992.

Contents

PART I
Politics and Economy

PART V
Globalization

PART VI
Media and Popular Culture

PART VII
Science and Health

PART VIII
Social Movements

PART IX
Thinking About Change and Working for Change

Preface

We both strongly believe that humans have come to a turning point in terms of our destruction of ecological resources and endangerment of human health. A daily look at the major newspapers points, without fail, to worsening environmental problems (and sometimes, but not often enough, a hopeful solution). Humans created these problems, and we have the power to resolve them. Naturally, the longer we wait, the more devastating the problems will become; and the more we ignore the sociological dimensions of environmental decline, the more our proposed solutions will fail.

Out of our concern for and dedication to bringing about a more sustainable future, we have both worked hard to develop environmental sociology courses that not only educate students about environmental issues but also show them their potential role as facilitators of well-informed change. This reader results in large part from our commitment to the idea that sociology can be a starting point for social change and we have sought to include in it work that reflects our vision.

We actively looked for readings that interest, motivate, and make sense to an undergraduate audience. We shortened most of the pieces (deletions are indicated in the text by ellipses, except at the beginnings of chapters) in the interest of including a fairly large number of selections on a variety of topics. Choosing which selections to include has been exciting and thought-provoking but not without a few dilemmas. We do not include works published before about 1990. That choice sprang from our observation that undergraduate students tend to be more interested in current work. In addition, several other good edited volumes and readers include the "classics," so we did not see a need to reinvent the wheel. One of our most difficult decisions was to leave out many "big name" researchers who have profoundly influenced the field. Some of this work represents a dialogue with a long and intertwined body of thought and research. Understanding such a dialogue would require reading the lineage of research leading up to it. In addition, much of the theoretical work in environmental sociology (as in most of our subdisciplines) engages important, but very specialized, issues.

As a way of providing students with a beginning understanding of this lineage, our introductory chapter presents a brief overview of the field with extensive footnotes for students wishing to explore specific theoretical perspectives in greater depth. This introductory chapter is balanced by the works in the reader itself—recent articles and book chapters that illustrate a wide variety of ways that sociologists might address environmental questions.

We also wanted this reader to be accessible to a maximum number of instructors, whether or not they are specialists in environmental sociology. Most sociologists and social scientists we know speak the language of inequalities, political economy, and social constructionism; we tend not to be as fluent in the biological and mechanical details of energy production, watershed management, or climate change. Thus, we organized our reader not by environmental issue but by sociological perspective. The reader frames the issues in terms of sociological concepts (e.g., race, political economy, social movements, etc.) and seeks to show students how sociologists might go about examining environment-related issues. We do want to emphasize, though, that in developing the reader's conceptual blocks, we were careful to cover a broad range of topics—from coal mining to lead poisoning to food production.

Ultimately, we think the most important feature of the reader is not the topics we chose or how we decided to organize the different pieces into categories; rather, it is that woven throughout the collection in the choice of material is the connection between power and environmental decision making. All of the pieces address either systems of power relations (e.g., the relations between the globalization of Western culture and the impact of the automobile infrastructure on lower-income countries) or individual levels of power (e.g., empowering consumers to buy "green"). We believe that good environmental decision making must incorporate sociological perspectives and we hope that increasing numbers of activists, policymakers, and academics will benefit from these frameworks.

We would like to thank Brian Romer, our editor at Rowman & Littlefield. We also thank Terri Boddorf for all her work obtaining permission to use the pieces included in this reader. Thanks to Krista Harper, Daniel Faber, Angela Halfacer, Tracy Burkett, and several anonymous reviewers for comments and suggestions on various parts of the text. Our research assistant, Elisabeth Gish, worked tirelessly with us on this project—thank you, Lizzie!

Introduction
Environmental Problems Require Social Solutions
Deborah McCarthy and Leslie King

What Is Environmental Sociology?

The short answer involves exploring two ideas: sociology and environment. Sociology is, above all else, a way of viewing and understanding the social world. It allows us to better understand social organization, inequalities, and all sorts of human interaction. Sociology is a multifaceted discipline that theorists and practitioners use in diverse ways, and, along with many others (e.g., Feagin and Vera 2001), we think it has the potential to help us create a more just world. *Environment*, like *sociology*, can be an elusive term. Is the environment somewhere outside, "in nature," untouched by humans? Or are humans part of the environment? Does it include places where you live and work and what you eat and breathe? Or is it more remote: the rolling valleys of the Blue Ridge Mountains, the pristine waters of Lake Tahoe, the lush rainforests of Brazil?

For environmental sociologists the answer is that the "environment" encompasses the most remote regions of the earth as well as all the bits and pieces of our daily lives—from the cleaners we use to wash our carpets to the air we breathe on our way to work each day. We sociologists assume, first and foremost, that humans are part of the environment and that the environment and society can only be fully understood in relation to each other. We build on this understanding to point to fissures that are developing in the relationship between humans and nature. These are problems that humans have both con-tributed to and are feverishly attempting to solve. Our lack of understanding about the human/nature relationship has led to some of our worst environ-mental problems—climate change, ozone depletion, and so on—and has lim-ited our ability to solve those problems.

In fact, some of our attempts to solve environmental problems have actu-ally made them worse. The Green Revolution is an example. The goal of the Green Revolution was to increase world food yields through the transfer of Western agricultural techniques, knowledge, and equipment to lower-income countries. This shift in agricultural methods massively reorganized agricultural production on a global scale. Although the global rates of food production increased, the revolution, with its focus on export crop production, for-profit rather than sustainable agriculture, mechanization, and heavy pesticide and fer-tilizer use, contributed to the destabilization of social, political, and ecological systems in many regions of the world (Dowie 2001: 106–40). For example, farm-ers must now engage in a money economy in order to pay for pesticides and herbicides, and as a result, many small-scale, less affluent farmers have lost their land (Bell 1998). What is more, as Peter Rosset and colleagues (2000) point out, an increase in food production does not necessarily lead to a decrease in hunger. In their words, "narrowly focusing on increasing production—as the Green Revolution does—cannot alleviate hunger because it fails to alter the tightly con-centrated distribution of economic power, especially access to land and pur-chasing power. . . . In a nutshell—if the poor don't have the money to buy food, increased production is not going to help them." With a better understanding of how humans and nature interact and a willingness to learn from past mis-takes and misunderstandings, we can prevent such problems from occurring again and build a more socially and environmentally sustainable future. This collection of readings represents a broad sample of work by many writers and researchers who are attempting to do just that.

Environmental Problems Are Social Problems

Sociologists, by focusing their research on questions of inequality, culture, power and politics, the relationship between government and economy, and other soci-etal issues, bring a perspective to environmental problem solving that is quite different from that of most natural and physical scientists. Take the following examples: the catastrophe in Bhopal, India in 1984, the 1989 *Exxon Valdez* oil spill in Alaska, the massive contamination of a community in Niagara Falls, New York in the late 1970s—usually referred to as Love Canal—the ongoing air pol-lution problems in the United States, and global climate change. Why are those not uniquely "natural" science issues? The answer is that these "disasters" are not out of our control; they are not "accidents" as we sometimes hear them referred to in the news media. In each of these cases (briefly summarized below), and many other cases too numerous to mention, it was social organization—a series of identifiable managerial steps, collection of beliefs, set of regulations, or other social structures—that led to the environmental problems.

- Bhopal, India: In 1984, the Union Carbide pesticide plant in Bhopal leaked forty tons of lethal methyl isocyanate (MIC) into the low-income communities of Bhopal. Varying estimates indicate that between 6,600 and 16,000 people died immediately and at least 70,000 more were permanently injured. This was far from a mere "technological accident." As many social science and journalistic studies attest, the mismanagement of the Bhopal plant almost guaranteed a disaster. In short, Union Carbide, in an attempt to cut costs, took advantage of a repressed and desperate workforce to construct and manage the plant in an unsafe manner.[1]
- *Exxon Valdez*: The story of the *Exxon Valdez* disaster along Alaska's shoreline is also, sadly, the story of a preventable "accident." Exxon, as well as other members of the oil industry, waged a successful public relations campaign to convince Congress that requiring tankers to utilize double hulls would be too expensive and therefore dangerous to the industry. Exxon also spent the 1960s and 1970s working to delay and avoid Alaska's attempts to raise taxes on tankers entering the Prince William Sound that didn't comply with the state's regulations. The groundwork for the largest ever crude oil disaster in U.S. history had been successfully laid; in 1989, the *Exxon Valdez* tanker got stuck on Bligh Reef and spewed over 38,000 tons of oil into the Prince William Sound (BBC News 2002), contaminating approximately 700 miles of coastline and a national forest, four national wildlife refuges, three national parks, five state parks, four state critical habitat areas, and a state game sanctuary. In just one example of the impact of the spill on wildlife, it killed an estimated 3,500 to 5,500 sea otters and 300,000 to 675,000 birds (Greenpeace 1999). The spill also nearly devastated the local fishing and seafood packaging industries, not to mention making a serious impact on the national insurance industry.[2]

The *Exxon Valdez* and Bhopal events represent crises that occurred in a flash and left immediately visible human and ecological tragedies in their wakes. Environmental disasters also occur in slow motion and inflict damages that are harder to detect but no less severe. This is the kind of disaster that occurred at Love Canal in Niagara, New York, and it is the kind of disaster that we all are watching as air pollution and climate warming continue to worsen. Like Bhopal and the *Exxon Valdez*, these "slow motion" environmental problems also have their roots in human decision making and social structure.

- Love Canal: In the case of Love Canal, tragedy was created by corporate and governmental decisions that were based on maximizing profits rather than social welfare. The first chapters of the Love Canal horror were written by a powerful corporation that attempted to reduce costs by selling its contaminated land to a local government for use as housing and then, when serious problems were discovered years later, acted to deny liability and subvert the regulatory functioning of local and federal government. In 1953, Love Canal, which had operated as a chemical waste dump since the 1920s, was covered and sold by the Hooker Chemical Company to the city of Niagara for one dollar. Subsequently, homes and a school were built at the site. By the 1970s, the residents began noticing the frightening consequences of living on a former toxic waste dump, including: the emergence of corroding waste disposal drums in residents' backyards after a heavy rain; the death of foliage, birds, and pets; the appearance of strange chemical smells which were often concentrated in residents' basements; the appearance of burns and rashes on children's hands and faces; and the increased rate of miscarriages, birth defects, cancers, and central nervous system disorders. In 1980, after much resistance on the part of local government and even the Environmental Protection Agency, massive activism on the part of community members led to a $15 million federal fund that was used to evacuate almost 900 families in the contaminated area.[3]
- Air pollution: The ongoing danger presented by a host of air pollution problems in the United States has roots in many social processes, one being the inability of governmental policies and laws to regulate industry amid the rise of neoliberal style economics, with its emphasis on deregulation and corporate rights, and the increasing monetary and political power of corporations. The 1970 Clean Air Act (and its 1977 and 1990 amendments), sets standards for air pollution levels, regulates emissions from stationary sources, approves state implementation plans for achieving federal standards, and sets emissions standards for motor vehicles (Rosenbaum 2002). This behemoth-size package of regulations, however, has failed to address many of our known air pollution problems. For example, two kinds of air pollution—ground-level ozone and fine particulates, which are both created by cars and trucks—continue to persist at dangerous levels in the United States. Ground-level ozone, a component of smog, exacerbates many respiratory problems including asthma. Health standards for ozone are exceeded at 95 percent of monitoring sites in the United States. Fine particulates also continue to pose serious health problems and are estimated to kill twice as many people as automobile accidents each year (REHN 2001).
- Global climate change: Climate warming is caused by the increasing amounts of greenhouse gases, produced in large part by the burning of

fossil fuels. Greenhouse gases trap heat, and are predicted to warm the earth by 1 to 11 degrees Fahrenheit by the year 2100. Most scientists believe climate change will bring about rising sea levels, increased flooding, increased drought, extremely dangerous temperature increases in some regions of the world, and more severe weather (IPCC 2001; EPA 2004). Climate warming continues to worsen, in part, because of national and international political practices and regulatory structures that place the interests of corporations over the interests of citizens. For instance, though the United States produces approximately one-third of all global carbon dioxide emissions and though polls show repeatedly that the majority of the U.S. public supports a global climate treaty (Sussman 2001; Cushman 1997), in 2001 President George W. Bush nevertheless rejected the Kyoto Climate Treaty.[4] In addition, national governmental policies and many international treaties, such as the North American Free Trade Agreement (NAFTA), exacerbate the release of fossil fuels into the atmosphere by subsidizing road construction, rather than public transportation, and by encouraging long distance trade in which businesses produce goods in one place only to transport them across the globe for consumption in another.

Some sociologists follow the crumb trail of social facts left behind by Love Canal, climate warming, and other ecological stories, in order to envision for the future a more just and sustainable environment. Why should we assume that human ingenuity begins and ends with the invention of a deadly, yet extraordinarily powerful pesticide, such as Bhopal's methyl isocyanate? We have put our minds together to invent automobiles (which contribute to sprawl, asthma, and greenhouse gas production), polyvinyl chlorides (which are carcinogenic), and nuclear energy (which has led to the nuclear waste problem). Can we not put the same creative optimism and energy into preventing global climate change, regulating polyvinyl chlorides and other carcinogens, and developing safe forms of renewable energy?

The answer is "Yes, but . . . ." On the one hand, the twentieth century brought us the passage of several environment-related laws,[5] as well as the explosion of environmental activism and legislation that marked the 1970s as the "environmental decade" in the United States. This one decade alone, for example, is witness to the successful passage of the 1970 National Environmental Policy Act (NEPA), the 1970 Clean Air Act, the 1972 Water Pollution Control Act (Clean Water Act), the 1976 Toxic Substances Control Act, and the 1977 Surface Mining Control and Reclamation Act, to name a few.[6] There is little doubt that ecological problems would be much worse absent these environmental laws and the current system of regulation. On the other hand, it is increasingly apparent, as noted earlier, that many of our environmental problems have worsened and several new environmental problems have appeared since the "environmental decade."[7] A good deal of our failure can be linked to a lack of understanding of what seemingly separate issues—like the Bhopal industrial leak and global climate change—have in common. The linkages among environmental abuse, poverty, inequality, racism, lack of democracy, and the increasing concentration of power within corporations form a complex social nexus that allows environmental degradation to continue and that prevents meaningful policy from being enacted. In almost all parts of the world, environmental laws have been poorly enforced, and when they are enforced, it is usually after the damage has already been done. In other words they focus on controlling pollutants/toxins after they have been released rather than preventing them before they pose a threat.[8]

Environmental sociologists contend that environmental problems are inextricably linked to societal issues (such as inequality, democracy, and economics). Endocrine-disrupting hormones, bioengineered foods, ocean dumping, deforestation, asthma, and so on, are each interwoven with economics, politics, culture, television, religious worldviews, advertising, philosophy and a whole complex tapestry of societal institutions, beliefs, and practices. Environmental sociology teaches us that a pesticide, such as Bhopal's methyl isocyanate, is not inherently "bad." Rather, the social organization of the pesticide industry is problematic. Blaming methyl isocyanate for the death toll in Bhopal's factory leak is like cursing the chair after you stub your toe on it. Humans invent pesticides, and humans decide how to manage those chemicals once they have been produced. Who decides whether to ban or regulate a toxin? How do we decide this? Do some citizens have more say than others? How do the press, corporate advertising, and other forms of media shape our understanding of a dangerous chemical or pollutant? These are just a whisper of the chorus of questions sociologists are asking about environmental problems and their solutions.

While science and technology can help us to invent "safer" pesticides, they cannot tell us what is safe. Is it safe enough that a new pesticide causes 1 in 100,000 of those exposed to develop a form of cancer? What about 1 in 1,000,000? What do you do when the scientific studies contradict each other? Who decides which ones represent the "facts," and how do they decide? Or, finally, as a colleague recently commented, "We scientists might be able to tell you how much of a toxic chemical you can dump in the ocean without killing all the fish; but the important question—and one that the natural sciences can't answer—is 'Why are we dumping that chemical in the first place?'"

Sociology can also help us see that environmental concerns are not merely about individuals. Some people disregard environment-related problems, explaining that "life itself is a risky business and the issue is ultimately about choice and free will." Along this line, the argument is that just as we choose to engage in any number of risky activities (like downhill skiing without a helmet or eating deep-fried Twinkies), we also choose to risk the increased cancer rates that are associated with dry-cleaning solvents in order to have crisp, well-hung suits and dress shirts. A sociologist, however, would begin by pointing out that not everyone is involved in that choice and not every community experiences the same level of exposure to that toxin. As Steingraber (reading 20) argues, the person who lives downstream from a solvent factory may not be the same person who chooses the convenience (and the risk) of dry-cleaned clothing.[9]

A Brief History of Environmental Sociology

In this section, we provide a broad and, by necessity, partial overview of the emerging subdiscipline of environmental sociology. Let's start at the beginning: Auguste Comte first coined the term *sociology* in 1838. However, it wasn't until the late 1960s and the 1970s (the environmental decade) that a significant number of sociologists began studying the impacts of natural processes on humans and its reverse—the social practices that organize our development and use of the products and technologies that impact ecological *and* human communities. Certainly many people in the early part of the twentieth century were interested in a wide range of environmental issues—from the conservation of vast expanses of wilderness to the improvement of urban spaces (see Taylor 1997). So why the long wait for an environment-focused sociology?

Part of the answer lies in the efforts of sociologists to establish the discipline of sociology as separate from other areas of study, especially the natural sciences (Hannigan 1995). Sociology provides an important counterpoint to the natural sciences by showing how social interaction, institutions, and beliefs shape human behavior—not just genetics, physiology, and the natural environment. In addition, the sociological perspective has been a crucial tool for dismantling attempts to use the natural sciences to justify ethnocentrism, racism, sexism, and homophobism. Sociologists have traditionally been reluctant, therefore, to venture outside the study of how various social processes (e.g., politics, culture, and economy) interact to look at human/nature interactions.[10] Writing about this trend in the discipline, Riley Dunlap and William Catton argued in the 1970s that sociologists should claim the study of the environment and not leave the "natural" world to natural and physical scientists (Catton and Dunlap 1978; Dunlap 1997). Catton and Dunlap thought environmental sociology ought to examine how humans alter their environments and also how they are affected by their environments. They developed a "new ecological paradigm," which represented an initial attempt to explore society-environment relations.

During the first decade of environmental sociology, researchers focused primarily on the same issues that the emerging environmental movement highlighted, including air and water pollution, solid and hazardous waste dumping, litter, urban decay, the preservation of wild areas and wildlife, and fossil fuel dependence. These problems were easy to measure and see: think of polluted rivers catching on fire, visible and smelly urban smog, ocean dumping of solid and hazardous wastes, and the appearance of refuse along the side of the road. Most early sociological studies focused on people's attitudes toward problems and the impacts of those problems on demographic trends (e.g., trends in health and mortality).[11]

As public and academic awareness of environmental issues has intensified, so has our understanding of the complexity of environmental problems. Potential and currently existing hazards that are socially, politically, and technologically complex, difficult to detect, potentially catastrophic, sometimes long range in impact, and attributable to multiple causes (e.g., environmental racism, acid rain, rainforest destruction, ozone depletion, loss of biodiversity, technological accidents, and climate change)[12] began to attract the attention of the nascent environmental sociology community in the 1980s.[13] Environmental sociologists of the late twentieth and early twenty-first centuries are studying a broad range of issues—from environmental racism[14] to strip mining[15] to lead poisoning.[16]

Social scientists have developed a number of specialist lenses to explore the increasingly complex relationships between environments and societies. Allan Schnaiberg (1980), for example, developed the concept of the "treadmill of production," which emphasizes the tendency of capitalist production to constantly seek to expand. According to Schnaiberg and his colleagues, this emphasis on growth leads to increased resource consumption as well as the increased generation of wastes and pollutants (both from the by-products of production and from consumption). Thus, according to this perspective, capitalist production, by its very nature, is at odds with efforts to clean up or improve the environment.

James O'Connor, a key figure in the development of the ecosocialist perspective, also argues that environmental devastation and resource exhaustion are inevitable consequences of capitalist accumulation.[17] Other conceptualizations of the relationship between capitalism and the environment can be found in the work of Marxist-oriented scholars such as Juan Martinez Alier (1997), Paul Burkett (1998), Daniel Faber (1993), John Bellamy Foster (2002), Barbara Rose Johnston (2003), Joel Kovel (2003), Ariel Salleh (1995), and many others.[18] For instance, Faber (1993) traces environmental destruction in Central America first to Spanish colonialism and, more recently, American imperialism. Faber documents the restructuring, in recent decades, of Central America's agricultural system into export-based and pesticide-reliant crop production at the expense of the health of Central America's people and land. The environmental devastation became especially profound in Nicaragua during the Nicaraguan revolution as the U.S.-backed contras waged war not only on the Sandinista revolutionaries but also on nature itself. The Sandinistas had brought about significant environmental improvements during the early years of the revolution, and so parks and environmental projects became targets for the contras.

Social ecologists, like ecosocialists, also draw on critiques of capitalism to direct attention to the intersection of ecodestruction with issues of class, race, and gender. Social ecologists, among them Murray Bookchin, believe that environmental destruction and human oppression are ultimately rooted in social hierarchy and that capitalist accumulation processes represent only one problematic manifestation of this hierarchy. Social ecologists, therefore, critique capitalist accumulation indirectly by way of their explicit critiques of hierarchy, including all types of state-, national-, and international-level governance. The only viable solution, according to social ecologists, is the creation of a decentralized society in which citizens engage in direct democracy at the local and city levels and where production and industrial processes would be tailored to the resources of specific regions (Bookchin 1990).

A number of scholars are less critical of existing social and economic structures. Arthur P. Mol and others (e.g., Mol 1997; Spaargarten and Mol 1992) have contributed to the development of a theoretical perspective known as *ecological modernization*, which calls our attention to the ways in which environmental degradation may be reduced or even reversed within our current system of institutions. Theorists working within this framework reject the notion that there is a fundamental opposition between the environment and the capitalist system of production and consumption. They believe that our institutions are transforming themselves through the use of increasingly sophisticated technologies and that production processes in the future will have fewer negative environmental consequences.

Yet another important set of theoretical perspectives to emerge in the realm of environmental sociology addresses the intersections of science and risk analysis. Ulrich Beck (1999), for example, has written that people in modern times feel increasingly at risk, due in large part to environmental degradation. Beck developed the concept of the *risk society*, which Michael Bell describes as "a society in which the central political conflicts are not class struggles over the distribution of money and resources but instead non-class-based struggles over the distribution of technological risk" (Bell 1998: 193). According to Beck (1999: 72), we are now at "a phase of development in modern society in which the social, political, ecological and individual risks created by the momentum of innovation increasingly elude the control and protective institutions of industrial society." Unlike authors such as Mol, who see science and technology as potentially improving environmental conditions, Beck examines how risks, and especially the social stresses associated with our perceptions of risks, are fostering a deterioration in our quality of life. In Beck's words, the *risk society* concept "describes a stage of modernity in which the hazards produced in the growth of industrial society become predominant" (74).

Some environmental sociologists pay particular attention to the health consequences of environmental degradation. Phil Brown and his colleagues (see, e.g., Brown and Mikkelson 1990) document and promote the idea of "popular epidemiology"—a process whereby nonscientist citizens become active producers and users of scientific data. Brown and Mikkelson's book *No Safe Place* describes how residents of Woburn, Massachusetts, identified and sought remediation for a cancer cluster caused by toxic chemicals in the water supply of certain parts of the town. Popular epidemiology requires more than the involvement of citizens in epidemiological decision making. In the words of Brown and Mikkelson, it "goes further in emphasizing social structural factors as part of the causative chain of disease and in involving social movements, in utilizing political and judicial remedies, and in challenging basic assumptions of traditional epidemiology, risk assessment, and public health regulation" (126).

Theoretical perspectives such as the treadmill of production, ecosocialism, social ecology, ecological modernization, the risk society, and popular epidemiology have been developed specifically to help us better understand the human/environment nexus. Many researchers, including those summarized here, also draw on more traditional sociological perspectives to study environment-related issues. For example, working within the tradition of sociological research that has long been interested in the creation and perpetuation of inequalities, some recent work applies theories of gender, race, and class to environmental injustices. The study of environmental justice, a major recent development that derives from a combination of social scientists' long-standing interest in inequalities and social movement activism, seeks to remedy environmental inequities. Another theoretical mainstay of sociology, the political economy perspective, is being used to examine the environmental implications of different types of relationships between business and politics.

Finally, several authors (e.g., Greider and Garkovich 1994; Hannigan 1995; Burningham and Cooper 1999; Scarce 2000; Yearley 1992) use a social constructionist framework to examine environment-related questions. Social constructionism emphasizes the process through which concepts and beliefs about the world are formed (and reformed) and through which meanings are attached to things and events. Environmental sociologists often draw a distinction between the "realists," who prefer not to question "the material truth of environmental problems" (Bell 1998: 3), and the "constructionists," who emphasize the creation of meaning—including the meaning of "environment" and "environmental problems"—as a social process (see Bell 1998; Lidskog 2001). However, it is important to note that, as Burningham and Cooper (1999) argue, all sociology is, in some ways, constructionist. Thus, to some extent, the distinction between realists and constructionists is the extent to which the authors *emphasize* the process through which meanings are created.

A Look at What's Included in the Reader

The roots of environmental sociology run deep and wide for such a relatively new field of study. The terrain of environmental sociology at the dawn of this new century could perhaps best be described as a collage rather than a cohesive structure of theory and research. This collage reflects the increasingly complex and global nature of our environmental challenges. Our intent in creating this reader is to provide a buffet of issues and perspectives on environmental sociology so that you can sample a wide variety of pieces. Later, when you know what you like, go to the library for seconds!

As mentioned, often the theories used by sociologists and other social researchers to study environment-related topics build on the same theories sociologists use to study most anything else. The pieces presented in our reader reflect that theoretical lineage. We include several works that use a political economy perspective, examining how, for example, corporations gain government support for various initiatives or whether concepts such as sustainable development might better be termed "sustainable capitalism." For instance, in the first piece, John Bellamy Foster asks whether the excesses of capitalism (with its attendant over-consumption, overuse of resources, unequal distribution of both wealth and environmental destruction, and often environmentally damaging production techniques) can be rendered more environmentally accountable.[19] Julia Fox examines the coal industry's control over both the elected governmental officials and regulatory agencies in West Virginia and the consequent problems with water contamination, deforestation, and other forms of environmental degradation (reading 2). The pieces by Luiz Barbosa (reading 3) and Oriol Pi-Sunyer and R. Brooke Thomas (reading 4) also provide critiques of environmental inequalities from political-economic perspectives. Allan Schnaiburg and Kenneth Gould (reading 5) argue that there is a fundamental conflict between capitalism's need for constant expansion and the protection of the environment.

Other authors examine the relationship between social inequality and environmental degradation—an issue central to many recent sociological studies of the environment. These studies ask questions regarding how capitalism, racism, sexism, and so forth, foster environmental injustices or inequalities. This investigation of the intersection of environment and inequality is relatively new to both the sociological study of environmentalism and environmental sociology. William Shutkin's piece (reading 6), for example, reveals how transportation planning and policy may both exacerbate and reduce environmental inequalities. Dorceta Taylor (reading 7) provides an historical overview on the relationships of people of color, women, and middle- and working-class whites to various environment-related movements in the United States. The piece by Lois Bryson, Kathleen McPhillips, and Kathryn Robinson (reading 8) examines the gender and class dimensions of Australian lead pollution policies and public health projects. Robert Bullard and Glenn Johnson (reading 9) provide a chronology of the key policy successes and challenges of environmental justice activists and supporters over several decades of struggle.

We have also included research on the intersections of environment, work, and corporations. William Freudenberg, Lisa Wilson, and Daniel O'Leary (reading 10) investigate the "jobs versus the environment" debate, analyzing the case of logging in the Northwest. David Pellow (reading 11), meanwhile, describes workplace conditions in the recycling industry. Thomas Beamish (reading 12) uses organizational theories to illuminate "accidents" within a large corporation and Gerald Markowitz and David Rosner (reading 13) provide a historical perspective on corporations and environmental health.

The process of globalization has profound implications for local and global environments as well as for social movement activists working to improve environmental quality. This reader features several pieces on globalization, such as the piece by Peter Freund and George Martin on the spread of the automobile culture and infrastructure across the globe and its impacts on lower-income nations (reading 15), the work by Deborah Barndt (reading 14) that examines transnational processes in agriculture, and the study by David Pellow, Adam Weinberg, and Allan Schnaiberg (reading 16), which discusses the international trade in toxic waste.

Researchers are increasingly beginning to study the way media and popular culture both reflect and shape our understanding of the human/nature relationship. Susan Davis (reading 17), for example, explores the creation of "nature" and "nature experiences" in a theme park, while Christopher Podeschi (reading 18) looks at conflicting depictions of nature in science fiction films. Robin Andersen (reading 19) examines how corporate advertising uses images of pristine nature to sell products that ultimately contribute to the destruction of many of those environments.

Because science plays an important role in environmental policymaking, we have included research on the sociology of science. Sandra Steingraber (reading 20) provides a very personal account of her experience with cancer and discusses the role of scientific knowledge as it relates to human health. The piece by Connie Ozawa (reading 21) examines how science has been used by various kinds of stakeholders in a wide range of environmental conflicts.

A growing body of literature examines environmental movements that seek to effect various types of environment-related change. The pieces we include explore social movement challenges to corporate and governmental power. Thomas Shriver (reading 22) asks why citizens of Oak Ridge, Tennessee, home to the U.S. government's Oakridge Nuclear Reservation, have not been able to organize a successful opposition to nuclear pollution and dangers. Boris Holzer, by contrast, provides an example of a relatively successful effort by social movement activists against the Mitsubishi Corporation (reading 23). The Mik Moore piece (reading 24), which provides a case study of conflicts over the proposed New Los Padres Dam in California, discusses how the traditional Native American philosophy acted as a crucial symbolic resource for both the native communities and the environmental organizations that were struggling against the dam.

The role of identity in the character and resolution of environmental conflicts is an emerging and increasingly important theme in the literature. Our reader features the work of Thomas Dunk (reading 25), who explores how stakeholders utilized a conception of an endangered "white male identity" in an attempt to challenge Canadian bear-hunting restrictions.

As students become aware of environmental/social inequalities and injustices many naturally want to use their sociological knowledge to change society. Some find it frustrating that while sociology reveals many grave societal problems, it does not tell them what to do about those problems. Indeed, there is no one, grand, sociological recipe for change. To some extent, just learning more about specific issues is the beginning of change. We have to know about and understand issues like the overproduction of organochlorines in order to even think about doing something about that problem. We believe that in almost every sociological study, one can find implied ideas—or "minirecipes"—for how to work toward solving some of the dilemmas illuminated by environmental sociologists. However, sometimes it is valuable to have specific and explicit examples of how change might occur. The concluding section of this reader, therefore, features pieces that provide good examples of these minirecipes— some put into practice, some at the proposal stage.

The piece by Juliet Schor (reading 28) focuses on the types of changes we can each make as individual consumers (with a focus on the garment industry). The Bill McKibben article (reading 26) looks at the power of the consumer as well and challenges readers, especially those in wealthy parts of the world, to take responsibility for the impact that their individual levels of consumption have on the whole of the earth's resources. Myron Glazer and Penina Glazer (reading 29) discuss, among other things, the importance of social support (social capital) in the lives of environmental activists. Peter Rosset (reading 27) presents a case study of a successful society-wide environmental change in the agricultural sector of Cuba's domestic economy.

A Final Word

There are two main reasons we wanted to edit this book. First, we know that people are increasingly interested in ecological change. Environmental interest and support is especially evident in academic institutions and new courses and programs relevant to environmental studies appear each year.[20] Universities and colleges not only offer environmental courses but are becoming more active as positive models of environmental change. According to a 2001 study by the National Wildlife Federation, 20 percent of U.S. campuses currently recycle up to 40 percent of their waste, and 5 percent currently recycle 70 to 100 percent of their waste. This trend is growing as 49 percent of institutions surveyed report that they have programs in place to encourage increased recycling (McIntosh et al. 2001).[21] Both of us, and many of our students, are a part of this ongoing dialogue and activity on our own campuses. We see this book as another way we can contribute to the work we find so important.

Our second reason for doing this book is that we hope to attract more sociologists (both instructors and students) to the subdiscipline. We have spent many hours on our own and, later, in collaboration with each other, to design environmental sociology courses that reflect our values and teaching philosophies. We believe that this is a tremendously important field of research. As sociologists continue to learn about the interconnectedness of our ecological problems with other social problems (e.g., toxic waste and racism), it becomes ever more clear that our ecological health is dependent on our understanding of broader social issues (e.g., democracy, inequality, economics, mass media, etc.) and vice versa. The degree to which we progress in one area (or regress) will echo into every other area of social thought and change. Environmental problems involve power relations; they derive from cultural and institutional practices and they are unequally distributed among populations. Environmental problems require social solutions.

Notes

1. For more detailed accounts of the Bhopal disaster, see Dinham and Sarangi (2002); Jasanoff (1995); Dembo, Morehouse, and Wykle (1990); Weir (1987); and Everest (1986).

2. For a brief summary of the *Exxon Valdez* disaster and the social conditions leading up to it, see Karliner (1997: 180–182).

3. For an in-depth description of the Love Canal disaster, see Gibbs and Levine (1982); for a brief description of the Love Canal disaster, see Gibbs (1995: 64–67).

4. The 1997 treaty sets binding emission targets for industrialized nations and encourages developing nations to join in voluntary emissions reductions. In the case of the United States, it would mean the reduction of greenhouse emissions to 7 percent below 1990 levels by 2010 (Rosenbaum 2002).

5. Some laws include the 1938 Food, Drug, and Cosmetic Act; the 1947 Federal Insecticide, Fungicide, and Rodenticide Act; and the 1965 Solid Waste Disposal Act. For a complete list of environmental laws up to 1990, see the Environmental Protection Agency website (www.epa.gov).

6. For a complete list of environmental laws up to 1990, see the Environmental Protection Agency website (www.epa.gov). Also, for a comprehensive history of environmental legislation in the United States, see Rosenbaum (2002) and Rothman (1998).

7. The Worldwatch Institute produces several yearly publications that document key environmental trends, including *State of the World* and *Vital Signs*. This introduction draws generally from three Worldwatch yearly publications, including Worldwatch Institute (2004); Gardner, Bright, and Starke (2003); and Abramovitz (2003).

8. See in this reader the excerpt by Steingraber (reading 20) that demonstrates this point.

9. See the following piece in this reader for a discussion of dry-cleaning solvents: Steingraber (reading 20).

10. It is important to note that sociological thought has always, to some degree, included the consideration of ecological factors. For example, some scholars at the Institute for Social Research, in Frankfort, worked on environment-related issues, such as the question of domination of nature, in the mid–twentieth century; see Merchant 1999 for a review. Brief summaries of the history of sociological thought on nature and environment can also be found in Hannigan (1995) and Buttel and Humphrey (2002: 35–36).

11. Sample titles of articles from sociological journals during this time include "Support for Resource Conservation: A Prediction Model" (Honnold and Nelson 1979); "The Costs of Air Quality Deterioration and Benefits of Air Pollution Control: Estimates of Mortality Costs for Two Pollutants in 40 U.S. Metropolitan Areas" (Liu 1979); "The Public Value for Air Pollution Control: A Needed Change of Emphasis in Opinion Studies" (Dillman and Christenson 1975); and "The Impact of Political Orientation on Environmental Attitudes and Actions" (Dunlap 1975).

12. Sample titles of articles from sociological journals during the 1980s include "The Social Ecology of Soil Erosion in a Columbian Farming System" (Ashby 1985); "Manufacturing Danger: Fear and Pollution in Industrial Society" (Kaprow 1985); "Cultural Aspects of Environmental Problems: Individualism and Chemical Contamination of Groundwater" (Fitchen 1987); "Blacks and the Environment" (Bullard and Wright 1987); and "Exxon Minerals in Wisconsin: New Patterns of Rural Environmental Conflict" (Gedicks 1988).

13. For a summary of the history of environmental sociology and the change in focus of environmental studies over time, see Dunlap, Michelson, and Stalker (2002).

14. For examples in this reader, see Bullard and Johnson (reading 9) and Shutkin (reading 6).

15. For an example in this reader, see Fox (reading 2).

16. For an example in this reader, see Bryson, McPhillips, and Robinson (reading 8).

17. For an early overview of this body of theory, see O'Connor (1988: 11–38). For a collection of O'Connor's work, see the 1998 book titled *Natural Causes: Essays in Ecological Marxism*. The ecosocialist perspective, as well as other perspectives relating to the politics/economy of environmental problems, continues to be explored and revised in the pages of the interdisciplinary journal *Capitalism, Nature, and Socialism*.

18. A great variety of perspectives converse and conflict within the broad field of Marxist-related environmental sociology, including, but not limited to, the voices of social ecologists, ecosocialists, ecofeminists, and Marxist-ecologists.

19. With a few notable exceptions (e.g., Spaargarten and Mol 1992), sociologists have strong reservations about the long-term sustainability of capitalism.

20. A quick search on the Yahoo Environmental Studies Directory reveals a list of seventy-one environmental studies programs and departments across the globe. The editors

work at Smith College and the College of Charleston (both of which have environmental studies programs); since neither college made it to this already lengthy list, we can only assume that this is an underestimation of the numbers of academic programs (not to mention courses) available to students interested in these issues.

21. More than 250 colleges/universities worldwide have signed the Talloires Declaration in support of ten principles for environmental sustainability (calling for raised environmental awareness, active engagement in environmental research, teaching, policy formation, and literacy). The Talloires Declaration specifically urges academic institutions to set an environmental example in their day-to-day campus operations. You can view this declaration at www.ulsf.org/about/tallosig.html. Also see recent books on green campuses (e.g., Creighton 1998; Barlett and Chase 2004).

References

Abramovitz, Janet N. 2003. *Vital Signs 2003: The Trends That Are Shaping Our Future* (Worldwatch Institute). New York: Norton.

Ashby, Jacqueline A. 1985. "The Social Ecology of Soil Erosion in a Columbian Farming System." *Rural Sociology* 50(3) (Fall): 377–96.

Barlett, Peggy F. and Geoffrey W. Chase (eds.). 2004. *Sustainability on Campus: Stories and Strategies for Change.* Cambridge, MA: MIT Press.

BBC NewsWorld Edition. "Comparing the Worst Oil Spills." November 19, 2002. http://news.bbc.co.uk/2/hi/europe/2491317 (retrieved on December 30, 2003).

Beamish, Thomas. 2002. *The Silent Spill: The Organization of an Industrial Crisis.* Cambridge, MA: MIT Press.

Beck, Ulrich. 1999. *World Risk Society.* Malden, MA: Polity.

Bell, Michael M. 1998. *An Invitation to Environmental Sociology.* Thousand Oaks, CA: Pine Forge.

Bookchin, Murray. 1990. *Remaking Society: Pathways to a Green Future.* Boston: South End.

Brown, Phil, and Edwin J. Mikkelsen. 1990. *No Safe Place: Leukemia and Community Action.* Berkeley: University of California Press.

Brulle, Robert J. 2000. *Agency Democracy and the Environment: The U.S. Environmental Movement from the Perspective of Critical Theory.* Cambridge, MA: MIT Press.

———. 1996. "Environmental Discourse and Social Movement Organizations: A Historical and Rhetorical Perspective on the Development of U.S. Environmental Organizations." *Sociological Inquiry* 66(1): 58–83.

Bullard, Robert D. 1993. "Anatomy of Environmental Racism and the Environmental Justice Movement." Pp. 15–39 in Robert Bullard (ed.), *Confronting Environmental Racism: Voices from the Grassroots.* Boston: South End.

Bullard, Robert D., and Beverly Hendrix Wright. 1987. "Blacks and the Environment." *Humboldt Journal of Social Relations* 14(1–2) (Fall–Summer): 165–84.

Burkett, Paul. 1998. "Analysis of Capitalist Environmental Crisis." *Nature, Society, and Thought* 11(1): 17–51.

Burningham, Kate, and Cooper, Geoff. 1999. "Being Constructive: Social Constructionism and the Environment." *Sociology* 33(2): 297–316.

Buttel, Frederick H., and William L. Flinn. 1974. "The Structure of Support for the Environmental Movement, 1968–1970." *Rural Sociology* 39(1) (Spring): 56–69.

Buttel, Frederick H., and Craig R. Humphry. 2002. "Sociological Theory and the Natural Environment." Pp. 32–69 in Riley E. Dunlap and William Michelson (eds.), *Handbook of Environmental Sociology*. Westport, CN: Greenwood.

Catton, William R., and Riley E. Dunlap. 1978. "Environmental Sociology: A New Paradigm," *American Sociologist* 13: 41–49.

Cole, Luke, and Sheila Foster. 2001. *From the Ground Up: Environmental Racism and the Rise of the Environmental Justice Movement*. New York: New York University Press.

Creighton, Sarah Hammond. 1998. *Greening the Ivory Tower: Improving the Environmental Track Record of Universities, Colleges, and Other Institutions*. Cambridge, MA: MIT Press.

Cushman, John. 1997, November 11. "Polls Show Public Support for Treaty." *New York Times*.

Dembo, David, Ward Morehouse, and Lucinda Wykle. 1990. *Abuse of Power: Social Performance of Multinational Corporations: The Case of Union Carbide*. New York: New Horizons Press.

Dillman, Don A., and James A. Christenson. 1975. "The Public Value for Air Pollution Control: A Needed Change of Emphasis in Opinion Studies." *Cornell Journal of Social Relations* 10(1) (Spring): 73–95.

Dinham, Barbara, and Satinath Sarangi. 2002. "The Bhopal Gas Tragedy 1984 to ? The Evasion of Corporate Responsibility." *Environment and Urbanization* 14 (1): 89–99.

Dowie, Mark. 1995. *Losing Ground: American Environmentalism at the Close of the Twentieth Century*. Cambridge, MA: MIT Press.

———. 2001. "Food." Pp. 106–40 in *American Foundations: An Investigative History*. Cambridge, MA: MIT Press.

Dunlap, Riley E. 1975. "The Impact of Political Orientation on Environmental Attitudes and Actions." *Environment and Behavior* 7(4): 428–54.

———. 1997. "The Evolution of Environmental Sociology: A Brief History of the American Experience." Pp. 21–39 in Michael Redclift and Graham Woodgate (eds.), *The International Handbook of Environmental Sociology*. Northampton, MA: Elgar.

Dunlap, Riley E., and Angela G. Mertig (eds.). 1992. *American Environmentalism: The U.S. Environmental Movement, 1970–1990*. Philadelphia: Taylor & Francis.

Dunlap, Riley E., William Michelson, and Glenn Stalker. 2002. "Environmental Sociology: An Introduction." Pp. 1–32 in Riley E. Dunlap and William Michelson (eds.), *Handbook of Environmental Sociology*. Westport, CN: Greenwood.

EPA (Environmental Protection Agency). 2004. "Climate." Environmental Protection Agency website, http://yosemite.epa.gov.oar/globalwarming.nsf/content/climate/html (retrieved January 1, 2004).

Everest, Larry. 1986. *Behind the Poison Cloud: Union Carbide's Bhopal Massacre*. Chicago: Banner.

Faber, Daniel. 1993. *Environment under Fire: Imperialism and the Ecological Crisis in Central America*. New York: Monthly Review Press.

Faber, Daniel (ed.). 1998. *The Struggle for Ecological Democracy: Environmental Justice Movements in the United States*. New York: Guilford.

Feagin, Joe, and Hernán Vera. 2001. *Liberation Sociology*. Boulder, CO: Westview Press.

Fitchen, Janet M. 1987. "Cultural Aspects of Environmental Problems: Individualism and Chemical Contamination of Groundwater." *Science, Technology, and Human Values* 12(2): 1–12.

Foster, John Bellamy. 2002. *Ecology against Capitalism*. New York: Monthly Review Press.

Gardner, Gary T., Chris Bright, and Linda Starke. 2003. *State of the World 2003* (Worldwatch Institute). New York: Norton.

Gedicks, Al. 1988. "Exxon Minerals in Wisconsin: New Patterns of Rural Environmental Conflict." *Wisconsin Sociologist* 25(2–3): 88–103.

Gibbs, Lois Marie. 1995. *Dying from Dioxin: A Citizen's Guide to Reclaiming Our Health and Rebuilding Democracy*. Boston: South End.

Gibbs, Lois Marie, and Murray Levine. 1982. *Love Canal: My Story*. Albany: State University of New York Press.

Gilbert, Michael J., and Steve Russell. 2002. "Globalization of Criminal Justice in the Corporate Context." *Crime, Law & Social Change* 38: 211–38.

Gottlieb, Robert. 1993. *Forcing the Spring: The Transformation of the American Environmental Movement*. Washington, D.C.: Island.

Greenpeace. 1999. "Remembering the Exxon Valdez." www.greenpeaceusa.org/features/exxontext.htm (retrieved on December 30, 2003).

Greider, Thomas, and Lorraine Garkovich. 1994. "Landscapes: The Social Construction of Nature and the Environment." *Rural Sociology* 59(1): 1–24.

Hannigan, John. 1995. *Environmental Sociology: A Social Constructionist Perspective*. New York: Routledge.

Harper, Charles L. 1996. *Environment and Society: Human Perspectives on Environmental Issues*. Upper Saddle River, NJ: Prentice Hall.

Harr, Jonathon. 1995. *A Civil Action*. New York: Random House.

Herbert, Josef. March 30, 2001. "Bush Rejects Warming Treaty." Environmental News Network, www.enn.com/news/wire- stories/2001/03/03302001/ap_bush_42805.asp?P=2 (Retrieved January 1, 2004).

Honnold, Julie A., and Lynn D. Nelson. 1979. "Support for Resource Conservation: A Prediction Model." *Social Problems* 27(2): 220–34.

Humphrey, Craig, and Frederick Buttel. 1982. *Environment, Energy and Society*. Belmont, CA: Wadsworth.

IPCC (Intergovernmental Panel on Climate Change). 2001. *Climate Change 2001: Synthesis Report*. www.grida.no/climate/ipcc_vol4/english (Retrieved December 31, 2003).

Jasanoff, Sheila (ed.). 1995. *Learning from Disaster: Risk Management after Bhopal*. Philadelphia: University of Pennsylvania Press.

Johnston, Barbara Rose. 2003. "Political Ecology of Water: An Introduction." *Capitalism, Nature, and Socialism* 14 (Fall): 73–90.

Kaprow, Miriam Lee. 1985. "Manufacturing Danger, Fear and Pollution in Industrial Society." *American Anthropologist* 87(2): 342–56.

Karliner, Joshua. 1997. *The Corporate Planet: Ecology and the Politics in the Age of Globalization*. San Francisco: Sierra Club Books.

Kovel, Joel. 2003. "Racism and Ecology." *Socialism and Democracy* 17, 1(33) (Winter–Spring): 99–107.

Kroll-Smith, Steve, Valerie Gunter, and Shirley Laska. 2000. "Theoretical Stances and Environmental Debates: Reconciling the Physical and the Symbolic." *American Sociologist* 31: 44–61.

Lidskog, Rolf. 2001. "The Re-Naturalization of Society? Environmental Challenges for Sociology." *Current Sociology* 49(1): 113–36.

Liu, Ben-Chieh. 1979. "The Costs of Air Quality Deterioration and Benefits of Air Pollution Control: Estimates of Mortality Costs for Two Pollutants in 40 U.S. Metropolitan Areas." *American Journal of Economics and Sociology* 38(2): 187–95.

Lynch, Michael, and Paul Stretesky. 2000. "Media Coverage of Chemical Crimes, Hillsborough County Florida, 1987–97." *British Journal of Criminology* 40 (1).

Martinez Alier, Juan. 1997. "Environmental Justice (Local and Global)." *Capitalism, Nature, and Socialism* 8 (Spring): 91–107.

McIntosh, Mary, Kathleen Cacciola, Steven Clermont, and Julian Keniry. 2001. *State of the Campus Environment: A National Report Card on Environmental Performance and Sustainability in Higher Education.* National Wildlife Federation survey. Princeton, NJ: Princeton Survey Research Associates.

Merchant, Carolyn (ed.). 1999. *Key Concepts in Critical Theory: Ecology.* Amherst, NY: Humanity Books.

Mertig, Angela G., Riley E. Dunlap, and Denton E. Morrison. 2002. "The Environmental Movement in the United States." Pp. 448–81 in Riley E. Dunlap and William Michelson (eds.), *Handbook of Environmental Sociology.* Westport, CN: Greenwood.

Michelson, William. 1976. *Man and His Urban Environment: A Sociological Approach.* Reading, MA: Addison-Wesley.

Mol, Arthur P. 1997. "Ecological Modernization: Industrial Transformations and Environmental Reform." Pp. 138–49 in Michael Redclift and Graham Woodgate (eds.), *The International Handbook of Environmental Sociology.* Northampton, MA: Elgar.

O'Connor, James. 1988. "Capitalism, Nature, and Socialism: A Theoretical Introduction." *Capitalism Nature and Socialism* 1 (Fall): 11–38.

———. 1998. *Natural Causes: Essays in Ecological Marxism.* New York: Guilford.

O'Riordan, Timothy. 1971. "The Third American Conservation Movement: New Implications for Public Policy." *American Studies* 5(2):155–71.

Perrow, Charles. 1997. "Organizing for Environmental Destruction." *Organization and Environment* 10(1).

REHN (Rachel's Environment and Health News). 2001, November 22. "Environmental Trends, Part 2"; can be found in the online archive at Rachel.org (retrieved December 31, 2003).

Rosenbaum, Walter A. 2002. *Environmental Politics and Policy.* Washington, D.C.: CQ Press.

Rosset, Peter, Joseph Collins, and Frances Moore Lappe. 2000. "Lessons from the Green Revolution." *Tikkun Magazine,* March 1.

Rothman, Hal K. 1998. *The Greening of a Nation? Environmentalism in the United States Since 1945.* New York: Harcourt Brace.

Salleh, Ariel. 1995. "Women, Labor, Capital: Living the Deepest Contradiction." *Capitalism, Nature and Socialism* (March): 21–39.

Scarce, Rik. 2000. *Fishy Business: Salmon, Biology, and the Social Construction of Nature.* Philadelphia: Temple University Press.

Shabecoff, Phillip. 1993. *A Fierce Green Fire: The American Environmental Movement.* New York: Hill & Wang.

Shutkin, William A. 2000. *The Land that Could Be: Environmentalism and Democracy In the Twenty-First Century.* Cambridge, MA: MIT Press.

Schnaiberg, Allan. 1980. *The Environment: From Surplus to Scarcity.* New York: Oxford University Press.

Schnaiberg, Allan, and Kenneth A. Gould. 1994. *Environment and Society: The Enduring Conflict.* New York: St. Martin's.

Simon, David R. 2000. "Corporate Environmental Crimes and Social Inequality." *American Behavioral Scientist* 43(4): 633–45.

Spaargarten, Gert, and Arthur P. Mol. 1992. "Sociology, Environment, and Modernity: Ecological Modernization as a Theory of Social Change." *Society and Natural Resources* 54(4): 323–444.

Sussman, Dalia. 2001, April 17. "Global Warming Trend: Six in 10 Say U.S. Should Join Kyoto Treaty." ABCNews.com, http://abcnews.go.com/sections/DailyNews.

Taylor, Dorceta E. 1993. "Chapter 3, Environmentalism and the Politics of Inclusion." Pp. 53–61 in Robert Bullard (ed.), *Confronting Environmental Racism: Voices from the Grassroots.* Boston: South End.

———. 1997. "American Environmentalism: The Role of Race, Class and Gender in Shaping Activism 1820–1995." *Race, Gender & Class* 5(1):16–62.

Thiele, Leslie Paul. 1999. *Environmentalism for a New Millennium: The Challenge of Coevolution.* New York: Oxford University Press.

Weir, David. 1987. *The Bhopal Syndrome.* San Francisco: Sierra Club Books.

Worldwatch Institute. 2004. *State of the World 2004: Progress Towards a Sustainable Society.* London: Earthscan.

Yearley, Steven. 1992. *The Green Case.* London: Routledge.

PART

Politics and Economy

The Vulnerable Planet

John Bellamy Foster

In the first five selections, the authors examine the intersections between politics, economics, and the environment. The sociological study of politics and economics challenges our tendency to take the current systems for granted and therefore represents an important cornerstone of environmental sociology. Many of us tend to assume, for instance, that capitalist trade is "natural," that competition is the most efficient method of distributing resources, and that a free economy is the same as a free citizenry. Sociological works on politics and economics help us see that our current system of production and consumption is not an imperative. Our political/economic system takes the shape that it does precisely because human society "made" it that way.

In this first reading, John Bellamy Foster argues that as corporations have grown and made use of new scientific methods and technologies since World War II, the relationship between economy and environment has become increasingly problematic. Specifically, Foster argues that while we can invent more green technologies, our hopes for a truly sustainable economy and environment are limited under our current economic system. The key to profit is growth. Foster argues that the level of growth required by a "healthy" capitalist economy contradicts the balanced level of inputs and outputs that would be required of a healthy environment.

In the period after 1945 the world entered a new stage of planetary crisis in which human economic activities began to affect in entirely new ways the basic conditions of life on earth. This new ecological stage was connected to the rise, earlier in the century, of monopoly capitalism, an economy dominated by large firms, and to the accompanying transformations in the relation between science and industry. Synthetic products that were not biodegradable—that could not be broken down by natural cycles—became basic elements of industrial output. Moreover, as the world economy continued to grow, the scale of human economic processes began to rival the ecological cycles of the planet, opening up as never before the possibility of planet-wide ecological disaster. Today few can doubt that the system has crossed critical thresholds of ecological sustainability, raising questions about the vulnerability of the entire planet.

The Scientific-Technical Revolution

During the heyday of the Industrial Revolution, the individual firm had only a small impact on the economy as a whole. The concentration and centralization (or growth and merger) of individual capitals altered this situation dramatically, changing forever the relation between firm and economy at both the national and international levels. Today over 60 percent of all U.S. manufacturing assets are owned by two hundred corporations. These changes in the economic character of the system, along with the growing internationalization of capital, constitute the essence of what has been called the monopoly stage of capitalism. Today monopoly capitalism is turning into what might be termed globalized monopoly capitalism, as a handful of multinational corporations rule over the production and finance of the entire world.[1]

Much of the concentration and centralization associated with the rise of monopoly capitalism was made possible by changes at the level of production. The most significant of these was the incorporation of scientific research and scientific management into the industrial process. "Science," according to the modern theorist of labor and technology Harry Braverman,

> is the last—and after labor the most important—social property to be turned into an adjunct of capital. . . . The contrast between science as a generalized social property incidental to production and science as capitalist property at the very center of production is the contrast between the Industrial Revolution, which occupied the last half of the eighteenth and the first third of the nineteenth centuries, and the scientific-technical revolution, which began in the last decades of the nineteenth century and is still going on.[2]

Although science played a large role in the early years of the Industrial Revolution, the relationship between science and industry remained indirect and diffuse. This changed in the period of the scientific-technical revolution, primarily as the result of advances in five fields: steel, coal-petroleum, chemicals, electricity, and the internal combustion engine. Germany played a leading role in these changes. As one industrial historian has noted, "It was Germany which showed the rest of the world how to make critical raw materials out of a sandbox and a pile of coal. . . . IG [Farben] changed chemistry from pure research and commercial pill-rolling into a mammoth industry affecting every phase of civilization."[3]

In the United States, corporate research laboratories arose more or less in tandem with monopoly capitalism. The first research laboratory systematically organized for invention was set up by Thomas Edison in 1876. By the turn of the century, large corporations such as Eastman Kodak, B. F. Goodrich, General Electric, Bell Telephone, Westinghouse, and General Motors had each established scientific research organizations (or acquired previously independent laboratories). By 1920, there were around 300 such corporate laboratories and by 1940 some 2,200.[4] These scientific laboratories provided a whole range of synthetic products, based on the development of new molecular arrangements, out of the essentially limitless number of those theoretically possible. This resulted in new forms of matter, many of which were created with commercial purposes in mind—from a new way of coloring fabric to a new way of killing bacteria. Unfortunately, this progress in physics and chemistry was not accompanied by an equally rapid expansion in the knowledge of how such substances might affect the environment.[5]

It is crucial to emphasize that what drove this revolution in science and technology was not simply the accumulation of scientific knowledge but the "transformation of science itself into capital."[6] This transformation was aimed at extending both the division of labor and the division of nature, and in the process both were transformed. Applied directly to the worker, this took the form of the scientific management of the labor process, the main purpose of which was to remove control over the job from the laborer and give it to management. In other words, scientific management altered labor's relation to the production process as the laborer was systematically reduced to the status of an instrument of production.

The principles of scientific management were most clearly enunciated in the early twentieth century by Frederick Winslow Taylor. Taylor was concerned with devising the theoretical tools for smashing worker resistance on the shop floor. The approach that he devised and advocated in such works as *Principles of Scientific Management,* published in 1911, centered on three principles, summarized by Braverman as: (1) the "dissociation of the labor process from the skills of the workers," (2) the "separation of conception from execution," and (3) the "use of this monopoly of knowledge to control each step of the labor process and its mode of execution."[7] In other words, complex, highly skilled labor was to be reduced to its simplest, most interchangeable—and hence cost-efficient—parts. The end result was the growing commodification of human labor and the destruction of human productive and cultural diversity.

As labor became more homogeneous, so did much of nature, which underwent a similar process of degradation. For example, as science was increasingly applied to the management of forests by profit-making businesses, the natural complexity of forest habitats was replaced by the artificial simplicity of industrial tree plantations. The goal of the scientific forester, as one authority has explained, is to simplify the forest by channeling

> the maximum amount of nutrients, water, and solar energy into the next cut of timber. He cleans up the diversity of age and size classes that are less efficient to cut, skid, process, and sell. He eliminates slow-growing and unsalable trees, underbrush, and any animals that might harm his crop. He replaces natural disorder with neat rows of carefully spaced, genetically uniform plantings of fast growing Douglas-firs. He thins and fertilizes to maximize growth. He applies herbicides and insecticides and suppresses fires to protect this crop against the ravages of nature that must be fought and defeated.

The result is a loss of biological and genetic diversity (the destruction of species and of the genetic varieties within species): industrial tree plantations are biological and genetic deserts when compared to the rich complexity of natural forests. The streams that flow through these plantations contain few fish. The variety of plants, animals, insects, and fungi is minimized. The floor of an old-growth forest is a lush carpet of vegetation; the floor of a tree plantation is almost barren by comparison. The trees themselves, which are viewed as mere commodities (i.e., so many board feet of standing timber), are "genetically improved" to allow for a lower rotation time and hence for profit maximization. Natural diversity is destroyed in the same proportion as profits are promoted.[8]

Such efficiency in the division of nature and human labor is accompanied, paradoxically, by the incorporation of useless inputs into the business process. Thus one of the primary tendencies of monopoly capitalism is "the interpenetration of the production and sales efforts," as advertising costs, superficial product changes, unnecessary packaging, and the costs of marketing in general became increasingly incorporated into the production costs of a commodity.[9]

The Synthetic Age

"We know that *something* went wrong in the country after World War II," Barry Commoner wrote of the United States in his bestseller *The Closing Circle* (1971), "for most of our serious pollution problems either began in the postwar years or have greatly worsened since then." That something, Commoner suggested, was "the sweeping transformation of productive technology since World War II . . . productive technologies with intense impacts on the environment have displaced less destructive ones. The environmental crisis is the inevitable result of this counterecological pattern of growth." Increased throughput of energy and materials creates enormous problems associated with the depletion of resources and the treatment and disposal of wastes. But these problems have been magnified many times over by the replacement of the products of nature with synthetics. This "technological displacement" of nature can be seen in the substitution of artificial fertilizers for organic fertilizers; the development of pesticides to replace biological forms of insect control; the use of synthetics and plastics instead of materials occurring in nature, such as cotton, wool, wood, and iron; and the replacement of soap by detergents with high phosphate content.[10]

What transpired in the post–World War II period was thus a qualitative transformation in the level of human destructiveness. Some of history's most harmful pollutants were only introduced in the 1940s and 1950s. Photochemical smog made its debut in Los Angeles in 1943. DDT first began to be used on a large scale in 1944. Nuclear fallout dates from 1945. Detergents began to displace soaps in 1946. Plastics became a major waste disposal problem only after World War II. Nuclear power and human-generated radioactive elements were a product of war industry and became industrial products in the 1950s.

During what might be termed the golden age of the postwar period (1946–1970), the physical output of food, clothing, fabrics, major household appliances, certain basic metals, and building materials (such as steel, copper, and brick) only grew at the rate of population increase, so that per capita production remained the same. For example, the per capita availability of food, whether measured in calories or protein intake, remained essentially unaltered in the U.S. over this period. Further, physical output in certain key areas actually declined over the period: cotton fibers, wool, soap, lumber, and work animal horsepower. Yet some (mainly synthetic) kinds of production increased dramatically.

While the production of basic needs—food, clothing, housing—has kept pace with the growth of population, the types of goods produced to meet these needs have changed dramatically. New technologies have replaced older ones. Synthetic detergents have replaced soap powder; synthetic fabrics have replaced clothing made out of natural fibers (such as cotton and wool); aluminum, plastics, and concrete have displaced steel and lumber; truck freight has displaced railroad freight; high-powered automobile engines have displaced the low-powered engines of the 1920s and 1930s; synthetic fertilizer has in effect displaced land in agricultural production; herbicides have displaced the cultivator; insecticides have displaced earlier forms of insect control.

It is therefore the *pattern* of economic growth rather than growth (or population) itself that is the chief reason for the rapid acceleration of the ecological crisis in the postwar period. In a 1972 study of the environmental impact of six pollutants (detergent phosphates, fertilizer nitrogen, nitrogen oxides, beer bottles, tetraethyl lead, and synthetic pesticides), Commoner pioneered in the introduction of the formula Environmental Impact = Population – Affluence – Technology (I = P – A – T) to assess the relative impact of the different environmental factors. He showed that P (Population) accounted for only 12 to 20 percent of the total changes in I (Impact) for these pollutants. In the case of nitrogen oxides and tetraethyl lead (both from automobile sources) around 40 percent of the changes in I were attributable to A (Affluence, defined by Commoner as economic goods/population); while in all cases other than automobile-based pollutants, A accounted for no more than 5 percent of the changes in I. T (Technology, defined as pollution/economic good) meanwhile accounted for 40 to 90 percent of all changes in I. The heightened environmental crisis of the 1970s, Commoner argued, was therefore attributable to a considerable degree to "counterecological" systems of production introduced in the postwar period.[11]

A major element in this counterecological trend has been the growth of the automobile complex. "In terms of high energy consumption, accident rates, contribution to pollution, and displacement of urban amenities," Bradford Snell stated in a famous report to a U.S. Senate committee, "motor vehicle travel is possibly the most inefficient method of transportation devised by modern man." Nevertheless, the central determinant of U.S. national transportation policy during most of the twentieth century has been a corporate strategy geared to the high profits associated with automobilization. Lavish federal funding for highways has been coupled with declining government subsidies for public transport. Moreover, at least some of the enormous present-day dependence of the United States on cars, which today account for 90 percent of all travel, can be traced to the deliberate dismantling of the nation's earlier mass transportation system. From the 1930s to the 1950s, General Motors (GM), the nation's top automobile manufacturer, operating in conjunction with Standard Oil and Firestone Tire, systematically bought up many of the nation's electric streetcar lines, converting them to buses. The number of streetcar lines dropped from 40,000 in 1936 to 5,000 in 1955. Meanwhile, GM used its monopolistic control of bus production and of the Greyhound Corporation, on the one hand, and its monopoly in the production of locomotives, on the other, to ensure the growing displacement of bus and rail traffic by private automobiles in intercity ground transport—essentially undercutting itself in intercity mass transit in order to make higher profits off increased automobile traffic. More than any other country, the United States has thus come to rely almost exclusively on cars and trucks for the ground transport of its people and goods, with disastrous consequences for the environment. *In the United States in 1988, the people/motor vehicle ratio was 1.3:1, the lowest in the world!*[12]

No less central as a counterecological force is the petrochemical industry, which creates a huge variety of synthetic products from a few starting materials, primarily petroleum and natural gas. Synthetic fibers, detergents, pesticides, and plastics are all products of the petrochemical industry, as are most toxic wastes. Today there are around 70,000 chemical preparations in use. About

400 of these have been found in the human organism, and most have never been tested for their toxic effects.[13]

The ecological impact of petrochemical production can be better understood if we take into account the way in which this industry has positioned itself in relation to both agriculture and manufacturing. On the one hand, the petrochemical industry markets agricultural chemicals. On the other hand, it produces the synthetic products that compete with farm output. Thus the giant corporations have at one and the same time molded the nation's farms into a "convenient market" and a "weakened competitor." Farming has been transformed from its ancient form, connected to ecological cycles, into a qualitatively new form of commercial enterprise known as "agribusiness."[14]

As Richard Lewontin, one of the world's leading geneticists, who teaches in the Harvard School of Public Health, and Jean-Pierre Berlan, director of research at the French National Institute of Agronomic Research, have explained, there is an "increasing differentiation between *farming* and *agriculture*. Farming is producing wheat; agriculture is turning phosphates into bread." Although its products are essential for our survival, farming itself now accounts for only about 10 percent of the average value added of agricultural products. Of the remaining 90 percent, 40 percent is accounted for by farm inputs (such as seeds, fertilizers, pesticides, and machinery) and 50 percent is added *after* the product leaves the farm, primarily in the form of marketing and distribution costs. The result is that a small number of large corporations, which monopolize the sale of farm inputs and the marketing and distribution of farm products, control the conditions of production in farming and reap the bulk of agricultural profits, even though farming itself is "spread over a large number of petty producers."[15]

At the heart of the Green Revolution carried out by agribusiness in the twentieth century has been the commodification of seed production. Biotechnology has produced hybrid corn and other seed varieties that are widely touted as producing superior high-yield crops. Some scientists, however, believe that the use of the hybrid method, rather than the direct selection of high-yielding varieties from each generation and the propagation of seeds from those plants, was motivated primarily by considerations of profitability. The reason is that the use of hybrid corn seeds makes farmers purchase new seeds each year, because to pursue the traditional farming method (selecting the best plants for seeds for the following year) would in the case of hybrids result in a sharp reduction in productivity (since hybrids do not breed true and their progeny will not produce the same yields). Hybrid corn, Berlan and Lewontin write, "the flagship of the successful innovations of twentieth century agricultural research," thus "expanded the sphere of commodity production by creating a new and extraordinarily profitable commodity."[16]

Perhaps even more important, in order to cultivate these new crop varieties (which are generally less suited than earlier varieties to naturally occurring ecological conditions), large quantities of inorganic fertilizers, herbicides, and pesticides are needed, along with mechanization. The new varieties, according to ecologists Yrjö Haila and Richard Levins, are bred to perform well only if accompanied by a whole technical package. Crops such as hybrid corn therefore become the entry point for an entire model of agribusiness, one that transforms traditional farmers into economic dependents of the major agribusiness corporations.

The Four Laws of Ecology and Economic Production

In order to understand the ecological impact of these trends, it is useful to look at what Barry Commoner and others have referred to as the four informal laws of ecology: (1) everything is connected to everything else, (2) everything must go somewhere, (3) nature knows best, and (4) nothing comes from nothing. The first of these informal laws, *everything is connected to everything else,* indicates how ecosystems are complex and interconnected. This complexity and interconnectedness, Haila and Levins write, "is not like that of the individual organism whose various organs have evolved and have been selected on the criterion of their contribution to the survival and fecundity of the whole." Nature is far more complex and variable and considerably more resilient than the metaphor of the evolution of an individual organism suggests. An ecosystem can lose species and undergo significant transformations without collapsing. Yet the interconnectedness of nature also means that ecological systems can experience sudden, startling catastrophes if placed under extreme stress. "The system," Commoner writes, "is stabilized by its dynamic self-compensating properties; these same properties, if overstressed, can lead to a dramatic collapse." Further, "the ecological system is an amplifier, so that a small perturbation in one place may have large, distant, long-delayed effects elsewhere."[17]

The second law of ecology, *everything must go somewhere,* restates a basic law of thermodynamics: in nature there is no final waste, matter and energy are preserved, and the waste produced in one ecological process is recycled in another. For instance, a downed tree or log in an old-growth forest is a source of life for numerous species and an essential part of the ecosystem. Likewise, animals excrete carbon dioxide to the air and organic compounds to the soil, which help to sustain plants upon which animals will feed.

Nature knows best, the third informal law of ecology, Commoner writes, "holds that any major man-made change in a natural system is likely to be *detrimental to that system.*" During 5 billion years of evolution, living things developed an array of substances and reactions that together constitute the living biosphere. The modern petrochemical industry, however, suddenly created thousands of new substances that did not exist in nature. Based on the same basic patterns of carbon chemistry as natural compounds, these new substances enter readily into existing biochemical processes. But they do so in ways that are frequently destructive to life, leading to mutations, cancer, and many different forms of death and disease. "The absence of a particular substance from nature," Commoner writes, "is often a sign that it is incompatible with the chemistry of life."[18]

Nothing comes from nothing, the fourth informal law of ecology, expresses the fact that the exploitation of nature always carries an ecological cost. From a strict ecological standpoint, human beings are consumers more than they are producers. The second law of thermodynamics tells us that in the very process of using energy, human beings "use up" (but do not destroy) energy, in the sense that they transform it into forms that are no longer available for work. In the case of an automobile, for example, the high-grade chemical energy stored in the gasoline that fuels the car is available for useful work while the lower grade thermal energy in the automobile exhaust is not. In any transformation of energy, some of it is always degraded in this way. The ecological costs of production are therefore significant.[19]

Viewed against the backdrop offered by these four informal laws, the dominant pattern of capitalist development is clearly *counter*ecological. Indeed, much of what characterizes capitalism as an ecohistorical system can be reduced to the following counterecological tendencies of the system: (1) the only lasting connection between things is the cash nexus; (2) it doesn't matter where something goes as long as it doesn't reenter the circuit of capital; (3) the self-regulating market knows best; and (4) nature's bounty is a free gift to the property owner.

The first of these counterecological tendencies, *the only lasting connection between things is the cash nexus,* expresses the fact that under capitalism all social relations between people and all the relationships of humans to nature are reduced to mere money relations. The disconnection of natural processes from each other and their extreme simplification is an inherent tendency of capitalist development. As Donald Worster explains,

> Despite many variations in time and place, the capitalistic agroecosystem shows one clear tendency over the span of modem history: a movement toward the radical simplification of the natural ecological order in the number of species found in an area and the intricacy of their interconnections. . . . In today's parlance we call this new kind of agroecosystem a *monoculture* meaning a part of nature that has been reconstituted to the point that it yields a single species, which is growing on the land only because somewhere there is a strong market demand for it.[20]

The kind of reductionism characteristic of "commercial capitalism," Indian physicist and ecologist Vandana Shiva states, "is based on specialized commodity production. Uniformity in production, and the unifunctional use of natural resources, is therefore required." For example, although it is possible to use rivers ecologically and sustainably in accordance with human needs, the giant river valley projects associated with the construction of today's dams "work against, and not *with,* the logic of the river. These projects are based on reductionist assumptions [of uniformity, separability, and unifunctionality] which relate water use not to nature's processes but to the processes of revenue and profit generation."[21]

All of this reflects the fact that cash nexus has become the sole connection between human beings and nature. With the development of the capitalist division of nature, the elements of nature are reduced to one common denominator (or bottom line): exchange value. In this respect it does not matter whether one's product is coffee, furs, petroleum, or parrot feathers, as long as there is a market.[22]

The second ecological contradiction of the system, *it doesn't matter where something goes unless it re-enters the circuit of capital,* reflects the fact that economic production under contemporary capitalist conditions is not truly a circular system (as in nature) but a linear one, running from sources to sinks—sinks that are now overflowing. The "no deposit/no return" analogue, the great ecological economist Nicholas Georgescu-Roegen has observed, "befits the businessman's view of economic life." The pollution caused by production is treated as an "externality" that is part of the costs to the firm.[23]

In precapitalist societies, much of the waste from agricultural production was recycled in close accordance with ecological laws. In a developed capitalist society, in contrast, recycling is extremely difficult because of the degree of division of nature. For instance, cattle are removed from pasture and raised in feedlots; their natural waste, rather than fertilizing the soil, becomes a serious form of pollution. Or, to take another example, plastics, which have increasingly replaced wood, steel, and other materials, are not biodegradable. In the present-day economy, Commoner writes, "goods are converted, linearly, into waste: crops into sewage; uranium into radioactive residues; petroleum and chlorine into dioxin; fossil fuels into carbon dioxide. . . . The end of the line is always waste, an assault on the cyclical processes that sustain the ecosphere."[24]

It is not the ecological principle that *nature knows best* but rather the counterecological principle that *the self-regulating market knows best* that increasingly governs all life under capitalism. For example, food is no longer viewed chiefly as a form of nutrition but as a means of earning profits, so that nutritional value is sacrificed for bulk. Intensive applications of nitrogen fertilizer unbalance the mineral composition of the soil, which in turn affects the mineral content of the vegetables grown on it. Transport and storage requirements take precedence over food quality. And in order to market agricultural produce effectively, pesticides are sometimes used simply to protect the appearance of the produce. In the end, the quality of food is debased, birds and other species are killed, and human beings are poisoned.[25]

Nature's bounty is a free gift to the property owner, the fourth counterecological tendency of capitalism, expresses the fact that the ecological costs associated with the appropriation of natural resources and energy are rarely factored into the economic equation. Classical liberal economics, Marx argued, saw nature as a "gratuitous" gain for capital. Nowhere in establishment economic models does one find an adequate accounting of nature's contribution. "Capitalism," as the great environmental economist K. William Kapp contended, "must be regarded as an economy of unpaid costs, 'unpaid' in so far as a substantial portion of the actual costs of production remain unaccounted for in entrepreneurial outlays; instead they are shifted to, and ultimately borne by, third persons or by the community as a whole." For example, the air pollution caused by a factory is not treated as a cost of production internal to that factory. Rather it is viewed as an external cost to be borne by nature and society.[26]

By failing to place any real value on natural wealth, capitalism maximizes the throughput of raw materials and energy because the greater this flow—from extraction through the delivery of the final product to the consumer—the greater the chance of generating profits. And by selectively focusing on minimizing labor inputs, the system promotes energy-using and capital-intensive high technologies. All of this translates into faster depletion of nonrenewable resources and more wastes dumped into the environment. For instance, since World War II, plastics have increasingly displaced leather in the production of such items as purses and shoes. To produce the same value of output, the plastics industry uses only about a quarter of the amount of labor used by leather manufacture, but it uses ten times as much capital and thirty times as much energy. The substitution of plastics for leather in the production of these items has therefore meant less demand for labor, more demand for capital and energy, and greater environmental pollution.[27]

The foregoing contradictions between ecology and the economy can all be reduced to the fact that the profit-making relation has become to a startling degree the sole connection between human beings and between human beings and nature. This means that while we can envision more sustainable forms of technology that would solve much of the environmental problem, the development and implementation of these technologies is blocked by the mode of production—by capitalism and capitalists. Large corporations make the major decisions about the technology we use, and the sole lens that they consider in arriving at their decisions is profitability. In explaining why Detroit automakers prefer to make large, gas-guzzling cars, Henry Ford II stated simply "minicars make miniprofits." The same point was made more explicitly by John Z. DeLorean, a former General Motors executive, who stated, "When we should have been planning switches to smaller, more fuel-efficient, lighter cars in the late 1960s in response to growing demand in the marketplace, GM management refused because 'we make more money on big cars.'"[28]

Underlying the general counterecological approach to production depicted here is the question of growth. An exponential growth dynamic is inherent in capitalism, a system whereby money is exchanged for commodities, which are then exchanged for more money on an ever increasing scale. "As economists from Adam Smith and Marx through Keynes have pointed out," Robert Heilbroner has observed, "a 'stationary' capitalism is subject to a falling rate of profit as the investment opportunities of the system are used up. Hence, in the absence of an expansionary frontier, the investment drive slows down and a deflationary spiral of incomes and employment begins." What this means is that capitalism cannot exist without constantly expanding the scale of production: any interruption in this process will take the form of an economic crisis. Yet in the late twentieth century there is every reason to believe that the kind of rapid economic growth that the system has demanded in order to sustain its very existence is no longer ecologically sustainable.[29]

Notes

From Foster, John Bellamy. 1994. "The Vulnerable Planet" in *The Vulnerable Planet: A Short Economic History of the Environment*. New York: Cornerstone Books.

1. *Statistical Abstract of the United States*, 1986 (Washington, D.C.: U.S. Government Printing Office, 1986), p. 524.

2. Braverman, *Labor and Monopoly Capital*, p. 156; also Marx, *Capital*, vol. I, p. 1035.

3. Richard Sasuly, *IG Farben* (New York: Boni and Gaer, 1947), p. 19.

4. Ibid., pp. 163–66; Robert Bruce Lindsay, *The Role of Science in Civilization* (New York: Harper and Row, 1963), pp. 214–19.

5. Commoner, *Closing Circle*, pp. 128–30.

6. Braverman, *Labor and Monopoly Capital*, pp. 166–67.

7. Ibid., pp. 112–21; Frederick Winslow Taylor, *The Principles of Scientific Management* (New York: W. W. Norton, 1947).

8. Elliott A. Norse, *Ancient Forests of the Pacific Northwest* (Washington, D.C.: Island Press, 1990), pp. 152–60.

9. Paul Baran and Paul Sweezy, *Monopoly Capital* (New York: Monthly Review Press, 1966), p. 132.

10. Commoner, *Closing Circle*, pp. 138, 175.

11. Schurr, ed., *Energy, Economic Growth, and the Environment*, pp. 44–62.

12. Bradford Snell, "American Ground Transport," in U.S. Senate, Committee on the Judiciary, *Industrial Reorganization Act*, Hearings Before the Subcommittee on Antitrust and Monopoly, 93rd Congress, 2nd Session, Part 4a (Washington, D.C.: U.S. Government Printing Office, 1974), pp. A-26, A-47; Glenn Yago, "Corporate Power and Urban Transportation," in Maurice Zeitlin, *Classes, Class Conflict, and the State* (Cambridge, MA: Winthrop Publishers, 1980), pp. 296–323; Motor Vehicles Manufacturers Association of the United States, *MVMA Facts and Figures* (Detroit, MI: MVMA, 1990), p. 36.

13. Rolf Edberg and Alexi Yablokov, *Tomorrow Will Be Too Late* (Tucson: University of Arizona Press, 1991), pp. 92, 101.

14. Barry Commoner, "Preface," in Michael Perelman, *Farming for Profit in a Hungry World* (Montclair, NJ: Allanheld, Osmun & Co., 1977), p. vii.

15. R. C. Lewontin and Jean-Pierre Berlan, "Technology, Research, and the Penetration of Capital," *Monthly Review* 38, no. 3 (July–August 1986), pp. 21, 26–27.

16. Jean-Pierre Berlan and R. C. Lewontin, "The Political Economy of Hybrid Corn," *Monthly Review* 38, no. 3 (July–August 1986): 35–47; R. C. Lewontin, *Biology as Ideology* (New York: HarperCollins, 1991), pp. 54–57.

17. Commoner, *Closing Circle*, pp. 29–42; Edberg and Yablokov, *Tomorrow Will Be Too Late*, p. 89; Haila and Levins, *Humanity and Nature*, pp. 5–6. Although Commoner refers to the fourth law as "there's no such thing as a free lunch," the Russian scientist Yablokov has translated this more generally as "nothing comes from nothing."

18. Commoner, *Closing Circle*, pp. 37–41; and *Making Peace*, pp. 11–13. Commoner's third law should not be taken too literally. As Haila and Levins write, "The conception that 'nature knows best' is relativized by the contingency of evolution." Haila and Levins, *Humanity and Nature*, p. 6.

19. Ibid., pp. 14–15; Herman E. Daly and Kenneth Townsend, eds., *Valuing the Earth* (Cambridge, MA: MIT Press, 1993), pp. 69–73.

20. Donald Worster, *The Wealth of Nature* (New York: Oxford University Press, 1993), pp. 58–59.

21. Vandana Shiva, *Staying Alive* (London: Zed Books, 1989), pp. 23–24, 186.

22. Haila and Levins, *Humanity and Nature*, p. 201.

23. Nicholas Georgescu-Roegen, *The Entropy Law and the Economic Process* (Cambridge: Harvard University Press, 1971), p. 2.

24. Commoner, *Making Peace*, pp. 10–11.

25. Haila and Levins, *Humanity and Nature*, p. 160.

26. Georgescu-Roegen, *Entropy Law*, p. 2; K. William Kapp, *The Social Costs of Private Enterprise* (Cambridge, Mass.: Harvard University Press, 1971), p. 231.

27. Chandler Morse, "Environment, Economics and Socialism," *Monthly Review* 30, no. 11 (April 1979): 12; Commoner, *Making Peace*, pp. 82–83; and *The Poverty of Power* (New York: Alfred A. Knopf, 1976), p. 194.

28. Ford and DeLorean quoted in Commoner, *Making Peace*, pp. 80–81.

29. Robert Heilbroner, *An Inquiry into the Human Prospect* (New York: W. W. Norton, 1980), p. 100.

2

Mountaintop Removal in West Virginia: An Environmental Sacrifice Zone

Julia Fox

In the next piece, Julia Fox explores how the newest technologies that have made mountaintop removal possible are not only decimating the environment in West Virginia but also costing the residents jobs, as workers are being replaced by machines. The few jobs that remain provide little financial security to the workers. Though the industry generates great profit for its owners, little of this wealth goes back to the community. This essay also shows how coal companies exercise enormous control over the state (and sometimes national) government. Therefore, regulations on the industry are weak, those regulations that are in place are poorly enforced, and the destructive mining practices and exploitative labor conditions continue.

Historically, West Virginia is a state that has been controlled by coal interests. The natural resources of the state and the labor power of the coal miners have been exploited to achieve a high level of profit for companies engaged in this extractive industry. Humans and nature are, in this respect, external to the logic of the coal market. As the great ecological economist

K. William Kapp (1971) demonstrated, such unpaid social costs are the very essence of the capitalist economy. Coal mining communities, in particular, have been devastated by a for-profit system that has promoted the extraction of coal in order to satisfy an insatiable need for energy that is disproportionately consumed by large corporations and the well-to-do—all the while denying the wider social and environmental costs. The vested interests who gained control of the coal resources have used technological innovations to displace coal miners. As a consequence, the unemployed coal miners and members of the coal mining communities have become redundant populations. The rapid increase in the scale and intensity of coal production that is characteristic of mountaintop removal has accelerated the social and environmental devastation already characteristic of the region.

A dependent periphery within the continental U.S. economy, the West Virginian economy, and its coal industry, have long been controlled by absentee owners. As the world market increased the demand for highly-efficient, low-sulfur coal, the rate of extraction of these coal resources also increased. Today, the major external force that drives mountaintop removal is the increased demand for cheap electricity. The enormous and growing demand for energy is rooted in the specific class relations in which the flow of energy, especially fossil fuel, disproportionately supports a few concentrated economic interests. The rate of demand for electricity on the part of corporations and the wealthy greatly exceeds that of the poor and working class in the United States. Since then, the coal mining companies control the technology and the profit from these operations; technological innovations like the massive draglines, which are geared to the removal of the largest amount of coal at the least cost for the corporations, are adopted without any real consideration of the consequences for local communities or the natural habitat. The scale of the mountaintop removal projects and the introduction of these massive machines allow for more rapid resource depiction with less workers. Yet, behind this apparent increase in efficiency lie human and ecological costs that are incalculable.

These conditions have created a number of internal contradictions that have sparked an organized struggle against mountaintop removal. This case study indicates the importance of creating political alliances between workers, environmentalists, and impacted members of the coal mining communities—indeed, all sectors of the community outside the privileged economic interests—to contest these brutal social and ecological relations. Given the relative power of the coal monopolies, a central proposition of the following argument is that the traditional regulatory regime is quite limited in its capacity to deal with the external (i.e., not recognized by the market) social and ecological costs associated with this kind of development. In fact, such regulatory regimes create the legalistic rules that permit increased ecological and social damage in the coal mining communities. Mountaintop removal must therefore be placed in the context of the logic of profit maximization. In the unending quest for a greater supply of fossil fuel, to generate higher profits, humans and nature are commodified and exploited—humans displaced, nature depleted—reflecting the alienation of humanity from the very conditions of human existence.

The Morphology of Destruction

Although mountaintop removal is a coal mining method that has been used since the 1970s, there was a precipitous increase in its use for extracting coal in the 1990s. Since the 1990s, West Virginia has been the major site for mountaintop removal. Between 1995 and 1998, the West Virginia Division of Environmental Protection has authorized 27,000 acres of new mountaintop mining. In 1997, the state granted permits for 20 new projects covering 20 square miles. The largest mine is expected to cover more than 20,000 acres. West Virginia's southern mountains contain the most efficient coal in the nation, producing 50% more energy per pound than Western coal (Purdy, 1998; Warrick, 1998). West Virginia coal has the added advantage of being the least polluting low-sulfur coal. Ironically, one of the major environmental regulations, the 1990 Clean Air Act, increased the demand for low-sulfur coal (Loeb, 1997). According to the Office of Surface Mining Reclamation and Enforcement, West Virginia had more mine repositories than any other state, with 45,458 listed repositories in 1997 (U.S. Department of the Interior, Office of Surface Mining, 1997, p. 4). West Virginia is thus being turned into an environmental sacrifice zone, subjected to horrendous environmental destruction to provide cleaner, less polluting coal for the nation.

Although strip mining removes seams of coal at the surface, mountaintop removal involves removing 500 or more feet of a mountain to gain access to the veins of coal. As Loeb (1997) points out, "an aerial inspection suggests that 15 percent of the mountain tops in the south-central part of the state and—perhaps 25 percent in some places . . . are being leveled" (p. 26) in the massive mountaintop removal strip mining operation. The coal in West Virginia's mountains is difficult to extract because it is embedded in narrow horizontal seams between the rock inside steep mountains. The extraction requires the use of dynamite. As Loeb (1997) points out, "blasts are made with the same mixture of ammonia nitrate fertilizer and fuel oil in the bomb that killed 168 people in Oklahoma City . . . but the mining explosions are 10 to 100 times stronger." After the coal is exposed by a complex process of removing topsoil, clear-cutting forests, and blasting the rock above the coal seams (the "overburden"), a massive machine, a dragline, is used to remove the rock. The dragline is a $100 million machine that is as high as a 20-story building and weighs 8 million pounds. The dragline's bucket has the capacity to hold 26 Ford Escorts and can remove 110 cubic yards of earth with a single scoop. The savings in labor are immense.

In one mountaintop removal site in Wyoming, 450 workers (including supervisors) mined 1.2 tons of coal per second—40 million tons per year—compared to 3 years earlier when 500 workers were able to produce only 33 million tons of coal from the same mine. Such surface mining, including mountaintop removal as its main means, produces 6 tons of coal per labor hour nationally, compared to 2½ tons per hour for underground miners (Berman & O'Connor, 1996, pp. 152–153).

After the coal is separated out, rock trucks, which can haul 380 tons of rock per load, dump the rock into surrounding valleys. This process, which is called a valley fill, may be composed of waste piles of rock and earth that are up to 1,000 feet wide and 500 feet deep. In the past, the valley fills were composed of low terraced steps; however, with the massive amounts of rock and earth that has been removed, these fills are now dumped over the mountains and nearby valleys (Loeb, 1997).

Mountaintop removal has been called a "muscled-up form of strip mining" (Warrick, 1998, p. A I). The removal of entire mountaintops is more ecologically damaging than strip-mining. During the initial process of mountaintop removal, forests are clear cut and the topsoil is removed. The topography and landscaping around the mountains are transformed into "moon-scape rock and dirt" (Warrick, 1998). . . .

As Tom Rodd, a West Virginia attorney who won a landmark case against the coal industry, recently pointed out, "West Virginia is the latest of the country's energy sacrifice zones, surrendered to keep power cheap. It isn't the first, and it may not be the last" (Purdy, 1998, p. 33). It is not just energy, however, that is being depleted nor simply is it energy that is being provided. Rather, West Virginia has become an environmental sacrifice zone, providing low-sulfur (i.e., low-polluting) coal at the expense of the rapid degradation of its own environment.

West Virginia Coal Mining and the Uneven Nature of Capitalist Development

The rapid growth of mountaintop removal in West Virginia must be placed in a theoretical and historical context. Although West Virginia has historically been a major site for coal mining and bloody protracted struggles between the coal companies and the coal miners, the rapid increase in this form of strip mining is due to a number of social and economic developments. The major external force that has expanded the level of production of mountaintop removal projects has been the increased demand for electricity. In the past 20 years, the U.S. demand for electricity increased 70% (Loeb, 1997). Coal as a source of fuel is now largely invisible, although production and profits are soaring. A great deal of the coal is now burned at remote mine-mouth generating plants owned by electrical utilities. The electricity is then generated at an extremely high voltage for hundreds of miles. Innumerable protests across the United States regarding the health dangers from high-voltage electromagnetic fields have occurred (Berman & O'Connor, 1996, p. 152). West Virginia's low-sulfur coal accounted for 57% of the electricity used in the United States. The amount of coal extracted in West Virginia has increased from 131 million tons in 1986 to 174 million tons in 1996. West Virginia now ranks number one in coal production in the United States. In 1996, the value of the coal was $4.4 billion (Loeb, 1997; Warrick, 1998).

The increased external demand for cheap electricity has intensified the pace of mountaintop removal. However, West Virginia has become the major site for this type of strip mining because of two other major factors as well: (a) the supply of low-sulfur and highly efficient coal reserves and (b) the internal class relations that have created the conditions for unfettered production that allow for the scale and intensity of degradation associated with mountaintop removal.

Historically, West Virginia has been an economically peripheral state whose natural resources have been controlled by coal monopolies, linked to external capital located in urban centers such as London, New York, Philadelphia, Boston and Baltimore. In 1900, absentee landowners owned more than 90% of Mingo, Logan, and Wayne counties and 60% of Boone and McDowell Counties. By 1923, nonresidents of West Virginia owned more than half the state and controlled four fifths of its total value. West Virginia coal was described by geologists as "some of the best in the world," and the mining operations in the state could be undertaken with "relatively small capital investments" (Conley, 1960; Corbin, 1981, pp. 4–51). The coal production in West Virginia increased from 489,000 tons in 1867 to 89,384,000 tons in 1917. By 1920, West Virginia was second only to Pennsylvania as the nation's leading coal producer (Corbin, 1981, p. 5).

This historic pattern of external ownership of West Virginia's resources continues today. The coal industry owns 75% of the land in West Virginia's major coal-producing counties (Miller, 1974; Purdy, 1998). Given the relative power of the coal monopolies in West Virginia, the internal political machinery has created some of the most lax environmental and labor regulation in the United States. Although capitalism, as Marx (1967) demonstrates, is a system in which humans and nature are exploited and commodified for profit in the market, the specific historical class relations in West Virginia create an uneven nature of capitalist development in this poor state—an almost Third World pattern of development.

Based on Marx's theoretical analysis of uneven development of capitalism, James O'Connor (1991) has argued that some of the "worst human and ecological disasters as a rule occur in the Third World" and the poorest regions of the First World (p. 257). O'Connor (1991) argues that two of the effects of uneven development are resource depletion and the rapid exploitation of fossil fuels. This is well illustrated by West Virginia because the major forces that have generated the ecological crisis there are the energy monopolies that maintain external control over the resources and technology. As the world market for cheap energy sources increases the demand for natural resources, the energy monopolies use new technologies to increase the rate of depletion, and as a consequence, the ecological devastation and social disruption is greatest in a poor state like West Virginia.

Given the disproportionate power exercised by the coal monopolies in West Virginia, the legal fetters of environmental and labor regulations have been less constraining in this state. West Virginia's historical class relations created conditions extremely favorable for mountaintop removal. The accelerated exploitation of the earth and of the workers associated with mountaintop removal is a consequence of the logic of capitalism's creative destruction. Applying Marx's critique of capitalism, Foster (1994; see also Sweezy, 1989) explains this "creative drive is the seemingly infinite ability to produce new commodities by combining materials and labor in new ways, and the destructive drive is the systematic degradation, transformation, and absorption of all the elements of existence outside of the system's own orbit" (p. 32). As a natural resource, like fossil fuel, becomes a precious commodity to be sold on the market, then humans and nature are commodified to transform this commodity into a profit, as Marx (1967) demonstrated. Because the coal companies own and control the resources and the technology, the enormous machines increase the rate of the rapid depletion of the resources and displacement of workers. In the past 10 years, the level of productivity for the average coal miner has tripled (Purdy, 1998). This is the precisely the relationship that Marx (1967) described as the most effective way to increase profit rates by increasing the productivity rate of workers and therefore lowering the unit costs of labor. The result is massive social dislocation and the creation of redundant populations. Mountaintop removal employed only 4,317 workers in West Virginia in 1997. This is less than 1% of the state's entire work force. Mining employment in West Virginia has declined from 130,000 in 1950 to 22,000 in 1996 (Loeb, 1997). As a local newspaper commented, "The industry has eliminated 100,000 West Virginia jobs, replacing miners with giant machines. Nearly all profits from West Virginia coal go to out-of-state owners" ("Shear Madness," 1997). The rapid developments in technological innovations have allowed the coal companies to extract more coal with less workers. Hence, many coal miners have been relegated to the surplus army of labor as the technological innovations displace the miners. The process of mountaintop removal has created massive social dislocation as the inhabitants around the coal fields have become an expendable population that stands in the way of a voracious demand for a precious commodity that is extracted for profit in the market.

This is precisely the type of massive social dislocation that Marx (1967) described as capitalist production penetrating every sphere of social life that comes "dripping from head to foot, from every poor, with blood and dirt" (p. 760). The extreme conditions of exploitation of the natural and human environment, the existence of large surplus populations of former workers seeking employment, and the rising levels of poverty all give West Virginia, in the age of mountaintop removal, a Dickensian character in which relations of exploitation of both human beings and the natural environment are extremely transparent despite the fact that all of this is taking place under the mantle of economic and ecological modernization.

Although the coal companies have made massive profits from the extraction of coal, the prosperity has not trickled down to the workers or the residents of West Virginia. Despite the vast natural resources, West Virginia is the second-poorest state in the United States. As the most recent census data indicate, the coal-producing states are the poorest. In West Virginia, the poverty rate was 23%, the per capita income $9,326, and the unemployment rate was 14.5% ("The Poverty of the Coal Fields Counties," 1998, p. 10).

In one of the poorest areas of West Virginia, Cabin Creek, the coal companies extracted $100 million of coal a year while the "poor children play in raw sewage" (Radmacher, 1998). Indeed, "if coal helps the local economies so much" one West Virginian journalist has asked, "Why are the southern coalfields the most impoverished part of the state?" (Radmacher, 1998, p. 4A). Because the coal companies control the natural resources, the profit from the coal production accrues to the companies at the expense of the coal miners, the members of the West Virginian communities, and the environment. As Foster (1994) points out, the "throughput (the flow of energy/material input) that supports a wealthy individual, class, or nation is obviously much greater than the throughput that supports a poor individual, class or nation" (p. 31). This can also be stated in the language of "ecological footprint" analysis, devised by ecological economists Mathis Wackemagel and William Rees (1996). The ecological footprint (the impact on the environment measured in terms of units of land needed to support the demands of a given individual, nation, etc.) is much greater for a rich individual than a poor one, for a wealthy nation as opposed to an impoverished one, for a society relying more on coal-based electricity than one relying on solar power generation (Wackemagel & Rees, 1996).

Coal companies, which are part of a much larger energy-industrial complex, control vast natural and economic resources and have the capacity to apply any technology that will intensify the rate of production, regardless of the social and environmental costs to society as a whole. The price of a cheap energy source such as coal does not include the "negative externalities" of mountaintop strip-mining. Indeed, the reduction of all such costs to externalities that do not enter into the balance sheets of corporations makes both human beings and natural externalities of the market valueless to the bottom line. Given the class relations of this production, the benefits that do accrue from this exploitation of the earth and human beings go mainly to the coal companies and the high demand energy consumers at the expense of the miners, the West Virginia communities, and nature.

Although it is the world market for cheap energy sources that drives the demand for more coal production, the coal companies control the resources and the technology that are used to intensify this production by driving down costs to further increase demand. Mountaintop removal is a manifestation of the creative destruction that applies technological innovations to increase the rate of exploitation of labor and nature, producing more coal with less workers. As a consequence, the massive social dislocations and ecological destruction are most pronounced at the site of extraction, and although transparent in West Virginia, they are invisible to those outside of this impoverished state.

The Political Power of the Coal Companies in West Virginia

Given the historical economic and political relations of the coal companies in West Virginia, the state government constitutes little more than the internal political machinery that lubricates the production process. In this sense, corporations and the state operate in a close "partnership" typical of capitalism (Miliband, 1991, pp. 30–34)—but more closely linked than usual in the West Virginian context, where it is perhaps misleading to talk about the relative autonomy of the state. The specific historical relations in West Virginia have created the conditions for greater relative power of the coal monopolies to control the internal political machinery. The lax environmental and labor regulations in this state provided the most favorable conditions for the acceleration of the scale and intensity of the mountaintop removal form of strip mining. . . .

Not content to control the state system from afar, the coal monopolies have long exercised extensive control over the state's internal political machinery. Historically, the coal companies were able to gain control of the state legislature and dominate legislative committees. In addition, the U.S. senators from West Virginia were either coal owners or men directly associated with the coal industry (Corbin, 1981). One of the most extensive studies of the power of the coal industry in West Virginia found that every governor and president of the state senate between 1950 and 1970 worked in the coal industry (Purdy, 1998). Coal companies were free from safety regulations. Until 1904, there was not a single prosecution in West Virginia, and there were none after 1912 for safety regulations. West Virginia was noted "in the mining world for insufficiency of proper laws" (Corbin, 1981, p. 16). The state received the dubious distinction of a major violator of child labor laws. West Virginia's child labor laws had so many loopholes that the National Child Labor Committee ranked West Virginia 34th (of the then 46 states in child labor) (Corbin, 1981). . . .

West Virginia's Weak Environmental Regulatory Regime

The ability of the coal companies to translate their economic clout into other forms of power is most evident in the weak regulatory regime governing the industry. West Virginia has had some of the weakest environmental laws and most lax regulatory rules in the United States. Recently, Governor Underwood signed one of the most controversial mountaintop mining bills in April 1998. The bill made a major change in the number of acres that coal companies can fill with the debris from mountaintop removal before paying damages. The new law allows companies to increase the amount of valley fill from 250 acres to 480 acres before paying damages to impacted streams and water sheds. Although the bill increased the penalty for land and water damages marginally from $200,000 per acre to $225,000 per acre, it made an important provision that allows state officials to monitor the impact of mountaintop removal projects in an attempt to cut out the federal Environmental Protection Agency (EPA). Hence, the EPA indicated that the bill violated the agencies' responsibilities under the Clean Water Act (Kennedy, 1998). . . .

There are a number of cases in which West Virginia's environmental regulatory agencies have violated federal laws themselves. The state regulators, for example, have made exemptions to the required 100-foot buffer zone around perennial streams. State regulators have allowed permits for mountaintop removal although the coal companies have failed to submit plans for stream reclamation. In July 1998, the West Virginia Highlands Conservancy filed a lawsuit claiming that the permit process violates federal laws and that debris from the mountaintop valley fills are "burying the state's headwater's at an alarming rate."

Mining companies are required to seek special permission for projects that significantly alter the landscape of a region. The 1977 Surface of Mining Control and Reclamation Act requires mining companies to restore the land to "approximate original contour" (AOC). However, exceptions are permitted if the landscape is leveled for future development (e.g., housing, schools, airport). A recent review of the of the permit process indicated that three quarters of the active mountaintop removal mining permits in West Virginia did not receive the required exemptions ("Operators Not Getting Required Exemptions," 1998, p. 1C). Based on a 6-month study by the U.S. OSM [Office of Surface Mining], the federal agency found that "for more than 20 years, mountaintop removal mines have been illegally permitted by the state without plans for developing the stripped land" (Ward, 1998, p. 1A). The federal report concluded that

> most of the AOC variance permits did not contain sufficient documentation to grant those variances. Few, if any, contained the required agency

approvals or the specific plan and assurances demonstrating a need for and supporting the proposed post-mining land use. Among the 11 permits that did receive AOC variances, OSM inspectors found that what the DEP accepted as constituting AOC varied widely. (Ward, 1998, p. 7A)....

The Limits of Regulation: Mountaintop Removal as an Egregious Example

... Given the tremendous economic and political power of the coal monopolies, regulation is not a sufficient means to constrain the corporations. As Tokar (1997) points out, "regulations are, in fact, highly mutable, subject as any other law to changes in the political climate" and "enforcement is often selective.... Officials can easily bend the rules on behalf of powerful economic interests" (p. 57)....

The regulation of strip mining in West Virginia is a classic example of how formal legal equity conceals the actual economic and political power of the coal interests and how this regulation became extraordinarily mutable when the coal interests exerted their power. At the outset, West Virginia's regulatory laws included many exclusions and innovative interpretations. Field inspectors, for example, were given the "local version" of the federal law (Corbin, 1981; Purdy, 1998, p. 30). A West Virginia law professor and environmental litigator, Patrick McGinley, points out that in the 1980s he challenged the illegally issued permits in West Virginia and found that "virtually everything the law required was not being done" (Purdy, 1998, p. 30).

Recently, the director of the West Virginia Division of Forestry, Bill Maxey, resigned his position because Governor Underwood tried to stifle his opposition to mountaintop removal ("Repulsed by Mining Method," 1998). Maxey argued that mountaintop removal had destroyed 250,000 acres of forest. However, he was ordered to issue a rebuttal statement and was asked to "tone down his position." His revised statement was "300,000 acres had not been disturbed" ("Repulsed by Mining Method," 1998). Maxey said, "I had to, against my will, really, say that it could be properly reforested.... That isn't what I really wanted to say, that's what I was told to say" ("Repulsed by Mining Method," 1998). Maxey also pointed out that he was pressured by the state DEP and the federal OSM to approve a phrase that "would justify leveling mountains." The agency proposed the phrase be included in specifications written by the Division of Forestry for voluntary reclamation of mines into woodlands. The phrase, which was included in the 1997 state surface mining regulation says, "flat or gently rolling land on a site reclaimed to hardwood is essential for the operation of mechanical harvesting equipment" ("Repulsed by Mining Method," 1998).

In addition to problems with the West Virginia's lax enforcement and local interpretations of the federal law, there were many loopholes in the 1977 federal

Surface Mining Control and Reclamation Act. When the law was written, mountaintop removal was intended to be the exception rather than the rule. However, by 1990, mountaintop removal became the dominant type of mining in West Virginia. As Warrick (1998) points out, "changes in technology and mining practices have left the OSM and states at odds on how to interpret the . . . Surface Mining Control and Reclamation Act" (p. A8). As Kathy Karpan, the director of the Office of Surface Mining, told a journalist for the Washington Post, "the law says we can have mountaintop removal but it doesn't tell you how many valley fills is too many" (Warrick, 1998, p. A8). . . .

One of the major loopholes in the mining law is the definition of the approximate original contour (AOC). West Virginia environmental regulators say they do not push mine operators to restore the land because the "legal definition of 'approximate original contour' is vague" ("Operators Not Getting Required Exemptions," 1998). John Ailes, chief of the DEP Office of Mining and Reclamation, indicated he has no idea what AOC may mean and that "nobody has ever defined it." In addition, West Virginian mountains are steeper than specified by federal laws for the restoration of mountains to their original contour. Therefore, "restoring mountains to their original dimensions would mean following the law on one hand and breaking it on another." Ben Green, a former state regulator who is currently a lobbyist for the coal industry, pointed out "you do not re-create that mountain. . . . That has never been anybody's idea or intention" ("Operators Not Getting Required Exemptions," 1998). . . .

Technological changes and the scale of mountaintop removal have created even more loopholes in the Surface Mining Control and Reclamation Act, demonstrating that such artificial regulations are helpless against the dynamic of accumulation. The president of the West Virginia Mining and Reclamation Association conceded that the "regulations were written with 20-year-old technology, and smaller blasts in mind" (Loeb, 1997, p. 29). The draglines are permitted to operate within 300 feet of houses. Although West Virginian residents filed more than 287 blast complaints in 1996, the mining explosions are within the legal limit. As one of the residents from Blair, Carlos Gore pointed out, "we're dying and they (the coal companies) are in compliance" (Forman, 1997, p. 12). . . .

The limits of the regulation of the coal industry are especially apparent in the case of the fines for the violation of state and federal laws. As Loeb (1997) points out, "an examination of violation and complaint files for all West Virginia mines shows most to be low, even for seemingly serious violations" (p. 34). The average fine is about $800. Although the maximum fine is $5,000 per incident, "nearly 80 percent of the fines recommended by inspectors are reduced by DEP assessment officers" after the mining companies protest (Loeb, 1997, p. 34).

Organized Opposition

The residents of West Virginia have mounted a major organized campaign to contest mountaintop removal. Similar to other environmental justice movements, the residents have developed an understanding of the economic and political power of the coal companies and the limits of environmental regulation. The West Virginia activists discovered that it was necessary to file lawsuits against the state and federal agencies for failing to implement mining regulations that these agencies developed. A number of environmental organizations including the Ohio Valley Coalition, the West Virginia Citizens Action Group, the West Virginia Environmental Council and the West Virginia Highlands Conservancy have filed several suits against the regulatory agencies ("Environmental Group," 1998). . . .

[However] [t]he West Virginian environmental organizations and activists have also discovered what Brian Tokar (1997) has pointed out, that "many regulatory programs simply codify the terms by which corporations are granted permits to pollute" (p. 80). The loopholes in the Surface Mining Control and Reclamation Act, the vague definitions of AOC, and the permitting process for mountaintop removal are all classic examples of how the coal companies are permitted to pollute. . . .

The relative power of the coal companies to control the internal political machinery of the state creates lax regulations that are comparable to some Third World countries. . . . If West Virginia is to cease to be an environmental sacrifice zone, more fundamental changes in social and environmental relations are needed.

Note

Fox, Julia. 1999. "Mountaintop Removal in West Virginia: An Environmental Sacrifice Zone." *Organization & Environment* 12(2): 163–183.

References

Berman, D. M., & O'Connor, J. T (1996). *Who owns the sun? People, politics, and the struggle for a solar economy.* White River Junction, VT: Chelsea Green.

Conley, R. (1960). *History of the coal industry of West Virginia.* Charleston, WV: Educational Foundation.

Corbin, D. (1981). *Life, work and rebellion in the coal fields.* Urbana: University of Illinois Press.

Environmental group, citizens intend to sue DEP. (1998, April 20). *Herald-Dispatch,* p. 9C.

Forman, L. (1997, November). West Virginians meet with new federal OSM director. *"E" Notes,* pp. 11–12.

Foster, J. B. (1994). *The vulnerable planet.* New York: Monthly Review Press.

Kapp, K. W. (1971). *The social costs of private enterprise.* New York: Schocken.

Kennedy, N. (1998, April 9). Underwood signs mountaintop removal bill. *Herald-Dispatch*, p. 2C.

Loeb, P. (1997, October 13). Shear madness. *U.S. News & World Report*, pp. 26–34.

Marx, K. (1967). *Capital* (Vol. 1). New York: Vintage.

Miliband, R. (1991). *Divided societies*. New York: Oxford University Press.

Miller, T. (1974). *Who owns West Virginia?* Huntington, WV: Herald-Dispatch.

O'Connor, J. (1991). Capitalism, uneven development and the ecological crisis. In B. Berberoglu (Ed.), *Critical perspectives in sociology* (pp. 256–264). Dubuque, Iowa: Kendall/Hunt.

Operators not getting required exemptions. (1998, August 9). *Herald-Dispatch*, p. 1C.

The poverty of coal fields counties. (1998, January/February). *Appalachian Reader*, p. 10.

Purdy, J. (1998, November/December). Rape of the Appalachians. *American Prospect*, pp. 28–33.

Radmacher, D. (1998, June 19). Coal hurts those it pretends to help. *Charleston-Gazette*, p. 4A.

Repulsed by mining method, forestry chief resigns. (1998, November 1). *Herald-Dispatch*, p. 4C.

Shear madness: Strip mine ravages. (1997, August 7). *Charleston-Gazette*, p. 9A.

Sweezy, P. M. (1989). Capitalism and the environment. *Monthly Review*, 41, 2.

Tokar, B. (1997). *Earth for sale: Reclaiming ecology in the age of corporate greenwash*. Boston: South End.

U.S. Department of the Interior, Office of Surface Mining Reclamation and Enforcement. (1997, November). *Mine map repositories*. Washington, DC: Government Printing Office.

Wackemagel, M., & Rees, W. (1996). *Our ecological footprint: Reducing human impact on the earth*. Philadelphia: New Society.

Ward, K. (1998, August 16). Feds confirm mining problems. *Charleston-Gazette*, pp. 1A, 7A.

Warrick, J. (1998, August 3 1). Mountaintop removal shakes coal state. *Washington Post*, pp. A1, A8.

The People of the Forest against International Capitalism

Systemic and Anti-Systemic Forces in the Battle for the Preservation of the Brazilian Amazon Rainforest

Luiz C. Barbosa

Drawing on world systems theory, which focuses on the unequal power between the wealthy "core" countries and the poorer countries of the "periphery," Barbosa examines how development projects of the mid–twentieth century wrought environmental destruction in the Amazon region of Brazil. These development projects were linked to international "systemic forces," including international lending institutions, large corporations, and governments. In the 1980s, "anti-systemic forces"— indigenous peoples, rubber tappers living in the forest, and environmentalists— gained political leverage and began to not only resist the often socially and environmentally harmful projects but also to contest the racist and colonialist idea that they needed to be "developed."

This paper is about the battle for the preservation of the Brazil's Amazon rainforest.[1] It is about the expansion of the frontier of global capitalism into Amazonia and resistance thereto. It is about the battle between government planners, organizations like the World Bank and the International Monetary Fund (IMF), and multinational corporations (the systemic[2] or capitalist forces) on one side, and environmentalists, indigenous people (the people of the forests), and grassroots social movements on the other, the anti-systemic forces.[3] I contend that this conflict is intertwined with the dynamics of an evolving capitalist world-system.[4] Here, the conflict is a byproduct of the expansion of capitalism in Amazonia, an expansion largely dependent on the collaboration of international capital. The systemic forces still have the upper hand. The anti-systemic forces have long resisted as best they could, for example, the centuries-old Indian resistance (Hemming 1987), but did not have the power to stop the expansion. Global changes and changes within Brazil itself in the 1980's have favored the anti-systemic forces. These changes and their consequences are the theme of this paper. My argument is that changes in the ecopolitics of the world-system[5] beginning in the mid-1980's have given political leverage to Brazil's anti-systemic forces, strengthening their resistance to capitalist penetration. This resistance, coupled with a "greening" of global ecopolitics, explained below, threatens the collaboration of international capital in developing Amazonia. It is first important to provide a world-systemic analysis of ecopolitics prior to the mid-1980's and its effects on Amazonia.

Ecopolitics before the Mid-1980's

The ecopolitics that prevailed in the world-system until the mid-1980's was that inherited from the evolution of capitalism itself. Of critical importance was the expansion of European capitalism to dominate other parts of the world, a process which began in what Immanuel Wallerstein calls "the long-sixteenth century" (1450–1650). This expansion allowed Europeans not only to exploit the natural resources of the regions they found but also the people who occupied them. . . .

The onset of capitalism involved a process of commodification, turning everything into marketable goods. With capitalism, what once was not perceived as commodity, for example land, became so. According to Wallerstein (1987:15), "Historical capitalism involved . . . the widespread commodification of processes—not merely exchange processes, but production processes—that had previously been conducted other than via the 'market.' And, in the course of seeking to accumulate more and more capital, capitalists have sought to commodify more and more of these social processes in all spheres of economic life." Nature, of course, provided the raw materials for this commodification. Nature itself, however, became commodified. In capitalism nature does not have value for its own sake. It only acquires value when it becomes a commodity (Marx 1906:661). Industrialization intensified this view. Goods were produced on a grand scale. This meant that the industrial apparatus of Europe hungered for raw materials, energy, and new markets. The economic and political systems of Third World countries, as well as their physical environments, were adulterated in order to fulfill these needs.

Later, the apparent economic success of Western countries made these economies models for the new emerging countries of the post-colonial era. Commodification, mechanization, and industrialization were equated with modernization, development, and progress. The message was that third-world countries should develop along the Western model. Tropical rainforests were defined as "undeveloped" regions that eventually would be developed (Barbosa 1993a; 1993b).

It is important to note that Western countries promoted and financed their model of development in the Third World. They created key international organizations (e.g., the World Bank and the IMF), that guaranteed the implementation of the model. Poor ex-colonies wishing to develop had to adhere to the guidelines of these institutions to receive assistance. . . .

Pre-Mid-1980's Global Ecopolitics and Amazonia

Brazilians inherited the West's vision of economic development via the diffusion of ideas in the world-system, including through the diffusion of academic works on development. They believed that, due to its size and natural resources, Brazil would one day enter the small club of developed countries. The natural wealth of Amazonia they thought would be Brazil's ticket to the developed world.[6] The perception of wealth and demographic emptiness in Amazonia also created a fear that it could be invaded by foreign powers. . . .

The desire to integrate Amazonia economically with the rest of Brazil for fear of foreign takeover has been an important feature of Brazilian history. The Republican Constitution of February 24, 1891, already called for the transfer of the capital from coastal Rio de Janeiro to the Central-West region of the country for the purpose of developing the "interior" (Kubitscheck 1975). This feat was not accomplished until April 21, 1960, when the new capital, Brasilia, was inaugurated. The building of an entire city was a costly enterprise, as has been the expansion of the frontier in general. The building of Brasilia was the first step in the process of expanding capitalism into Amazonia.[7] It created a network of roads that allowed penetration into the forest. Roads were constructed connecting it to the major costal cities of Brazil, such as Rio de Janeiro and Belo Horizonte, allowing migration from these cities to the interior. The new capital in the heartland of Brazil was also connected to the city of Belem at the mouth of the Amazon river with the 1,900 kilometer, all-weather Belem-Brasilia highway, completed in

1964 (Mahar 1989:12–13). It became the focal point of regional deforestation in recent decades. When the military took over in 1964, the whole process of capitalist expansion accelerated. Military leaders were obsessed with the possibility of a foreign takeover of the region. They wanted to integrate Amazonia with the rest of Brazil quickly, so allied themselves with big capital in this endeavor. All sorts of tax incentives were given to businesses to settle in the region along the Belem-Brasilia highway. Foreign capital was easily obtained from international lending institutions, such as the World Bank, because the military coup had the approval of the U.S. government, which supported the overthrow of the leftist government of President Joao Goulart. The U.S. feared that Brazil might transform itself into a South American Cuba (see Black 1977).[8] The price of U.S. support, however, was the opening of the Brazilian market to foreign investors. For example, U.S. assets in Brazil grew from US$846 million in 1966 to US$2.033 billion in 1973 (Evans 1979:168). Brazil's links with the capitalist world-economy were thus strengthened. Brazil followed a model which Peter Evans (1979:10) called "dependent development": development dependent on the infusion of international capital.

From the beginning, the military saw big business as the most viable way of implementing its *Programa de Integracao Nacional* (Program of National Integration/PIN) in Amazonia. It provided seductive tax incentives for corporate ventures.... The most common form of investment has been cattle ranching. However, many of the ranches are idle; they were created simply to take advantage of tax incentives. The biggest of these ranches were owned by major multinational corporations. Volkswagen, for example, bought a 139,392 hectare ranch in the district of Santana do Araguia in the state of Para. Swift-Armour-King (Detect International Packers) and Liguigas (an Italian multinational) owned 72,000 and 560,000 hectare ranches respectively (Branford and Glock 1985; Campuzano 1979; Davis 1977; Irwin 1977). Some of these ranches received substantial financial and technical assistance from international organizations. The Swift-Armour-King ranch received US$11 million from USAID and Suia-Missu, the Liguigas ranch, US$32 million (Campuzano 1979:32), despite the environmental devastation associated with cattle ranching.

In addition to opening Amazonia to large investors, the Brazilian military devised a series of grand projects for the region.... I address four of these major projects here: the Transamazon highway, and the Jari, Carajas, and Polonoroeste projects.

The Transamazon (Br-320) project called for the construction of a 4,960 Km highway crossing the Amazon from east to west: from Recife and Joao Pessoa on the Atlantic Coast of Brazil to Cruzeiro do Sul near the Peruvian border (Campuzano 1979). The project was never completed due to difficulties of the terrain, lack of capital, and the failure of colonization projects along its extent. The rationale given by the Brazilian government for starting it was that it would connect "men without land with land without men;" in fact, allowing the government to avoid the explosive issue of land reform in the Northeast. Critics called the Transamazon the road that went from nowhere to nowhere (Bunker 1985). And their criticisms were well-founded. The colonization schemes along the road failed to live up to expectations. By 1974, only 5,700 families had settled, less than 10% of the target set by the government (Mahar 1990). The World Bank and the Inter-American Development Bank contributed to the project by lending US$400 million to the Brazilian National Highway Department (DNER), whose function was "to form a unified network of highways in which both civil and military interests regarding national integration would be taken into account" (Davis 1977:63). It is important to note that "these loans represented the largest grants ever made to any country for highway construction in the history of the World Bank . . . " (Davis 1977:64; see also Campuzano 1979). Multinational corporations also contributed substantially to Project Radam, a project using radar to map the natural resources of Amazonia (Campuzano 1979:31). The end result of this highway project was an increase of foreign debt for the Brazilian people, the invasion of the lands of at least five Indian tribes (the Jurunas, Arara, Parakana, Asurini, and Karao), and the needless deforestation of thousands of hectares of virgin forest.

Another project that shows the alliance between the Brazilian government and foreign capital is the Jari Project. Jari was a 1,618,800 hectare, US$1 billion project devised by American tycoon Daniel K. Ludwig (Kinkead 1981:102–103). Forecasting a shortage of fiber in the world economy by the year 2000, Ludwig intended Jari to fulfill this need. Parts of the project were also devoted to cattle and rice production. In his search for a quick-growing plant, he imported *Gmelina arboria* from Asia. It proved disastrous. As a non-native species, *Gmelina* was not adapted to local conditions. . . . Ludwig unloaded the failed project to a consortium of 22 Brazilian companies in 1982. The project was renamed "New Jari" (Fearnside 1988:83). It received a loan of US$60 million (1964) for cattle raising and US$ 76 million for beef production (1967–1972) from the World Bank. The project intruded into the lands of nine Apalai Indian villages (Campuzano 1979:33).

Perhaps the greatest influence of foreign capital in Amazonia has been in the mining sector. From 1982 to 1986, foreign capital owned more than 38% of mining rights (subsolo) in Amazonia.[9] This percentage declined substantially after the passage of the 1988 Constitution which imposed limitations on foreign ownership of Brazilian land. By 1990, direct ownership of Amazonia land was limited to only six percent. Brazilian companies, allied with foreign capital, held another 10.5% (Brasiliense 1991). Despite a decline in direct ownership of land, the presence of foreign loans in the mining industry has been critical. The best example of this is the Carajas Project.

Carajas is immense. It occupies 800,000 square kilometers, close to 11% of the Brazilian territory. It will involve an estimated US$62 billion in investments.

Mining is the backbone of the project, even though it also entails cattle raising, forestry, and farming. The area is extremely rich in mineral ores. In addition to gold, nickel, copper, manganese, and bauxite, it contains 18 billion tons of iron ore with the highest iron content ever found (Hall 1987:533; Hoge 1981). . . .

Another project worth noting is the Northwest Brazil Integrated Development Program or *Polonoroeste*. While formulated prior to the mid-1980's, the international financing for this project was affected by the greening of global ecopolitics. Before discussing this project, I will finish the discussion of the impact of capitalist penetration before the mid-1980's on the people of the forest.

The People of the Forest prior to the 1980's

The original people of the forest were the Native Brazilians. Their population at the time of the arrival of the Portuguese in 1500 has been estimated at anywhere between one and five million. The diseases and abuses of contact reduced their numbers to a current 220,000 (Treece 1990:265; Worcman 1990:51). Non-native people also depend on the forest. *Caboclos*,[10] for example, survive by typically combining horticulture, extraction, hunting, and fishing in various proportions (Bunker 1980:17). The Brazilian census of 1980 also counted 68,000 rubber-tapper families living in the forest (Fearnside 1989:387).[11] They are the descendants of people who migrated to the region primarily from the Northeast of Brazil during the rubber boom of the late 19th and early 20th century. There are also an estimated half million people in Amazonia who make a living by gathering, not only latex, but also Brazil nuts and other forest products (Hyman 1988:24).

Rivalries have always existed among the people of the forest. Indians fought among themselves prior to the arrival of the Portuguese. *Caboclos* and rubber tappers saw the Indians as a threat to themselves and were thus frequently hostile to them. But, these people had to put their differences aside due to the dimensions of the threat they faced. They could only survive by forging alliances. . . .

The military's Program of National Integration accelerated the expansion of the frontier to previously unknown levels. The "non-Indians" began to collide once again with the Indians. Conflicts were in many ways worse than earlier ones. The expansion was orchestrated and heavily funded, and disastrous for the Indians. Many tribes disappeared. Those who survived had their lands stolen and their cultural traditions corrupted.

The government created the *Fundacao Nacional do Indio* (National Indian Foundation/FUNAI) to be in charge of Indian affairs. As happened with the U.S. Bureau of Indian Affairs, FUNAI was in reality not created to defend Indian interests but to assimilate or "civilize" Indian populations. It embodied the ethnocentric view of the government, and of Brazilian society as a whole: that the Indians were primitive savages and that they should be prepared ("civilized") for life in modern society. . . . FUNAI thus functioned as a systemic agent of destruction from the very beginning: an agent of internal colonialism.

With each new boom, with each new project, with each new road, the Indians were pressed harder against the wall. The ventures in Amazonia brought in more and more landless peasants from other regions of Brazil.

The Indians were not the only people affected by capitalistic expansionism. Even though themselves a product of capitalist expansion, *caboclos* and rubber tappers also came to live in harmony with the forest they depended on for their livelihoods. They became people of the forest, and the penetration of capitalism also clashed with their way of life. The expanding frontier meant deforestation, the end of their extractive resources. The incoming rich landowners and the landless peasants were not interested in preserving the forest, but in clearing it for agriculture and cattle ranching.

The rich landlords presented a special problem. Due to the poor soils of Amazonia, the clearing of virgin forest for new pasture is necessary if cattle growing is to persevere. Thus, powerful landowners were constantly encroaching on forests claimed by people like the rubber tappers. Those who resisted the encroachment were often killed. . . .

The people of the forest resisted but they faced major problems. First, rich and corporate landowners and large government projects, such as *Carajas,* had the financial backing of the Brazilian government and international lending organizations. Second, the people of the forest themselves were politically divided and not yet politically organized. Third, potential allies like environmentalists and nongovernmental organizations, did not yet have the political leverage they now exercise on a global scale. Changes in the ecopolitics of the world-system in the mid-1980's would empower these groups and they, in turn, would help empower the Brazilian anti-systemic groups.

The Greening of Global Ecopolitics in the Mid-1980's

Global ecopolitics changed significantly beginning in the mid-1980's. The world entered a period of discovery about the state of the environment. The hole in the atmosphere's ozone layer, the greenhouse effect, the disaster at Bhopal, and the Exxon Valdez spill, all helped to sensitize public opinion about the need to preserve the environment. . . . Perhaps of most importance in changing global ecopolitics was the possibility of atmospheric changes caused by the greenhouse effect and the hole in the ozone layer. The economic consequences of such a change, coupled with pressure from environmentalists, forced politicians in the U.S. to focus on the destruction of tropical rainforests in the Third World. Tropical rainforests were being burned at alarming rates in countries such as Brazil, thus aggravating the greenhouse problem through carbon dioxide emissions.[12] Organizations that lent money to development projects in the Third

World, notably the World Bank, suddenly came under criticism for contributing to the problem.

The World Bank had been the most powerful force for environmental destruction in the Third World. It invested heavily in large development projects such as hydroelectric dams, and highways. It was assumed that the benefits of these projects would "trickle down" to the poor. But critics pointed out that World Bank financing benefitted primarily the elites and the multinational corporations operating in the Third World, particularly by providing them with infrastructure necessary to continued development. . . . The trickle-down policy of the Bank came under severe attack. Environmentalists argued that large projects often harmed the poor, for example, by displacing them. These projects were also viewed as conducive to environmental destruction (Subcommittee on International Economic Policy and Trade 1989).

. . . [T]he greening of the Bank really began in 1987. Pressured by the criticism of environmentalists, then-president Barber Conable announced a sweeping reorganization. Even though 500 positions were eliminated, 40 new environmental positions were created (Hopper 1988:769; Holden 1988:1610). The Bank also became stricter about requiring environmental impact studies before the approval of loans. . . .

The Greening of the World-System and Brazilian Ecopolitics

Jose Sarney's Brazilian government (1985–1990) reacted with anger to the new environmental restrictions. Brazil's international allies for development projects were turning their backs. Sarney resisted the pressure and called the criticisms of Brazil "unjust, defamatory, cruel, and indecent" (Linden 1989). He argued that the Brazilian economy needed to grow in order for Brazil to pay its staggering foreign debt. He also argued that the U.S. had no right to criticize Brazil when it spewed more pollutants than any other country (Linden 1989; see also Barbosa 1990). The anger of the Brazilian government is perhaps best exemplified by *Polonoroeste*, the Northwest Brazil Integrated Development Programme. *Polonoroeste* was created to coordinate and expand the spontaneous migration and development already taking place in the states of Acre, Rondonia, and parts of Mato Grosso, an area of 410,000 Kms (Mahar 1990:64). The conflict between the Brazilian government and its former allies, the international lending institutions, involved the extension and paving of Highway BR-364. The paving of this road to the state of Rondonia brought in an estimated 160,000 migrants a year. Environmentalists and the people of Acre insisted that the extension of the road to their state would bring about similar migration and the resulting environmental devastation. The road was eventually supposed to reach the border with Peru and thus connect Brazil with the Pacific Ocean. Environmentalists feared that this would accelerate the process of destruction, allowing, for example, the export of Amazonian timber to Japan.

Due to pressure from environmentalists, the World Bank stopped its financial support to Polonoroeste in 1985. Brazil then attempted to obtain loans from the Inter-American Development Bank but failed because the growing political leverage of anti-systemic forces made it politically costly for the Bank.

The People of the Forest after the Mid-1980's

The acquisition of political power by the people of the forest began with the re-democratization of Brazilian society beginning in the late 1970's, a process known as *Abertura Democratica*, or Democratic Opening. After decades of political repression, Brazilian anti-systemic forces were unleashed: media censorship was abolished, opposition parties were allowed, and civil disobedience was tolerated. This process allowed not only the people of the forest to be heard, but it also allowed an open debate on the Amazonian question. The new political freedom also allowed the people of the forest to form coalitions. The Amazonian Alliance of the People of the Forest was created in the state of Acre in 1986 "to unite the National Council of Rubber-Tappers and the Union of Indian Nations in their fight to defend their forests and ways of life . . ." (Treece 1990:285).

Political leverage increased substantially when coalitions with international environmentalist organizations were formed. The campaign to stop the expansion of BR-364 is very illustrative of this mutual help. When the Brazilian government tried to secure a loan from the Inter-American Bank for the paving of BR-364, Chico Mendes was flown to the United States by American environmental organizations to lobby top executives of the World Bank and other American politicians. The plight of the rubber tappers received considerable media attention as a result. The loan was never approved. Native Brazilians were also brought to the U.S. by these organizations for lobbying and publicity purposes, receiving the support of media personalities like the rock singer Sting.

The publicity received abroad helped make Brazil an environmental villain in the 1980's. This image contributed to international loans drying up or being conditioned by environmental clauses. Because Brazil is development-dependent, it is vulnerable to pressure. In 1988, two extractive reserves were established in the state of Acre. Four more were proposed (see Fearnside 1989).[13] Pressure was also mounting within Brazil and abroad for the demarcation of Indian lands. Despite Chapter VIII in the New Constitution of 1988 which is dedicated to Indian rights and land ownership (Brazil 1988:132–133), *de facto* victories for the Indians would come only with the election of Fernando Collor de Mello in 1990.

Collor became the first president elected by the people in 27 years. His political platform was in line with American and European conservative ideology:

small government, less government spending, and privatization. Collor also promoted the idea that since Brazil is one of the world's top ten economies, it should work to become a developed country. It should align itself not with Third World countries, but with the First World where it strives to belong. This desire made Collor's administration more receptive to foreign complaints of environmental destruction in Amazonia, an attitude which can be seen as an attempt to normalize Brazil's relations with international lending institutions. This pro-environmental position was also part of a package to renegotiate Brazil's mounting foreign debt, now US$120 billion. Collor was quite aware of Brazil's dependent status in the world economy. Instead of antagonizing foreign institutions on the environmental issue, his administration played along with their new concerns. . . .

Major victories for the Indians during the Collor administration included the demarcation of Kayapo and Yanomami lands, in November, 1991.[14] The Kayapo reservation comprises an area of 4.9 million hectares, slightly larger than Switzerland,[15] and the Yanomami, 9.4 million hectares, roughly the size of Portugal. In one month, Collor protected half again as much Indian land as was set aside in the proceeding 80 years (Inter-American Development Bank, United Nations Development Programme, and Amazon Cooperation Treaty 1992).[16] This would have been impossible without changes within Brazil and in the ecopolitics of the world-system.

It is also important to note that the demarcation of Indian reservations does not necessarily mean that Indians are protected within their boundaries. For example, in August 1993, the Yanomamis were subjected to a massacre in their reservation by invading gold prospectors. It has been difficult to ascertain the number of Indians killed because Yanomami have the cultural tradition of cremating their dead. Estimates range from 20 to 120. Reports of the massacre by the Brazilian and the international media prompted the Brazilian government to investigate the case (*"Pouca luz na selva"* 1993).

Conclusions

Often, individual countries are taken as the unit of analysis in discussions of rainforest destruction. Even at macro levels, the economic connections between poor and rich countries are often missing. The Brazilian case discussed here suggests the fallacies of this method. Changes in the ecopolitics of the world-system beginning in the mid-1980's accelerated changes in the ecopolitics of Brazil. These changes benefitted anti-systemic environmentalist forces everywhere, but were especially beneficial for the people of the forest in Amazonia. Amazonian forest people benefitted from the disruption in the alliance of the Brazilian government with international organizations, especially the World Bank. Given the relatively futile centuries-old resistance of the Amazonian

Indians, it is doubtful that without these changes in global ecopolitics the people of the forest would have their voices heard to such an extent.

Taking world-systemic processes into consideration does not diminish the importance of internal or national processes. Throughout their history, Brazilians developed their own attitude about Amazonia as a reservoir of raw materials and a ticket to the future. They still fear its internationalization. This fear is a critical component in understanding ecopolitics in Brazil. The policies of the Brazilian state to develop Amazonia is directly related to the possibility of a foreign takeover of the region.

Notes

Barbosa, Luiz C. 1996. "The People of the Forest against International Capitalism: Systemic and Anti-Systemic Forces in the Battle for the Preservation of the Brazilian Amazon Rainforest." *Sociological Perspectives* 39 (2): 317–31.

1. The area defined as legal Amazonia in Brazil is just under 5.0 million km^2 or 58% of the Brazilian territory (Mahar 1990:59). It remained relatively undisturbed until the mid-1970's. Landsat survey data show that in 1975 0.6% (28,595.25 km^2) of legal Amazonia had been cleared. This figure, however, increased to 1.5% (77,171.75 km^2) by 1978. By 1980 the destruction increased to 2.5% (125,107.8 km^2) and by 1988 to 12% (598,921.5 km^2) (Fearnside 1986:5; Mahar 1989:6), an almost fivefold increase in the rate of deforestation in eight years.

2. Systemic forces are capitalist-motivated. They are represented by profit-oriented organizations, such as private banks or others which seek to advance the cause of global capitalism (e.g., the International Monetary Fund/IMF). The modus operandi of these organizations has been conducive to the destruction of tropical rain forests. They have functioned under the principle that nature is either a source of raw materials for development or a source of profits.

3. The anti-systemic forces do not necessarily oppose the profit motive of capitalism. They do oppose predatory capitalism; that is, the idea of profits and development at all cost. They also defend the rights of the people of the forest. The goal is either total preservation or at least environmentally sustainable economic development.

4. According to world-system theory, since the "long-sixteenth century" (1450–1650) the European capitalist economy expanded to engulf the local economies of other parts of the world. This expansion created a world-system, stratified into core, periphery, and semi-periphery. The Western European core was joined in the 20th century by the United States and Japan. The periphery is what we think today as the Third World. The relationship between the core and the periphery has been one of exploitation. The resources of the periphery—natural resources, labor, etc.—are exploited for the benefit of the core. The shifting semi-periphery is the world's middle class. It is composed of declining core countries or of the few countries that managed to break periphery status (see Chirot and Hall 1982).

5. The term *ecopolitics* is used in this paper in two ways. First, as the general position of governments and international organization on environmental issues, as reflected in policies and documents. Second, as political maneuvering or conflict between groups in

favor of developing ecosystems and those in favor of preservation. The ecopolitics of the world-system (or global ecopolitics) is the predominant environmental view in the system, reflected largely by such international organizations as the World Bank and international public opinion.

6. The Amazon rainforest has always been surrounded by a myth of wealth. Its lushness proved to be not from the fertility of its soil, as previously believed, but from the natural recycling of the forest itself. Surveys show that 75% of the Amazon region is composed of nutrient poor, acid Oxisols and Utisols. A further 14% of the region is occupied by poorly drained alluvial, flood plain and palm swamp areas (Furley 1990: 310). On the other hand, parts of Amazonia have been wealthy in minerals, for example, the *Carajas* region.

7. We say "expanding" because Amazonia was already linked to the capitalist world-economy through the extractive economy of the region.

8. It is important to note that the justification for the development of Amazonia by the military was grounded on the National Security Doctrine. According to Jose Goldemberg and Eunice R. Durham (1990:30), this doctrine was a byproduct of the Cold War. It "was created by the United States to insure its hegemony in Latin America and was adopted by the large sectors of Brazil's armed forces. Geopolitical conceptions in that doctrine characterized all governmental policies during that period, furthering a real militarization of the Amazon question."

9. Brazilian state companies owned 30% and the Brazilian private sector 32%. These figures do not show, however, the partnership between Brazil's privately-owned companies and foreign capital (Brasiliense 1991).

10. *Caboclo* is the general term for the members of the rural lower class in Amazonia. The term is often applied to people of Indian-"white" ancestry who live alone in isolated areas along rivers (Fearnside 1986:232).

11. It is important to note that rubber-tapper organizations dispute this figure. They claim that the number of tappers is much higher than was registered (Fearnside 1989:387).

12. While the problem of deforestation indeed was urgent, the fact that the developed countries were the greatest emitter of carbon dioxide was often put aside.

13. Fearnside reports nine existing extractive reserves comprising area over 5,531 km^2 in 1989, when the article was published. Eleven other reserves were also being proposed comprising together an area of 16,630 km^2. A total of 2,290 families occupied the total area of over 22,161 km^2.

14. There were other minor victories as well. For example, Collor legalized the 35,922 hectares of Caigangue Indian lands in the Southern State of Rio Grande do Sul.

15. The demarcation of Kayapo lands clearly shows the importance of international allies in helping the Indian cause. US$1.2 million were raised to help with the demarcation by the rock singer Sting, who in the late 1980's and early 1990's promoted the cause of Brazilian Indians. It is ironic that when the Kayapos began selling their trees to lumber companies the alliance with Sting ended. The Brazilian magazine, *Veja,* called it "O fim do romanticismo" or "The End of Romanticizing [the Indian as a "noble" savage] ("O fim do romanticismo" 1993).

16. Severe problems still persist in the Yanomami reservation. Yanomami lands continue to be threatened by invading gold prospectors. The thickness of the forest and lack of equipment and staff make the task of patrolling the area extremely difficult.

References

Barbosa, Luiz. 1993a. "The 'Greening' of the Ecopolitics of the World-System: Amazonia and Changes in the Ecopolitics of Brazil." *Journal of Political and Military Sociology* 21(1): 107–134.

Barbosa, Luiz. 1993b. "The World-System and the Destruction of the Brazilian Rain Forest." *Review* (A Journal of the Fernand Braudel Center for the Study of Economies, Historical Systems, and Civilizations) 16(2): 215–240.

Barbosa, Luiz. 1990. "Dependencia, Environmental Imperialism and Human Survival: A Critical Essay on the Global Environmental Crisis." *Humanity and Society* 14: 329–344.

Black, Jan Knippers. 1977. *United States Penetration of Brazil*. Philadelphia: University of Pennsylvania Press.

Branford, Sue, and Oriel Glock. 1985. *The Last Frontier*. London: Zed Books Ltd.

Brasiliense, Ronaldo. 1991. "Subsolo da reserva ianomami e cobicado por 25 empresas." *Jornal do Brasil*, December 8, 1991, 1º Caderno, p. 14.

Brazil. 1988. *Nova Costituicao do Brasil*. Rio de Janeiro: Grafica Auriverde, Ltda.

Bunker, Stephen G. 1985. *Underdeveloping the Amazon*. Urbana, IL: University of Illinois Press.

Bunker, Stephen G. 1980. "Forces of Destruction in Amazonia." *Environment* 22: 14–42.

Campuzano, Joaquin Molano. 1979. "As multinacionais na Amazonia." *Econtros com a Civilizacao Brasileira* 11: 21–34.

Chirot, Daniel, and Thomas D. Hall. 1982. "World-System Theory." *Annual Review of Sociology* 8: 81–106.

Davis, Shelton H. 1977. *Victims of the Miracle*. Cambridge, MA: Cambridge University Press.

Evans, Peter. 1979. *Dependent Development*. Princeton, NJ: Princeton University Press.

Fearnside, Philip. M. 1989. "Extractive Reserves in Brazilian Amazonia." *BioScience* 39: 387–393.

Fearnside, Philip M. 1988. "Jari aos dezoito anos: licoes para os planos silviculturais em Carajas." *A Amazonia Brasileira em Foco* 17: 81–101.

Fearnside, Philip M. 1986. *Human Carrying Capacity of the Brazilian Rainforest*. New York: Columbia University Press.

Furley, Peter A. 1990. "The Nature and Sustainability of Brazilian Amazon Soils." Pp. 309–358 in *The Future of Amazonia*, edited by David Goodman and Anthony Hall. New York: St. Martin's Press.

Goldemberg, Jose, and Eunice Ribeiro Durham. 1990. "Amazonia and National Sovereignty." *International Environmental Affairs* 2: 22–39.

Hall, Anthony. 1987. "Agrarian Crisis in Brazilian Amazonia: The Grande Carajas Programme." *The Journal of Development Studies* 23: 522–552.

Hemming, John. 1987. *Amazonian Frontier*. Cambridge, MA: Harvard University Press.

Hoge, Warren. 1981. "Big Amazon Project Unfolds." *The New York Times*, November 18.

Holden, Constance. 1988. "The Greening of the World Bank." *Science* 240(4859): 1610.

Hopper, David W. 1988. "The Seventh World Conservation Lecture The World Bank's Challenge: Balancing Economic Need with Environmental Protection." *The Environmentalist* 8: 165–175.

Hyman, Randall. 1988. "Rise of the Rubber Tappers." *International Wildlife* 18: 24–28.

Inter-American Development Bank, United Nations Development Programme, and Amazon Cooperation Treaty. 1992. *Amazonia Without Myths.* (No place of publication or publisher given.)

Irwin, Howard. 1977. "Coming to Terms with the Rain Forest." *Garden* 1: 29–33.

Kinkead, Gwen. 1981. "Trouble in D.K." *Fortune*, April 20.

Kubitscheck, Juscelino. 1975. *Por que Construi Brasilia.* Rio de Janeiro: Edicoes Bloch.

Linden, Eugene. 1989. "Playing with Fire." *Time*, September 18: 76–79.

Mahar, Dennis J. 1989. *Government Policies and Deforestation in Brazil's Amazon Region.* Washington, D.C.: The World Bank.

Mahar, Dennis J. 1990. "Policies Affecting Land Use in the Brazilian Amazon." *Land Use Policy* 7: 59–69.

Marx, Karl. 1906. *Capital.* New York: The Modern Library.

"O fim do romantismo." *Veja*, April 28, 26(17): 74–75.

"Pouca luz na selva." *Veja*, September 1, 26(35): 28–29.

Subcommittee on International Economic Policy and Trade of the Committee on Foreign Affairs, House of Representatives, U.S. Government. 1989. "Environmental Impact of World Bank Lending, Volumes I and II." Washington, D.C.: U.S. Government Printing Office.

Treece, David. 1990. "Indigenous People in Brazilian Amazonia and the Expansion of the Economic Frontier." Pp. 264–287 in *The Future of Amazonia* edited by David Goodman and Anthony Hall. New York: St. Martin's Press.

Wallerstein, Immanuel. 1987. *Historical Capitalism.* New York: Verso.

Worcman, Nira Broner. 1990. "Brazil's Thriving Environmental Movement." *Technology Review* October: 42–51.

Tourism, Environmentalism, and Cultural Survival in Quintana Roo

Oriol Pi-Sunyer and R. Brooke Thomas

This article was written by two anthropologists interested in the cultural, economic, and environmental consequences of tourism in Mexico's Yucatán Peninsula. Tourism is promoted as a means of economic development in many parts of the world (and more recently has been advocated as a green alternative to more polluting industries). The authors reveal how tourist development, by constructing natural environments as pristine and uniquely beautiful, and therefore desirable to consumers, often destroys the very uniqueness of those places. In addition, tourist development has profound consequences for the people living in tourist destinations. Many of us will be tourists at some point in our lives, and we should be aware of some of the hidden negative consequences of tourism.

In the course of the last 25 years, the state of Quintana Roo, which occupies the eastern portion of the Yucatán peninsula, and which until 1974 had been a thinly settled federal territory, has been experiencing the massive penetration of tourism and collateral processes. This has transformed the area, and particularly the coast, from one of the most isolated regions of Mexico to a veritable tourist factory. Quintana Roo, for all the vast sums of money poured into the construction of resorts and supporting infrastructure, continues to offer beautiful beaches and reefs, Mayan pyramids and tropical forest environments—all a short jet-flight from Florida and the eastern United States. Cancún has become an architectural fantasy city with its grandiose hotels and full range of distractions and attractions vying for tourists' attention and money. Sun-and-sea resorts have now have spread southward to almost every available lagoon and beach; archaeological tours, including New Age pilgrims in search of ancient knowledge, disgorge bus after bus at Maya sites; and ecotourism has sought out remote locations deep in the interior forests.[1]

By standard economic measures, tourism development in this part of Mexico has been an unqualified economic success for the Mexican government and major investors, international and national; the Cancún model has even been replicated in other parts of the country and beyond. This tourist-driven transformation has been minimally studied, and it is not clear how it is affecting the Maya: their health and diet, the environment and resources upon which they depend, their social organization and collective identity. We should recall that these are the native people of the Yucatán, the descendants of the builders of the pre-Hispanic cities and ceremonial centers. Furthermore, although the Maya now find themselves a numerical and ethnic minority in Quintana Roo, the Maya-speaking people constitute one of the largest indigenous groups in Mesoamerica. These collectives are pressing—as has become very evident in Chiapas—for cultural recognition, autonomy, dignity, and social and economic justice.[2] The history of resistance stretches back centuries. Since colonial times, the forests of Quintana Roo have served as a refuge area and region of resistance for the Maya as they fled the exploitation of Spanish and Mexican landowners and administrators. Here, in relative isolation, they lived in small communities dependent on slash-and-burn agriculture, beekeeping, hunting, and the utilization of forest resources.

Today, as Mexican and Mayan history collide once again along the Caribbean coast, this way if life is being severely challenged. The customs, rituals, and language that have held Mayan communities together, binding them to their ancestors, the spirit world, their land, and one another, are being challenged externally by agents of change, and internally by members of their own communities. A grid of all-weather roads crisscrosses the state, bringing with it a growing degree of demographic, economic, and political penetration. Tourism, government services, and commercial activity directly and indirectly touch all Maya settlements. Flowing out of the hamlets and villages are young men and women in search of work on the coast. Ideas of "progress" return with them, following the lines of electrification and the Coca-Cola trucks. Generations in the community seem split as the elderly attempt to hold on to traditional usages while younger people confront modernity. The questions "Who are we, where are we going, and what will become of us?" are at the core of a continuous discourse that takes many specific forms.

Our research attempts to give voice to these concerns, placing them within the broader political economic context of tourist development and the encompassing dependency relationships that characterize this industry. Tourism in locations such as Quintana Roo faces enormous competition from similar destinations, and tourism policy is only partially within the control of host countries. In a very real sense, Quintana Roo can be thought of as constituting a "double periphery": it lies at the margins of a national state that is itself highly dependent on its powerful neighbor to the north, the United States. Within this structure of dependency, indigenous peoples—and certainly the Maya of the Yucatán—are the most subordinated.

The perspectives and opinions of the Maya concerning such issues remain largely unheard as they move through the tourist world as silent and invisible waiters and domestic workers, groundskeepers, construction workers, petty vendors, and hammock weavers. Experience, age, gender, and levels of education all play a role in influencing their hopes and perceptions. For the most part there is little outright resistance to the technological and material aspects of consumer society; however, there appears to be a growing awareness of the costs of shifting from subsistence agriculture to a cash economy. Most obviously, the wages of the average Maya worker are pitifully low. As a higher percentage of the population experiences some formal education, and television becomes the entertainment of choice in the majority of Maya households, attitudes toward modernity become marked by ambivalence. Money—or lack of it—is a factor, but also the caution and resentment of becoming a marginalized people in their own land. And with what or whom can they identify in the mass-mediated messages they receive through television? Fundamentally, their actions reflect patterns of accommodation and resistance appropriate to very confusing times, when the institutions of the state are under siege. And what it means even to be Maya is constantly being negotiated, as everyone in Mexico is poised between the uncertain present and a future filled with hope and risk. In listening to the Maya, one learns to examine many things—including tourism—from the bottom up. ...

The Structure of the Touristic System

Is tourism a "life and death matter" as we approach the end of the millennium? Perhaps not in a literal sense, but for many inhabitants of Third World societies, and especially for indigenous groups in such countries, tourism represents a particular face of both development and forced social change. Fundamentally, this externally driven process works to dismantle corporate identities and reduces the ability of individuals and communities to survive economically and culturally.

For some groups it is thus not an overstatement to speak of a "crisis" engendered by tourism: a crisis variously of identity, employment, and security.

Tourism has received limited attention from anthropologists and other social scientists. There is little in the classical tradition of either anthropology or sociology that discusses the phenomenon, and even today it is sometimes difficult for academics to take the subject seriously.[3] Yet, as Erik Cohen observed more than a decade ago, tourism deserves to be treated as something other than an exotic and marginal topic derived from a superficial activity.[4] The economic case can be argued with little difficulty. Travel is now the single-largest item in international trade. Some 425 million visitors generate US$230 billion in expenditures per year.[5] Since the 1960s tourism has grown at the rate of some 10 percent per year, a trajectory that is likely to continue for the rest of the century.[6] A phenomenon of this type—global in scope, touching societies of different types, manifesting itself in sundry forms and scales—is not inherently all good or all evil. Much depends on the context, and in our opinion the key analytical question is the degree to which mass tourism may work to reinforce and perpetuate relations of dependency and inequality. Dependency, as we use the concept, is a historical process that changes the internal functioning of economic and social systems in such a manner that local economies are weakened and thereby forced to serve the needs of external markets.[7]

Tourism in Mexico has become a major source of foreign exchange, a new kind of tropical export product. But tourism is also a highly competitive industry, since tourists and tour operators have numerous alternative destinations at their disposal.[8] Consequently, attracting foreign tourists and foreign capital is a difficult process that has meant austerity (particularly for the poor), a labor force under increasingly rigid discipline, and other "belt-tightening" measures. What is harder to understand is how the state itself has become a partner in this process of exploitation. At least part of the answer must be sought in the influence of a general myth of development (shared by Mexican and international technocrats alike) that holds that the poverty of poor countries arises from an inadequate spread of market forces. It follows that the recommended solution to poverty is increased capitalist penetration. An economic transformation along these lines typically entails massive inputs of capital from rich countries and a restructuring of local-level economic relations. The guiding myth is also—indeed, *has to be*—uncritically optimistic: any demonstrated failures are attributed to local aberrations, not to the model, and real social costs are always played down.

Writing of what he refers to as the ideology and practice of "developmentalism," Arturo Escobar observes that:

> Wherever one looked, one found the repetitive and omnipresent reality of development: governments designing and implementing ambitious development plans, institutions carrying out development programs in city and

countryside alike, experts of all kinds studying underdevelopment and producing theories ad nauseam.[9]

Escobar could easily have been discussing the plans, first formulated in the late 1960s, drawn up to develop the coast of Quintana Roo into a major tourist destination dominated by the new resort city of Cancún.

The putative rewards of tourism are many, including international and cultural understanding, but the tangible benefits, it is claimed, are chiefly economic and easily demonstrable. Tourism, as we have noted, is essentially a sort of internal export industry. What are sold are the various "attractions" of the country, be these natural or cultural. In the process, a flow of foreign exchange is generated and one can balance the money coming in against the money going out. In the case of Mexico, the figures are strongly favorable: in 1988, the 5.69 million foreign visitors entering the country brought in $1.35 billion more than Mexicans spent abroad, a balance that had risen to $1.79 billion by 1992.[10] Furthermore, the claim is often made that tourism is the ideal "smokeless" industry—a relevant selling point in this era of ecological anxiety.

Tourism differs from other export sectors in that production and consumption both take place at home—an appealing feature for countries such as Mexico that suffer high rates of unemployment and underemployment. Again, to turn to specifics, some estimates for Mexico assert that the hospitality industry is responsible for about 3 percent of the gross domestic product and generates almost 2 million direct and indirect jobs.[11]

Tourism not only employs many people, but it does so in ways significantly different from most other industries. Many tourism-related occupations call for little formal education or specialized training. In addition, tourism in general, and resort tourism in particular, often hires people from surrounding semirural environments, thus presumably helping to stem migration to the overcrowded cities.[12]

Finally, a number of indirect advantages are commonly attributed to the expansion of this sector. Given that tourism in less-developed countries caters to many guests from wealthier societies, a whole series of improvements in services and infrastructure can be expected to benefit both visitors and residents. Roads must be built and maintained, systems of sanitation and potable water installed where these are not present, and access made available to health providers, communications systems, and the like. This piggyback effect again seems unproblematic to planners and bureaucrats in distant offices.

A related argument is linked to the needs of tourists. Not only do they consume "services" in the abstract, but they must be fed and housed, and provided with all those tangible tokens—the straw hats, the textiles, the carvings—that make the visit memorable. Tourist expenditures, therefore, should bring direct benefits to local shopkeepers, farmers, craft workers, and even prostitutes. The tourist dollar, once spent, is assumed to touch and benefit many people through what is termed the multiplier effect.

The Other Side of the Tourist Coin

This is necessarily a brief overview of the most common arguments presented in favor of the type of touristic development that has taken place in Quintana Roo, but anyone with minimal experience of tourism in the Third World is bound to recognize that it does exact a price, although it is harder to say what form this price takes. At the level of popular discourse and media reporting, the costs of tourism are commonly expressed in cultural and aesthetic terms. Tourists are often perceived as altering host societies in ways that are negative, generally by bringing about some loss of "authenticity." Sometimes it seems that local life was inevitably better in the past, that villagers were more contented, less mercenary, more "traditional." Blame for this turn of events is put on the deluge of visitors and, in lesser measure, on natives who, somehow, should not have permitted themselves to be seduced by the forces of modernity.

The cultural costs of tourism are often painful and deserve a nuanced and sympathetic analysis. It is nevertheless true that much of the discussion of "loss" probably reflects, more than anything else, a contemporary western distress that small-scale societies and "remote" peoples will, in the words of Errington and Gewertz, become "demoralized and disoriented with the shattering of their formerly beautiful customs by modernity."[13] Such concerns may tell us more about "our" fears of anomie and disorientation than about "their" priorities in the contemporary world.[14]

Granted that there is often a cultural price to pay for tourism, the most powerful argument against the type of tourism we are discussing is a variant of the critique of development.[15] Not only have Mexican economic policies in general failed to narrow the enormous gap between the rich and the poor, but also one can reasonably claim that the expansion of tourism has further reinforced the dualistic structure of Mexican society.[16] Today, income distribution in Mexico is so skewed that the top 10 percent of the population accounts for 39.5 percent of the national income—leaving the bottom 20 percent to manage with a share of 4.1 percent.[17]

Our position is not simply that mass tourism is likely to reinforce preexisting socioeconomic structures, but rather that inequalities will impact most severely on the lives and prospects of just those populations with whom tourists come into contact, people who are often characterized by extreme material poverty and minimal political influence.[18] Clifford Geertz, discussing economic development in Southeast Asia, observes that modernizing states "do not bring all their citizens equally with them when they join the contemporary world of capital flows, technology transfers, trade balances, and growth rates," a problem that was already apparent a generation ago.[19] In 1974, the British ambassador to Panama wrote in a dispatch to the Foreign Office:

> Unless the Kuna can perform a miracle unique in our world, of teaching themselves to accommodate to and manage a 20th Century explosion of tourists, industrial development and cash nexus, it seems impossible that they will not be obliterated.[20]

Fundamentally, when peripheral regions, such as Quintana Roo, are brought into the process of modernization, local populations (commonly regarded as "backward" by the national elites) are assigned very subordinate roles in the development process sometimes to the point of invisibility. A recognition of these asymmetries helps us to engage a whole range of questions and issues. For example, it is significant that tourism development in Mexico commonly takes the form of distinct enclaves, resorts such as Acapulco, Puerto Vallarta, and Cancún. More recently, the shift has been to packaged tours and self-contained "gated" complexes that provide accommodation and recreation within highly controlled environments. Fundamentally, what we have in all such settings are pieces of the First World ensconced in Third World environments. There is no process of "convergence"; on the contrary, arrangements of this type work to keep apart the worlds of tourist and native, a gap likely to be filled by various forms of "staged authenticity."[21]

Leaving aside the question of what might have been done for local people with even some of the resources earmarked for hotels, roads, and airports, there is no question that the majority of those who work in the tourist sector live under very difficult conditions and enjoy little in the way of job security. To this we can add the phenomenon of an ethnic division of labor. Concretely, in Quintana Roo the Maya are defined as cheap manual labor, quite unsuited to positions of responsibility and authority.[22] This results in Mexicans and foreigners occupying the public relations positions and managerial roles. In conclusion, the realities of luxury tourism in Quintana Roo can be summed up as a system in which the poor subsidize the rich. "Tourism," in the words of Cynthia Enloe, "is not just about escaping work and drizzle; it is about power, increasingly internationalized power."[23]

Quintana Roo

A generation ago, Quintana Roo was one of the most inaccessible locations in Mexico, a frontier where the institutions of the state were barely represented. In 1950, the total population numbered only 26,967 inhabitants, a figure that would increase to 50,169 by 1960.[24] Thus, until the advent of mass tourism in the early 1970s, it remained little changed from what it had been for centuries:

an extensive region with a very low population density inhabited by Maya farmers in the interior and small mestizo populations in a few coastal locations.

From the 16th century to the 20th, Quintana Roo was what Aguirre Beltran[25] has so aptly termed a "zone of refuge," first for Indians fleeing Spanish control and secondly, following independence, for Maya villagers no more inclined to come under Mexican rule. The 1847 native uprising in Yucatán known as the Caste War, and the other insurrections that followed, further isolated the region. Campaigns by the Mexican army well into this century reinforced a Maya distrust of state authority.

It was not until the late 1960s that this high degree of isolation—and autonomy—began to change. First came the roads, then the airports; and with them a mounting influx of outsiders, both Mexicans and foreigners. What is particularly interesting about these changes is that up to two or three decades ago the region had hardly registered on the mental maps of urban Mexicans. When they thought about the territory at all, it was usually in terms of a backward area associated with the questionable activities of smugglers, chicle gatherers, and forest-dwelling Indians.

With the expansion of tourism the region, and particularly its coast, rapidly became redefined as a tropical paradise, a land of broad, sandy beaches and pristine forest. Tourism propaganda often contrasts the modernity of Cancún and other resorts with the theme of a changeless natural world of "still crystal-clear waters and virgin beaches," a land "that still belongs to nature, where ocelots, kinkajous and spider monkeys still roam wild in their native habitat."[26]

The reality is rather different. In the 1960s, Cancún was a village of some 600 people. Today it is not only a city of more than 300,000 inhabitants, but it is growing at an annual rate of 20 percent, a truly frightening statistic. This huge resort with over 140 hotels attracts more than 2 million visitors a year, of which 1.5 million are foreigners.[27] More than this, the whole coast from Cancún south to Tulum has been converted into a string of resorts, theme parks, and hotels, effectively transforming what had been space of public access into the highly controlled world of the managed resort.

Cancún dominates Quintana Roo, and it is very much a child of 1960s development thinking, a mega-project requiring massive investment for a colossal plan of construction. It has certainly become a major international tourist destination, one particularly attractive to North Americans.[28] But it would be wise to remember the cyclical and uncertain character of global tourism, and particularly the decline of Acapulco, once the major Mexican playground for the rich. By the 1970s, a number of factors, including the opening up of new resort areas with more modern facilities, a deterioration in the Acapulco physical environment, and the fact that the resort was no longer seen as an elite location, led to a marked drop in Acapulco's popularity and profitability.[29] Some observers claim that Cancún is already confronted by similar problems of image and security.

Ecological Anxiety

It is perhaps the realization that Cancún's long and massive line of hotels no longer quite responds to contemporary needs and tastes that has helped to bring about a diversification of the market in the direction of environmental and cultural tourism. We don't find the romanticization of a simpler, more "authentic," way of life such as has been described for cultural tourism in Mexico's state of Chiapas and in Guatemala. More than anything else, alternative tourism in Quintana Roo plays on the theme of the grandeur that was the Maya civilization (now safely in the past) and the quest for the particular kind of dream world associated with jungles and exotic animal life. Not surprisingly, one encounters many latter-day Indiana Joneses indulging their fantasies, as well as more serious ecological tourists. The impressive Maya sites of the area, particularly Tulum and Cobá, receive a large influx of visitors—and Tulum is showing clear signs of wear and tear.

A major problem related to all forms of tourism stems from the inherent fragility of the biogeographical system on which the touristic system has been erected. The tourist boom has had both anticipated and unforeseen consequences on environmental quality, consequences that, unless remedied, are bound to impinge on the profitability of the industry. At the same time, the growing environmental awareness—as well as an emerging market for ecologically friendly tourism—has directed attention to the uniqueness of the peninsula and to the crisis it faces. But these issues—problems posed by the limits of growth—hardly form part of the traditional experience-perceptions of the Maya. Prior to the advent of tourism they had seen no need to undertake programs of environmental preservation, nor had they had disputes surrounding land use practices or the utilization of their natural resources.

A Fragile, Living Rock

Generally unnoticed by the tourist, and certainly little heeded by developers, is the fragile nature of Quintana Roo's linked environments. The 16th century Spanish bishop Diego de Landa described this land thus: "Yucatán is the country with the least earth I have ever known, since all is a living rock." Contemporary Maya farmers describe much of this terrain as *zekel*: "land very stony or full of stones."[30] Precipitation from tropical storms falls on the interior forest percolating through thin pockets of topsoil; it then rapidly makes its way into the porous calcite bedrock that underlies the peninsula. From here it flows through a complex network of underground rivers to the coast where it surfaces in freshwater vents in the mangrove swamps behind the beaches, in cuts or lagoons, or beyond the beach itself. In many cases, resorts along the tourist zone are built on a lens of beach built up between the swamp and the sea, and are protected under calm conditions by the offshore barrier reef running the length the peninsula's eastern coast. Connecting these terrestrial and marine ecosystems is the underground flow of water that makes its way across the limestone platform.

In the area there has been remarkably little land-use planning or environmental impact analysis for most projects, and that which follows is a brief review of some of the more apparent consequences. Cancún now suffers from problems of urban congestion, including a decline in air quality; there has also been a substantial deterioration of the lagoon behind the beach resorts as a consequence of erosion and water pollution. The surrounding mangroves, which might counteract these processes, are being filled in for further development, and a golf course encroaches on the city's most important archaeological site.

The main highway south from Cancún to Tulum (136 km long) is lined with an assortment of billboards advertising resorts with such alluring names as Secret Beach and Shangri-La. The tourist is invited to visit Puerto Aventuras, touted as "A new civilization on the Maya Caribbean" combining experiences that are both "*primitiva*" and "*sofisticado*." These supposedly remote and beautiful locations are often the source of very tangible environmental problems. The complexes consume great quantities of potable water and release nutrient-rich waste materials that in due course reach the fragile reefs.

Debris, seemingly from cruise ships, litters unsupervised beaches, while sea turtles seeking beaches on which to lay their eggs are distracted by resort lights. Likewise, cuts and lagoons that gave refuge to fishers and even an occasional manatee, have been transformed into marinas and charter-boat anchorages. Conch have largely disappeared from the inshore waters, while sharks, around which a short-lived industry was formed in the 1980s, have become a rare catch. A lively underground trade exists in exotic animals, and resorts frequently display caged birds and melancholy spider monkeys as if to verify their rarely-seen existence.

Finally, cattle ranching has become a prestigious enterprise, and with it has come the extensive clearing of forest. This change of the natural cover often degrades the quality of the soil (already poor), as it simultaneously reduces habitats and hence biodiversity. In contrast, the small-scale farming of forest patches—"making *milpa*"—is much less disruptive, owing to the fact that once the plot is abandoned the forest rapidly regenerates. Cattle is a cash crop, reared by those with connections to urban markets, and consequently a product that does not normally benefit the average Maya.

In conclusion, an obvious contradiction exists between the theme of unmarred nature and the reality of an increasingly impacted environment. A number of sincere and meaningful efforts are under way to counteract these trends, but all too often ecology has been co-opted by the market. It takes only paint and a little imagination to turn standard transportation into "ecobuses" and "ecotaxis," a marketing technique that does little more than gesture at ecological sensitivity. These problems and shortcomings are inevitable in an industry that measures success by its ability to fill airplanes, hotel rooms, and the rosters of tour operators. The underlying reality is that both nature and people—including the visiting anthropologist—are constantly asked to define everything in terms of price: "How much does it cost to fly to Miami?"; "How much is that camera in dollars?"; "Is that an expensive backpack?" The visitor off the plane engages in a rather similar discourse: "Beer, cold beer, *cerveza, cuanto*?" This mediation of relationships by the cash nexus is not simply a consequence of tourism but also reflects the several parallel processes that have functioned to incorporate the region into a world where everything seems to have a price.

Some final observations. All tourism, it has often been pointed out, is a form of play and fantasy. Surely, it will be argued, if tourists want to identify with jungle adventurers (ecological or otherwise), this is strictly their business. It would be that, except that the shift toward environmental and "adventure" tourism, as well as the transformation of archaeological sites into tourist attractions, carries material and cultural consequences for local people. Most obviously, whatever land and resources are taken out of local control reduces local economic options. Cultural options are precluded as well when sacred spots such as *cenotes* (water-filled sinkholes), caves, and ancient temples (where Maya still go to pray and converse with the spirits) become tourist attractions. Even when the Maya are not barred from using these sites by admission fee, they explain that the spirits are reluctant to come—because of the intrusion of alien influences. . . .

Differing Faces of Tourism

Sun and Sea Tourism

As sun-and-sea resorts spread down the coast, service communities developed in their vicinity. These are composed of people drawn from towns and villages across the Yucatán peninsula, and increasingly from other parts of Mexico. Since few inhabitants have access to agricultural land, they are completely dependent on stores and vendors for basic needs.

Akumal was selected because of its dependency on resort employment and the pressures faced by its residents. The 400 villagers live directly outside the boundaries of a resort and recreation complex of the same name, and the contrast between the resort and the service community could not have been starker. The resort is pleasant, spacious, and well-maintained, and it faces a sandy beach. It is filled with contented visitors and the owners of holiday properties. The village houses of the service community were built of natural materials, scrap or cinder blocks—more shantytown than tropical hamlet. Although electricity was available, the community had to manage without garbage removal, a sewage system, or potable water. While the main water line to the resort passed through the village, residents had no access to it and were forced to rely on a series of shallow and unsanitary wells.

Already in 1993, pressure was mounting to evict the people of Akumal. The resort management and condo owners declared it an eyesore and an actual or potential health hazard. They also hypothesized that runoff from the village might damage the lagoon or reef (we only heard a few of these people express concern for the health of the villagers). By the summer of 1994, approximately half the homes in the community had been razed: the remains of the demolished houses are still visible. Finally, in the summer of 1996, assisted by a hurricane that damaged many of the structures, the government ordered all houses and businesses vacated. This action was facilitated by the circumstance that the Akumal people lacked titles to their house lots and were technically squatters on federal land.

It is unclear whether the vacated land will become an ecopark or, as is now rumored, be turned into a tastefully landscaped extension of the resort. Whatever the outcome and in spite of the efforts by the Akumal management and others to relocate residents to more "suitable" locations—resentment and resistance to these actions ran high. Akumal provided low-cost housing for service workers and, equally important, a space of relative autonomy from the growing controls of the industry and government.

Ciudad Chemuyil, the second service community that we looked at, is a new development that, at the time of the study, contained 250 houses. This community was established by a consortium of hotels and the government in order to concentrate, and presumably the better to control, the growing labor force attracted by the expanding tourism. Its location, several miles inland from the resort area, appears designed to segregate workers from clients, a hypothesis reinforced by the fact that nearly half of the inhabitants of Ciudad Chemuyil had been evicted (some left voluntarily) from Akumal and other "irregular settlements."

Promotional literature describes Ciudad Chemuyil as a "new Maya city," and plans for an eventual expansion to 250,000 inhabitants. Even if this target is not reached, such a massive development is bound to transform the human geography of the coast and to reinforce racial and ethnic separation. The new settlement (presented as superior, healthier, and more modern) consists of small, poorly ventilated, cement block houses—all identical—on minimal lots. Company minibuses pick up maids, gardeners, waiters, and cooks in the morning, take them to their places of work, returning them in the evening. Most employees work a grueling seven-day week for four weeks and then receive a break of five days. Ciudad Chemuyil is essentially a company town, totally dependent on resort employment, lacking natural resources of its own, and with a population tied to 30-year mortgages underwritten by employers. It is hard to escape the conclusion that this is an arrangement designed to sap self-reliance

and meant to transform hotel workers into the modern equivalent of plantation labor.

Archaeological Tourism

Unlike Akumal and Ciudad Chemuyil, the village of Cobá did not come into
being as a result of tourism. Settled some 50 years ago by a small group of
Yucatán farmers, Cobá has grown over the years to approximately 900 inhabitants. What makes Cobá a tourist destination is its location at the edge of one
of the largest archaeological complexes in Mesoamerica: the Classic Maya site
from which the community takes its name.

The development of the ruins into an attraction for the "adventurous"
tourist followed the completion in the 1970s of a road connecting Cobá to the
coastal highway. This link brought not only tourists, but also in due course electricity, easy delivery of commercial goods, potable water, a health clinic, and a
school. By the time of our study, the men and women of Cobá were increasingly
dependent on different types of wage work, including resort construction. As
our colleague Ellen Kintz notes, this growing dependence on wage labor "provides cash but strains social ties."[31]

Today, some 60,000 annual visitors, mostly foreigners, make their way to
that archaeological site and the protected forest trails that surround it. Tourists
arrive by car and bus and, except for those booked into one of the two small
hotels (one of which is a Club Med "Villas Arqueológicas"), most leave by the
end of the day. Although tourism has stimulated the sale of handicrafts and
the establishment of a handful of stores and restaurants, most townspeople
complain that the tourist trade leaves them nothing, and many insist that it
represents a net loss.

From the very beginning, tourism was placed in contention, since the
newly designated archaeological zone contained some of the community's
prime agricultural land. This initial loss of resources was followed by other
destabilizing changes, including a heightened awareness of class and ethnic
inequalities, plus pressures to privatize the *ejido* communal holdings (this tends
to split the community generationally). The strains also manifest themselves
in the religious sphere. A generation ago, religious practice was a variant of
folk syncretism that joined elements of Catholic ritual to a Mayan cosmology in
which the milpa, the forest, and the community represented a model of the universe, and the task of ritual practitioners was to maintain these components in
proper harmony.

Something of this belief remains, but Cobá is now an arena for competing
congregations, and about two-thirds of the population identify with different
Protestant and Evangelical denominations. Sectarian balkanization has reinforced individualism and undermined the old ties of reciprocity sanctioned by

custom and ritual. The new religious ideology tends to privilege the individual over the community, much as the market constructs the individual as consumer and autonomous economic agent.

Ecotourism

At the opposite pole of dependency typical of the communities along the Cancún–Tulum corridor is the hamlet of Punta Laguna. This is a forest community deep in the interior, composed of a handful of interrelated families and subsisting almost entirely from its milpas, kitchen gardens, and the products of the surrounding forest. Unlike the situation in nearby Cobá, ritual practices remain largely intact, and community bonds, essentially links of kinship, remain strong. Honey represents one of the few cash crops, and this helps explain why the residents have long protected a large stand of the canopy forest. The forest, in turn, offers food and shelter to a colony of spider monkeys.

Tourists come in small numbers to see the monkeys and are escorted by teenage boys who act as guides. These youths speak with enthusiasm about the local flora and fauna and use tracking skills to locate the monkeys. Most tourists are fascinated by the experience, even if initially disconcerted by a certain looseness of organization. Still, having to wait some minutes while a small child searches for an available guide may increase the value of the experience; it certainly reflects the homespun nature of Punta Laguna ecotourism.

Interesting as the experience is likely to be for the visitor, such tourism does not contribute much to the local economy. The tips collected by guides put little extra money into circulation, and the guest waiting in the entrance shed is invited to buy postcards and honey. In recent years, the environmental organization Pronatura has provided funds for improving trails and offered other forms of support, but perhaps its most valuable contribution has been to underscore the importance of protecting the environment.

Punta Laguna, although not many miles from the main road, remains very quiet and traditional. It lacks telephone service or electricity, and harbors no resident outsiders. All our ethnographic visits had to be negotiated well in advance and required community approval. In our interviews we found that most people did not think that the Maya language was in danger of disappearing, but that they were worried about possible challenges to religious practices and a decline in the systems of reciprocal assistance. An important concern, and certainly the chief concern of the young, was a lack of money. This little hamlet is about as self-sufficient in basic needs as any place in the state, but people feel poor and young men are beginning to leave to search for work on the coast. However, the jobs open to clearly rural, and often monolingual, Maya-speaking workers are the hardest and most menial—"slave labor," as one young man expressed it. . . .

Recommendations

Obviously, tourism in Quintana Roo is part of a much bigger picture, and any "recommendations" that we offer are conditioned by two recognitions: that one can hardly change the part without addressing the whole, and that certainly there are no quick fixes. One statistic stands for many others: excluding Africa, Mexico now ranks sixth from the bottom of the world's nations in income inequality.[32] Today, seven out of ten Mexican wage earners—9 million people— earn $300 or less each month.

A whole literature has merged analyzing and explaining the Mexican crisis, and while a discussion of their work falls outside of our purview, our first recommendation is one of greater awareness.[33] It is in the nature of tourism that those who travel abroad leave their cares and concerns at home. But Quintana Roo, exotic though it may appear, is basically an economic extension of the United States. The American dollar functions as a parallel currency; hotel rooms, airfares, and items in luxury stores are dollar denominated. Maids, waiters, taxi drivers, beach attendants, and other service personnel prefer to receive tips in dollars—particularly following the devaluations of recent years.

This relationship is well known to all working in Quintana Roo. One of our middle-class Mexican informants commented that the state had much more in common with South Florida than it did with Mexico City. Fundamentally, what we suggest is that economically Quintana Roo is akin to northern Mexico. It is equally dependent on U.S. trade, and the bulk of the area's investment is from the United States. The *maquiladoras*, the 2,500 border factories, produce almost exclusively for the American market and are in physical proximity to the United States. As such, they have been the subject of numerous studies and plans. People in the United States, we believe, should also be made aware of the role of multinationals in developing tourism among our neighbors to the south, something that the tourism industry has no interest in doing.

It is, we believe, possible to exert considerable pressure on the industry (and the Mexican government) to meet minimal standards of employment and labor practice. The model here is one that already is having some impact on corporations that engage in cut-throat "outsourcing" in the Third World: apparel companies, toy manufacturers, and so forth. What is being outsourced in Quintana Roo (and Mexico as a whole) is a particular kind of commodity: fantasy and leisure. The working conditions of those who make these breaks possible should certainly concern different kinds of organizations, including trade unions and pension funds. Similar tactics can be applied to environmental issues and questions of indigenous rights as well.

The goal of such efforts is not to destroy or cripple tourism—which is bound to prevail as the mainstay of the state's economy—but rather to help defend those who have become dependent on it. This is a responsible agenda that ensures that the conditions attracting tourists in the first place are protected and reinforced. Among these conditions is social tranquility, an absolutely indispensable component of any successful tourist policy. Our argument is that the long-term prospects of the industry are intimately tied to the well-being and security of the local populations, and that these can only be realized when ways are found to redistribute some of the immense profits earned by the sector.

It should also be possible to increase Maya involvement in tourism. The current structure of tourism—the massive wall of hotels along the Cancún beachfront and the gated complexes further down the coast—denies opportunities to small businesses (wood carvers, blouse-and huipil-makers, roadside stands, handicraft shops, and so forth). Such enterprises would bring a little more money into the village economy, but, equally important, they would signal a much-needed *Maya* involvement in tourism. For the same reasons, community-run operations such as those at Punta Laguna should be supported. None of this entails large-scale investment. The ideal sponsors for such work are specialized foundations, nongovernmental organizations, and environmental groups.

For the same reasons, it is vital to overcome the current employment policies that presently relegate the Maya to the lowest-paying jobs and place Mexicans and foreigners in virtually every position that entails contacts with the customer. We have been told by managers that the issue is one of skills and education, not discrimination; this seems a very questionable explanation. Even a generation ago, it might have been reasonably suggested that the Maya would make their own decisions on how to articulate with the modern world. Today, though, options have become reduced and the poor people of Quintana Roo—Maya and non-Maya alike—must find solutions in solidarity with other groups, other organizations. . . . As anthropologists, we see our task as dual: first, to work as interpreters of cultural systems; second, to act as advocates, to our best ability, for a people that badly needs help and support. We have not appropriated a role, yet still have been asked to fill it.

Notes

Pi-Sunyer, Oriol and R. Brooke Thomas. 1997. "Tourism, Environmentalism, and Cultural Survival in Quintana Roo" in Barbara Rose Johnston (ed.), *Life and Death Matters: Human Rights and the Environment at the End of the Millennium*. Walnut Creek, Calif.: AltaMira.

1. Veronica Long, "Tourism Development, Conservation and Anthropology," *Practicing Anthropology* 14:2 (1992), 14–17.

2. Gary H. Gossen, "Maya Zapatistas Move to the Ancient Future," *American Anthropologist* 89:3 (1996), 528–538.

3. Malcolm Crick, "Representations of International Tourism in the Social Sciences: Sun, Sex, Sights, Savings and Servility," *Annual Review of Anthropology* 18:44 (1989), 309.

4. Erik Cohen, "The Sociology of Tourism: Approaches, Issues, and Findings," *Annual Review of Sociology* 10 (1984): 373–392.

5. Valene L. Smith and William R. Eadington (eds.), *Tourism Alternatives, Potentials and Problems in the Development of Tourism* (Philadelphia: University of Pennsylvania Press, 1992). These figures are taken from a 1991 World Travel Organization report and, as is often the case with WTO statistics, they fail to distinguish between leisure and business travel. Increasingly, though, much business travel includes a tourist itinerary.

6. "Third-World Tourism," *The Economist,* March 11, 1989, p. 19.

7. For a discussion of tourism and dependency, see Colin Michael Hall, *Tourism and Politics, Policy, Power and Place* (Chichester and New York: John Wiley, 1994), 122–132. A whole literature on dependency theory, Marxist or neo-Marxist in orientation, emerged out of the Latin American experience in the 1960s. See, for example, Andre Gunder Frank, *Development and Underdevelopment in Latin America* (New York: Monthly Review Press, 1968); Samir Amin, *Unequal Exchange* (New York: Monthly Review Press, 1976); Fernando Enrique Carduso and Enzo Faletto, *Dependency and Development in Latin America* (Berkeley: University of California Press, 1979); Ronald H. Chilcote and Joel C. Edelstein, *Latin America: The Struggle with Dependency and Beyond* (New York: Schenkman Publishing Company, 1974).

What is important for our purpose is the stress on the inequality of relationships in international trade between First World and Third World countries. This critique very seldom addressed questions of cultural or ethnic domination related to the development process. The premises of developmentalism did not go entirely unchallenged in anthropology. In 1967, one of us drew attention to the dangers of designating a particular economic model as the goal of research or policy, and cautioned against drawing almost exclusively on the "experience of the capitalist countries of the Western world" and taking these as "ideal forms and events. See Oriol Pi-Sunyer, *Zamora: A Regional Economy in Mexico* (New Orleans: Middle American Research Institute, Tulane University, 1967), 170.

8. Charles L. Geshekter, "International Tourism and African Underdevelopment: Some Reflections on Kenya," in Mario D. Zamora et al., eds., *Tourism and Economic Change* (Williamsburg, Va.: Dept. of Anthropology, 1978), 57–88.

9. Arturo Escobar, *Encountering Development* (Princeton: Princeton University Press, 1995).

10. Secretaria de Turismo, *El Turismo en México Durante 1992* (Supplemento de La Gaceta del Sector Turismo, vol.7, año 2, Mexico City, 1993), 3.

11. Daniel Hiernaux Nicolis and Rodriguez Woog, *Tourism and Absorption of the Labor Force in Mexico. Economic Development.* Working Paper 34 (Washington, D.C.: Commission for the Study of International Migration and Cooperative, 1990).

12. The population pressures in Mexico are real enough. The workforce is expanding at 3 percent per year, and more than one-third of Mexicans are under 15 years of age; a full 80 percent are under 40. United Nations projections indicate that by the year 2015, metropolitan Mexico City will have a population of 18.7 million.

13. Frederick K. Errington and Deborah B. Gewertz, *Articulating Change in the "Last Unknown"* (Boulder, Colo.: Westview Press, 1995), 162.

14. The myth of the shattered Eden is often juxtaposed with one that emphasizes the inevitability of change and "modernization." Fundamentally, "traditional societies" are depicted as *resisting* change, transformation that can only come about when pressure is applied. In its 19th century form, it underpinned colonialism and explained human diversity; more recently it was (and continues to be) a key component of development ideology. Especially in the 1950s and 1960s planners insisted that economic progress was

impossible without painful adjustments that would dismantle established systems of belief and community organization.

15. Oriol Pi-Sunyer, "The Cultural Costs of Tourism," *Cultural Survival Quarterly* 6:3 (1992) 7–12.

16. Crick, op. cit., note 3, p. 44; Emanuel de Kadt, "Making the Alternative Sustainable: Lessons from Development for Tourism" in Smith and Eadington, op. cit., note 5, p. 54.

17. *World Development Report 1994* (Washington, D.C. and New York: World Bank and Oxford University Press), 221.

18. John E. Kicza, ed., *The Indian in Latin American History, Resistance, Resilience and Acculturation* (Wilmington, Del.: Jaguar Books, 1993).

19. Clifford Geertz, "Life on the Edge," *New York Times Review of Books,* April 7, 1994, pp. 3–4.

20. Julian Burger, *Report from the Frontier, The State of the World's Indigenous Peoples* (London: Zed Books, 1987), 2.

21. See Dean MacCannell, "Staged Authenticity," *American Journal of Sociology* 79:3 (1975), 589–603; and *Empty Meeting Grounds: The Tourist Papers* (London: Routledge, 1992).

22. We would argue, on the basis of both observation and literature, that tourism tends to reinforce prejudice and racism, particularly when the objects of the tourist gaze are exoticized "others." That this prejudice is typically rendered in the language of paternalism is hardly surprising.

23. Cynthia Enloe, *Bananas, Beaches and Bases* (Berkeley and Los Angeles: University of California Press, 1989), 40.

24. Dirección General de Estadística Octavo Censo General de la Población (Resumen General) (México, D.E: Estados Unidos Mexicanos, Secretaría de Economía, 1962), 7.

25. Gonzalo Aguirre Beltran, *Regions of Refuge,* Monograph 12 (Washington, D.C.: Society for Applied Anthropology, 1979).

26. *Passport Cancun,* 20th ed. (Cancún: Apoyo Promocional, 1994), 9.

27. Secretaría de Turismo, "El Turismo en México Durante 1992," Supplemento de La Gaceta del Sector Turismo, Mexico City (vol. 7, año 2, 1993).

28. Michael Collins, *Ecotourism in the Yucatan Peninsula of Mexico* (Syracuse, N.Y.: SUNY, ESF-IEPP, 1991).

29. Gustavo Lins Riberiro, and Flávia Lessa de Barros, *A Corrida por Paisagens Autênticas: Turismo, Meio Ambiente e Subjetividade na Contemporaneidade* (Brasilía D.E: University of Brasilía) Anthropology Series No. 171, 1994, p. 8.

30. Rayfred L. Stevens, "The Soils of Middle America and Their Relation to Indian Peoples and Cultures," in *Natural Environment and Early Cultures,* Robert C. West, ed., *Handbook of Middle American Indians,* 1. Robert Wauchope, gen. ed. (Austin: University of Texas Press, 1964), 265–315; see p. 303.

31. Ellen Kintz, *Life under the Tropical Canopy* (Fort Worth, Texas: Holt, Reinhart and Winston, 1980), 80.

32. Jorge Castañeda, "Mexico's Circle of Misery," *Foreign Affairs* 75:4 (1996), 92–105; see p. 63.

33. Ibid. See also Andres Oppenheimer, *Bordering on Chaos: Guerrillas, Stockbrokers, Politicians and Mexico's Road to Prosperity* (Boston: Little Brown, 1996).

5

Treadmill Predispositions and Social Responses

Population, Consumption, and Technological Change

Allan Schnaiberg and
Kenneth Alan Gould

The "treadmill of production" is a concept that focuses our attention on institutions and social structure. It emphasizes that we are all part of a system that must continue to grow—to continually produce more and create consumers to consume that which is produced. This process requires ever more energy and resources and causes industrial and consumer wastes to be constantly generated. In this excerpt from their book, Environment and Society: The Enduring Conflict, *Allan Schnaiberg and Kenneth Gould describe technological changes that have allowed for the expansion of production and argue that population growth plays only a very limited role in that expansion.*

Directions of and Motives for Technological Change

... In its simplest form, the treadmill of production [see table 5.1] is built around the interaction of two processes. The first process is an expansion of technological capacity.... Associated with expanded technological capacity, especially in

TABLE 5.1.   **The Logic of the Treadmill of Production**

1. Increasing accumulation of wealth, through ownership of economic organizations that successfully use ecological resources to expand production and profits.
2. Increasing movement of workers away from self-employment, into positions of employees who must rely on expanded production to gain jobs and wages.
3. Increasing allocations of the accumulated wealth to newer technologies in order to replace labor with physical capital, thereby generating more profits for wealth-holders, in order to sustain and expand their ownership in the face of growing competition from other wealth-holders.
4. Increasing activities of governments to facilitate expanded accumulation of wealth for "national development," on the one hand, and "social security," on the other.
5. The net result of these processes is an increasing necessity for ever greater ecological withdrawals and additions in order to sustain a given level of social welfare.
6. The ecological obverse of no. 5 is the increasing likelihood of an industrial society creating ecological disorganization, as economic pressures push toward greater extraction of market values from ecosystems.
7. Extending no. 6, societies become increasingly *vulnerable* to socioeconomic disorganization, as their ecological "resource base" itself becomes disorganized.

modern industrial societies, is an expanded necessity of using this capacity in order to provide economic support for the population. The population of modern societies is increasingly made up of a small number of very large productive organizations and the concomitant growth of the number of wage employees of these and other productive organizations (Granovetter 1984). Both categories are committed to the treadmill but for somewhat different reasons. Investors and managers of the large production organizations seek to survive in a globally competitive milieu . . . and need to grow in order to cumulate wealth. . . .

The second process entails the dominance of economic growth preferences, even when many decision makers know that ecosystem disorganization is a likely outcome. This projected ecosystem *disorganization* is often dismissed as merely a temporary *disruption*. . . .

The role of government is often viewed as that of policing the relationship of economic structure to ecosystem disruption (Williamson 1985). It is also seen as that of mediating many relationships between workers and investors/owners/managers of economic organizations. But neither of these ideal roles can be actualized, since governments are not impartial mediators between dominant treadmill values and alternative values (Reich 1991). For modern governments have as their constituencies both major groups of treadmill interests. For political and administrative agents of government, economic growth offers some gains for both constituencies (investors and workers). Moreover, all government expenditures, which can be allocated to either constituency (e.g., subsidies for investors, or social welfare for employees and dependents), arise as a tax or levy on national income. Thus, growing national income, which is typically generated by speeding up the treadmill, simultaneously makes it easier to collect a given level of tax revenue and allows governments to allocate more revenue to its two constituencies.

Thus, the treadmill tends to be the dominant value of capital owners, workers, and governments alike (Granovetter 1979). . . .

Expanding Production as a Social Choice

Woven through the treadmill of production is one particular form of economic growth: the expansion of production (Reich 1991). . . .

. . . [T]his chapter examines how profits are linked to technological changes and how such technological changes are in turn tied to the expansion of production. The competitive framework . . . has been present in skeletal form since early forms of trading existed, with horseback transportation of goods. The pace of competition accelerated somewhat with more advanced means of transportation, since it permitted more firms to send their products to more distant markets. Thus, sailing ships and especially steamships permitted industrial country producers to expand their markets in other industrial and underdeveloped countries. At the same time, railroads permitted intrasocietal competition in the industrial countries. In contrast, in the underdeveloped countries, most of this transportation improvement merely facilitated the expansion of monopoly corporations. . . . Improved transportation enabled them to exploit these populations and ecosystems on a grander scale, extracting still more ecological resources from these countries (e.g., through mining and forestry especially and, later, through plantation agriculture, with the growth of refrigerated ships).

Improvements in transportation were part of a growing attentiveness to technologies in all forms of production. While "tinkering" was a widespread form of improving and adapting production technologies for several centuries, the twentieth century began to alter this social and economic process (Schnaiberg 1980: ch. 3). Many different historical processes combined to generate these substantial changes. These processes are listed in table 5.2 . . . in different forms and degrees across societies and production arenas.

These diverse historical factors helped pave a general path to expanded production, especially from the periods following World War II (though the impacts of some of these factors increased as early as the mid–nineteenth century or before, in some cases). From an ecological perspective, we should distinguish two different forms of production expansion. The first is a quantitative form: here factories and mills primarily got larger and more numerous, and withdrew *more withdrawals* and added *more additions*. In the twentieth century, however, much greater qualitative changes in production were created: in these qualitative

TABLE 5.2. **The Roots of Expanded Production**

1. The increasing portability of production, as sources of energy became more separable from their applications in production (e.g., stream power versus steam, and then diesel and electrical power).
2. The portability of instruments of finance, increasing the "circulation" speed of production and distribution, by establishing national and international banking networks.
3. Improvements in transportation, allowing for more rapid and easier movement of raw materials, sources of energy, products, workers, and distribution agents.
4. Improvements in communication, leading to a reduction in the transaction costs of producing for distant and less familiar markets, and increasing the pace of economic activity.
5. Increased availability of credit from financial institutions, and insurance from such institutions and government agencies in order to permit taking longer distance risks.
6. The rise of technological specialties (e.g., applied scientists or engineers), which would have skills at redesigning capital equipment in production.
7. The rise of financial-organizational specialties (e.g., systems analysts and industrial engineers), which permit reallocation of financial and human resources.
8. Political control over international and domestic trade, through the use of government instruments, ranging from patent and property rights to the use of "gunboats" and armies, as well as trade consuls and small business administrations.

changes, we often had *new withdrawals* or *new additions,* which were previously never extracted from or dumped into ecosystems. [. . .] When the source of power for mills changed from water to steam, we had primarily quantitative changes: they became larger, and they simply impacted on more ecosystems in the production process, as well as in the mining process (which could be farther away, with steam-powered trains to ship the coal). New types of coal were mined, with steam-powered equipment and new explosives. In addition to the quantitative effects of this transformation, some qualitative changes also occurred. New forms of air pollution arose from burning new types of coal, and perhaps some new forms of land and water pollution also occurred near factories, with fly-ash and other effluents from the use of the new coal. This early transformation of production is strongly associated with *changing energy sources.* Moreover, much of the ecological impact of this path to expanding production was associated with withdrawals of fossil fuels, which we termed depletion or resource exhaustion.

When factories changed from coal to fuel oil, and later to electricity, however, the results were often more pronounced qualitatively as a result of further industrial processing. Some quantitative impacts of energy use and transformation occurred. Indeed, some were even seen as positive in environmental terms, such as burning clean oil versus dirty coal, which had left more visible pollutants in the form of particulates made up of fly-ash. But the availability of large supplies of portable energy (especially electricity), coupled with new scientific resources, led to "better living through chemistry" in many fields. This change led to many problems of *additions* to ecosystems, or what we have come to call "pollution," a situation that accelerated in the post-1945 period.

The modern economic system faces technologically induced problems of both withdrawals and additions (Schnaiberg 1980: chs. 2 and 3). Among industrial countries, there has been a historical withdrawal of and depletion of petroleum (on land) and some mined resources. But in recent decades in such countries, additions such as toxic wastes, threatening human health as well as animal species, have become more prominent through the rise of synthetic chemicals in production. By the late 1980s the United States alone produced 275,000 tons of industrial waste per year (OECD 1991). Included in that waste were more than 48,000 different chemicals registered with the U.S. Environmental Protection Agency (National Wildlife Federation 1991). In contrast, in the underdeveloped countries, deforestation and desertification have proceeded from quantitative shifts, including increased extraction of firewood by individuals, and massive timber cutting and some plantation agriculture by multinational companies. Between 1981 and 1990 about 40 million acres of tropical forest were destroyed (Rudel & Horowitz 1993; World Resources Institute 1992). Within both sets of societies, as well as the Eastern Bloc states, there is a wide range of both depletion and pollution, arising from intensified use of traditional power sources such as soft coal (producing toxic air pollution and acid rainfall).

Global problems of warming arise primarily from the combustion of fossil fuels (carbon dioxide) and the waste products of modern agriculture and industry (methane gas), heavily influenced by the level of national industrialization. Ozone depletion is even more skewed to the production of industrial societies, which use chlorofluorocarbons in both industrial and consumer products (e.g., refrigeration and air-conditioning)....

Why then does the treadmill system favor expanding production, thereby increasing ecological withdrawals and additions and the threat of ecosystem disorganization? In an earlier period, before worldwide competition was so intense, why did production expand? The usual answer relates to the growth of population: in the next section we treat this as a real but very incomplete explanation for the expansion of production. Our interpretation of the work of historians, economists, and sociologists suggests that a tension exists within industrial systems, especially but not exclusively in capitalistic ones. In the early days of the twentieth-century treadmill, the speed was slower, but the directions of modern production systems were already laid out. The process is a relatively straightforward one. For any given producer who is successful in extracting ecosystem resources and in marketing them, a profit is generated. Some portion of the profit or "surplus" is likely to be allocated to new physical capital. This capital is then used to expand production, often thereby reducing the per unit costs of products and improving the company's competitive position and/or the size and share of the total market it supplies.

A first question can be raised: why can't a corporation's level of production simply remain static rather than expand? One answer is that some family-owned enterprises have indeed remained fairly stable and static over protracted periods. These owners typically were satisfied with their rates of return, since they didn't need or want more profits. Often, even in such firms, the next generation of owners may be larger, or seek new challenges, and thereby seek to modernize and expand their operations. (The son or daughter faces an economic organization that his or her parent had the satisfaction of *building*.) When we move beyond family-owned businesses, we encounter variants of the situation. . . .

A company that enters a public market which provides it with operating or expansionary funds must compete with other firms for such funds (Williamson 1981). This produces both . . . resistance to environmental protection regulations . . . and the need to expand profits (Williamson 1985). From these arguments, we can state that expansion of profits is more of a goal in this public marketplace, while reshaping production systems may be a more important goal in family-owned businesses. Somewhere in the middle of these two positions is the case of family-owned businesses in which the next generation is larger and thereby needs more profits to support more owners. Such owners then expand the pie, rather than just fighting each other for a fixed profit level. In other cases, inheritors of businesses have no skills or desires to continue the business, and they sell off the firm to a market actor, thereby moving the firm onto the "conveyor belt" of the treadmill of production (Granovetter 1979, 1984).

The second question posed has to be: why do owners invest more in new equipment, rather than in adding more workers (Noble 1979)? A conventional answer is that this path results in greater capital efficiency (defined as lower costs of production per unit produced), permitting more rapid expansion of profits through expansion of market size. This path differs from the model in which a social reorganization of workers permits more profits from a given level of production by increasing labor efficiency. Interestingly, most of the examples of a labor-relations path seem to arise within family businesses, where owners are also managers. They retain social relationships to their employees, rather than seeing employment as a business transaction (Williamson 1985; cf. Granovetter & Tilly 1988). In sharp contrast to both of these models, the theory of rationalization or social control suggests that owners of capital and their managers can predict and control their transactions with machinery and physical technology rather than their transactions with workers. Whether this difference in control is always or usually actualized, the *belief* of investors and managers that these transaction costs will be predictably reduced itself creates pressures to develop new technologies to supplant more human labor. Unlike the first two models, the last model treats the employment relationship as only a transaction, peripheral to the major tasks of the production organization (Williamson 1981). It assumes enduring conflict between owners and workers, which agrees with both neo-conservative and neo-Marxist views of the firm and the economy. . . .

Population and Production: The Complex Connections

It is tempting to explain the contemporary growth in production by referring to the growth of the human population in each society. At every level of analysis, however, this inference is misleading. Starting with the world as a whole, those societies with the highest rates of economic growth and the highest rates of ecological withdrawals and additions, are precisely those societies with the lowest rates of population growth in the twentieth century. Indeed, the proportion of the global population represented by these high-economic growth nations is shrinking rather rapidly. As we move from this global perspective down to a nation-state and regional level, this disjuncture between population growth and ecological withdrawals and additions becomes more evident—but also more complex. At one extreme, we cannot assume that population growth is irrelevant to both temporary ecological disruption and enduring ecological disorganization. However, the size of a population appears to be most relevant to ecological disruption where populations are closer to subsistence production rather than industrial production. In recent decades, the world has become more aware of drought, starvation, and their connection to the press of populations on limited ecosystems. Desertification in sub-Saharan Africa has been increasingly linked to the attempts of regional nomadic peoples to support larger herds of animals by grazing them in areas of marginal moisture and productivity. Here and elsewhere, this process is accelerated by the collection of firewood through cutting down younger and younger trees, which in turn makes soil more vulnerable to sandstorms and loss of topsoil, when moisture is low.

In establishing any form of environmental protection, we must attend to two features of this population-induced ecological disorganization. First, many of these local and regional ecological problems are "solved" in ways such that ecosystems pass on these pressures to all animal species: members of the population gain less nutrition, become ill and disabled, and die in much larger than usual numbers. . . .

. . . The form and scale of later industrialization created productive mechanisms in societies whose expansion vastly exceeded the stimulus of population growth itself. That is, it was the growth of industrial production systems rather than the growth of human reproduction that increased societal impacts on ecosystems. To expand production, industrial investment of capital "overcame" local ecological limitations, largely by moving resource extraction activities farther afield to many more and to more distant ecosystems. As we noted earlier, this shift depended on the revolution in transportation technologies. Improved transportation provided ease of movement both of fuels and raw materials from distant resource areas and of the finished products to distant consumers.

To understand the limited role of population growth, we need merely to turn to the South. In the twentieth century, and especially since World War II, Southern societies have experienced the sharpest rates of population increase in part because of food stabilization and in part because of public health controls that together lowered mortality in this period. As a consequence, they have become the youngest societies and thus capable of even more population growth in the future, as the vast numbers of children become parents. While rising population has indeed led to many local shortages of arable land, fuel, and water, the major impact has been malnutrition, disease, and death. This problem has limited actual withdrawals from and additions to ecosystems, making them well below the rates of such ecological disruptions in industrial countries, some of which even have negative rates of population growth.

In a socially realistic analysis, production expansion has complicated relationships to population growth. Quite different combinations of the population's savings and consumption patterns can produce highly variable patterns of productive investment and production expansion. Thus, for example, the U.S. population consumes more of its per capita income than does Japan, which invests more of its income in Japanese industries (either directly or through government intervention). Thus, the effect of population on ecological demands will also differ in the two societies: more per capita consumer impacts exist in the United States than in Japan, but potentially more industrial impacts can and will occur in Japan in the absence of governmental and private sector environmental protection. . . . Whereas in all industrial societies, a population represents a potential market for goods and services produced by the economic sectors, with modern flows of investment and goods, markets are increasingly multinational and international, as are flows of capital. . . . This further weakens the connections among a society's population size, level of production, and withdrawals and additions within domestic ecosystems. . . .

[T]he failure of growing world production to meet basic human needs is additional powerful evidence that no clear link exists between production expansion and population growth. It indicates that the biomass quality of human actors is less significant for predicting their ecological impacts than is the socioeconomic quality of human actors. One way of demonstrating this idea is by reference to social expectations about the size of children and the linkage of family size to nutritional adequacy. In the North, over the past century individuals have grown substantially taller than their ancestors. Yet this growth of biomass actually represents a significant drain on ecosystems, requiring higher food and raw material inputs (e.g., for clothing, bedding, and housing) and increasing human wastes per capita. Once more, it is our social values that take precedence over our biological necessities. The opposite end of this scenario is the *shrinking* size of individuals under conditions of malnutrition and starvation in the South. Contrasting these two sets of conditions (higher weight and diffused obesity in the North versus lower weight and malnutrition in the South), we find that the per capita ecological impact of individual members of the two types of societies as biological demands on ecosystems is higher in the North. Yet we decry population growth as an ecological *hazard* in the South.

A second example of the logical flaws in a crude population-based explanation of expanded production is found in our recent attention to the disappearing rainforests (Rudel & Horowitz 1993). There is often a confusion in our public rhetoric about South American rainforests, since many South American countries have high rates of population growth. But the major forces influencing the cutting of rainforests are industrial investors, who want the land for mining, cattle grazing, and plantation agriculture. First, such investors act to remove the population by killing or forcibly moving them and then they cut the rainforest trees, which they also sell for large profits (Rudel & Horowitz 1993). Some of the more socially and ecologically progressive movements in industrial societies to find alternative, sustainable uses of rainforests by preserving the tree cover may be unaware of actual efforts along these lines by indigenous leaders of local populations. These include groups of rubber-tappers and others involved in more sustainable forms of production. Some of them, like Chico Mendez, have died in the process of opposing miners, cattle ranchers, and other industrial investors (Gore 1992). . . . [H]owever, these groups face serious pressures from industrial and other economic elites, who look to industrial societies for models and money to expand high-technology production, with all its attendant ecological risks.

In summary, from a local and global ecological perspective, population growth represents a clear and present danger. Contrary to some analysts, we see little social or ecological advantages to a growing population in most societies. At the same time, however, the historical record is clear: it is not population growth that has created the national and global environmental disruption we confront today. Rather . . . it is the expansion of profits or surplus that has led to an enormous expansion of production in the past century, especially in the last half-century. With the modern potential for global diffusion of money, physical production equipment (technology) and information, goods and services, to both producers and consumers, the role of population growth in expanding production becomes attenuated. Instead, the generation of surplus and the allocation of this capital influence production expansion. Both domestically and internationally, not only are production decisions controlled disproportionately by a small number of decision makers, but also consumption capacity is increasingly being allocated to a smaller share of the world's consumers (Galbraith 1992; Reich 1991).

Note

Schnaiberg, Allan, and Kenneth Alan Gould. 1994. "Treadmill Predispositions and Social Responses: Population, Consumption, and Technological Change." In *Environment and Society: The Enduring Conflict*. New York: St. Martin's.

Selected References

Galbraith, John Kenneth
  1992 *The Culture of Contentment.* New York: Houghton Mifflin.
Gore, Senator Al
  1992 *Earth in the Balance: Ecology and the Human Spirit.* Boston: Houghton Mifflin.
Granovetter, Mark
  1979 "The idea of 'advancement' in theories of social evolution and development." *American Journal of Sociology* 85 (3): 489–515.
  1984 "Small is bountiful: Labor markets and establishment size." *American Sociological Review* 49 (June): 323–334.
  1985 "Economic action and social structure: The problem of embeddedness." *American Journal of Sociology* 91 (3): 481–510.
Granovetter, Mark, & Charles Tilly
  1988 "Inequality and labor processes." Pp. 175–221 in Neil Smelser, editor, *Handbook of Sociology.* Newbury Park, CA: Sage Publications.
National Wildlife Federation
  1991 *The Earth Care Annual.* Washington, DC: National Wildlife Federation.
Noble, David F.
  1979 *America by Design: Science, Technology, and the Rise of Corporate Capital.* New York: Alfred A. Knopf.
Organization of Economic Cooperation (OECD)
  1991 *Environmental Indicators.* Paris: OECD.
Reich, Robert B.
  1991 *The Wealth of Nations: Preparing Ourselves for 21st Century Capitalism.* York: Alfred A. Knopf.
Rudel, Thomas K., & B. Horowitz
  1993 *Tropical Deforestation: Small Farmers and Land Clearing in the Ecuadorian Amazon.* New York: Columbia University Press.
Schnaiberg, Allan
  1980 *The Environment: From Surplus to Scarcity.* New York: Oxford University Press.
  1992 "The recycling shell game: Multinational economic organization vs. political ineffectuality." Working paper WP-92-16, Center for Urban Affairs & Policy Research, Northwestern University, Spring.
Schnaiberg, Allan, N. Watts, & K. Zimmermann, editors
  1986 *Distributional Conflicts in Environmental-Resource Policy.* Aldershot, England: Gower Publishing.
Williamson, Oliver
  1981 "The economics of organization: The transaction cost approach." *American Journal of Sociology* 87 (November): 548–577.
  1985 *The Economic Institutions of Capitalism.* New York: Free Press.
World Resources Institute
  1992 *The Environmental Almanac.* Washington, DC: World Resources Institute.

PART

Environmental Justice: Race, Class, and Gender

Oakland's Fruitvale Transit Village

Building an Environmentally Sound Vehicle for Neighborhood Revitalization

William A. Shutkin

The following four selections explore how race, class, and gender intersect with environmental issues. In this first selection, William Shutkin presents a case study of transit planning in a California neighborhood to illustrate several important ways that an environment-related issue, transportation, is interwoven with social equity. The environmental issues at stake include highways that often slice through neighborhoods creating physical barriers, noise from automobiles, particulate matter that may cause health problems such as asthma, and dangers to pedestrians from traffic. The excerpt also shows how the costs of environmental degradation are often not borne by those who create the problem or those who benefit from its production. Instead, people who bear the costs of environmental problems are often lower-income people and people of color. Finally, this reading reveals the importance of citizen action for developing workable policies and creating positive environmental change. To learn more about Fruitvale's project, visit the Unity Council website at http://www.unitycouncil.org/index.html.

Environmental Quality, Community, and Transportation in the Bay Area

... Due east across the bay from San Francisco lies the city of Oakland. Its rise as the region's second largest city, less dramatic than its neighbor to the west, Oakland was first inhabited by the Ohlone Indian tribe and later became a vast land holding of the Mexican general Peralta into the nineteenth century. The city began as a series of farms and estates established by squatters, who eventually got title to most of Peralta's property by the mid-1800s. Not as prone to the fog that engulfs the San Francisco peninsula to the west in the warmer months of summer and early fall, the flat plains and undulating hills of Oakland and the East Bay afforded settlers sunny planting grounds for orchards of all kinds. With local access to the Western and Southern Pacific Railroads, Oakland farmers could transport their produce to markets well beyond the East Bay.

This merchantilist prowess spawned the Fruitvale district. . . . Fruitvale was the seat and commercial center of Alameda County until it was annexed by the city of Oakland in 1909. Located in the flatlands south of Oakland's central business district, Fruitvale was settled primarily by German immigrants and was long considered Oakland's second downtown because of its vibrant business and civic culture. . . . With World War II came an economic boom and an influx of war industry workers, including large numbers of African Americans and Hispanics.

Following the war, the suburbs around the Bay Area burgeoned, owing largely to the freeway projects and mortgage subsidies that enticed middle-class urban residents to leave the increasingly crowded and polluted central cities of San Francisco and Oakland for the towns and subdivisions farther out. Meanwhile, urban neighborhoods like Fruitvale fell on hard times. Wartime factories closed, and workers were laid off. The canneries that once employed local residents began leaving the area in the 1960s.

In 1970, Fruitvale was still a majority white community; by 1980, it was majority black; by 1990, Latinos joined blacks as the dominant racial and ethnic groups, comprising almost two-thirds of the population, followed by Asians, whites, and a substantial Native American population. With each decade, the neighborhood experienced greater disinvestment and white flight. Jobs, housing, tax revenues, and other essential ingredients of community life dwindled, displaced to the suburbs to Oakland's east and south. Today more than 20 percent of Fruitvale families live below the poverty line, and almost a third of the residents are under the age of eighteen.[1]

Over the past three decades, urban sprawl resulting from the abandonment of central cities in the Bay Area has eaten away at the region's social and environmental stability from the inside out. In California as a whole, sprawl has resulted in a shortage of affordable housing, inaccessible jobs, and notorious traffic jams, leading to some of the nation's worst air pollution hot spots. In

1995, California-based Bank of America issued a report entitled *Beyond Sprawl* that declared, "As we approach the 21st century, it is clear that sprawl has created enormous costs that California can no longer afford."[2]

The primary cause of much of the Bay Area's sprawl has been the automobile. As with the rest of the country, suburbanization and the withering away of central city neighborhoods have been made possible by the development of massive roadways and the proliferation of cars that began in the 1950s. Transportation policy in the Bay Area, as elsewhere, has for decades favored cars over alternative forms of transportation such as mass transit, walking, and bicycling. As suburban commuters sit in traffic, lurching day in and day out toward the workplace or home on expensive highways paid for by public dollars, residents of inner-city neighborhoods contend with the noise, air pollution, and pedestrian hazards associated with that traffic, often without access to quality mass transit service or other offsetting amenities.

In Oakland, several major freeways, including Interstates 580, 680, 880, and 980, and hundreds of heavily traveled, high-capacity roadways pass through and above neighborhoods like Fruitvale. Motor vehicles going fast generate significant noise. For example, from sixteen yards away, a car traveling at 56 miles per hour makes ten times as much noise as it would going 31 miles per hour.[3] Noise from vehicular traffic typically drowns out all other ambient sound in congested urban neighborhoods and contributes to the same sense of congestion and unease that drivers themselves experience.

Further, the air pollution caused by cars, trucks, and buses driving through urban neighborhoods contains a deadly mixture of gases and particles. Transportation emissions, which include vehicle exhaust, fuel and paint vapors from gas stations and auto body shops, and tire dust, affect human health, especially in densely settled urban neighborhoods, where emissions tend to occur close to where people live and work. Also, because of frequent idling and stop-and-go traffic, urban transportation produces considerably more pollution per mile than does smooth-flowing traffic, when engines operate most efficiently. A lack of air pollution sinks or filters such as street trees and other vegetation can exacerbate the adverse effects of air pollution from motor vehicles.

People of color, who live in cities to a far greater extent than whites, are disproportionately exposed to urban air pollution. According to a California study, people of color are nearly three times as likely as whites to be exposed to harmful fine-particulate matter, emitted by sources that burn fossil fuels. Fine particles lodge in the lungs and, because of their size, cannot be expunged, resulting in respiratory and cardiovascular disease. The California study, which relied on readings from 161 air pollution monitors in different communities, found that 54 percent of the monitors in communities of color showed readings above the 1997 U.S. EPA health standard for fine particles. Only 19 percent of the air monitors in white communities had readings above the standard.[4]

In addition to particulate matter, motor vehicles that run on gas or diesel emit carbon monoxide and carbon dioxide, nitrogen oxides, and volatile organic compounds, resulting in environmental hazards such as ozone smog and greenhouse gases that cause global climate change. Transportation sources account for roughly one-third of greenhouse gas emissions. On the ground, these pollutants contribute to asthma, a serious and sometimes fatal illness that especially affects children. Asthma rates among children have risen 75 percent between 1982 and 1994, and young African Americans are four to six times more likely to die of asthma than are young whites.[5]

Beyond noise and air pollution, motor vehicles kill and injure urban pedestrians at alarming rates. In Boston, New York City, and San Francisco, half of the people killed in car accidents are pedestrians. Most at risk are children and the elderly, who depend most heavily on walking and bicycling for transportation.

Transportation infrastructure such as highways also usually comes at the expense of accessible open space for urban residents. Multilane highways invariably separate people from places like waterfronts or parks. Whether elevated or at groundlevel, such highways form impenetrable barriers of concrete, steel, and asphalt, shunting people from the precious few natural areas that remain in most cities. In Oakland, high-capacity surface roadways, including several interstates, criss-cross the city; the massive I-880 stands between the bulk of the city's residential areas and the Oakland waterfront, historically an industrial corridor that, in the Fruitvale and San Antonio sections, abuts a sleepy estuary separating the city from Alameda Island.

Urban residents have borne the brunt of the environmental harms associated with transportation, and most have not enjoyed access to quality mass transit services. Between the late 1960s and early 1990s, the average annual number of vehicle miles traveled by Americans doubled, reflecting increased suburbanization and highway construction and a concomitant decline in mass transit services. Yet in cities like Oakland, mass transit remains the preferred, if not sole, means of getting around for most people of color. Blacks and Hispanics account for almost half of transit riders in the United States, are four times as likely as Americans in general to take public transit to work, and are one and half times as likely to walk to work.[6] Access to affordable, reliable transit services means access to jobs, shopping centers, parks, and other basic building blocks of a healthy community. In Oakland, the Church Community Jobs Commission has tried to help local African Americans gain employment with Bay Area employers such as AT&T. However, because most jobs are located in distant suburban communities, Oakland job seekers who do not own cars have no way of getting to work.[7]

Notwithstanding the importance of mass transit to urban communities of color, most state and local governments have failed to invest in transit infrastructure improvements. Instead, they have subsidized projects that serve suburban commuters, such as highways and commuter rail lines. Until recently in Los Angeles, for example, the Metropolitan Transit Authority (MTA) spent 70 percent of its discretionary funds on a rail system that served only 6 percent of the passengers; the MTNs buses, serving 94 percent of the passengers (many of whom are lower income and 81 percent of whom are people of color), got only 30 percent of the funds.[8]

In Oakland, residents of the city's West Side battled Caltrans, the state highway agency, to stop the rebuilding of the Cypress Freeway after it was damaged by the Loma Pleta earthquake in 1993. The Church of the Living God Tabernacle, along with the Clean Air Alternative Coalition, claimed that "freeways have caused high cancer rates in the communities alongside them and high incidences of lead in the brains of our children living in these communities." The area abutting the freeway is 92 percent people of color.[9] The community groups filed a civil rights lawsuit, alleging that the proposed freeway reconstruction violated Title VI of the Civil Rights Act of 1964, which prohibits discrimination in any federally funded project. "The brunt of the proposed project's negative social, human health and environmental impacts," the groups argued, "including those associated with noise and air pollution, the dislocation of persons, the condemnation of homes and businesses, the chilling of economic development, as well as the disruption of the life of the community—will be borne by minority residents of West Oakland."[10]

That the residents of Oakland's West Side had to resort to civil rights laws to protect their environment underscores the sometimes pernicious limits of traditional environmental regulations. The unchecked development of massive highway projects, proliferation of cars, and lack of access to open space and the Oakland waterfront demonstrate that the urban environment has been the orphan of traditional environmental law and policy. Although California's air quality regulations are the strictest in the nation due to the state's unique and persistent air pollution problems, the environmental impacts of highway projects meant to serve regional commuters have traditionally been reviewed according to regional air quality conditions, not local hot spots. As in most other jurisdictions, California's health-based regulations look at air quality on a parts-per-million, pollutant-by-pollutant basis and tend to discount the cumulative, compounded health and environmental effects of many different kinds of pollutants in the atmosphere, especially as they affect people living or working close to the pollution sources.

Moreover, air pollution from mobile sources such as cars is regulated with standards that dictate how much pollution a single car can emit from its tailpipe and how many miles per gallon the car's engine must achieve. California's air quality regulations, like those in every other state, say nothing about how many cars can be produced or driven or how many highways can be built to accommodate them. Thus, air quality gains from tougher emissions and efficiency standards can be wiped away by the proliferation of more and more automobiles riding on ever-widening roadways.

Meanwhile, the California Environmental Quality Act, the state's environmental impact review statute, examines projects in isolation, one at a time. Each highway development in Oakland has been dealt with on its own terms and ultimately has been driven by the need for greater roadway capacity to handle the regular increases in Bay Area commuter traffic. This piecemeal review has resulted in both increased air pollution from highway traffic and the isolation of most Oakland residents from the waterfront and the loss of open space. With little in the way of comprehensive land use planning or growth management in Bay Area communities, the neighborhoods in Oakland's flats have been left to deal with the adverse environmental consequences.

As civil rights leader and U.S. congressman John Lewis explains, "Even today, some of our transportation policies and practices destroy stable neighborhoods, isolate and segregate our citizens in deteriorating neighborhoods, and fail to provide access to jobs and economic growth centers."[11] Transportation is the vehicle by which a community develops; it is a critical enabling mechanism that allows individuals within a community to engage in professional, recreational, cultural, and other essential activities. . . .

Transportation thus exercises a profound influence on the quality of life of urban neighborhoods. It is a critical part of the fiber, the infrastructure, that holds urban communities together and enables urban denizens to pursue the variety of daily activities that make up their lives and livelihoods. . . .

It seems only natural that efforts to revitalize a hard-hit urban neighborhood might start with transportation as a means of bringing people together, literally and figuratively, and helping to restore a safe, healthy environment. In Oakland's Fruitvale neighborhood, where residents have seen the worst of what automobiles, pollution, and disinvestment can do to a community, that is exactly what is happening.

The Road to Recovery: Fruitvale's Transit Village

Along Fruitvale Avenue, the crowded boulevard that runs southwest from the Oakland hills toward the San Francisco Bay, forming the northern border of the Fruitvale district, the offices of the Unity Council (UC, formerly the Spanish-Speaking Unity Council) bustle with activity. Formed in 1964 by activists seeking to create a forum to address issues of concern to the growing Latino community, UC began as a civil rights organization, later evolving into a community development corporation focused exclusively on Fruitvale.[12]

UC's director, Arabella Martinez, began her career as an Alameda County welfare case worker after completing graduate school at the University of California, Berkeley. She was among the group of activists who founded UC in

1964. Her commitment to the community stems from her sense of place. "I certainly have a sense that this is my community," she explains, "I've always had a place here. I'm rooted here."[13]

Under Martinez's leadership, UC designed a comprehensive approach to dealing with Fruitvale's challenges that emphasizes the neighborhood's strengths and assets rather than its weaknesses. Focusing on Fruitvale's rich tradition of commercial activity and ethnic diversity, UC has been able to bring together residents, community organizations, and businesses in delivering successful community projects such as affordable housing, child care and senior programs, and community centers. As well, unlike many other lower-income neighborhoods, Fruitvale has been able to maintain a vibrant business district along the neighborhood's main commercial thoroughfare, East Fourteenth Street, considered a regional destination for Latino shoppers. The Fruitvale Community Collaborative, a consortium of local groups spearheaded by UC in 1991, helps organize local residents and businesses in establishing block associations and conducting community cleanups and tree plantings, traffic-calming campaigns, and flea markets.

Yet as is so often the case in hard-hit urban neighborhoods, no matter how well organized or successful, community residents and organizations must constantly contend with unwanted development. A case in point was a 1991 proposal by the Bay Area Rapid Transit authority (BART) to build a 500-car parking facility at its Fruitvale station, in the middle of the area's residential and commercial center. . . .

When BART officials approached the neighborhood in 1991 with plans for a multilevel concrete parking structure on a parcel abutting the station, UC and neighborhood residents came out in force against the proposal, arguing that the facility, a massive eyesore, would not only exacerbate existing air quality and traffic problems but would act as a dangerous barrier, severing pedestrian access to the East Fourteenth Street business district. . . . Eventually BART dropped the proposal, citing concerns about potential adverse air quality impacts due to increased traffic.

Notwithstanding its successful opposition to the BART proposal, UC recognized that the BART station itself was a significant neighborhood asset, providing a convenient, low-cost alternative for the many area residents who do not own cars and making the neighborhood accessible to residents from other Bay Area communities. In 1991, Rich Bell, a graduate planning student at Berkeley, working with Martinez and UC staff, saw the parking garage proposal as an opportunity for UC to present its own ideas for development of the Fruitvale BART station, which would take advantage of the station as a neighborhood asset. In his courses at Berkeley, Bell had learned about the new planning concept called "transit-oriented development," which sees mass transit as a lever for neighborhood revitalization and environmental improvement. Transit-oriented development attracts ridership (thus reducing traffic and air quality impacts) not by expanding parking infrastructure on site (which merely increases vehicle miles traveled through the neighborhood), but enhancing existing amenities and increasing residential development near transit stations. Traditionally, BART emphasized parking capacity as the key to attracting riders. As a result, most of BART's thirty-four stations, like Fruitvale, are surrounded by sprawling surface parking lots.

At the heart of the new approach is the concentration of mixed-use development—high-density housing, shops, and public spaces—within a quarter-mile radius of transit stations. Concomitantly, transit-oriented development discourages density development elsewhere in the area. In essence, development is driven by pedestrian access to transit, which makes transit more appealing to riders.

In California, the push toward transit-oriented development was driven largely by the state's stringent air quality regulations, which require sharp reductions in pollutants like carbon dioxide emitted by automobiles. Research has shown that transit-oriented development can reduce auto trips by as much as 18 percent, and several reports have demonstrated dramatically higher BART ridership among people living within a quarter-mile of transit stations than among the general public. Moreover, these commuters tend to walk to the station, thus avoiding the automobile engine cold start, a major contributor to regional air pollution. Also, commuters driving one-half mile to BART stations cause almost the same amount of pollution as commuters who drive ten or fifteen miles to work.[14]

Using this new policy concept, Martinez, Bell, and UC were able to transform their opposition to the BART parking garage into a sophisticated plan for community-based economic revitalization, enlisting BART officials in their effort as codeveloper with the Fruitvale Development Corporation, UC's economic development wing. "This [was] not the usual planning process," Martinez recalls, "It came from the community and the people that live here. That's new. . . . By working with us and meeting the needs of the community, the costs of development [would] be much less because we [wouldn't] be fighting all the time. We really do want development here, but we want it done in terms of what this neighborhood will support."[15]

Martinez and her colleagues reckoned that most new housing, office, and retail space is built on the fringes of cities, in suburbs. They understood that this kind of development results in region-wide environmental and economic problems. For example, suburban development destroys prime farmland, open space, and plant and wildlife habitat, and it causes traffic congestion and air pollution. In addition, it draws investment away from central cities, further distancing inner-city residents from much-needed jobs and economic opportunities. For BART officials, UC's transit-oriented approach was just what they were looking for. "Fruitvale for 20 years has had various programs for revitalization, none of which has gone anywhere," explains BART director Michael Bernick. "This is the first program that tries to use the transit station both as the center of and the spur to development. . . . This is a very important step."[16] Before Fruitvale, every large-scale transit-oriented development project undertaken in the United States had been located in a new development or wealthy suburb. This would be the nation's first such project in a lower-income, inner-city community.

Immediately UC began to engage community residents in a visioning and planning process to flesh out the parameters of what was soon being called the Fruitvale Transit Village. Oakland's mayor in 1992, Elihu Harris, having been elected partly on his promise of changing the focus of city government from downtown to neighborhoods like Fruitvale and responding to a strong turnout by Fruitvale voters, supported the effort by earmarking $185,000 in city Community Development Block Grant funds to jump-start the planning process. UC raised an additional $290,000 from government and foundation sources, including the federal Intersurface Transportation Equity Act (ISTEA).

Beginning in 1992, UC acted as a facilitator and convener, bringing together different stakeholder groups from around the community and disseminating the results of the planning process. In the spring of 1993, working with the planning department from Berkeley, UC sponsored a community design symposium, or charette, to review and comment on various design proposals for the transit village. UC then held a series of community planning meetings, based on feedback from the charette, to begin to develop a comprehensive redevelopment plan.

The village would be located on the current BART parking lot, a nine-acre site bounded by Fruitvale and Thirty-Seventh avenues on the west and east, respectively, and East Twelfth Street and the BART station on the north and south, respectively. The lot accommodates over a thousand vehicles. As required by BART policy, replacement parking would have to be integrated into the project's design so as to maintain existing capacity. In addition to the environmental benefits that would result from increased ridership, early planning discussions embraced a vision of a transit village where pedestrians and bicyclists could enjoy easy access to local businesses, a health care clinic, a child care facility and community center, UC offices, affordable housing, and even a Fruitvale museum showcasing the area's Latino heritage. The center of the village would be a capacious, elegant pedestrian plaza, lined by trees and filled with people around the clock.

By the spring of 1993, the community planning effort began to receive national attention. In April, U.S. Transportation Secretary Federico Peña visited the Fruitvale BART station, promising to push for federal aid for the project. He returned in August to report that the transit village planning effort would receive a $470,000 grant from the Federal Transit Administration. By this stage in the planning process, money had become a critical factor. As the scale of the project unfolded, each month grander than the last, the need for funding to match the vision increased. In the summer of 1995, as UC convened a series of community planning meetings to develop final project goals and an overall design, and prepared to enter the technical phase of the development process, including environmental review, traffic studies, and economic modeling, the projected cost of the transit village had reached tens of millions of dollars.

Hoping to receive $20 million in federal Empowerment Zone funds, UC learned in October 1995 that it would get only a fraction of that amount: $3.3 million in grants and $3.3 million in loans. Also in October, BART officials informed UC that money it expected BART to contribute to the project as a result of the aborted parking garage, $15 million, would likely not be made available. Suddenly the need for dollars was driving the dream, diminishing and reshaping it.

In an effort to raise the funds necessary to make the Fruitvale Transit Village a reality, UC hired Chris Hudson, an urban planner by training, as project manager in the late summer of 1995. As Hudson explains, his job "was to make the money match the vision."[17] Undaunted by October's setbacks, Hudson, Martinez, and their UC colleagues geared up for some creative fund raising, and not a moment too soon, for the project was moving from the planning phase into action, with environmental review about to begin. . . .

During the years 1996–1997, the transit village took on a definitive shape and scale. After over three years of community involvement in planning and design, the community, led by UC, embraced the transit village as the key to transforming Fruitvale's commercial and transportation core into a vital, healthy place to live, work, and shop. . . .

The heart of the project is a pedestrian plaza adjacent to the existing BART station entrance, lined with small shops and restaurants. It will be a venue for neighborhood festivals and concerts. Vendors will sell food and other goods from pushcarts, while neighbors commune on tree-shaded benches. Surrounding the pedestrian plaza will be approximately 146,000 square feet of new space for commercial and nonprofit use. There will be a health care clinic, La Clinica de la Raza, the Fruitvale-San Antonio Senior Center, a day care center, a branch of the Oakland Public Library, and offices for UC and other community service providers. Also near the plaza will be close to 18,000 square feet of existing retail space. Additionally, the transit village will have approximately fifteen residential units. Three hundred twenty-five new parking spaces will be created, as well as a parking structure for relocation of 979 existing BART surface parking spaces. Finally, East Twelfth Street will be narrowed from two lanes to one in each direction.

The design of the transit village reflects the core goals and principles underlying UC's revitalization efforts:

- *Strengthening existing community institutions.* The transit village will allow community members and institutions to work together more easily and efficiently by housing them in a single, convenient location. . . .

- *Providing a stable source of jobs and income for the community.* Currently 90 percent of Fruitvale's workers have to go outside the neighborhood for work. With new nonprofit and commercial ventures, the transit village will be a source of jobs, for which workers will not need to own a car. UC estimates the transit village will create over seven hundred new jobs. As well, UC hopes to incubate many small businesses and microenterprises within a public marketplace on site, providing technical assistance and preserving the locally owned, small business flavor of Fruitvale. . . .
- *Increasing the variety of retail goods and services available in the community.* As with employment opportunities, many local residents frequently leave the neighborhood to buy many kinds of goods and services. By broadening the types of retail shops and services available, local income and taxes will remain in the community.
- *Beautifying a blighted area.* The transit village will convert sprawling parking lots, brownfield sites, and traffic-congested streets into an attractive community and commercial space centered around a pedestrian plaza. With street and facade improvements as part of the development, the transit village will serve as a catalyst for community-wide beautification.
- *Increasing real and perceived safety.* . . . With housing, commerce, and a steady stream of pedestrians, the transit village will ensure that crime is minimized, if not eradicated altogether. From improved lighting to traffic-calming measures, the transit village will reduce the opportunities for crime and demonstrate that Fruitvale residents care about and are invested in their neighborhood.
- *Providing high-quality affordable housing.* As the quality of life in a neighborhood improves, invariably it becomes more expensive to live in. Already, Oakland and the Bay Area generally are among the nation's most expensive housing markets. The transit village will alleviate some of this pressure by providing lower-income residents affordable rental housing, with access to nearby jobs, shops, and other community resources.
- *Encouraging and leveraging public and private investment.* Because the transit village will improve the look and sense of the neighborhood, highlighting its resources and assets, it will attract outside investment and encourage local home owners and businesses to invest in improvements.
- *Increasing BART ridership and reducing traffic and pollution.* The genesis and core of the transit village lies in the notion that, by reducing the number of cars on neighborhood streets, the neighborhood will be made safer, cleaner, quieter, and healthier. Environmental improvement thus becomes the catalyst for community-based problem solving addressing the full range of neighborhood concerns, from jobs and housing to health care and crime. . . .

With the project scaled down somewhat from earlier plans (for instance, no museum and fewer housing units), UC was able to secure funding in the fall of 1997 to develop the replacement parking facility that was critical to building the rest of the project; without it, BART could not allow the project to proceed. Subsequently, additional funding commitments came through from private foundations and public sources, enabling UC to move forward with its predevelopment activities. The groundbreaking took place in September 1999, and the project is on schedule to be completed by the end of 2000.

Going One Step Further: Restoring Public Access to Oakland's Waterfront

Because of the momentum and visibility of the transit village project, UC has been able to leverage its success to benefit other efforts aimed at improving the quality of life in the neighborhood. For example, UC has raised funds to help local businesses not directly associated with the transit village undertake facade improvements as part of the National Trust for Historic Preservation's Main Street program. According to Chris Hudson, "The transit village project has provided an opportunity to look at the community as a whole and to see the connections. . . . The project is a catalyst for positive change in ways we might not have anticipated going into it."[18]

Feeding off the civic energy created by the transit village effort, UC and several community partners formed the Fruitvale Recreation and Open Space Initiative (FROSI) in early 1997. Working with Oakland's Office of Parks, Recreation and Cultural Services, the University–Oakland Metropolitan Forum (a community development partnership between the University of California at Berkeley and the city of Oakland), the California State Coastal Conservancy, the Friends of the University of California at Berkeley Crew Team, and the Trust for Public Land, UC and community residents are trying to increase Fruitvale's and the adjacent San Antonio neighborhood's environmental assets by converting a degraded nine-acre vacant lot on Oakland's waterfront into a flagship community park, called Union Point Park.

. . . As with the BART station, UC and community residents recognized that the Oakland waterfront was a highly underused and undervalued community asset. With nineteen miles of shoreline, Oakland has one of the longest Bay Area shorelines, yet few people think of Oakland as a waterfront city. Historically, the Oakland estuary has been inaccessible to the public because of a host of industrial and military facilities located on the water's edge. "Oakland and the people that live here have felt cut off from the waterfront. . . . Right now, the waterfront is not a very friendly place to go to. You don't want to get out of your car in some places," says Virginia Hamrick, a member of the Oakland Estuary Advisory Committee, the group charged with implementing the Oakland

Estuary Plan, a long-range planning effort to determine the fate of Oakland's waterfront.[19] ...

Civic Environmentalism and the Fruitvale Transit Village

In a 1996 report, *Blueprint for a Sustainable Bay Area*, the Bay Area environmental group Urban Ecology spelled out the challenge that now confronts Bay Area communities: "The San Francisco Bay Area has reached a pivotal point in its history. ... Because of past patterns of planning and development the region is losing the special qualities that have made it a desirable place to live and work for generations. The time to start changing these destructive patterns and chart a new course for the future is now."[20]

The Fruitvale Transit Village and Union Point Park represent a powerful civic environmental response to this challenge. UCs commitment to participatory planning and a proactive, asset-based approach to community development and environmental protection, defines the transit village and park projects and accounts for their success to date. ...

Despite occasional setbacks, resulting in the scaling back of the original transit village design, UC's ability to engage community residents, professionals, and BART in a long-term, iterative planning process proved decisive. Recognizing in local air quality and traffic problems an opportunity for neighborhood improvement and exploiting the state's push for increased mass transit ridership, UC took the high road in opposing the 1991 proposed BART parking facility and became, in effect, the lead agency, a term usually reserved for government actors, in promoting the transit village concept.

Thus, what was at first merely another case of environmental "us versus them" was transformed into a dynamic, collaborative process that went well beyond environmental and land use issues to incorporate a community-wide revitalization strategy. As evidenced by the Union Point Park project, the momentum generated by the transit village spawned new opportunities for change, which created exciting synergies between the transit village project and other neighborhood development initiatives.

Moreover, as both Chris Hudson and Michael Rios emphasize, consistent community involvement allowed a large, educated constituency to develop, who understood the connection between environmental improvements and overall community change and who were prepared to capitalize on opportunities for action. ...

Fruitvale residents, most of them lower-income people of color, carried the banner of environmental justice in their campaign to improve transportation and overall environmental quality. Having put up with traffic, air pollution, and lack of access to quality parks and open space for decades, UC and the Fruitvale community took a stand, and in the process transformed their environmental problems into a strategy for economic, social, and environmental revitalization. . . .

Notes

Shutkin, William A. 2000. "Oakland's Fruitvale Transit Village: Building an Environmentally Sound Vehicle for Neighborhood Revitalization." In *The Land That Could Be: Environmentalism and Democracy in the Twenty-First Century.* Cambridge, Mass.: MIT Press.

1. A. Haupt, "Union Point Park: Waterfront Access, Recreation, and Equity," *Urban Ecologist,* no. 4 (1997): 10.

2. Bank of America, *Beyond Sprawl: New Patterns of Growth to Fit the New California* (Feb. 1995).

3. S. Burrington and B. Heart, *City Routes, City Rights: Building Livable Neighborhoods and Environmental Justice by Fixing Transportation* (Boston: Conservation Law Foundation, June 1998), 13.

4. Ibid.

5. Ibid., 14–15.

6. Ibid., 19.

7. J. Holtz Kay, *Asphalt Nation: How the Automobile Took Over America, and How We Can Take It Back* (New York: Crown Publishers; 1997), 40.

8. Ibid., 21.

9. Ibid., 46 (quoting C. Hayes).

10. Ibid., 47.

11. R. Bullard and G. Johnson, eds., *Just Transportation: Dismantling Race and Class Barriers to Mobility* (Philadelphia: New Society Publishers, 1997), xi (quoting J. Lewis).

12. Much of the information regarding UC's history comes from my August 20, 1998, interview with Chris Hudson, project manager at the Fruitvale Development Corporation.

13. D. Kim, "Neighborhood Thinks Transit Villages Work," *Oakland Tribune* (quoting A. Martinez).

14. M. Bernick, "Can't Walk to Work? Then Walk to the Train," *Los Angeles Times,* May 4, 1993.

15. B. O'Brien, "This Time We Make the Plans," *Oakland Express,* Jan. 29, 1993, 2, 20 (quoting A. Martinez).

16. B. Wildavsky, "Transport Official Offers Oakland Help," *San Francisco Chronicle* (quoting M. Bernick).

17. Interview with Chris Hudson, Aug. 20, 1998.

18. Ibid.

19. K. Kirkwood, "Renewing the Waterfront, Better Days Ahead for Oakland Estuary," *Oakland Tribune,* Apr. 26, 1998, 1.

20. Urban Ecology, *Blueprint for a Sustainable Bay Area* (San Francisco: Urban Ecology, Nov. 1996), 9.

7

American Environmentalism
The Role of Race, Class and Gender in Shaping Activism 1820–1995
Dorceta E. Taylor

Dorceta Taylor's contribution provides a sweeping history of the intersection of race, gender, and class with the rise of American environmentalism. Taylor argues that the white middle-class environmental movement failed to recognize and address the needs of working-class whites, people of color, and some middle-class communities. Taylor shows that though people of color, women, and other marginalized communities were largely excluded from participation in and benefits of traditional environmentalism, they did, nonetheless, make significant contributions to environmental change throughout the nineteenth and twentieth centuries. She concludes that the potential for bringing about a future that is both socially and environmentally just depends on the development of a more inclusive, culturally sensitive, and broad-based environmental agenda in the United States.

People's relationship to the environment has to be understood in the context of historical and contemporary class, race and gender relations. American environmentalism has been profoundly shaped by a unique set of social, political and economic factors arising from the period of conquest and subsequent industrialization. It can be argued that American environmental activism evolved over the last 175 years, and is characterized by four distinct periods of mobilization: the pre-movement era—1820s–1913, the post–Hetch Hetchy era—1914–1959, the post-Carson era—1960–1979, the post–Love Canal/ Three Mile Island era—1980–present (Taylor, 1996). . . .

. . . This [essay] tries to explore how environmental activism is shaped by one's race, class position, gender, societal experiences, ideology, political power, and social networks.

The Pre-Movement Era (1820s–1913)

White, middle class, outdoor- and wilderness-oriented, elite males influenced by cultural nationalism or Romanticism and Transcendentalism began espousing pro-environmental ideas and started to publicize the natural wonders of the country during the first half of the 19th century. . . .

The work of these visionaries influenced successive generations of white middle class men to embrace outdoor pursuits and to undertake actions to preserve the environment. From the 1850s–1913, John James Audubon, George Perkins Marsh, Henry David Thoreau, and John Muir were the leading naturalists, ecologists and advocates of wilderness preservation. [These activists] called attention to the destruction and domination of nature, and advocated compassion for other species, the harmonious coexistence of humans and nature, government protection of wild lands, and a return to a simpler lifestyle. These activists were also influential in preserving some of the earliest national parks and forests. . . .

Because these men were financially secure, they were free to embark on outdoor expeditions at will. They sought out the wilderness as an antidote to the ills of the urban environment. They did not include issues relating to the workplace or the poor in their agenda. They were basically middle class activists procuring and preserving environmental amenities for middle class benefits and consumption.

What were the conditions of the working class throughout the period the middle class white males were exploring the wilderness and building the foundations of the early reform environmental movement? During the 1800s, the white working class worked in deplorable conditions. They worked long hours for little wages. The long hours and intensive labor practices combined to give the U.S. one of the highest industrial accident rates in the world. . . .

Housing conditions were abominable. Workers lived in crowded, unsafe, unsanitary, over-priced housing. Unemployed and homeless people lived in parks and undeveloped lots. . . . Twenty percent of the population of Boston and New York City were said to be living in distress, and one in ten New Yorkers were buried in paupers graves in Potters Field (Dubofsky, 1996:27). The little free time workers had was spent in local pubs or in the streets. Because of overcrowding, private personal and family activities (like courtship, sexual encounters, drinking, and socializing) often spilled onto the streets. This rankled the middle class who set out to Americanize and acculturate immigrants and curb what they saw as morally bankrupt, uncivilized behavior. Not surprisingly, interactions between both groups grew increasingly tense (Rosenzweig, 1983, 1987; Peiss, 1986; Beveridge and Schuyler, 1983; Dickason, 1983). Some of these tensions were fought out around the issues of access to and utilization of urban open space.

Not all the middle class white males interested in the environment and open space issues concentrated their efforts on wilderness and wildlife. Some, like Andrew Jackson Downing, explored their recreational interests in the city. Downing sought to bring pastoral bliss to squalid cities like New York by building elaborate, European-inspired, picturesque parks in urban centers. . . . While the afore-mentioned wilderness-oriented activists separated their environmental activism, and recreation from those of urban dwellers and the working class, the urban park builders worked alongside the poor coming face to face with working class concerns. Though the park builders were often insensitive to working class concerns (Olmsted, for instance, fired his Central Park work crew who struck for better wages and working conditions, and dismissed those who missed work due to illness or injuries sustained on the job), they built parks where the middle and working classes interacted—even if on a limited basis (Beveridge and Schuyler, 1983; Roper, 1973; Gottlieb, 1993; Rosenzweig, 1983, 1987; Olmsted, 1860: November 13). These activities, and the extent to which they believed that one could find solace in urban green space, distinguished them from their wilderness-oriented counterparts.

. . . These parks provided free or cheap leisure for the working class and soon became the focal point of environmental and political activism. They were the sites of labor unrest, bread riots and political rallies (Peiss, 1986; Piven and Cloward, 1979:43–44; Beveridge and Schuyler, 1983:15; Olmsted, 1860: December 8; *New York Times*, 1857:6, 17, 21, 25). In addition, the working class pointing out the environmental inequalities they faced (lack of sanitation, clean water, public open space, overcrowding, ill health and diseases) organized and lobbied for neighborhood parks. The working class also influenced the design of these urban parks by lobbying for open space designed for more active recreation like ball fields (Rosenzweig, 1983, 1987).

During the 1800s and early 1900s, white middle class females found themselves in a peculiar position. On the one hand, they attended elite colleges; read extensively about Romanticism and Transcendentalism; ecology; botany; and natural history; and attended lectures on the environment; but on the other, they were not expected to act on their impulses to explore the wilderness (Muir, 1924). . . .

Nonetheless, some middle class white women broke this mold and were able to combine their interest in ecology, the environment, health, moral upliftment, cultural enlightenment, and civic improvement with political activism and a desire to help the poor.

... As crowding reached unbearable levels in the cities in the mid to late 1800s, the streets became the social and recreational space of the working class. Children roamed the streets and were often jailed for playing or loitering in the streets (Rosenzweig, 1983, 1987). During the Progressive Era (1880–1920s), upper-middle class women—the wives of wealthy industrialists—sought to remedy the situation by building small neighborhood playgrounds and "sand gardens." One prominent group involved in this effort was the Massachusetts Emergency and Hygiene Association (MEHA). Explicating the basic ideology of the male urban park builders, the female park builders and recreation planners believed that recreation would improve the health, moral and cultural outlook of the children. They believed that recreation should be provided in a structured environment, because their efforts to acculturate working class children was the most effective means of improving the lives of the working class. They occupied a niche ignored by the male park builders who focused on grandiose urban parks or park systems designed with a bias towards passive recreation. . . .

People of color lived in inhumane conditions in the U.S. during the 1800s and the early 1900s. When Jamestown was established in 1607, an estimated 1–18 million Native Americans lived in the U.S.; however, by 1890 when the last of the Native American wars ended, there were about 250,000 surviving Native Americans (Stiffam, 1992:23–28; Wax, 1971:17; McNickle, 1973). Under the 1830 Indian Removal Act, Native Americans were driven from their land and forced onto desolate reservations. They lost their traditional hunting, gathering and fishing grounds, and sacred sites. About 20,000 Indians died on the trek west (Lurie, 1982:131–144; Josephy, 1968; Debo, 1970). Native Americans, forced to live in some of the most inhospitable areas of the country (Lenarcic, 1982:137–139), used some of their traditional knowledge and practices to develop resource management techniques to sustain themselves on the reservations. . . . As DeLoria (1994:4) argues, from the 1890s to the 1960s Indians were the "Vanishing Americans" because most people thought Native Americans had been exterminated.

By the early 1800s, the northern states began to industrialize, but the southern states remained largely agrarian. Slavery had been in place for about two hundred years and African American labor was used to exploit the resources of the south. . . .

The system of sharecropping further reinforced the inequality of African Americans. By 1910, more than half of all employed African Americans worked in agriculture (Geschwender, 1978:169). . . . As the first wave of African Americans started working in the urban centers of the North, they were recruited as strike breakers and were offered the most dangerous factory jobs. Because of rigid segregation, they lived in the most dilapidated, crowded, unsanitary and unsafe housing. They earned less wages and paid higher rents than whites (Hurley, 1995; Morris, 1984: Tuttle, 1980; Dubofsky, 1996:12).

In the first half of the 1800s, territory (which later became Texas, New Mexico, Arizona, and California) was appropriated from Mexicans living in the southwest. The area had a regional economy based on farming and herding; an elite class of wealthy Mexican landowners dominated the affairs of the region (Cortes, 1980:697–719). . . .

The construction of the railroads and the expansion of agriculture stimulated demand for low-wage Mexican labor. Mexican workers were employed chiefly in the cultivation, harvesting, and packaging of fruits, vegetables and cotton in California and sugar beets in Colorado and California. At the turn of the century Mexican workers were used in the sugar-beet industry of central California to curb the demands of Japanese workers and in Colorado to help control German and Russian workers. They also worked in New Mexico and Arizona in copper, lead and coal mines (Jiobu, 1988:21–22; Dubofsky, 1996:3). Until the 1920s, movement across the border between Mexico and the U.S. was informal and largely unrestricted. Throughout the century, Mexican Americans have been offered some of the worst jobs for little wages. They are often paid less to do the same jobs as Anglos. This split labor market has been further divided by gender; Mexican American women are assigned worse jobs than men and received lower wages (Takaki, 1993:318–319; Dubofsky, 1996:13). Mexican Americans often toiled in "factories in the fields" where about 2,000 men, women and children worked in 100+ degree-heat, had no drinking water, shared 8 outdoor toilets, and slept among the insects and vermin (Dubofsky, 1996:24).

From the 1820s onwards, the Chinese and later the Japanese and Filipinos, migrated as laborers to help in the opening and exploitation of the West—in the Gold Rush, mining, railroad construction, farming, fisheries, sugar plantations, and factories. For example, in 1860, in Butte County, California (gold mining area), there were 2,200 Chinese (Jiobu, 1988:34). In the immigration were young men traveling without wives, or other family members. Asians worked long hours under dangerous working conditions. They were subject to severe discrimination and quite often were stripped of some of the basic rights accorded the white working class (Friday, 1994:2–7, 51; Lenarcic, 1982:140; Carlson, 1992:67–84; Anderson and Lueck, 1992:147–166). Chinese were hired to do menial tasks and as the claims played out, white gold miners sold their old claims to Chinese miners. As the California Gold Rush ran its course, the percentage of Chinese miners steadily increased from one percent of all miners in 1850 to more than 50 percent in 1920 (Jiobu, 1988:34).

In 1862 when Congress authorized the construction of the transcontinental railroad, Central Pacific had labor problems. There was not a large pool of white laborers in the Pacific region who were willing to work on the railroads, so Central Pacific recruited Chinese laborers from the gold fields, farms, cities, and from China. Most were acquired through labor contracting firms. At $35 per month (minus the cost of food and housing), Chinese laborers were paid two-thirds of what white laborers earned. With the hiring of Chinese, whites moved up the occupational ladder to hold whites-only jobs such as foremen, supervisors, and skilled craftsmen. The Chinese were left to tunnel through the mountains, dig through dangerously deep snow banks, and dangle in baskets along sheer cliff faces while they planted explosives. Many Chinese lost their lives building the tracks across the sierras (Jiobu, 1988:34–35)....

Because of the conquest and domination of people of color, at some points men and women of color from the respective racial groups shared common environmental experiences. However, they were separated spatially and occupationally at times, thereby having very different experiences. For instance, Native American men and women were moved onto reservations together and African American men and women were enslaved and subjected to the same kind of work conditions. However, in the late 1800s and early 1900s the African American men would migrate to distant cities or rural areas to work in factories and mines, etc. Similarly, Latinos would also migrate long distances for jobs leaving Latinas behind.

Asian families were also torn apart because of immigration rules—young men migrated in search of jobs leaving their families behind in Asia (Almquist, 1979:430–450; Kitano, 1980:563; Chan, 1991:66; Duleep, 1988:24). Employers in the West played a role in splitting up families. They preferred to hire single men because it cost less to feed, clothe and house one male worker than whole families....

The Post–Hetch Hetchy Era (1914–1959)

The Hetch Hetchy controversy was the catalyst that resulted in the formation of the environmental movement. The controversy, which was at its most intense from 1910 to 1913, arose when the city of San Francisco proposed a dam on the Toulumne River (Hetch Hetchy) Valley in Yosemite. According to some, Hetch Hetchy was as spectacular as the Yosemite Valley which was already designated a national park. The opponents of the Hetch Hetchy dam, John Muir and the Sierra Club, claimed the land should be preserved and not used in a utilitarian manner. It should be left untouched for future generations. The proponents of the dam, the City of San Francisco, Gifford Pinchot (the founder of the U.S. Forest Service) and many Western Congresspersons, argued that Muir and his colleagues were being selfish. They argued that the land should be conserved or used wisely or sustainably. They argued that the good of the many (providing water for a parched city) should prevail over the good of a few (saving land for itself and for future generations). The controversy, therefore, pitted the preservationists against the conservationists (Nash, 1982).

The controversy thrust environmental issues onto the public stage. For the first time, citizens who were not a part of the small elite group of preservationists, conservationists and outdoor enthusiasts got involved in environmental debate by writing letters, newspaper articles and participating in public debates (Bramwell, 1989; Fleming, 1972; Fox, 1985; Nash, 1982; Oelschlaeger, 1991; Paehlke, 1989; Pepper, 1986; Taylor, 1992). White, middle class wilderness-oriented activists, realizing that they needed to organize and coordinate their efforts more effectively, formed numerous environmental groups focused on protecting wilderness, wildlife, outdoor recreation, habitat restoration, and water pollution (as it affected fish, waterfowl and habitats). For the most part, these environmental groups were segregated by race and class.

It is not surprising that a movement sprang out of one of the first major public environmental controversies. A study of 1,053 organizations listed in the 1993 and 1994 Conservation Directory and the 1992 Gale Environmental Sourcebook (National Wildlife Federation, 1993; Hill and Piccirelli, 1992), shows that by the turn of the century many of the elements of what would become the environmental movement were already in place. A substantial number of the organizations were formed before or during the time of Hetch Hetchy. Seventy-eight national and regional organizations existed by 1913, and in the decade after the Hetch Hetchy decision, 43 new environmental organizations were formed. Overall, 214 organizations were formed in the post–Hetch Hetchy period (1914–1959). Prior to Hetch Hetchy these institutions and organizations did not coordinate their activities extensively, but Hetch Hetchy focused attention on one issue and that allowed for the necessary coordination and communication—so vital to movement building—to take place.

From the turn of the century onwards, middle class women who were formerly confined to home, urban and/or community activism, began embarking on strenuous outdoor expeditions in greater numbers than ever before. Though women joined and participated actively in organizations like the Sierra Club, males dominated the leadership and the agenda of these organizations. The agenda centered on wilderness and wildlife preservation, hunting, fishing, bird-watching, hiking, and mountaineering. From 1914 to 1959, men, many of whom were businessmen or had significant business ties, sought to consolidate this agenda by establishing and reinforcing contacts with government, influential policy groups and industry. They decided to espouse a brand of environmentalism that sought to make small incremental changes or reforms in the existing system by working with both government and industry. This laid the groundwork for reform environmentalism (McCloskey, 1992:77–82). However, by the 1930s, the newly formed movement began to stagnate as the political activities and issues being tackled failed to capture the imagination of large sectors of the population. This period of malaise continued through the 1940s (Bramwell, 1989; Fleming, 1972; Fox, 1985; Nash, 1982; Oelschlaeger, 1991; Paehlke, 1989; Pepper, 1986; Taylor, 1992; Gottlieb, 1993).

During this time period, middle class white women still maintained their interest in natural history, garden clubs, local ecology, but they shifted the focus of other aspects of their activism. Groups like MEHA placed less emphasis on playground construction and supervision as these tasks were turned over to cities and other government entities with the capital and human resources to fund and operate them more effectively. In addition, with a reduction in immigration, improved living and working conditions, and with many immigrant groups forming their own ethnic organizations and social networks, there was less need for the acculturation, morality and hygiene lessons from the upper class. As the Progressive Era drew to a close, some of the concerns (worker rights) addressed by activists of the era were being tackled by labor unions.

From the 1920s onwards, the urban agenda of park building continued with city and state governments being responsible for building, maintaining and supervising these parks, and setting the standards for the equipment found in them. The concern for health and sanitation evolved into concerns over environmental quality and quality of life issues. Again, city and state governments having taken over the role of providing basic services, citizens' groups adopted the role of monitoring the government's performance and lobbying for improved or expanded services. These issues transcended the urban domain, and became a part of the suburban and rural agenda also.

While Hetch Hetchy motivated middle class whites to join reform environmental organizations, Taylorism impelled the working class to unionize. Between 1880 and 1920, employers introduced Taylorism (scientific management or Fordism) to the factories. As a result, assembly lines moved faster, workers lost control over their work, the owners got richer, and the workers saw little material benefits for their increased output. Union rolls swelled: from 1897 to 1904, union membership rose from 447,000 to 2,073,000. . . . (Dubofsky, 1996:94–95, 102–103, 118–119). However, as workers pinned their hopes to the unions during the 1900s, some of the issues like hazard reduction in the workplace and community were downplayed at the expense of creating and maintaining jobs (especially during the Depression), securing wage increases and improving benefits. However, from 1914 to 1959, with or without the support of the union, workers expressed their discontent about poor working conditions by participating in organized and wildcat strikes, protests, and by demanding safety equipment on the job (Hurley, 1995). Again, there was considerable overlap in the interests of working class men and women. Since both groups toiled under dangerous conditions and lived in polluted communities, family and community health and safety concerns were salient.

By the early 1900s, unionized people of color began expressing their concerns about deplorable working conditions. They experienced on-the-job discrimination—they were stuck in the most dangerous, dirty and hazardous work sites with little or no chance of promotion, and for lower pay than whites. People of color had unskilled jobs which meant they worked longer hours than skilled laborers (whites). In 1920, skilled laborers worked an average of 50.4 hours per week. In comparison, unskilled laborers worked an average of 53.7 hours per week (Dubofsky, 1996:24). So, by the early 1930s, people of color began to take strike action. For instance, African Americans in unionized factories and steel mills like Gary Works staged wildcat strikes/work stoppages to improve work conditions (Hurley, 1995).

To understand the position of workers of color and the likelihood that they would adopt pro-environmental positions, one has to understand the role of race and class in structuring their response. During the 1800s and early 1900s, oppression among whites vis-à-vis the work place and access to environmental amenities amounted to white-on-white class and ethnic oppression. That is, American born whites of northern European descent comprised the middle class that discriminated against the immigrant and/or Southern and Eastern Europeans. The latter group comprised a large portion of the working class. However, much harsher forms of (racial) discrimination distinguished whites from people of color. The white–nonwhite relationship, marked by enslavement, forced relocations, appropriation of land, internment and deportations, continued through the twentieth century with rigid occupational, educational and residential segregation.

This meant that while the white working class were able to start advocating a radical working class environmental agenda at the turn of the century, people of color saw their biggest problem in the community and in the workplace as racial oppression. This is not to say that they were unable to perceive other forms of discrimination—they did—but they had to overcome the racial oppression in the work place in order to relieve occupational discrimination (Hurley, 1995). When both the employers and the unions reinforced patterns of *de jure* and *de facto* discrimination and segregation, workers of color were left to their own devices or social networks to resolve their problems.

On the job workers of color had to deal with the class oppression of unsafe, hazardous work and the racial and gender oppression of being permanently assigned to such jobs for the lowest wages. People of color were quite aware that these jobs were the least likely to be cleaned up and made safe. The white worker, because of his or her race, knew that with time he or she would be moved to safer jobs. Workers of color knew such opportunities did not exist for them. The interlocking and multiple sources of oppression (racial, class and gender) led workers of color to support occupational safety improvements, but demand racial equality at the same time. . . .

The Post-Carson Era (1960–1979)

During the 1960s and 1970s, there was an unprecedented level of mobilization around environmental issues. The mobilization was spurred, in part, by [Rachel] Carson who linked her concern for wildlife and nature with questions about the effects of pesticides on humans and wildlife. She used the injustice frame to question the immorality and danger of widespread spraying of pesticides, and argued that people had a right to a safe environment. She argued that if they were harmed by the deliberate acts of others, the victims had a right to compensation (Carson, 1962). Carson focused on home, community, nature, and the wilds. She linked urban and rural concerns, and publicized an issue that affected everyone in the country.

Carson's work led to an immense public outcry over pollution and chemical contamination that launched the birth of the modern environmental movement. Reform environmental organizations benefited from the mobilization. Membership skyrocketed in leading organizations like the Sierra Club, the National Audubon Society and the Wilderness Society in the 1960s. A study of eight of the major environmental organizations found that membership went from about 123,000 in 1960 to about three-quarters of a million members in 1969 (Fox, 1985:315). The mass mobilization drive resulted in cleaner air, rivers and lakes for many Americans. Towards the end of the decade many young and radical environmentalists—students, former civil rights activists and antinuclear activists—joined the movement (Zinger, Dalsemer and Magargle, 1972:381–383). Some of these youthful environmentalists joined the leading environmental organizations while others formed their own organizations. This brought new constituencies into the reform environmental movement. The concerns broadened to include more issues relating to the urban environment, community, home, and humans. More attention was paid to environmental hazards and industry was scrutinized more heavily.

The second surge of mobilization in the post-Carson era came in 1970, before and after Earth Day. Between 1970 and 1979 membership in the eight major environmental organizations mentioned above went from 892,100 to 1.583 million. More environmental groups were formed in the post-Carson era than at any other period in environmental history; 469 or 45 percent of the 1,053 environmental groups studied were formed between 1960 [and] 1979. However, the mobilization of the 1960s and 1970s was largely a white middle class mobilization. Surveys of the membership of leading environmental groups and of environmental activists nationwide in the late 1960s and early 1970s demonstrate this point. A 1969 national survey of 907 Sierra Club members indicated that the organization had a middle class membership. Seventy-four percent of the members had at least a college degree; 39 percent had advanced degrees. Ninety-five percent of the male respondents were professionals, and five percent occupied clerical and sales positions, were owners of small businesses or unskilled laborers. Fifty-eight percent of the respondents said their family incomes was over $12,000; 30 percent reported family incomes over $18,000 per year (Devall, 1970:123–126)....

...A 1972 study of 1,500 environmental volunteers nationwide showed that 98 percent of the members of the environmental organizations were white and 59 percent held a college or graduate degree. Forty-three percent held professional, scientific-technical, academic or managerial jobs. A half of the respondents had family incomes of more than $15,000, 26 percent had between $10,000 and $15,000 and the remainder earned less than $10,000 (Zinger,

Dalsemer and Magargle, 1972). In general, studies find that environmentalists are highly educated, older, urban residents who are political independents. In addition, education, and to a lesser extent, income is associated with naturalistic values and environmental concern. (Harry, Gale and Hendee, 1969:246–254; Devall, 1970:123–126; Hendee, Gale and Harry, 1969:212–215; Buttel and Flinn, 1978:433–450, 1974:57–69; Cotgrove and Duff, 1980:333–351; Dillman and Christenson, 1972:237–256; Faich and Gale, 1971:270–287; Lowe, Pinhey and Grimes, 1980:423–445; Harry, 1971:301–309; Tognacci, et al., 1972:73–86; Wright, 1975; Martison and Wilkening, 1975).

There was also a major ideological shift during the post-Carson era. During the 1960s and 1970s, the Romantic environmental paradigm gave way to a broader vision of environmentalism—the new environmental paradigm (NEP). Building on the basic ideological framework of the REP, the NEP expanded on the environmental dialogue and articulated a bold new vision that critiqued the development of high (large, complex, energy-intensive) technology like the nuclear industry, encouraged population control, pollution prevention, risk reduction, energy, recycling, environmental clean-ups and espoused post-materialist values (Inglehart, 1992).

During this era, the environmental movement enjoyed strong public support. Opinion polls show what could be described as a Carson effect and an Earth Day effect. There was a steady increase in concern over pollution through the latter part of the 1960s and a sharp increase in levels of concern in 1970. For instance, in 1965, 17 percent of the respondents in a Gallup survey said they wanted the government to devote most of its attention to reducing air and water pollution. However, by 1970, 53 percent of the respondents wanted the government to devote most of its time to these issues (Gallup, 1972:1939). Other polls showed that 46–60 percent of the respondents indicated they were "very concerned" about reducing water and air pollution, 50–61 percent of the respondents thought too little was being spent on the environment; 58 percent thought we must accept a slower rate of economic growth in order to protect the environment. In addition, most respondents were not willing to relax environmental standards to achieve economic growth, did not think that pollution control requirements had gone too far and did not think we had made enough progress on cleaning up the environment to start limiting the cost of pollution control (State of the Nation, 1972–1976; NORC, 1973–1980; ORC, 1977–1978). A distinction should be made between concern and support for the environment and environmental activism. Not all people who are concerned about the environment or support environmental actions become environmental activists (Taylor, 1989).

Though the environmental activities of the sixties and seventies heightened awareness of environmental issues among the working class, and though some joined outdoor recreation organizations like the Izaak Walton League, by and large, the working class did not flock to preservationist environmental organizations.

They intensified their efforts to strengthen the traditional working class agenda. Pressuring and working through their unions and newly formed working class environmental organizations, they sought to improve working conditions, bring the issues of worker health and safety into the national consciousness, pushed for safety equipment, etc. Using the Occupational Safety and Health Administration's (OSHA) guidelines, they reported environmental violations, filed complaints and forced companies to comply with the regulations. In addition, they used collective bargaining strategies to ascertain general environmental improvements and to establish safety committees at the workplace. They negotiated the "right to refuse hazardous work" clauses, hazard pay and safety equipment as part of their union contracts (Hurley, 1995; Robinson, 1991).

The working class was also concerned about health of residents and the environment outside the factory gates. During this period working class environmental groups were formed to reduce pollution in the community. Focus was on air and water pollution, factory emissions and improved sanitation (illegal dumping, garbage removal). Even though the broadened emphasis of the reform environmental movement included a focus on reducing pollution, collaboration between middle and working class activists was still limited and strained. Because the middle and upper classes no longer lived in parts of the cities close to the sights, sounds and smell of the factories, middle class environmental groups did not lend much support to working class environmental struggles. The middle class focused on preventing the degradation of their communities and improving the environmental amenities close by.

The working class, having more free time and income at their disposal, intensified their interests in outdoor recreation pursuits like hunting, fishing, visits to parks, etc. Working class groups pushed for access to fishing and hunting grounds, improved quality of recreation sites, and larger number of parks, etc. Conflicts arose between the working and middle classes when the middle class tried to restrict the use of recreational areas in their communities to prevent overcrowding or use by the working class and people of color. In addition, as middle class communities passed zoning ordinances to restrict development, the middle class perceived these actions as efforts to preserve the environment while the working class and the unions viewed them as saving the environment at the expense of working people's jobs and livelihoods (Hurley, 1995).

However, towards the end of the post-Carson era, two major environmental disasters in lower middle class and working class communities precipitated an unprecedented level of activism and mobilization in these communities. The nuclear accident at Three Mile Island (TMI) and toxic contamination of Love Canal came as a wake-up call to communities nationwide. Though they did not receive the media attention of TMI and Love Canal, many communities of color had similar problems. Prior to Love Canal, what people perceived as a local, isolated case of toxic contamination emerged as a national problem of immense proportions. This resulted in the formation of working class, grassroots environmental groups all over the country. While some of these groups organized to clean up the toxic contamination in their communities (reactive mode), like the middle class environmentalists before them, many of these groups became more proactive. They organized to halt the development of noxious or nuisance facilities or other locally unwanted land uses (LULUs). . . .

The Post–Three Mile Island/Love Canal Era (1980–Present)

Throughout the 1980s, white, middle class, reform environmentalism continued to dominate the environmental landscape. Environmental organizations grew increasingly big, bureaucratized, hierarchical and distant from local concerns and politics. They focused on national and international issues, lobbied Congress and business, and developed close ties with industry (through funding, negotiations and board representation). Grassroots organizing had long given way to direct-mail recruiting, and direct-action political strategies were rarely used. . . .

Males dominate the top leadership positions in reform environmental organizations. A 1992 nationwide study conducted by the Conservation Leadership Foundation found that of the 248 chief executive officers (CEOs) and top leaders surveyed, 79 percent of the respondents were male. Their mean age was 45 years, and 50 percent had a bachelor's degree, 28 percent a master's and 21 percent a doctorate (Snow, 1992:48–49). . . . The Conservation Leadership Foundation's national study of environmental volunteers also found that males dominated the voluntary sector of the reform environmental movement. . . . The profile of the reform sector in the 1990s is similar to profile of the membership of the organizations during the late 1960s–early 1970s. . . .

. . . [I]n 1992, wildlife, wilderness and waterway protection dominated reform organizations' agendas both in terms of the presumed importance of the issue and the percentage of the organization's resources that was spent on the issue. While 46 percent of organizations' budgets were spent on fish, wildlife, land preservation, and wilderness, and 16 percent on water conservation, only 8 percent of the budgets were being spent on toxic waste management and 4 percent on land use planning. This pattern of environmental perception, problem definition and spending was in place at a time when communities all over the country were struggling with environmental health issues, toxic contamination, urban sprawl, pollution, and solid waste disposal issues. There seemed to be a disconnect between the priorities of the reform environmental organizations many local communities.

Although the reform environmental agenda continued to dominate environmental politics, it was the emergence of many persistent, radical grassroots organizations that have profoundly changed the nature of the environmental movement during the 1980s. . . .

. . . [T]oxics have remained in the news and have continued to be a mobilizing factor because what people thought was the worst-case scenario when the story first captured national attention in the late 1970s turned out to be just the tip of the iceberg. In addition, many huge corporations—some long-time providers of jobs and supporters of local civic organizations and events—were found to be the source of the contamination. Many people felt that the trust or social compact between host communities and corporations was broken. People were further dismayed to find that the government, policy makers, scientists and other experts could not offer ready solutions or any solutions at all. In many instances nothing was done while the government, business, scientists remained bogged down in long delays due to legal or technical skirmishes. In some cases people discovered there were no safeguards to protect the balance of power and to ameliorate the situation (Habermas, 1975; Pusey, 1993:92–110; Ingram, 1987:155–160). Some of these radical groups participate in the environmental justice movement.

It was the image of a silent spring—a spring silent of bird song that motivated thousands of middle class whites to become active in the reform environmental movement in the 1960s. In the early 1990s, another image motivated people of color to form the environmental justice movement. They were aroused by the specter of toxic springs—springs so pervasive and deadly that no children sang. Soon after the publication of Carson's *Silent Spring*, middle class whites relocated to pristine areas, cleaned up and slowed or prevented the degradation of their communities. Long before the phenomenon known as NIMBYism (Not In My Backyard) was labeled, middle class residents skillfully used zoning laws, legal challenges and every other means available to them to control and maintain the integrity of the communities they lived in. Their success left developers and industry flustered; but only temporarily.

Because effective community resistance is costly, industry responded to the challenges of middle class white communities by identifying the paths of least resistance (Blumberg and Gottlieb, 1989; Cerrell Associates, 1984; Trimble, 1988). By the 1980s, working class white communities recognized that that path went through their communities also. As working class communities organized to stop the placement of LULUs in their neighborhoods, industry quickly adjusted to the new political reality. The path of least resistance became a expressway leading to the one remaining toxic frontier (in the U.S., that is)— people of color communities. By the 1990s, people of color communities were characterized by declining air and water quality, increasing toxic contamination, health problems, and declining quality of life. Since the 1970s there have been isolated efforts to mobilize communities of color around environmental issues; such efforts started to pay off during the late 1980s (Hurley, 1995). From 1987 through the early 1990s, the book read in people of color communities was *Toxic Waste and Race* (United Church of Christ, 1987). This study did for people of color and the environmental justice movement what *Silent Spring* did for middle class whites in the 1960s.

Toxic Waste and Race, other books and newspaper articles that began appearing shortly after this study was published, had an immediate impact. These publications made an explicit connection between race, class and the environment. Using the injustice frame, they articulated the issues in terms of civil and human rights, racism and discrimination. They identified the widespread perception that people of color communities were viewed as politically impotent and were either unwilling or unable to take control of and change the social and environmental conditions in their communities.

This framing was the obvious bridge that transformed the previous attempts of people of color to articulate their environmental concerns in a way that linked their past and present experiences in an effective manner. Environmental justice embodied all these concerns and experiences. The Environmental Justice Movement sought to: (1) recognize the past and present struggles of people of color; (2) find a way to unite in the various struggles; (3) organize campaigns around fairness and justice as themes that can interest a variety of people—these are also themes that all people of color had built a long history of community organizing around; (4) build a movement that linked occupational, community, economic, environmental, and social justice issues; (5) build broad class and racial coalitions; (6) strive for gender equity; (7) employ a combination of direct action and non-direct-action strategies; and (8) educate, organize and mobilize communities of color. Since many people of color still live, work and play in the same community, the environmental justice agenda made explicit connections between issues related to workplace and community, health, safety, environment, and quality of life. . . .

People of color environmental activism and the Environmental Justice Movement grew out of an awareness of the increasing environmental risk people of color faced and a dissatisfaction with the reform environmental agenda. The advent of the Environmental Justice Movement marks a radical departure from the traditional, reformist ways of perceiving, defining, organizing around, fighting, and discussing environmental issues; it challenges some of the most fundamental tenets of environmentalism that have been around since the 1800s. It questions some of the basic postulates, values and themes underlying reform environmentalism. The environmental justice movement also questions the ideological hegemony of reform environmentalism and the tendency to marginalize or dismiss other perceptions. In addition, environmental justice questions the racial and class homogeneity of the environmental movement, male dominance, racism in the movement (as manifested through publications, hiring practices, the composition and recruitment of the membership and the board), the relationship with industry, strained relationship with people of color communities, and the practice of ignoring the environmental issues that plague people of color communities. . . .

Conclusion

As the above discussion shows, race, class and gender have profound impacts on a person's experiences, which in turn have significant impacts on political development, ideology, and activism. . . . The history of American environmentalism presented by most authors is really a history of white, middle class male environmental activism. The tendency to view all environmental activism through this lens has deprived us of a deeper understanding of the way in which class, race and gender relations structured environmental experiences and responses over time. . . .

. . . White working class grassroots and environmental groups differ from those of white middle class reform groups in the emphasis the former groups place on workplace and community experiences. So while occupational health and safety and jobs are still minor or non-existent parts of reform environmental organizations' agendas, they are major issues for white grassroots and people of color environmental justice groups. In addition, issues relating to toxics, the urban environment, and environmental risks and burdens are more prominent on the agendas of working class grassroots and environmental justice groups than on reform environmental organizations' agendas. Environmental justice groups differ from white working and middle class groups in their utilization of networks involved in past social justice struggles and religious groups. They use the injustice frame to identify and analyze racial, class and gender disparities, and emphasize improved quality of life, autonomy and self determination, human rights, and fairness.

The environmental movement is a powerful social movement, however, the movement faces enormous challenges in the future. Among the most urgent is the need to develop a more inclusive, culturally sensitive, broad-based environmental agenda that will appeal to many people and unite many sectors of the movement. To do this the movement has to re-evaluate its relationship with industry and the government, re-appraise its role and mission, and develop strategies to understand and improve race, class and gender relations.

Note

Taylor, Dorceta E. 1997. "American Environmentalism: The Role of Race, Class and Gender in Shaping Activism 1820–1995." *Race, Gender & Class* 5(1):16–62.

Bibliography

Anderson, T. L, & Lueck, D. (1992). "Agricultural Development and Land Tenure in Indian Country." In T. L. Anderson, ed., *Property Rights and Indian Economies*. Lanham, MD: Rowman & Littlefield Publishers.

Almquist, E. M. (1979). "Black Women and the Pursuit of Equality." In Jo Freeman, ed., *Women: A Feminist Perspective*. Palo Alto, CA: Mayfield.

Beveridge, C., & Schuyler, D. (1983). *The Papers of Frederick Law Olmsted*, Volume III, *Creating Central Park, 1857–1961*. Baltimore, MD: Johns Hopkins University Press.

Blumberg, L., & Gottlieb, R. (1989). *War on Waste: Can America Win Its Battle with Garbage?* Covelo, CA: Island Press.

Bramwell, A. (1989). *Ecology in the 20th Century: A History*. New Haven: Yale University Press.

Buttel, F. H., & Flinn, W. I. (1978). "Social Class and Mass Environmental Beliefs: A Reconsideration." *Environment and Behavior*, 10:433–450.

———. (1974). "The Structure and Support for the Environmental Movement, 1968–70." *Rural Sociology*, 39(1):56–69.

Carlson, L. A. (1992). "Learning to Farm: Indian Land Tenure and Farming before the Dawes Act." In T. L. Anderson, ed., *Property Rights and Indian Economies*. Lanham, MD: Rowman & Littlefield Publishers.

Carson, R. (1962). *Silent Spring*. New York: Houghton Mifflin.

Cerrel Associates, Inc. (1984). "Political Difficulties Facing Waste-to-Energy Conversion Plant Siting." J. Stephen Powell, Senior Associate, Waste-to-Energy Technical Information Series, Chapter 3a. California Waste Management Board, Los Angeles.

Chan, S. (1991). *Asian Americans: An Interpretive History*. Boston: Twayne.

Cortes, C. (1980). "Mexicans." In Stephen Thornstrom, ed., *Harvard Encyclopedia of Ethnic Groups*. Cambridge: Harvard University Press.

Cotgrove, S., & Duff, A. (1980). "Environmentalism, Middle-Class Radicalism and Politics." *Sociological Review*, 28(2):333–351.

Debo, A. (1970). *A History of the Indians in the United States*. Norman: University of Oklahoma Press.

Deloria, V. (1994). *God Is Red: A Native View of Religion*. Golden, CO: Fulcrum Press.

Devall, W. B. (1970). "Conservation: An Upper-Middle Class Social Movement. A Replication." *Journal of Leisure Research*, 2(2):123–126.

Dickason, J. G. (1983). "The Origin of the Playground: The Role of the Boston Women's Clubs, 1885–1890." *Leisure Sciences*, 8(1).

Dillman, D. A., & Christenson, J. A. (1972). "The Public Value for Pollution Control." In William Burch et al., eds., *Social Behavior, Natural Resources, and the Environment*. New York: Harper and Row.

Dubofsky, M. (1996). *Industrialism and the American Worker 1865–1920*, 3rd ed. Wheeling, IL: Harlan Davidson.

Duleep, H. O. (1988). *Economic Status of Americans of Asian Descent*. Washington, DC: U.S. Commission on Civil Rights.

Faich, R. G., & Gale, R. P. (1971). "The Environmental Movement: From Recreation to Politics." *Pacific Sociological Review*, 14(2):270–287.

Fleming, D. (1972). "Roots of the New Conservation Movement." *Perspectives in American History*, vol. 6. Cambridge: Belknap Press of Harvard University Press.

Fox, S. (1985). *The American Conservation Movement: John Muir and His Legacy*. Madison: University of Wisconsin Press.

Friday, C. (1994). *Organizing Asian American Labor: The Pacific Coast Canned-Salmon Industry, 1870–1942*. Philadelphia: Temple University Press.

Gallup Poll. (1972). *The Gallup Poll: Public Opinion 1935–1971*. New York: Random House.

Geschwender, J. A. (1978). *Racial Stratification in America*. Dubuque, IA: William C. Brown.

Gottlieb, R. (1993). *Forcing the Spring: The Transformation of the American Environmental Movement*. Covelo, CA: Island Press.

Habermas, J. (1975). *Legitimation Crisis*. Translated by Thomas McCarthy. Boston: Beacon Press. Original title: *Legitimationsprobleme im Spatkapitalismus*, Frankfurt: Suhrkamp (1973).

Harry, J. (1971). "Work and Leisure: Situational Attitudes." Pacific Sociological Review, 14(July): 301–309.

Harry, J., Gale, R., & Hendee, J. (1969). "Conservation: An Upper-Middle Class Social Movement." *Journal of Leisure Research*, 1(2):255–261.

Hendee, J. C., Gale, R. P., & Harry, J. (1969). "Conservation, Politics and Democracy." *Journal of Soil and Water Conservation*, 24(Nov–Dec):212–215.

Hill & Picirelli. (1992). *Gale Environmental Sourcebook*. Detroit: Gale Research.

Hurley, A. (1995). *Environmental Inequalities: Class, Race, and Industrial Pollution in Gary, Indiana, 1945–1980*. Chapel Hill, NC: University of North Carolina Press.

Inglehart, R. (1992). "Public Support for Environmental Protection: Objective Problems and Subjective Values." Paper presented at the annual meeting of the American Political Science Association, Chicago.

Ingram, D. (1987). *Habermas and the Dialectic of Reason*. New Haven: Yale University Press.

Jiobu, R. M. (1988). *Ethnicity and Assimilation: Blacks, Chinese, Filipinos, Japanese, Koreans, Mexicans, Vietnamese, and Whites*. Albany, NY: State University of New York Press.

Josephy, A. (1968). *The Indian Heritage of America*. New York: Knopf.

Kitano, H. H. (1980). "Japanese." In Stephen Thronstrom, ed., *Harvard Encyclopedia of Ethnic Groups*. Cambridge: Harvard University Press.

Lenarcic, R. J. (1982). "The Moon of Red Cherries: A Brief History of Indian Activism in the United States." In Gerald R. Baydo, ed., *The Evolution of Mass Culture in America, 1877 to the Present*. St. Louis: Forum Press.

Lowe, G. D., Pinhey, T. K., & Grimes, M. D. (1980). "Public Support for Environmental Protection: New Evidence from National Surveys." *Pacific Sociological Review*, 23 (October).

Lurie, N. O. (1982). "The American Indian: Historical Background." In Norman Yetman and C. Hoy Steele, eds., *Majority and Minority*, 3rd ed. Boston: Allyn and Bacon.

Martinson, O. B., & Wilkening, E. A. (1975). "A Scale to Measure Awareness of Environmental Problems: Structure and Correlates." Paper presented at the Annual Meeting of the Midwest Sociological Society, Chicago, April.

McCloskey, M. (1992). "Twenty Years of Change in the Environmental Movement: An Insider's View." In Riley E. Dunlap and Angela G. Mertig, eds., *American Environmentalism: The U.S. Environmental Movement, 1970–1990*. Philadelphia: Taylor & Francis.

McNickle, D. (1973). *Native American Tribalism: Indian Survivals and Renewal*. New York: Oxford University Press.

Morris, A. (1984). *The Origins of the Civil Rights Movement: Black Communities Organizing for Change*. New York: The Free Press.

Muir, J. (1924). *The Life and Letters of John Muir*, 2 vols. Baltimore, MD: Johns Hopkins University Press.

Nash, R. (1982). *Wilderness and the American Mind*, 3rd. ed. New Haven: Yale University Press.

National Opinion Research Center. (1973–1980). *General Social Survey*. Chicago: University of Chicago Press.

National Wildlife Federation. (1993). *Conservation Directory*, 38th. ed. Washington, D.C.: National Wildlife Federation.

New York Times. (1857). November 6, 17, 21, 25.

Oelschlaeger, M. (1991). *The Idea of Wilderness: From Prehistory to the Age of Ecology*. New Haven: Yale University Press.

Olmstead, F. L. (1860). Letter to Charles Loring Brace, December 8.

———. (1860). Letter to Board of Commissioners of the Central Park, November 13.

Opinion Research Corporation. (1978, September). *Public Opinion Index*. Vol. 36(17) & (18). N.p.

———. (1977). Vol. 35. N.p.

Paehlke, R. (1989). *Environmentalism and the Future of Progressive Politics*. New Haven: Yale University Press.

Peiss, K. (1986). *Cheap Amusements: Working Women and Leisure in Turn-of-the-Century New York*. Philadelphia: Temple University Press.

Pepper, D. (1986). *The Roots of Modern Environmentalism*. London: Croom & Helm, Ltd.

Piven, F. F., & Cloward, R. A. (1979). *Poor People's Movements: Why They Succeed How They Fail*. New York: Vintage Books.

Pusey, M. (1993). *Jurgen Habermas*. New York: Routledge.

Robinson, J. (1991). *Toil and Toxics: Workplace Struggles and Political Strategies for Occupational Health*. Berkeley: University of California Press.

Roper, L. W. (1973). *FLO: A Biography of Frederick Law Olmsted*. Baltimore, MD: Johns Hopkins University Press.

Rosenzweig, R. (1987). "Middle-Class Parks and Working-Class Play: The Struggle Over Recreational Space in Worcester, Massachusetts, 1870–1910." In Herbert G. Gutman and Donald H. Bell, eds., *The New England Working Class and the New Labor History*. Urbana: University of Illinois Press.

———. (1983). *Eight Hours for What we Will: Workers and Leisure in an Industrial City, 1870–1920*. Cambridge: Cambridge University Press.

Snow, D. (1992). *Inside the Environmental Movement: Meeting the Leadership Challenge*. Covelo, CA: Island Press.

"State of the Nation." (1972–1976). Poll conducted by Potomac Associates.

Stiffam, L. A., & Lane, P., Jr. (1992). "The Demography of Native North America: A Question of American Indian Survival." In M. A. Jaimes, ed., *The State of Native America: Genocide, Colonization and Resistance*. Boston: South End Press.

Takaki, R. (1993). *A Different Mirror: A History of Multicultural America*. Boston: Little Brown.

Taylor, D. E. (1996). "Mobilizing for Environmental Justice in Communities of Color: An Emerging Profile of People of Color Environmental Groups." In Burch et al., eds., *Ecosystem Management: Adaptive Strategies for Natural Resources Organizations in the Twenty-First Century*. Washington, D.C.: Taylor & Francis.

————. (1992). "Can the Environmental Movement Attract and Maintain the Support of Minorities." In Bunyan Bryant and Paul Mohai, eds., *Race and the Incidence of Environmental Hazards*, Boulder: Westview Press.

————. (1989). "Blacks and the Environment: Toward an Explanation of the Concern and Action Gap between Blacks and Whites." *Environment and Behavior*, 21(2): 175–205.

Tognacci, L. N., Weigel, R. H., Wideen, M. F., & Vernon, D. T. A. (1972). "Environmental Quality: How Universal Is Public Concern?" *Environment and Behavior*, 4 (March): 73–86.

Trimble, L. C. (1988). "What Do Citizens Want in Siting of Waste Management Facilities?" *Risk Analysis*, 8(3).

Tuttle, W. M., Jr. (1980). *Race Riot: Chicago in the Red Summer of 1919*. New York: Atheneum.

United Church of Christ. (1987). *Toxic Waste and Race*. New York: United Church of Christ.

Wax, M. (1971). *Indian Americans: Unity and Diversity*. Englewood Cliffs, NJ: Prentice Hall.

Wright, S. (1975). "Explaining Environmental Commitment: The Role of Social Class Variables." Paper presented at the Annual Meeting of the Midwest Sociological Society, Chicago, April.

Zinger, C. L., Dalsemer, R., & Magargle, H. (1972). "Environmental Volunteers in America." Prepared by the National Center for Voluntary Action for the Environmental Protection Agency, Grant # R801243, Office of Research and Monitoring.

8

Turning Public Issues into Private Troubles

Lead Contamination, Domestic Labor, and the Exploitation of Women's Unpaid Labor in Australia

Lois Bryson, Kathleen McPhillips, and Kathryn Robinson

This reading examines government efforts to deal with contamination from lead smelters in some Australian provinces. Lead is a dangerous environmental toxin, which can cause an array of physical problems, from kidney damage to hyperactivity. Lead is especially dangerous to young children and can cause, among other things, mental retardation, brain damage, and behavior problems. However, rather than address the source of the problem (the smelters themselves), government policies have focused on individual behaviors, specifically on housecleaning, to minimize the amount of lead dust in homes. The authors, pointing to the fact that most of the homes near smelters are occupied by working-class families and most of the housecleaning is done by women, argue that the government policies are not only ineffective but are embedded in unequal social class and gender relations.

U sing a case study of state intervention in the industrial contamination of a residential community, this article offers a contribution to feminist analysis of the role of the state. Residents of three Australian lead smelter towns, with high levels of toxic pollution, have been given a strong message by state health authorities that their children's health could be irreparably damaged unless they adopt a strict regimen of housecleaning and child management to reduce the ingestion of lead particles by their children.

This case study of state action on a site that is classically women's domain provides insight into a "state gender regime" (Connell 1990) through examining the state "doing gender" (West and Zimmerman 1987). . . .

Unraveling the complexity of the state's role in developing and implementing its health strategy for dealing with the effects of lead pollution within smelter towns allows us to tease out some of the complexities of state intervention. How do they identify and address this health issue? Whose interests do the interventions serve? What are the impacts for different groups of women? Because the intervention is focused on mothering, we start by examining some relevant features of motherhood in contemporary Australia and its place within the wider scheme of gender relations.

Motherhood and the State Gender Regime

. . . While schools of feminist thought account in different ways for women's position and gender relations, they do not contest that motherhood involves a responsibility for family work, which falls unequally to women. Graham (1983) pointed to the bifurcated nature of caring as involving "caring about" and "caring for." In terms of parenthood, she suggested that fathers are expected to "care about" their children, and this may involve taking some responsibility for overseeing that care is provided. For mothers, the two aspects are firmly fused: They are expected to both "care about" and "care for."

Empirical studies in Australia and elsewhere of perceptions of motherhood clearly expose a dominating ideology that reflects such views of caring (Dempsey 1997; McMahon 1999; Russell 1983). . . .

Decades of research into time use confirm women's disproportionate share of the work involved in both domestic cleaning and child care relative to their male partners (Bittman and Matheson 1996; Bryson 1997; Fenstermaker Berk 1985). Women in Australia, as elsewhere, are far more likely to undertake cleaning chores and physical care of children. The Australian Bureau of Statistics (1994) found that women's share of laundry was 89 percent (90 percent if in full-time employment), their share of cleaning 82 percent (84 percent if in full-time employment), and of physical care of children 84 percent (76 percent if in full-time employment) (Bittman and Pixley 1997, 113). These tasks are central to the housecleaning regime devised by state authorities that we examine here. . . .

The Smelter Communities: Method and Background

We first became interested in the public health intervention processes in lead-affected towns when one of us (McPhillips) became caught up with the effects of lead contamination in her residential community, which bordered on Boolaroo, a smelter town in NSW [New South Wales]. McPhillips's personal experience of dealing with lead contamination became the subject of spirited debate and theorizing in our (then) shared work context—the sociology department of the local university. This led to the development of the research reported in this article.

Method

... In investigating the case study of Boolaroo, we collected qualitative data through direct engagement in the community; through participant observation in community activities; and through interviews with eight residents of Boolaroo, including female members of community groups (both for and against the smelter). We also interviewed personnel in the local health authority who had been involved in testing children's blood levels and in designing and implementing subsequent intervention measures.

Historical materials relating to the genesis of the political conflict over contamination in Boolaroo were available to us. These included reports in local newspapers (which had been systematically filed by the municipal library and collected by activists), reports and scholarly articles produced by the public health authorities, and a television documentary produced for the state-owned public broadcasting authority that critically recorded an intervention in 1991 intended to remove historic contamination from Boolaroo residences (Australian Broadcasting Corporation 1992). The smelter company produced regular community newsletters, which were available to us, as were the public health information brochures.

In interpreting the data relating to Boolaroo, we used a comparative perspective, drawing on the reports of similar interventions in the other Australian lead smelter towns. There are three major sites of lead production in Australia: Port Pirie in South Australia, and Broken Hill and Boolaroo in NSW [New South Wales]....

... The critical features of the populations in all three areas affected by the pollution is that they are of low socioeconomic status with manufacturing providing predominantly male employment.

The Household Cleaning Regime

Public health authorities have systematically turned to an approach dealing with lead effects that is focused on the ways in which it is ingested, particularly by children, rather than with stopping pollution. In the vicinity of the Boolaroo smelter, for example, an attractive poster was distributed to residents, with the title "Lead: Lower the levels & protect your child" (PHU n.d.). It mentions soil, food, household dust, old paint, and the lead worker as potential sources, identifying interventions that can be made, with most of them involving an intensification of domestic labor. To avoid exposure of children to lead in household dust, parents are advised to use a wet mop instead of a broom; keep dust from children's play areas, including under beds and in closets; and remember to dust corners, along sides of windows, and behind furniture and doors. Advice to the lead worker includes the following: Keep kit bags out of reach of children; keep children away from work clothes; and clean dust from inside and outside of car, especially if driven to work. To avoid lead in food, the parents are advised to intensify domestic labor by preventing children from sucking dirty hands, fingernails, or objects; wiping surfaces before preparing food; and covering food and utensils to prevent lead dust from settling on them. The problem of contaminated soil is addressed by advice to wash children's hands before eating, especially if playing outside; to provide clean soil or sand for children to play in; and to use a nail brush under nails. The poster does not represent the source of the lead contamination in any way.

The focus on domestic-based interventions deflects attention away from the source of the pollution, which is clearly in the interests of capital. We argue that the burden of activities to ameliorate the effects of pollution falls disproportionately to women and must be counted as an aspect of the state's gender as well as class regime. The communities, which already bear the burden of toxic contamination and its attendant health, social, and economic effects, are further burdened with the responsibility of "putting things right."

The spotlight in the three smelter towns in recent years has been very much on children, even though evidence suggests that lead is a health hazard for all age groups (Alperstein, Taylor, and Vimpani 1994; Centers for Disease Control and Prevention 1991). In earlier years, the focus of official programs was on workers and also in a manner that deflected attention away from the smelting companies' practices. . . .

The 1980s and 1990s: Developing the Domestic Cleaning Regime

In the early 1980s, the South Australian health department responded to U.S. research linking impaired intellectual development with lead by undertaking a survey of the factors implicated in elevated blood lead levels of young children living within the vicinity of the smelter at Port Pirie (Landrigan 1983). Using a control sample of children with low blood lead levels, researchers found that 5 of the 16 evaluated behavioral factors were significantly associated with high lead levels. These factors were a history of placing objects in the mouth, nail biting, dirty hands, dirty clothes, and eating lunch at home rather than at school

(Landrigan 1983, 8). The survey also found that higher lead levels in children were associated with the number of persons in the household working in the lead industry (cf. Donovan 1996).

However, the research also showed that none of the factors were as important as living in a contaminated environment and that living near the smelter was three times more important than anything else (Landrigan 1983, 9). Because past emissions still contaminate the atmosphere and the soil, reducing emissions was acknowledged as insufficient for dealing with the problem. More recently, public health officials recognized that the only effective way of dealing with lead contamination at smelter sites is relocating residents (Galvin et al. 1993, 377), although to our knowledge no public body has ever implemented a relocation program.

In 1984, the South Australian health department and the smelter management started an education program about the importance of personal hygiene and household cleaning. The focus was on "pathways" through which lead is ingested by children and the ways this can be minimized within the home. The general manager of smelter operations expressed the view that "given reasonable care and hygiene, then you can live with the levels of contamination from past emissions." Fowler and Grabosky (1989, 153–57) suggested that the company was "defensive and cautious" lest remedial action imply admission of legal responsibility. The company maintained public health to be a government responsibility and invested far less than the government. The South Australian government's response was restrained as well, illustrating the power of major capital by showing an unwillingness to "antagonise one of the state's largest employers" (Fowler and Grabosky 1989, 150). The government was prepared to burden working-class women, rather than business.

In 1986, another case control study examined children's blood lead levels in Port Pirie (Wilson et al. 1986). The study focused mainly on environmental issues and the implications for the smelter. It assessed behavioral differences between "cases" and "controls" and again pointed to the "pathways" for ingestion such as "biting fingernails" or "dirty clothing/hands at school."[1] In 1988, the public health program at Port Pirie also undertook external decontamination in the yards of 1,400 homes at highest risk, and adjacent vacant blocks and footpaths were sealed (Heyworth 1990, 178). Subsequently, a "partial decontamination" of the same area was instituted, in recognition that contamination is continuous. Such a program requires a permanent cycle of treatment of the homes and vacant blocks and a continuing awareness within the community of the need for vigilance in terms of personal and home hygiene (Heyworth 1990, 183).[2]

In 1993, after several months of remediation, including replanting tailings dumps and the implementation of housecleaning regimes, the blood lead levels had elevated in a part of the town deemed to be a nonrisk area and hence not subject to the campaign. Further investigation found the source to be stacks of lead ore left on the wharves in open piles, with dust being carried by the winds into residential areas not previously considered at risk from the smelter.

The South Australian Health Commission published a decade review of the Port Pirie program in 1993 that concluded, "Given the amount of lead contamination to which these households are exposed, changes in dust hygiene would not seem to be a realistic way to ensure lower exposures to lead by the child" (Maynard, Calder, and Phipps 1993, 25). This point, that domestic strategies are unsustainable in the long term (Maynard, Calder, and Phipps 1993, 5), had previously been made by Landrigan in 1983 and by Luke in 1991 after an extensive search of the world literature on the topic (Luke 1991, 161–67). Many people are prepared to modify their behavior in the short term, but sooner or later they revert to more comfortable habits. The report notes that because there is a constant process of recontamination, it is unrealistic to expect continuing voluntary participation of residents in programs for lead remediation. Another barrier to ongoing participation is the stigmatization that is involved if a child's blood levels are high. Parents are reluctant to have their child labeled as potentially intellectually impaired, and they themselves feel stigmatized by the implication that they have dirty houses.

Over time, parents tend to withdraw from participation in blood-monitoring programs. This can be seen as a form of resistance in situations where parents are expected to comply with surveillance of a problem that has a source external to the home and family, but they are expected to deal with the consequences. Their concerted efforts at housecleaning fail to bring the promised results, and they continue to suffer the stigmatization of the threats to their children's healthy development and the implication that they are poor housekeepers.

During the evaluations of cleaning regimes by health authorities, little attention was paid to the women involved as the cleaners, which indirectly provides us with some insight into the state gender regime. Apart from the effort and the responsibility, there is evidence that the cleaning process itself can be contaminating (Australian Broadcasting Corporation 1992; Luke 1991, 99).[3]

Fowler and Grabosky (1989, 148) suggested that the "regulatory orientation to pollution control has been characterized by negotiation and compromise, rather than strict enforcement." Governments have been reluctant to antagonize business, and business has been motivated by a "desire to avoid the loss of a marketable product rather than environmental concern" (Fowler and Grabosky 1989, 147). State recognition of business interests as more important than smelter community residents has ultimately resulted in greater recourse to the cleaning regime. Furthermore, it is unlikely that the public health care system has the long-term capacity to maintain the level of involvement and monitoring that is required. This suggests the program is "window dressing," a project of state legitimation, which rests on the state's class and gender regime and which masks the interests that are served by de facto tolerance of levels of industrial pollution damaging to a small and easily ignored segment of the population.

Domestic Cleaning Regimes—The 1990s

A similar household cleaning regime to that developed and applied in the 1980s by public health workers in the vicinity of the Port Pirie lead smelter was subsequently put in place in the NSW smelter towns of Boolaroo and Broken Hill. The regime was formalized in 1994 by a federal agency, the Commonwealth of Australia Environmental Protection Authority (EPA), and outlined in a document published by the EPA titled *Lead Alert: A Guide for Health Professionals* (Alperstein, Taylor, and Vimpani 1994).

Lead Alert set out the "steps to minimise exposure and absorption" of lead by children. Health professionals are told to give the following advice "to parents if a child's blood lead is more than 15½ g/dL" (Alperstein, Taylor, and Vimpani 1994, 17). Parents should ensure that children's hands and face are washed before they eat or have a nap, discourage children from putting dirty fingers in their mouths, encourage children to play in grassy areas, and wash fruit and vegetables. In terms of housecleaning, they are advised to wet dust floors, ledges, window sills, and other flat surfaces at least weekly or more often if the house is near a source point for lead; to clean carpets and rugs regularly using a vacuum cleaner; to wash children's toys, especially those used outside; and to wash family pets frequently and discourage pets from sleeping near children. The parent should ensure that the child does not have access to peeling paint or chewable surfaces painted with lead-based paint and that the child's diet is adequate in calcium and iron, which helps to minimize lead absorption. Children should be provided with regular frequent meals and snacks, up to six per day, because more lead is absorbed on an empty stomach (Alperstein, Taylor, and Vimpani 1994, 17).

The document was widely disseminated, not just in smelter communities but in situations where people had elevated blood lead levels from any source. The housecleaning regime has been promoted as a primary intervention strategy, despite the evidence from evaluations of the Port Pirie experience (as well as studies in other countries) that show it to be ineffective.

This extremely detailed cleaning regime involves an implicit threat for noncompliance, the threat of adversely affecting one's child's health and intellectual development. Because of the nature of the tasks, the responsibility for most of this work falls to mothers rather than all parents.

For communities in the vicinity of lead smelters in Australia, public health authorities recommend an even more stringent regime than that implied in *Lead Alert*. The regime suggests not vacuuming with a child in the room since the cleaning raises dust. Dusting should ideally cover what one Boolaroo mother described as "bizarre places" such as the fly wire in screen doors. Other suggested strategies include moving children's beds from under windows and putting away soft toys because they cannot be easily washed (Gilligan 1992, 4). To add to this intensification, some anxious parents intensify the regime further. The "wet dusting" or mopping over of all horizontal surfaces is done more than once a day by some mothers. In a television documentary on the subject of lead poisoning and children, women who had implemented the regime in both Boolaroo and Port Pirie expressed their frustration. "They tell you to run round with a washer after them. They are not allowed to put their fingers in the mouths"; "You do things that you just would not do"; "We must be the only housewives in New South Wales that do these tasks" (Australian Broadcasting Corporation 1992). Another mother verbalized the worry and guilt associated with such responsibility for her children's health, "You feel guilty if you just don't want to do the work . . . you think your child does not have a normal life . . . should you have more children?" (Gilligan 1992, 45).

The lead abatement program also involves the monitoring of the child's blood lead levels as the most significant means of measuring the levels of lead absorption. This means subjecting the child to frequent blood sampling and a constant measuring as to whether the family's efforts have been successful. There is constant contact for both mother and child with health and other professionals and often researchers (Australian Broadcasting Corporation 1992; Gilligan 1992, 4). The constancy of the monitoring creates the impression that the responsibility for the public health problem falls on the residents themselves (and especially on mothers), rather than the government or the corporation (see also McGee 1996, 14). The monitoring of blood lead levels also gives the impression that something is being done but actually does nothing to address the source of the contamination. In some cases, children's blood lead levels have risen after the implementation of household cleaning regimes.

A study of the effects of the lead issue on Boolaroo residents found that families of children with high blood lead levels experienced "feelings of guilt, stigma, anxiety, stress and powerlessness . . . that the difference may be due to or to be seen to be due to some action or inaction of them as parents" (Hallebone and Townsend 1993, 17). We found that in Boolaroo, health professionals regarded parental failure to present children for monitoring as evidence of irresponsibility. As with smelter workers in the 1920s, stigma is attached to those who do not conform to the cleaning standards. . . .

Not only are women enslaved by the domestic cleaning, but their children's psychological and/or emotional development is put in question. In the education booklet written by the education department for use in Boolaroo schools, the family unit is presented as the most important element in dealing with lead exposure: "Children who are supported and confident in their family unit will be better able to deal with any problems associated with the lead issue" (quoted in Mason 1992, 41).

Despite evidence of limited success in the long term, domestic cleaning regimes are becoming popular for dealing with lead contamination from gasoline.[4] As has occurred with the smelter communities, the effect again is to shift responsibility from the polluting source to the private sphere of the family and women.

Resistance to State Interventions

The focus on housecleaning regimes as a response to children's elevated blood lead levels has the effect of stigmatizing parents and calling into question their capacity to care for their children. As noted above, mothers who have implemented the regime, however, find it very oppressive and anxiety provoking, hence many cease doing it (McGee 1996, 14). There is also resistance to the surveillance of children's blood lead levels in situations where the health authorities are not offering an effective response to the problem. The proportion of smelter-town families who continue to present their children for blood testing has declined considerably over time (Isles 1993; Maynard, Calder, and Phipps 1993, 9; Stephenson, Corbett, and Jacobs 1992; Western NSW PHU 1994).

Nevertheless, women in smelter towns have responded with active as well as passive resistance. Following the PHU's 1991 revelation of elevated blood lead levels in Boolaroo children, local residents, mainly women, formed the North Lakes Environmental Action Defence Group (No Lead). It has campaigned for action by state and local government and by the smelter to counter the high levels of lead pollution. . . .

No Lead has addressed the issue of lead contamination by refocusing public attention on the responsibilities of the government to regulate the industry and the industry's responsibility to reduce toxic emissions. They have found themselves in direct conflict with the company's public relations strategists who have attempted to downplay their political significance as legitimate representatives of community interest. No Lead has been able to successfully work with environmental groups like Greenpeace and has achieved some success in refocusing government attention on Boolaroo. For example, a Parliamentary Select Committee set up in 1994 resulted in a management plan that placed more responsibility on the smelter and the local government in limiting the effects of plant emissions, although it reinforced the emphasis on housecleaning regimes (NSW Parliament 1994).

Other women in Boolaroo have actively resisted government interventions intended to ameliorate the exposure of children to lead, which they see as stigmatizing their children as intellectually limited or themselves as poor housekeepers. The most significant of these was the 1992 resistance by the Parents and Citizens Association to a temporary closure of the local school for remediation. The parents questioned whether the school was indeed contaminated or if the contamination was any more significant than that which they experienced in their homes. They rejected the implication that the school posed a threat to their children's intellectual development, citing cases of local children who had succeeded academically. The women involved in this public protest expressed a fear that the school would be closed forever, once the children had been relocated. McPhillips (1995, 48) commented that the issue brought out the "deep-felt suspicion that this section of the local community held for government bureaucracies." That is, they do not see the public authorities as acting in their interests and resist the authorities' efforts to intervene in the situation. In this case, the women's resistance was successful, and the remediation was carried out while the school was routinely closed for summer holidays.

Conclusion: Lead Levels, Public Health Strategies, and State Gender Regimes

The strategy promoted by public health authorities for smelter towns, rather than dealing with the source of the pollution, turns this public issue into a private family matter. In appearing to do something, the state selected a remediation strategy that has been repeatedly proven to be ineffective. Research from many countries (Luke 1991) persistently shows that the major issue is one of proximity to the smelter and the level of emissions, with past pollution also a critical factor. There is a continuing history of failure of the smelting companies to own the pollution problem they cause and to deal with it. Profit levels, rather than health concerns, have historically taken precedence, with the state mediating the corporations' interests. The state historically has facilitated the shifting of the focus of responsibility for the problem to the relatively powerless. Now the blame is laid on working-class women, whereas early in the twentieth century, it was directed toward recently arrived male immigrant workers, although women were indirectly implicated.

The official remediation strategy does not fall in a gender-neutral way on both parents. It relies for its implementation on additional daily caring labor being undertaken by mothers of children deemed "at risk" and thus on the basis of a traditional understanding of the responsibilities associated with motherhood. In addition, it relies for compliance on the mothers' emotional commitment to their children's health, a situation with great potential to engender feelings of guilt and to stigmatize those who "fail." The exploitation of these working-class women's unpaid caring labor is not only an example of the state "doing" gender but a recent form of "doing" class as well. Brown and Ferguson (1995, 161) noted that

> when activists discover that local industry values its bottom line or international reputation more highly than it does the health of children in the

community, this realization violates the trust that the women toxic waste activists have placed in the ideal of corporate citizenship and governmental protection.

Working-class mothers are the target of a burdensome housecleaning regime that absorbs their labor but at the same time effectively shifts responsibility for ensuring the pollution does not damage their children's health away from the corporation and the state. This strategy reflects a state gender regime that involves material exploitation in the form of reliance on women's labor in a manner that serves powerful interests and ideological exploitation through the manipulation of the women's sense of maternal responsibility.

The structural features of class and gender do not, of course, account for the whole story. Residents are not duped nor have they have been silent on the matter. The women as individuals resist or reject the recommended regime and have been at the forefront of organized community resistance strategies. Women organized into community-based pressure groups have had limited success in directing attention back onto the smelter as the source of the toxic pollution and away from their responsibility as domestic carers for the health of their children.

This case study raises specific questions about where the responsibility lies for the health of residents of smelter communities. But it also illuminates class and gender relations in contemporary Australia and the manner in which the state is engaged in the reproduction of class and gender difference.

Notes

Bryson, Lois, Kathleen McPhillips, and Kathryn Robinson. 2001. "Turning Public Issues into Private Troubles: Lead Contamination, Domestic Labor, and the Exploitation of Women's Unpaid Labor in Australia." *Gender & Society* 15(5): 755–72.

1. In terms of current standards, the controls also had elevated blood levels (with a mean of $17\frac{1}{2}$ g/dL), which casts doubt on the conclusions from the study. The study had the effect of normalizing the underlying high level of contamination (represented in the high blood levels) and, once again, the effect of focusing attention away from the source.

2. Over time, improvements have been made in levels of air emission. However, past emission levels ("historic contamination") remain part of the problem that needs to be dealt with by major decontamination programs that should establish buffer zones by buying up the houses closest to the smelter. (By 1992, 100 houses in Port Pirie had been purchased and demolished.)

3. This potential was belatedly recognized in the 1993 Port Pirie high-risk area program when the blood of both parents and children was tested after residents were encouraged to participate in the home contamination procedure (Maynard, Calder, and Phipps 1993, 9).

4. At the Newcastle conference on Lead Abatement and Remediation in 1994, Professor Bornschein, director of Epidemic Research at Cincinnati University, claimed that "full abatement is expensive and ineffective" and that good results can be obtained

from educating people about personal and home hygiene. He indicated that this was to be the focus of 29 programs to start in the United States in the near future (Maguire 1994).

References

Alperstein, G., R. Taylor, and G.Vimpani. 1994. *Lead alert: A guide for health professionals.* Canberra, Australia: Commonwealth Environmental Protection Agency.

Australian Broadcasting Corporation. 1992. *Living with lead.* Television documentary screened on *Four Corners*, 14 September.

Australian Bureau of Statistics. 1994. *How Australians use their time.* Canberra: Australian Bureau of Statistics.

Bittman, Michael, and George Matheson. 1996. *"All else in confusion": What time use surveys show about changes in gender equity.* SPRC Discussion Paper Series No. 72. Sydney: Social Policy Research Centre.

Bittman, Michael, and Jocelyn Pixley. 1997. *The double life of the family: Myth, hope, and experience.* Sydney: Allen and Unwin.

Brown, P., and F. Ferguson. 1995. "Making a big stink": Women's work, women's relationships, and toxic waste activism. *Gender & Society* 9:145–72.

Bryson, Lois. 1997. Citizenship, caring and commodification. In *Crossing borders: Gender and citizenship in transition*, edited by Barbara Hobson and Anne Marie Berggren. Stockholm: Swedish Council for Planning and Co-ordination of Research.

Centers for Disease Control and Prevention. 1991. *Preventing lead poisoning in young children.* Atlanta, GA: U.S. Department of Health and Human Services.

Connell, R. W. 1990. The state, gender, and sexual politics. *Theory and Society* 19:507–44.

Dempsey, Ken. 1997. *Inequalities in marriage: Australia and beyond.* Melbourne: Oxford University Press.

Donovan, John. 1996. *Lead in Australian children: Report on the national survey of lead in children.* Canberra: Australian Institute of Health and Welfare.

Fenstermaker Berk, Sarah. 1985. *The gender factory: The apportionment of working American households.* New York: Plenum.

Fowler, R., and P. Grabosky. 1989. Lead pollution and the children of Port Pirie. In *Lead pollution: Fourteen studies in corporate crime or corporate harm*, edited by P. Grabosky and A. Sutton. Milson's Point NSW, Australia: Hutchinson.

Galvin, J., J. Stephenson, J. Wlordarczyk, R. Loughran, and G. Wallerm. 1993. Living near a lead smelter: An environmental health risk assessment in Boolaroo and Argenton, New South Wales. *Australian Journal of Public Health* 17:373–78.

Gilligan, B., ed. 1992. *Living with lead: A draft plan for addressing lead contamination in the Boolaroo and Argenton areas, NSW.* Lake Macquarie, Australia: Lake Macquarie City Council.

Graham, Hilary. 1983. Caring: A labor of love. In *A labor of love: Women, work and caring*, edited by Janet Finch and Dulcie Groves. London: Routledge and Kegan Paul.

Hallebone, E., and M. Townsend. 1993. Who bears the weight of lead in society? A social impact assessment of proposed changes to the national guidelines for blood lead levels. Technical appendix 3. In *Reducing lead exposure in Australia: An assessment of impacts*, vol. 2, edited by Mike Berry, Jan Garrard, and Deni Greene. Canberra, Australia: Department of Human Services and Health.

Heyworth, J. S. 1990. *Evaluation of Port Pirie's environmental health program.* Master of Public Health diss., Adelaide University, Adelaide, Australia.

Isles, Tim. 1993. Boolaroo blood-lead tests show no change. *Newcastle Herald*, 13 July.

Landrigan, P. J. 1983. *Lead exposure, lead absorption and lead toxicity in the children of Port Pirie: A second opinion.* Adelaide: South Australian Health Commission.

Luke, Colin. 1991. A study of factors associated with trends in blood lead levels in Port Pirie children exposed to home based intervention. Master of Public Health diss., Adelaide University, Adelaide, Australia.

Maguire, Paul. 1994. Green buffer "best way" to cut lead poisoning in young children. *Newcastle Herald*, 3 June.

Mason, Chloe. 1992. Controlling women, controlling lead. *Refractory Girl* 43:41–42.

Maynard, E., I. Calder, and C. Phipps. 1993. *The Port Pirie lead implementation program: Review of progress and consideration of future directions (1984–1993).* Adelaide: South Australian Health Commission.

McGee, T. 1996. Shades of grey: Community responses to chronic environmental lead contamination in Broken Hill, NSW. Ph.D. diss., the Australian National University, Canberra.

McMahon, Anthony. 1999. *Taking care of men: Sexual politics in the public mind.* Cambridge, UK: Cambridge University Press.

McPhillips, K. 1995. Dehumanising discourses: Cultural colonisation and lead contamination in Boolaroo. *Australian Journal of Social Issues* 30:41–55.

NSW Parliament. 1994. *Report of the select committee upon lead pollution.* Vol. 1. December.

Public Health Unit. n.d. Hunter Area Health Service. Poster.

Russell, Graeme. 1983. *The changing role of fathers?* St. Lucia, Australia: University of Queensland Press.

Stephenson, J., S. Corbett, and M. Jacobs. 1992. Evaluation of environmental lead abatement programs in New South Wales. In *Choice and change: Ethics, politics and economies of public health. Selected papers from the 24th Public Health Association of Australia conference, 1992 Canberra,* edited by Valerie A. Brown and George Preston. Canberra: Public Health Association of Australia.

West, C., and D. H. Zimmerman. 1987. Doing gender. *Gender & Society* 1:125–51.

Western NSW Public Health Unit. 1994. *Risk factors for blood lead levels in preschool children in Broken Hill 1991–1993.* Dubbo, Australia: Western New South Wales Public Health Unit.

Wilson, D., A. Esterman, M. Lewis, D. Roder, and I. Calder. 1986. Children's blood lead levels in the lead smelting town of Port Pirie, South Australia. *Archives of Environmental Health* 41:245–50.

Environmental Justice

Grassroots Activism and Its Impact on Public Policy Decision Making

Robert D. Bullard and Glenn S. Johnson

Documentation accumulated over the last three decades provides ample evidence that people-of-color and low-income communities face a greater incidence of exposure to environmental hazards than other communities. In response, environmental justice activists across the country began fighting for the health of their families and communities. In 1991, activists across the country gathered for the First National People of Color Environmental Leadership Summit in D.C. (1991 Summit). As part of this process, they developed the seventeen "Principles of Environmental Justice," widely available on the internet (see www.ejrc.cau.edu/princej.html). Robert Bullard is a leading researcher and activist in the environmental justice movement. In the next selection, Bullard and his colleague Glenn Johnson show how, since the 1980s, grassroots activists have coordinated a nationwide effort to change governmental and corporate practices that endanger the health of people-of-color and low-income communities. These activists have experienced a wide range of successes, from resisting unwanted land uses at the local level to changing federal regulations at the national level.

Despite significant improvements in environmental protection over the past several decades, millions of Americans continue to live in unsafe and unhealthy physical environments. Many economically impoverished communities and their inhabitants are exposed to greater health hazards in their homes, on the jobs, and in their neighborhoods when compared to their more affluent counterparts (Bryant & Mohai, 1992; Bullard, 1994a). . . .

From New York to Los Angeles, grassroots community resistance has emerged in response to practices, policies, and conditions that residents have judged to be unjust, unfair, and illegal. Some of these conditions include (1) unequal enforcement of environmental, civil rights, and public health laws; (2) differential exposure of some populations to harmful chemicals, pesticides, and other toxins in the home, school, neighborhood, and workplace; (3) faulty assumptions in calculating, assessing, and managing risks; (4) discriminatory zoning and land use practices; and (5) exclusionary practices that prevent some individuals and groups from participation in decision making or limit the extent of their participation (Bullard, 1993b; C. Lee, 1992). . . .

Communities under Siege

. . . Elevated public health risks have been found in some populations even when social class is held constant. For example, race has been found to be independent of class in the distribution of air pollution, contaminated fish consumption, location of municipal landfills and incinerators, abandoned toxic waste dumps, cleanup of Superfund sites, and lead poisoning in children (Agency for Toxic Substances and Disease Registry, 1988; Bryant & Mohai, 1992; Commission for Racial Justice, 1987; Goldman & Fatten, 1994; Lavelle & Coyle, 1992; Pirkle et al., 1994; Stretesky & Hogan, 1998; West et al., 1990).

. . . Figures reported in the July 1994 *Journal of the American Medical Association* in the Third National Health and Nutrition Examination Survey (NHANES III) revealed that 1.7 million children (8.9% of children aged 1–5) are lead-poisoned, defined as having blood lead levels equal to or above 10 micrograms per deciliter. The NHANES III data found African American children to be lead-poisoned at more than twice the rate of White children at every income level (Pirkle et al., 1994). Over 28.4% of all low-income African American children were lead-poisoned, compared to 9.8% of low-income White children. During the time period between 1976 and 1991, the decrease in blood lead levels for African American and Mexican American children lagged far behind that of White children.

In 1992 in California, a coalition of environmental, social justice, and civil libertarian groups joined forces to challenge the way the state carried out its screening of poor children for blood lead levels. The Natural Resources Defense Council, the National Association for the Advancement of Colored People Legal Defense and Education Fund, the American Civil Liberties Union, and the Legal Aid Society of Alameda County, California, won an out-of-court settlement worth $15 million to $20 million for a blood lead-testing program. The lawsuit, *Matthews v. Coye*, involved the failure of the state of California to conduct federally mandated testing for lead on some 557,000 poor children who receive Medicaid (B. L. Lee, 1992). This historic agreement triggered similar lawsuits and actions in several other states that failed to live up to the federal mandates.

Federal, state, and local policies and practices have contributed to residential segmentation and unhealthy living conditions in poor, working-class, and people-of-color communities (Bullard & Johnson, 1997). Several recent California cases bring this point to light (B. L. Lee, 1995). Disparate highway siting and mitigation plans were challenged by community residents, churches, and the NAACP Legal Defense and Education Fund in *Clean Air Alternative Coalition v. United States Department of Transportation* (N.D. Cal. C-93-O721-VRW), involving the reconstruction of the earthquake-damaged Cypress Freeway in West Oakland. The plaintiffs wanted the downed Cypress Freeway (which split their community in half) rebuilt farther away. Although the plaintiffs were not able to get their plan implemented, they did change the course of the freeway in their out-of-court settlement.

The NAACP Legal Defense and Education Fund filed an administrative complaint, *Mothers of East Los Angeles, El Sereno Neighborhood Action Committee, El Sereno Organizing Committee, et al. v. California Transportation Commission, et al.* (before the U.S. Department of Transportation and U.S. Housing and Urban Development), challenging the construction of the 4.5-mile extension of the Long Beach Freeway in East Los Angeles through El Sereno, Pasadena, and South Pasadena. The plaintiffs argued that the state agencies' proposed mitigation measures to address noise, air, and visual pollution discriminated against the mostly Latino El Sereno community. For example, all of the planned freeway in Pasadena and 80% in South Pasadena will be below ground level. On the other hand, most of the freeway in El Sereno will be above ground. White areas were favored over the mostly Latino El Sereno in allocation of covered freeway, historic preservation measures, and accommodation to local schools (Bullard & Johnson, 1997; B. L. Lee, 1995).

Los Angeles residents and the NAACP Legal Defense and Education Fund have also challenged the inequitable funding and operation of bus transportation used primarily by low-income and people-of-color residents. A class action law-suit was filed on behalf of 350,000 low-income, people-of-color bus riders represented by the Labor/Community Strategy Center, the Bus Riders Union, Southern Christian Leadership Conference, Korean Immigrant Workers Advocates, and individual bus riders. In *Labor/Community Strategy Center v. Los Angeles Metropolitan Transportation Authority* (Cal. CV 94-5936 TJH Mcx), the plaintiffs argued that the Los Angeles Metropolitan Transit Authority (MTA) used federal funds to pursue a policy of raising costs to bus riders (who are mostly poor and people of color) and reducing quality of service in order to fund rail and other projects in predominately White suburban areas (Mann, 1996).

In the end, the Labor/Community Strategy Center and its allies successfully challenged transit racism in Los Angeles. The group was able to win major fare and bus pass concessions from the Los Angeles MTA. It also forced the Los Angeles MTA to spend $89 million on 278 new clean-compressed natural gas buses. . . .

Relocation from "Mount Dioxin"

Margaret Williams, a 73-year-old retired Pensacola, Florida, schoolteacher, led a 5-year campaign to get her community relocated from the environmental and health hazards posed by the nation's third largest Superfund site. The Escambia Wood Treating site was dubbed "Mount Dioxin" because of the 60-foot-high mound of contaminated soil dug up from the neighborhood. The L-shaped mound holds 255,000 cubic yards of soil contaminated with dioxin, one of the most dangerous compounds ever made (Olinger, 1996). Williams led Citizens Against Toxic Exposure (CATE), a neighborhood organization formed to get relocation, into battle with EPA officials, who first proposed to move only the 66 households most affected by the site (U.S. EPA, 1996). After prodding from CATE, EPA then added 35 more households, for a total cost of $7.54 million.

The original government plan called for some 257 households, including an apartment complex, to be left out. CATE refused to accept any relocation plan unless everyone was moved. The partial relocation was tantamount to partial justice. CATE took its campaign on the road to EPA's NEJAC and was successful in getting NEJAC's Waste Subcommittee to hold a Superfund relocation roundtable in Pensacola. At this meeting, CATE' s total neighborhood relocation plan won the backing of more than 100 grassroots organizations. EPA nominated the Escambia Wood Treating Superfund site as the country's first pilot program to help the agency develop a nationally consistent relocation policy that would consider not only toxic levels but welfare issues such as property values, quality of life, health, and safety.

On October 3, 1996, EPA officials agreed to move all 358 households from the site at an estimated cost of $18 million. EPA officials deemed the mass relocation as "cost efficient" after city planners decided to redevelop the area for light industry rather than clean the site to residential standards (Escobedo, 1996; Washington Post, 1996). This decision marked the first time that an African American community had been relocated under EPA's Superfund program and was hailed as a landmark victory for environmental justice (Escobedo, 1996).

From Dumping in Dixie to Corporate Welfare

The southern United States has become a "sacrifice zone" for the rest of the nation's toxic waste (Schueler, 1992, p. 45). A colonial mentality exists in Dixie through which local government and big business take advantage of people who are both politically and economically powerless. The region is stuck with a unique legacy: the legacy of slavery, Jim Crow, and White resistance to equal justice for all. This legacy has also affected race relations and the region's ecology.

The South is characterized by "look-the-other-way environmental policies and giveaway tax breaks" and as a place where "political bosses encourage outsiders to buy the region's human and natural resources at bargain prices" (Schueler, 1992, pp. 46–47). Lax enforcement of environmental regulations has left the region's air, water, and land the most industry-befouled in the United States.

Toxic waste discharge and industrial pollution are correlated with poorer economic conditions. In 1992, the Institute for Southern Studies' "Green Index" ranked Louisiana 49th out of 50 states in overall environmental quality. Louisiana is not a rich state by any measure. It ranks 45th in the nation in spending on elementary and secondary education, for example.

Ascension Parish typifies the toxic "sacrifice zone" model. In two parish towns of Geismar and St. Gabriel, 18 petrochemical plants are crammed into a 9.5-square-mile area. In Geismar, Borden Chemicals has released harmful chemicals into the environment that are health hazardous to the local residents, including ethylene dichloride, vinyl chloride monomer, hydrogen chloride, and hydrochloric acid (Barlett & Steele, 1998, p. 72).

Borden Chemicals has a long track record of contaminating the air, land, and water in Geismar. In March 1997, the company paid a fine of $3.5 million—the single largest in Louisiana history—for storing hazardous waste, sludges, and solid wastes illegally; failing to install containment systems; burning hazardous waste without a permit; neglecting to report the release of hazardous chemicals into the air; contaminating groundwater beneath the plant site (thereby threatening an aquifer that provides drinking water for residents of Louisiana and Texas); and shipping toxic waste laced with mercury to South Africa without notifying the EPA, as required by law (Barlett & Steele, 1998).

Louisiana could actually improve its general welfare by enacting and enforcing regulations to protect the environment (Templet, 1995). However, Louisiana citizens subsidize corporate welfare with their health and the environment (Barlett & Steele, 1998). A growing body of evidence shows that environmental regulations do not kill jobs. On the contrary, the data indicate that "states with lower pollution levels and better environmental policies generally have more jobs, better socioeconomic conditions and are more attractive to new business" (Templet, 1995, p. 37). Nevertheless, some states subsidize polluting industries in the return for a few jobs (Barlett & Steele, 1998). States argue that tax breaks help create jobs. However, the few jobs that are created come at a high cost to Louisiana taxpayers and the environment.

Nowhere is the polluter-welfare scenario more prevalent than in Louisiana. Corporations routinely pollute the air, ground, and drinking water while being subsidized by tax breaks from the state. The state is a leader in doling out corporate welfare to polluters (see table 9.1). In the 1990s, the state wiped off the books $3.1 billion in property taxes owed by polluting companies. The state's top five worst polluters received $111 million over the past decade (Barlett & Steele, 1998). A breakdown of the chemical releases and tax breaks includes

- Cytec Industries (24.1 million pounds of releases/$19 million tax breaks)
- IMC-Agrico Co. (12.8 million pounds/$15 million)
- Rubicon, Inc. (8.4 million pounds/$20 million)
- Monsanto Co. (7.7 million pounds/$45 million)
- Angus Chemical Co. (6.3 million pounds/$12 million)

TABLE 9.1. **Corporate Welfare in Louisiana**

The biggest recipients: Companies ranked by total industrial-property tax abatements, 1988–97

Company	Jobs Created	Total Taxes Abated
1. Exxon Corp.	305	$213,000,000
2. Shell Chemical/Refining	167	$140,000,000
3. International Paper	172	$103,000,000
4. Dow Chemical Co.	9	$96,000,000
5. Union Carbide	140	$53,000,000
6. Boise Cascade Corp.	74	$53,000,000
7. Georgia Pacific	200	$46,000,000
8. Willamette Industries	384	$45,000,000
9. Procter & Gamble	14	$44,000,000
10. Westlake Petrochemical	150	$43,000,000

The costliest jobs: Companies ranked by net cost of each new job (abatements divided by jobs created)

Company	Jobs Created	Cost per Job
1. Mobil Oil Corp.	1	$29,100,000
2. Dow Chemical Co.	9	$10,700,000
3. Olin Corp.	5	$6,300,000
4. BP Exploration	8	$4,000,000
5. Procter & Gamble	14	$3,100,000
6. Murphy Oil USA	10	$1,600,000
7. Star Enterprise	9	$1,500,000
8. Cytec	13	$1,500,000
9. Montell USA	31	$1,200,000
10. Uniroyal Chemical Co.	22	$900,000

Not only is subsidizing polluters bad business, but it does not make environmental sense. For example, nearly three-fourths of Louisiana's population—more than 3 million people—get their drinking water from underground aquifers. Dozens of the aquifers are threatened by contamination from polluting industries (O'Byrne & Schleifstein, 1991). The Lower Mississippi River Industrial Corridor has over 125 companies that manufacture a range of products, including fertilizers, gasoline, paints, and plastics. This corridor has been dubbed "Cancer Alley" by environmentalists and local residents (Beasley, 1990a, 1990b; Bullard, 1994a; Motavalli, 1998).

Winning in Court: The Case of *CANT v. LES*

Executive Order 12898[1] was put to the test in rural northwest Louisiana in 1989. Beginning that year, the Nuclear Regulatory Commission (NRC) had under review a proposal from Louisiana Energy Services (LES) to build the nation's first privately owned uranium enrichment plant. A national search was undertaken by LES to find the "best" site for a plant that would produce 17% of the nation's enriched uranium. LES supposedly used an objective scientific method in designing its site selection process.

The southern United States, Louisiana, and Claiborne Parish ended up being the dubious "winners" of the site selection process. Residents from Homer and the nearby communities of Forest Grove and Center Springs—two communities closest to the proposed site—disagreed with the site selection process and outcome. They organized themselves into a group called Citizens Against Nuclear Trash (CANT), which charged LES and the NRC staff with practicing environmental racism. CANT hired the Sierra Club Legal Defense Fund (which later changed its name to Earthjustice Legal Defense Fund) and sued LES.

The lawsuit dragged on for more than 8 years. On May 1, 1997, a three-judge panel of the NRC's Atomic Safety and Licensing Board issued a final initial decision on the case. The judges concluded that "racial bias played a role in the selection process" (Nuclear Regulatory Commission, 1997). A story in the *London Sunday Times* proclaimed the environmental justice victory by declaring "Louisiana Blacks Win Nuclear War" (1997). The precedent-setting federal court ruling came 2 years after President Clinton signed Executive Order 12898. The judges, in a 38-page written decision, also chastised the NRC staff for not addressing the provision called for under Executive Order 12898. The court decision was upheld on appeal on April 4, 1998.

A clear racial pattern emerged during the so-called national search and multistage screening and selection process (Bullard, 1995). For example, . . . African Americans comprise about 13% of the U.S. population, 20% of the southern states' population, 31 % of Louisiana's population, 35% of Louisiana's northern parishes, and 46% of Claiborne Parish. This progressive narrowing of the site selection process to areas of increasingly high poverty and African American representation is also evident from an evaluation of the actual sites that were considered in the "intermediate" and "fine" screening stages of the site selection process. . . . [T]he aggregate average percentage of Black population for a one-mile radius around all of the 78 sites examined (in 16 parishes) is 28.35%. When LES completed its initial site cuts and reduced the list to 37 sites within nine parishes, the aggregate percentage of Black population rose to 36.78%. When LES then further limited its focus to six sites in Claiborne Parish, the aggregate average percentage Black population rose again, to 64.74%. The final site selected, the LeSage site, has 97.10% Black population within a one-mile radius.

The LES plant was proposed to be built on Parish Road 39 between two African American communities—just one-quarter mile from Center Springs (founded in 1910) and one and one-quarter mile from Forest Grove (founded in the 1860s just after slavery). The proposed site is in a Louisiana parish that has per capita earnings of only $5,800 per year (just 45% of the national average of almost $12,800) and where over 58% of the African American population is below the poverty line. The two African American communities were rendered "invisible," since they were not even mentioned in the NRC's draft environmental impact statement (Nuclear Regulatory Commission, 1997).

Only after intense public comments did the NRC staff attempt to address environmental justice and disproportionate-impact implications, as required under the NEPA and called for under Executive Order 12898. For example, NEPA requires that the government consider the environmental impacts and weigh the costs and benefits of any proposed action. These include health and environmental effects, the risk of accidental but foreseeable adverse health and environmental effects, and socioeconomic impacts.

The NRC staff devoted less than a page to addressing environmental justice concerns of the proposed uranium enrichment plant in its final environmental impact statement (FEIS). Overall, the FEIS and Environmental Report (ER) are inadequate in the following respects: (1) they inaccurately assess the costs and benefits of the proposed plant, (2) they fail to consider the inequitable distribution of costs and benefits of the proposed plant to the White and African American population, and (3) they fail to consider the fact that the siting of the plant in a community of color follows a national pattern in which institutionally biased decision making leads to the siting of hazardous facilities in communities of color and results in the inequitable distribution of costs and benefits to those communities.

Among the distributive costs not analyzed in relationship to Forest Grove and Center Springs include the disproportionate burden of health and safety, diminished property values, fire and accidents, noise, traffic, radioactive dust in the air and water, and dislocation by closure of a road that connects the two communities. Overall, the CANT legal victory points to the utility of combining environmental and civil rights laws and the requirement of governmental agencies to consider Executive Order 12898 in their assessments.

In addition to the remarkable victory over LES, a company that had the backing of powerful U.S. and European nuclear energy companies, CANT members and their allies won much more. They empowered themselves and embarked on a path of political empowerment and self-determination. During the long battle, CANT member Roy Mardris was elected to the Claiborne Parish Jury (i.e., county commission), and CANT member Almeter Willis was elected to the Claiborne Parish School Board. The town of Homer, the nearest incorporated town to Forest Grove and Center Springs, elected its first African American mayor, and the Homer town council now has two African American members. In fall 1998, LES sold the land on which the proposed uranium enrichment plant would have been located. The land is going back into timber production, for which it was used before LES bought it.

Winning on the Ground: *St. James Citizens v. Shintech*

Battle lines were drawn in Louisiana in 1991 in another national environmental justice test case. The community is Convent and the company is Shintech. The Japanese-owned Shintech, Inc., applied for a Title V air permit to build a $800 million polyvinyl chloride (PVC) plant in Convent, Louisiana, a community that is over 70% African American; over 40% of the Convent residents fall below the poverty line. The community already has a dozen polluting plants and yet has a 60% unemployment rate. The plants are so close to residents' homes, they could walk to work. The Black community is lured into accepting the industries with the promise of jobs, but in reality, the jobs are not there for local residents.

The Shintech case raised similar environmental racism concerns as those found in the failed LES siting proposal. The EPA is bound by Executive Order 12898 to ensure that "no segment of the population, regardless of race, color, national origin, or income, as a result of EPA's policies, programs, and activities, suffer disproportionately from adverse health or environmental effects, and all people live in clean and sustainable communities." The Louisiana Department of Environmental Quality is also bound by federal laws (e.g., Title VI of the Civil Rights Act of 1964) to administer and implement its programs, mandates, and policies in a nondiscriminatory way.

Any environmental justice analysis of the Shintech proposal will need to examine the issues of disproportionate and adverse impact on low-income and minority populations near the proposed PVC plant. Clearly, it is African Americans and low-income residents in Convent who live closest to existing and proposed industrial plants and who will be disproportionately impacted by industrial pollution (Wright, 1998). African Americans comprise 34% of the state's total population. The Shintech plant was planned for the St. James Parish, which ranks third in the state for toxic releases and transfers. Over 83% of St. James Parish's 4,526 residents are African American. Over 17.7 million pounds of releases were reported in the 1996 *Toxic Release Inventory* (*TRI*). The Shintech plant would add over 600,000 pounds of air pollutants annually. Permitting the Shintech plant in Convent would add significantly to the toxic burden borne by residents, who are mostly low-income and African American.

After 6 months of intense organizing and legal maneuvering, residents of tiny Convent and their allies convinced EPA administrator Carol M. Browner to place the permit on hold. A feature article in *USA Today* bore the headline "EPA Puts Plant on Hold in Racism Case" (Hoversten, 1997). A year later, the Environmental Justice Coalition forced Shintech to scrap its plans to build the PVC plant in the mostly African American community. The decision came in September 1998 and was hailed around the country as a major victory against environmental racism. The driving force behind this victory was the relentless pressure and laser-like focus of the local Convent community.

Radioactive Colonialism and Native Lands

There is a direct correlation between exploitation of land and exploitation of people. It should not be a surprise to anyone to discover that Native Americans have to contend with some of the worst pollution in the United States (Beasley, 1990b; Kay, 1991; Taliman, 1992; Tomsho, 1990). Native American nations have become prime targets for waste trading (Angel, 1992; Geddicks, 1993). More than three dozen Indian reservations have been targeted for landfills, incinerators, and other waste facilities (Kay, 1991). The vast majority of these waste proposals have been defeated by grassroots groups on the reservations. However, "radioactive colonialism" is alive and well (Churchill & LaDuke, 1983).

Radioactive colonialism operates in energy production (mining of uranium) and disposal of wastes on Indian lands. The legacy of institutional racism has left many sovereign Indian nations without an economic infrastructure to address poverty, unemployment, inadequate education and health care, and a host of other social problems.

Some industry and governmental agencies have exploited the economic vulnerability of Indian nations. For example, of the 21 applicants for the DOE's monitored retrievable storage (MRS) grants, 16 were Indian tribes (Taliman, 1992a). The 16 tribes lined up for $100,000 grants from the DOE to study the prospect of "temporarily" storing nuclear waste for a half century under its MRS program.

It is the Native American tribes' sovereign right to bid for the MRS proposals and other industries. However, there are clear ethical issues involved when the U.S. government contracts with Indian nations that lack the infrastructure to handle dangerous wastes in a safe and environmentally sound manner. Delegates at the Third Annual Indigenous Environmental Council Network Gathering (held in Cello Village, Oregon, on June 6, 1992) adopted a resolution of "No nuclear waste on Indian lands."

Transboundary Waste Trade

Hazardous waste generation and international movement of hazardous waste pose some important health, environmental, legal, and ethical dilemmas. It is unlikely that many of the global hazardous waste proposals can be effectuated without first addressing the social, economic, and political context in which hazardous wastes are produced (industrial processes), controlled (regulations, notification and consent documentation), and managed (minimization, treatment, storage, recycling, transboundary shipment, pollution prevention).

The "unwritten" policy of targeting Third World nations for waste trade received international media attention in 1991. Lawrence Summers, at the time he was chief economist of the World Bank, shocked the world and touched off an international scandal when his confidential memorandum on waste trade was leaked. Summers writes: "'Dirty' Industries: Just between you and me, shouldn't the World Bank be encouraging *more* migration of the dirty industries to the LDCs?" (quoted in Greenpeace, 1993, pp. 1–2).

Consumption and production patterns, especially in nations with wasteful "throw-away" lifestyles like the United States, and the interests of transnational corporations create and maintain unequal and unjust waste burdens within and between affluent and poor communities, states, and regions of the world. Shipping hazardous wastes from rich communities to poor communities is not a solution to the growing global waste problem. Not only is it immoral, but it should be illegal. Moreover, making hazardous waste transactions legal does not address the ethical issues imbedded in such transactions (Alston & Brown, 1993).

Transboundary shipment of banned pesticides, hazardous wastes, toxic products, and export of "risky technologies" from the United States, where regulations and laws are more stringent, to nations with weaker infrastructure, regulations, and laws smacks of a double standard (Bright, 1990). The practice is a manifestation of power arrangements and a larger stratification system in which some people and some places are assigned greater value than others.

In the real world, all people, communities, and nations are *not* created equal. Some populations and interests are more equal than others. Unequal interests and power arrangements have allowed poisons of the rich to be offered as short-term remedies for poverty of the poor. This scenario plays out domestically (as in the United States, where low-income and people-of-color communities are disproportionately affected by waste facilities and "dirty" industries) and internationally (where hazardous wastes move from OECD states flow to non-OECD states).

The conditions surrounding the more than 1,900 maquiladoras (assembly plants operated by American, Japanese, and other foreign countries) located along the 2,000-mile U.S.-Mexico border may further exacerbate the waste trade (Sanchez, 1990). The maquiladoras use cheap Mexican labor to assemble imported components and raw material and then ship finished products back to the United States. Nearly a half million Mexican workers are employed in the maquiladoras.

A 1983 agreement between the United States and Mexico required American companies in Mexico to return their waste products to the United States. Plants were required to notify the U.S. EPA when returning wastes. Results from a 1986 survey of 772 maquiladoras revealed that only 20 of the plants informed the EPA that they were returning waste to the United States, even though 86% of the plants used toxic chemicals in their manufacturing process (Juffers, 1988). In 1989, only 10 waste shipment notices were filed with the EPA (Center for Investigative Reporting, 1990).

Much of the wastes end up being illegally dumped in sewers, ditches, and the desert. All along the Lower Rio Grande River Valley, maquiladoras dump their toxic wastes into the river, from which 95% of the region's residents get their drinking water (Hernandez, 1993). In the border cities of Brownsville, Texas, and Matamoras, Mexico, the rate of anencephaly—babies born without brains—is four times the national average. Affected families have filed lawsuits against 88 of the area's 100 maquiladoras for exposing the community to xylene, a cleaning solvent that can cause brain hemorrhages and lung and kidney damage.

The Mexican environmental regulatory agency is understaffed and ill-equipped to adequately enforce its laws (Barry & Simms, 1994; Working Group on Canada-Mexico Free Trade, 1991). Only time will tell if the North American Free Trade Agreement (NAFTA) will "fix" or exacerbate the public health and the environmental problems along the U.S.-Mexico border.

Conclusion

. . . The poisoning of African Americans in Louisiana's "Cancer Alley," Native Americans on reservations, and Mexicans in the border towns all have their roots in the same economic system, a system characterized by economic exploitation, racial oppression, and devaluation of human life and the natural environment. Both race and class factors place low-income and people-of-color communities at special risk. Although environmental and civil rights laws have been on the books for more than 3 decades, all communities have not received the same benefits from their application, implementation, and enforcement.

Unequal political power arrangements also have allowed poisons of the rich to be offered as short-term economic remedies for poverty. There is little or no correlation between proximity of industrial plants in communities of color and the employment opportunities of nearby residents. Having industrial facilities in one's community does not automatically translate into jobs for nearby residents. Many industrial plants are located at the fence line with the communities. Some are so close that local residents could walk to work. More often than not communities of color are stuck with the pollution and poverty, while other people commute in for the industrial jobs.

Similarly, tax breaks and corporate welfare programs have produced few new jobs by polluting firms. However, state-sponsored pollution and lax enforcement have allowed many communities of color and poor communities to become the dumping grounds. Louisiana is the poster child for corporate welfare. The state is mired in both poverty and pollution. It is no wonder that Louisiana's petrochemical corridor, the 85-mile stretch along the Mississippi River from Baton Rouge to New Orleans dubbed "Cancer Alley," has become a hotbed for environmental justice activity.

The environmental justice movement has set out clear goals of eliminating unequal enforcement of environmental, civil rights, and public health laws; differential exposure of some populations to harmful chemicals, pesticides, and other toxins in the home, school, neighborhood, and workplace; faulty assumptions in calculating, assessing, and managing risks; discriminatory zoning and land use practices; and exclusionary policies and practices that limit some individuals and groups from participation in decision making. Many of these problems could be eliminated if existing environmental, health, housing, and civil rights laws were vigorously enforced in a nondiscriminatory way.

The call for environmental and economic justice does not stop at the U.S. borders but extends to communities and nations that are threatened by the export of hazardous wastes, toxic products, and "dirty" industries. Much of the world does not get to share in the benefits of the United States' high standard of living. From energy consumption to the production and export of tobacco, pesticides, and other chemicals, more and more of the world's peoples are sharing the health and environmental burden of America's wasteful throwaway culture. Hazardous wastes and "dirty" industries have followed the path of least resistance. Poor people and poor nations are given a false choice of "no jobs and no development" versus "risky, low-paying jobs and pollution."

Industries and governments (including the military) have often exploited the economic vulnerability of poor communities, poor states, poor nations, and poor regions for their unsound and "risky" operations. Environmental Justice leaders are demanding that no community or nation, rich or poor, urban or suburban, Black or White, be allowed to become a "sacrifice zone" or dumping grounds. They are also pressing governments to live up to their mandate of protecting public health and the environment.

Notes

Bullard Robert D., and Glenn S. Johnson. 2000. "Environmental Justice: Grassroots Activism and Its Impact on Public Policy Decision Making." *Journal of Social Issues* 56(3): 555–78.

1. Executive Order 12898, "Federal Actions to Address Environmental Justice in Minority Populations and Low-Income Populations," was issued by President Bill Clinton on February 11, 1994. This order reinforces Title VI of the 1964 Civil Rights Act (which prohibits discrimination on the part of federally funded programs) in an attempt to respond to growing public concern and mounting scientific evidence regarding environmental injustices.

References

Agency for Toxic Substances and Disease Registry. (1988). *The nature and extent of lead poisoning in children in the United States: A report to Congress.* Atlanta, GA: U.S. Department of Health and Human Services.

Alston, D., & Brown, N. (1993). Global threats to people of color. In R. D. Bullard (Ed.), *Confronting environmental racism: Voices from the grassroots* (pp. 179–194). Boston: South End Press.

Angel, B. (1992). *The toxic threat to Indian lands: A Greenpeace report.* San Francisco: Greenpeace.

Austin, R., & Schill, M. (1991). Black, Brown, poor, and poisoned: Minority grassroots environmentalism and the quest for eco-justice. *Kansas Journal of Law and Public Policy,* 1(1), 69–82.

Barlett, D. L., & Steele, J. B. (1998, November 23). Paying a price for polluters. *Time,* pp. 72–80.

Barry, T., & Simms, B. (1994). *The challenge of cross border environmentalism: The U.S.–Mexico case.* Albuquerque, NM: Inter-Hemispheric Education Resource Center.

Beasley, C. (1990a). Of pollution and poverty: Keeping watch in Cancer Alley. *Buzzworm,* 2(4), 39–45.

Beasley, C. (1990b). Of poverty and pollution: Deadly threat on native lands. *Buzzworm,* 2(5), 39–45.

Bryant, B. (1995). *Environmental justice: Issues, policies, and solutions.* Washington, D.C.: Island Press.

Bryant, B., & Mohai, P. (1992). *Race and the incidence of environmental hazards.* Boulder, CO: Westview Press.

Bullard, R. D. (1993a). *Confronting environmental racism: Voices from the grassroots.* Boston: South End Press.

Bullard, R. D. (1993b). Race and environmental justice in the United States. *Yale Journal of International Law,* 18(1), 319–355.

Bullard, R. D. (1993c). Environmental racism and land use. *Land Use Forum: A Journal of Law, Policy & Practice,* 2(1), 6–11.

Bullard, R. D. (1994a). *Dumping in Dixie: Race, class and environmental quality.* Boulder, CO: Westview Press.

Bullard, R. D. (1994b). Grassroots flowering: The environmental justice movement comes of age. *Amicus,* 16 (Spring), 32–37.

Bullard, R. D. (1995). Prefiled written testimony at the *CANT v. LES* hearing, Shreveport, LA.

Bullard, R. D. (1996). *Unequal protection: Environmental justice and communities of color.* San Francisco: Sierra Club.

Bullard, R. D., & Johnson, G. S. (1997). *Just transportation: Dismantling race and class barriers to mobility.* Gabriola Island, British Columbia, Canada: New Society.

Center for Investigative Reporting. (1990). *Global dumping ground: The international traffic in hazardous waste.* Washington, DC: Seven Locks Press.

Chase, A. (1993). Assessing and addressing problems posed by environmental racism. *Rutgers University Law Review,* 45(2), 385–369.

Churchill, W., & LaDuke, W. (1983). Native America: The political economy of radioactive colonialism. *Insurgent Sociologist,* 13(1), 51–63.

Commission for Racial Justice. (1987). *Toxic wastes and race in the United States.* New York: United Church of Christ.

Cooney, C. M. (1999). Still searching for environmental justice. *Environmental science and Technology,* 33 (May), 200–205.

Council on Environmental Quality. (1971). *Second annual report of the Council on Environmental Quality.* Washington, DC: U.S. Government Printing Office.

Council on Environmental Quality. (1997). *Environmental justice: Guidance under the National Environmental Policy Act.* Washington, DC: Author.

Escobedo, D. (1996, October 4). EPA gives in, will move all at toxic site. *Pensacola News Journal,* p. A1.

Geddicks, A. (1993). *The new resource wars: Native and environmental struggles against multi-national corporations.* Boston: South End Press.

Goldman, B. (1992). *The truth about where you live: An atlas for action on toxins and mortality.* New York: Random House.

Goldman, B., & Fatten, L. J. (1994). *Toxic wastes and race revisited.* Washington, DC: Center for Policy Alternatives/National Association for the Advancement of Colored People/United Church of Christ.

Greenpeace. (1990). *The international trade in waste: A Greenpeace inventory.* Washington, DC: Greenpeace USA.

Greenpeace. (1993). *The case for a ban on all hazardous waste shipment from the United States and other OECD member states to non-OECD states.* Washington, DC: Greenpeace USA.

Hernandez, B. J. (1993). Dirty growth. *New Internationalist* (August).

Hoversten, P. (1997, September 11). EPA puts plant on hold in racism case. *US Today,* p. A3.

Institute for Southern Studies. (1992). *1991–1992 Green Index: A state-by-state guide to the nation's environmental health.* Durham, NC: Author.

Juffers, J. (1988, October 24). Dump at the border: U.S. firms make a Mexican wetland. *Progressive.*

Kay, J. (1991, April 10). Indian lands targeted for waste disposal sites. *San Francisco Examiner.*

Lavelle, M., & Coyle, M. (1992). Unequal protection. *National Law Journal,* 1–2.

Lee, B. L. (1992, February). *Environmental litigation on behalf of poor, minority children: Matthews v. Coye: A case study.* Paper presented at the annual meeting of the American Association for the Advancement of Science, Chicago.

Lee, B. L. (1995, May). *Civil rights remedies for environmental injustice.* Paper presented at Transportation and Environmental Justice: Building Model Partnerships conference, Atlanta, GA.

Lee, C. (1992). *Proceedings: The First National People of Color Environmental Leadership Summit.* New York: United Church of Christ, Commission for Racial Justice.

Louisiana Blacks win nuclear war. (1997, May 11). *Sunday London Times.*

Mann, E. (1991). *LA's lethal air: New strategies for policy, organizing, and action.* Los Angeles: Labor/Community Strategy Center.

Mann, E. (1996). *A new vision for urban transportation: The bus riders union makes history at the intersection of mass transit, civil rights, and the environment.* Los Angeles: Labor/Community Strategy Center.

Motavalli, J. (1998). Toxic targets: Polluters that dump on communities of color are finally being brought to justice. *E* (July/August), 28–41.

National Institute for Environmental Health Sciences. (1995). *Proceedings of the Health and Research Needs to Ensure Environmental Justice Symposium.* Research Triangle Park. NC: Author.

Nuclear Regulatory Commission. (1997). *Final initial decision-Louisiana Energy Services.* U.S. Nuclear Regulatory Commission, Atomic Safety and Licensing Board, Docket no. 70-3070-ML. May 1.

O'Byrne, & Schleifstein, M. (1991, February 19). Drinking water in danger. *Times Picayune*, p. A5.

Pirkle, J. L., Brody, D. J., Gunter, E. W., Kramer, R. A., Paschal, D. C., Flegal, K. M., & Matte, T. D. (1994). The decline in blood lead levels in the United States: The National Health and Nutrition Examination Survey (NHANES III). *Journal of the American Medical Association*, 272, 284–291.

Sanchez, R. (1990). Health and environmental risks of the maquiladora in Mexicali. *National Resources Journal*, 30(1), 163–186.

Schueler, D. (1992). Southern exposure. *Sierra*, 77 (November–December), 45–47.

Stretesky, P., & Hogan, M. J. (1998). Environmental justice: An analysis of Superfund sites in Florida. *Social Problems*, 45 (May), 268–287.

Taliman, V. (1992a). Stuck holding the nation's nuclear waste. *Race, Poverty, Environment Newsletter* (Fall), 6–9.

Taliman, V. (1992b). The toxic waste of Indian lives. *Covert Action*, 40(1), 16–19.

Templet, P. T. (1995). The positive relationship between jobs, environment and the economy: An empirical analysis and review. *Spectrum* (Spring), 37–49.

Tomsho, R. (1990, November 29). Dumping grounds: Indian tribes contend with some of the worst of America's pollution. *Wall Street Journal.*

U.S. Environmental Protection Agency. (1991). *Hazardous waste exports by receiving country.* Washington, DC: Author.

U.S. Environmental Protection Agency. (1992a). *Environmental equity: Reducing risk for all communities.* Washington, DC: Author.

U.S. Environmental Protection Agency. (1992b). Geographic initiatives: Protecting what we love. In *Securing our legacy: An EPA progress report 1989–1991.* Washington, DC: Author.

U.S. Environmental Protection Agency. (1993). *Toxic release inventory and emission reduction 1987–1990 in the Lower Mississippi River industrial corridor.* Washington, DC: EPA, Office of Pollution Prevention and Toxics.

U.S. Environmental Protection Agency. (1996). *Escambia Treating Company interim action: Addendum to April 1996 Superfund proposed plan fact sheet.* Atlanta, GA: EPA, Region IV.

U.S. to move families away from Florida toxic dump. *Washington Post,* October 6, 1996.

Wernette, D. R., & Nieves, L. A. (1992). Breathing polluted air. *EPA Journal,* 18(1),16–17.

West, P., Fly, J. M., Larkin, F., & Marans, P. (1990). Minority anglers and toxic fish con-sumption: Evidence of the state-wide survey of Michigan. In B. Bryant & P. Mohai (Eds.), *Race and the incidence of environmental hazards* (pp. 100–113). Boulder, CO: Westview.

Wright, B. H. (1998). *St. James Parish field observations.* New Orleans, LA: Xavier University, Deep South Center for Environmental Justice.

PART

Work

Forty Years of Spotted Owls?

A Longitudinal Analysis of Logging Industry Job Losses

William R. Freudenburg, Lisa J. Wilson, and Daniel J. O'Leary

The next two selections focus on the relationship between work and the environment. Often we see a "jobs versus the environment" dichotomy presented in the popular media. Many people believe that tougher environmental regulations mean fewer jobs. The role of sociologists is to question these kinds of assumptions—crack them open and see if they hold up to the light of scientific research. The authors of the next piece do exactly that in their quantitative analysis of national and regional employment data (from 1947 to 1989). Their findings challenge the widely held belief that environmental protection laws, specifically laws designed to protect the Northern Spotted Owl, have led to declines in logging and milling employment. The authors found no quantitative evidence of a statistically credible increase in job losses associated with the federal listing of the northern spotted owl as a threatened species.

While logging has been an important part of the Pacific Northwest economy ever since European Americans arrived in the region, only recently has the activity received increased attention from sociologists. Much of that attention has come in response to growing concerns over environmental regulations—specifically including efforts to list the Northern Spotted Owl as a "threatened" species—which have caused many loggers in the region to argue that they, not the owl, are "endangered" (Satchell 1990). The loggers' concerns have been echoed by many of the scholars who have examined the issue most closely (see especially Brown 1995; Carroll 1995; Lee 1993); in addition, government officials, timber industry executives, and the mass media (see Rice 1992; Fitzgerald 1992; Levine 1989) have joined the scholars and the loggers, predicting that spotted owl protection would lead to economic and social havoc in rural timber communities. In spite of the widespread nature of the concerns, however, it is not entirely clear just how many jobs have been lost, over just how long a period, due to the limitations imposed on logging through efforts to protect spotted-owl habitat and to provide other forms of environmental protection.

That is precisely the gap the present article is intended to fill. . . .

Existing Assessments

While scholarly writing on the spotted-owl issue is a relatively recent phenomenon, this recent writing both reflects and grows out of a much larger body of work that has dealt with society–environment relationships more broadly. Some of the best known analysts of society environment relationships, including Catton (1982), Schnaiberg (1980), and O'Connor (1988), have characterized economic growth as a major threat to environmental protection—and vice versa. More recently, the work of Schnaiberg and Gould (1994) has characterized the relationship between society and environment as involving an "Enduring Conflict," with the "major argument" of the book being that there is a "conflict between economic growth and environmental protection" (1994:94).

In many respects, the debates over logging in the Pacific Northwest would appear to provide a particularly clear empirical example of the environment-versus-economy expectation—a point that is underscored by the ways in which the spotted-owl issue has been discussed in the sociological literature to date. Lee (1993), for example, argues that efforts to preserve old growth forests and spotted owls have not only limited economic growth, but created "a severe economic and social impact" (1993:1; see also Greber et al. 1990; Conway and Wells 1993). Similarly, Humphrey et al. (1993:159) trace "the potential impoverishment" of forest products workers in the Pacific Northwest to "growing concern for old growth forests and their ecological structure," particularly given what these authors characterize as "the growing power of a national elite dedicated to environmentalism." While loggers were once seen as veritable folk heroes, Humphrey et al. (1993:161–62) argue, the newer views involve "images such as 'buffalo hunters,' 'tree murderers,' and 'rapers of the land'"—with the newer images being used to justify the exclusion of loggers from continued access to the trees and to their traditional basis for earning a living (see also Lee 1994; Carroll 1995).

Academic researchers, however, are only a small fraction of the people paying attention to the "enduring conflict" in the Pacific Northwest. Not surprisingly, a number of the more forceful statements have come from representatives of the timber and lumber industries (see Flynn 1991; Bland and Blackman 1990; Forest Industries 1991), who blame owl protection (and environmentalists) for job losses, timber shortages, and higher timber prices. Yet the tendency to blame owls and environmentalists also goes well beyond timber industry publications: In recent years, headlines in a variety of popular periodicals have spoken of "The Great Spotted Owl War" (Fitzgerald 1992), have noted claims that "The Spotted Owl Could Wipe Us Out" (Levine 1989), and have referred both to "The Endangered Logger" (Satchell 1990) and to a battle of "Owl vs. Man" (Gup 1990).

The message of the headlines is generally reinforced by the articles themselves. Fitzgerald (1992:93), for example, writes that, "[T]he wheels of government and the federal courts have been set in motion to protect the owl. The result has been havoc for people.". . .

. . . Despite the widespread claim that the mass media have an "anti-industry" bias (for a review of the relevant literature, see Freudenburg et al. 1996), it was the timber industry approach, rather than the environmentalists' approach, that was generally adopted by television reports. An outright majority of the news stories presented on the evening news broadcasts of the three major networks stressed the "jobs" side of the controversy over the environmental side, adopting and reinforcing the "jobs versus owls" frame of reference (Liebier and Bendix 1994:10).

While the periodicals of environmental groups have been more likely to state the case for saving the owls and their habitat, even these publications reflect the prevailing belief that efforts to protect owl habitat are likely to prove a major source of dislocation for the region's workers. In the official magazine of the Sierra Club, for example, Tisdale (1992), a self proclaimed "tree-hugger," describes posters asking "A spotted owl needs hundreds of acres to live—why can't I have some of that land to live on? Am I important?" (see also Mitchell 1990; Mitchell and Lamont 1991). A number of environmentally oriented analysts, however, have argued that the loss of jobs should instead be traced to the fact that so much lumber is being exported to other nations in the form of raw logs, rather than first being transformed into finished products such as furniture (see Anderson and Olson 1991; Brown 1995; Foster 1993; Glick 1995).

... While recent studies have begun to cast doubt on the expectation for a positive relationship between harvest levels and community stability (see, e.g., Machlis et al. 1990; Force et al. 1993; Heberlein 1994; see also Yoho 1965; Freudenburg 1992; Freudenburg and Gramling 1994b; Peluso et al. 1994), the expectation continues to be highly influential in policy discussions involving timber harvesting in the Pacific Northwest (see Lee 1993).

Still, despite the widespread agreement, there are at least two important problems, in empirical terms, with the tendency to blame the logging industry job losses and attendant socioeconomic disruptions on environmental protection efforts. The first has to do with the matter of turning points: Even if there is agreement that environmental protection is to blame for job losses, there is considerably less agreement about the time when this effect should be seen as having begun. While many analyses point to the 1989–1990 "listing" of the spotted owl as an officially "threatened" species, any number of authors have identified earlier starting points, with two dates having received particular attention. Many authors single out the importance of 1970, the time of the National Environmental Policy Act and the first "Earth Day" (see, e.g., Dunlap 1990; Freudenburg and Gramling 1994a; McCloskey 1992; Mitchell, Mertig and Dunlap 1992). Other writers have singled out the earlier turning point of September, 1964, when the Wilderness Protection Act was officially signed into law (Nash 1982:226; Runte 1987:240). Final Congressional action on this bill came only after years of debate—according to the Wilderness Society (1996), for example, the Act went through 66 rewrites and 8 years of legislative battles— largely due to the bitter objections of many senators and representatives from forest-dependent regions, particularly in the western U.S., who feared that the Act would "lock up" valuable forest lands and end the virtual "Golden Age" of booming timber demand that had characterized the first two decades after the end of World War II. . . .

Aside from the matter of differing views on turning points, the second problem is that, while agreement on the jobs-versus-environmental-protection assessment is clearly widespread, it is not a matter of complete consensus. At least within the research community, a number of respected authors have argued that the job losses should be seen as resulting from other factors, such as mechanization, the exporting of "raw" or unprocessed logs, or the fact that most of the giant old trees had already been cut before the spotted-owl issue erupted. These arguments are often overlooked in the noise of the ongoing debate, but they deserve careful consideration nevertheless. Roughly a decade before the spotted-owl issue came to widespread public awareness, for example, Young and Newton (1979) called attention to the widespread closing of sawmills in the Pacific Northwest, noting that the closures had been particularly extensive in the rural regions where the spotted-owl protests would later become the most intense. More broadly, any number of other analysts have expressed the concern that, in the words of Love's concise assessment of employment trends (1997:217), "[T]he owl controversy has been masking ongoing changes in the Northwest timber industry" (see also Beuter et al. 1976; Brunelle 1990). Even the figures used by the official "Forest Service Ecosystem Management Assessment Team" or FEMAT (1993) show annual timber industry employment in the spotted owl region to have declined significantly, from roughly 168,000 to roughly 151,000 jobs, between the early 1970s and the "pre-owl" late 1980s—despite the fact that wood exports from this region rose more than 50%, from about 3.2 billion to about 5 billion board feet, over the same period (FEMAT 1993: vi–23).

From Argument to Analysis

In short, despite the pervasive belief that the loss of logging and logging-related jobs in the Pacific Northwest can be traced to environmental concerns, there is less than full agreement over just when those environmental concerns—and which such set of concerns—should be seen as having begun to exercise an effect. This question, however, is an inherently empirical one, and it is to the answering of this question that we now turn.

Any such empirical examination needs to begin with the recognition that arguments about the negative impact of environmentalism on employment apply mainly to the logging and milling sectors of the timber industry, rather than to "forest-dependent" communities and populations more broadly....

National employment data have been drawn from the U.S. Bureau of Labor Statistics (BLS).... Employment data for the states of Washington and Oregon ... were obtained by contacting each state's employment agencies....

The best data on log exports are those from the Demand, Price and Trade Analysis Group of the U.S. Forest Service. Given that we were unable to obtain all of this group's original data reports back to 1947, we turned to publicly available sources that also report the relevant figures.... For data on Forest Service harvests, we turned to appropriate editions of Statistical Abstracts....

Operationalization and Analytical Approach

In the interest of avoiding any confusion, we have attempted to keep our analyses as straightforward as possible. Two sets of regressions were performed and [are] reported one, using national-level data and two, using data from the Washington/Oregon region.... In both sets of analyses, the dependent variable will be total employment in logging, saw mills and planing mills....

Both for analysis of national-level trends and of regional trends in the Pacific Northwest, we will consider the effects of each of the three potential turning point years [1964, 1969, and 1989]....

The two substantive independent variables are the U.S. Forest Service (USFS) "cut" (the total harvest of logs on U.S. Forest Service lands, in millions of board feet) and the net exports of raw or unprocessed logs (as measured in millions of cubic feet of roundwood equivalent).[1] . . .

. . . [T]here is no statistical evidence for a spotted-owl effect in any of the three sets of results. . . .

Discussion

Despite the widespread and apparently heartfelt conviction that the jobs of rural loggers and primary wood processors in the Pacific Northwest are being endangered by federal protection of the spotted owl, and by environmental protection more broadly, quantitative analysis provides a very different picture. Based on straightforward longitudinal regression analyses and time-series analyses of the best available data on employment in logging and milling, whether in the Pacific Northwest or in the nation as a whole, there is simply no credible evidence of a statistically believable job-loss effect. This conclusion holds whether the measurement is done at the national or at the regional level, and whether the focus is on the period associated with spotted-owl protection in 1989–1990 or on the period since "Earth Day" and the passage of the National Environmental Policy Act in 1969–1970.

By contrast, there is clear evidence of a major reduction in the prior rate of job losses—a change that is strongly significant statistically—beginning at about the time when U.S. National Forests began to be "locked up" under the Wilderness Act of 1964. While this Act is sometimes associated with what many proponents of resource extraction tend to see as the end of a "golden era" of logging—the exuberant era of booming home construction, subject to few significant environmental constraints, that stretched from the end of World War II to the time when the Wilderness Act was passed in 1964 (Hirt 1994; see Catton 1982)—the statistical effect of this turning point is in precisely the opposite of the "expected" direction: There is a 90% reduction in the prior rate of job losses at the national level, and roughly a two-thirds reduction at the regional level, even after controlling for other statistically significant effects. Indeed, had there been a continuation of the rates of job losses that existed before the passage of the Wilderness Act, total logging and milling employment would have dropped to zero, both nationally and in the Pacific Northwest, before 1990.

. . . Far from being a period in which the Pacific Northwest region has been "outta work for every American," the period since the listing of the spotted owl has actually been one of soaring job growth in the Northwest; as noted not just in environmentally oriented publications such as Glick (1995), moreover, but also in the mass media more broadly (see Egan 1994), at least some of the new jobs appear to have been attracted in part by the prospect of increased environmental protection in the region. Even in the forests themselves, as noted by Love (1997:215), "Signs are emerging that the market value of recreational uses of federal forests may exceed the market value of logging them."

All in all, if the regression coefficients from this study's analyses were to be taken literally, they would indicate that, after controlling for other statistically significant variables, the Wilderness Act was associated with an increase of roughly 500,000 logging and milling jobs in the nation as a whole, and more than 50,000 jobs in the states of Washington and Oregon, over the past 29 years. Clearly, we would not argue for such a literal interpretation; that would make nearly as little sense as have some of the past assessments that have blamed environmental protection measures for varying but apparently precise assertions about purported job "losses" (see Kazis and Grossman 1982; Freudenburg 1991). What may make the least sense of all, however, is to argue that the actual employment data should continue to be ignored, and that analyses and policy decisions should continue to be based on assertion instead of evidence.

Perhaps the most important questions to emerge from this analysis are two in number, and they are interrelated. The first has to do with what factors might do a better job of accounting for job losses than might environmental protection measures; the second has to do with how and why spotted owl protection could have been embraced so readily as the alleged cause of job losses, when the widespread assertions are so clearly at variance with the available empirical evidence. Based on the information available at present, the most reasonable answers would appear to be, first, that the job losses are best understood in terms of changes in the forest-products industry and in the natural resource base that remains available for supporting that industry, and second, that the failure to give fuller recognition to these well documented trends in the ongoing spotted-owl debates would appear to be due to a remarkable level of historical and perhaps also political naiveté.

Changes in Technology and Natural Resource Base

As will be clear to anyone who has spent a significant amount of time in timber-dependent communities or who has read the sensitive reports on the region by sociologists such as Brown (1995; see also Carroll 1995; Lee 1994), timber workers are suffering emotionally painful changes associated with the decline of the logging industry; importantly, moreover, so are their families, friends, neighbors, and communities. It will also be clear that the workers and their communities have a substantial level of entirely understandable anger about their plight. What is not so clear is whether that anger is directed at the actual causes of the pain.

Sociological authors such as Lee (1993, 1994), Carroll (1995) and Humphrey et al. (1993) have devoted a good deal of attention to the argument that loggers have recently been subjected to "moral exclusion"—being depicted in politicized debates as socially or morally unworthy of continued access to old-growth forests. Such arguments do provide an accurate depiction of the feelings in timber country, but as discussed at greater length in the review of this literature by Freudenburg and Gramling (1994b), the arguments do not appear to provide an accurate rendering of the ways in which loggers have actually been discussed, whether by the mass media in general or by environmental publications in particular. As indicated by the literature review at the start of this [essay], the vast majority of the actual discussions in environmental publications, as well as in the mass media more broadly, have portrayed the loggers and their families in a highly sympathetic light.

To the extent to which any portrayals of loggers have been unflattering, moreover, an examination of the historical record shows that such portrayals are anything but a recent phenomenon—and that they are by no means limited to the views that have been expressed from outside of the forest-products industry. An examination of forest-products publications such as *Journal of Forestry* or *Forest Products Journal* quickly reveals industry spokespersons to have expressed significant levels of concern about logging and milling employment since well before the emergence of concerns over the spotted owl, or even over the protection of wilderness areas. Most of the earlier expressions of concern, however, had to do with levels of employment that were considered to be too high, not too low. Authors such as Simmons (1947) and Compton (1956) fretted openly over what they saw as a tendency for timber-related jobs to be attracting a different (and in their view, "less desirable") sort of worker than had been available during the depression-era years before World War II. Compton (1956:19), for example, worried that "the sawmill owner or operator seems to have adopted a pessimistic attitude toward employees: that he will accept men of lower work capabilities ... [and] that men for the sawmill occupations need not be as intelligent, physically fit or socially acceptable as for other fields of endeavor." Others worried that, with the growing availability of alternative forms of employment, only the "less desirable" employees were being attracted to timber industry jobs; Simmons (1947:345), for example, saw the timber jobs as appealing mainly to the types of workers characterized by "a strong back and a weak mind." Such lines of argument appear to have virtually all of the characteristics that authors such as Lee describe as "moral exclusion," save for the fact that they were being expressed by some of the leading spokespersons inside of the forest products industry, not by the critics on the outside.

The concerns about "less desirable" workers appear to have been joined by a number of more tangible concerns about reducing costs and increasing the productivity of labor, as well as about easing "labor shortages" (Batori 1957; Kendrick 1961; for a more critical analysis, see Young and Newton 1979). Similar concerns continued to be expressed even during the late 1960s, during the period after the passage of the Wilderness Act but before the enactment of the National Environmental Policy Act; McConnen (1967:10), for example, noted that "the cost of forest labor has increased very rapidly," and Wambach (1969:108) noted that, while new capital investments in mechanization often proved expensive, "the investments can be justified by . . . lower labor requirements" and other factors (for a comparable recent assessment, see Greber 1993). The employment statistics considered in this paper, which show national employment to have dropped from 572,000 jobs in 1947 to just 342,000 jobs by 1964, can provide some indication of just how widely accepted such arguments evidently proved to be.

It is still possible, of course, that case studies will be able to identify specific locations where job losses can be blamed on specific environmental constraints; in fact, it would be remarkable indeed if absolutely no such cases could be identified. At least to date, however, most of the arguments about spotted owls, or about environmental protection more broadly, have not been depicted as involving a small scattering of isolated or atypical locations. Instead, not just in the mass media, but also in serious arguments by widely respected scholars, the preservation of old-growth forests has been depicted as posing threats such as "the potential impoverishment of at least 48,160 forest products workers in the Pacific Northwest" (Humphrey et al. 1993:159), and creating "a severe economic and social impact" throughout the communities of the region (Lee 1993:1), all while exemplifying an "enduring conflict" between economy and environment, more broadly (Schnaiberg and Gould 1994). Even for arguments about job losses in specific locations to be taken at face value, moreover, it may be necessary to ignore the inconvenient findings about broader trends that have been summarized in this article—as well as to ignore the fact that more detailed analyses of the Pacific Northwest have tended to mirror the types of findings that have been reported here. When Young and Newton (1979) did a more detailed analysis of the State of Oregon, for example, they found that over 33% of the state's large sawmills were closed during the period from 1948–1962, while among the smaller sawmills—which tended disproportionately to be located in the rural regions that have been at the focus of spotted-owl debates—over 85% of the mills closed during the same period of 1948–1962. All of these changes, in other words, were not merely underway, but had been largely completed, not just before the emergence of the spotted-owl controversy, but more than a quarter of a century earlier, before the passage of the Wilderness Act of 1964.

If we truly want to understand the ways in which the environment has affected logging industry job losses, accordingly, we may need a different analytical approach—one that includes attention to the physical environment itself, and not simply to the battles over the environment. One possibility that has received far too little attention, to be more specific, is that the job losses may ultimately have resulted not from "excessive" environmental protection efforts of the past several years, but from insufficient environmental protection over the past century or more.

To note a fact that is obvious enough to require little additional emphasis, while trees are a "renewable" resource, they are nevertheless not an infinite resource (see Dunlap and Catton 1979; Catton and Dunlap 1980). This point applies with particular force to old-growth forests of the sort that have been rapidly disappearing from the Pacific Northwest: While the U.S. Pacific Northwest may well have been the "logging capital of the world" during the past several decades, it is only the most recent of the many rural regions to have claimed such a distinction. To limit the illustrations simply to the United States, similar claims have previously been staked by the regions surrounding scores of other communities, including Bangor, Maine; Williamsport, Pennsylvania; Muskegon, Michigan; Eureka, California (Wood 1971) and Menominee/ Marinette, Wisconsin/Michigan (Connor 1978; Freudenburg, Frickel, and Gramling 1995). Even an examination of back issues of logging region newspapers (see ["Memories"] Aberdeen Daily World 1980) could show a widespread recognition of the fact that the sawmills being shut down in the region over the years have often been those that had been designed for large, old-growth forests—most of which had already disappeared well before concerns over spotted owls began to receive wider attention. Rather than being a reflection of efforts to protect the environment, ironically, a significant fraction of the long-term job loss in the rural areas of the Pacific Northwest may thus be due to nearly the opposite phenomenon—to the very speed and vigor with which the old-growth trees had been cut down in earlier years (see also Hirt 1994). In short, the problem may ultimately be traced to the fact that, across the decades preceding the spotted-owl issue, as noted in the title of one book (Van Syckle 1980), *They Tried to Cut It All*—and that by the late 1980s, "they" had so nearly succeeded in doing just that.

The Importance of Questioning the Taken for Granted?

Given the lack of credible quantitative evidence for job losses associated with environmental protection, plus the extensive and well-documented evidence of earlier, systematic efforts to reduce labor requirements and shut down the small sawmills in rural areas—not to mention the fact that all previous "logging capitals of the world" appear to have experienced similar fates—how is it that so little attention would have been devoted to the rate at which large corporations have been cutting down trees and laying off workers, relative to the rate at which small owls have led to the laying down of paperwork by federal agencies?

One important part of the answer would appear to reflect the continued relevance of Thernstrom's classic warning (1965) on "the dangers of historical naiveté." Social scientists have often learned, with good reason, to have high levels of respect for the insights and expertise of the citizens who live in areas we study, but as Thernstrom long ago pointed out, there is a significant difference between treating those views with respect and treating them as definitive—a distinction that proves to be particularly important when there is a need to understand historical antecedents of long-term trends. Affected local people are often highly knowledgeable not just about their own experiences, but also about the nature of the world they inhabit; at the same time, however, they may be no more likely to be infallible than are the social scientists who study them, whether in what they remember or what they forget. Where the existence of historical record makes it possible to double-check—whether on what is remembered or what is forgotten) the principle of prudence makes it imperative to do just that.

Another part of the answer may relate to the continuing relevance of the concerns expressed nearly two decades ago by "environmental sociologists" such as Dunlap and Catton (1979), among others—having to do with the excessive reluctance on the part of sociologists to deal explicitly with physical environmental variables, including in this case the rate at which the old-growth timber was being cut. Still, this problem may have been further aggravated by the fact that the early environmental sociologists, in turn, were often reluctant to consider the effects of "social construction" processes (see Berger and Luckmann 1986; for further discussion, see Freudenburg 1997; Gramling and Freudenburg 1996). In part, this reluctance is understandable, in that some earlier versions of social constructivism were so crude as to assert that "the physical characteristics of the environment may be ignored" (Choldin 1978:353). If "the physical characteristics of the environment" are actually ignored, however, there can be any number of unfortunate consequences—including, in the present context, the widespread failure to recognize the rapid rate of exhaustion of the old-growth timber, as well as the failure to analyze the obvious importance of this variable in the rapid disappearance of logging-related jobs over the past half-century.

Yet the tendency to ignore physical environmental variables can also lead to a number of oversights that are less immediately obvious. One example involves a point that is implicit in much of the above discussion, but that has also been made more explicitly by a number of environmental or natural resource sociologists, namely that resource exhaustion may actually be no less common for "renewable" resources than for nonrenewable ones (see Hamilton and Seyfrit 1994; Schurman 1993). A second example is suggested by Freudenburg (1992), who challenges the assumption that the extraction of nonrenewable resources should be expected to continue "until the resource is gone" (see Krannich and Luloff 1991). At least by the analysis of Freudenburg (1992), what may be more common—for renewable as well as for nonrenewable resources—is the phenomenon of the "cost-price squeeze," in which there are increases in the costs of specific extractive operations, over time, even though world-wide prices are less likely to rise, and may often decline. This phenomenon often causes extractive activities to stop "at an earlier stage—not when everything is gone, but at the point when the extraction is no longer profitable" (Freudenburg 1992:324). Yet there is an important catch:

> The catch is that there is no one such point. . . . In particular, extractive corporations have the option of using the ambiguity of exhaustion as an argument for extracting not just raw materials, but concessions from workers, who may be willing to accept wage and benefit cuts in exchange for a few months' or years' worth of employment, and from the political system, as in the form of the willingness to exempt facilities from environmental or health and safety regulations. (Freudenburg 1992:324)

In light of the experience in the Pacific Northwest, however, what should be emphasized is that the advantages of the concession-extraction process may be not merely financial; they can also be political. Efforts to extract concessions from workers, after all, can have very different implications than do efforts to extract environmental concessions. In 1986, for example, when one of the nation's largest lumber companies decided to squeeze wage concessions from its workers in the Pacific Northwest, "insisting that its 8,000 unionized employees accept cuts of 20% or more in wages and benefits" (Pollack 1986: D1), the effort made the front page of the "Business Day" section of The New York Times. While investors may have approved, the workers evidently did not; they were reported in that article to be "resisting the cutbacks," with the entire episode being characterized as a "key showdown" between workers and management (Pollack 1986: D1). The efforts to extract environmental concessions, by contrast, were often enthusiastically supported by the same workers. Rather than marching to protest the greed or callousness of their corporate employers, the workers often marched to protest the fact that the same corporations were not being allowed to cut down remaining old-growth forests—all while burning environmentalists in effigy.

The tendency to blame "environmentalists" for the loss of logging jobs, in other words, may not be simply a matter of faulty memory. As Freudenburg (1997) has noted, there may be a particular need for sociologists to question taken-for-granted assumptions—and perhaps especially those assumptions that also happen to provide convenient justifications for the prevailing distributions of political and economic power. If this suggestion is applied to the case of natural resources, in particular, it can lead to the recognition of one clear possibility involved in the concession–extraction process: Sooner or later, one of the parties being asked for a concession may be unwilling to provide it. At that point, at least if extractive interests and their allies are sufficiently skillful, they may be able to depict the party refusing to provide the concession—rather than the nonsustainable levels of prior extraction—as being responsible for the loss of extractive jobs (see Kazis and Grossman 1982). The recent experience in the Pacific

Northwest provides a striking degree of correspondence to the concession-extraction model, with the refusal to grant concessions being associated with particularly vigorous efforts to affix both attention and blame. As a result, it may not be simply a matter of happenstance, but also a sign of success in the relevant groups' efforts to construct an interpretation of the situation, that environmental groups and the federal government—rather than the logging industry itself—have ultimately wound up providing the targets for so much of the blame for the loss of logging-related jobs.

Whether or not such a shifting of responsibility was in fact a conscious intention of resource-extracting firms or their allies, of course, is a question of motivation for which the present study's empirical data on employment can provide little direct evidence. All that can be said is that the pattern of academic as well as popular discourse is quite consistent with such a possibility—while the actual employment trends are anything but consistent with the common convictions on this issue.

Despite the strength of the belief that the "endangered logger" of the Pacific Northwest has been suffering because of the habitat needs of small owls, rather than because of the tree-cutting and cost-cutting practices of large corporations, this common belief is remarkably devoid of empirical support. There is simply no quantitative evidence of any statistically credible increase in job losses associated with the federal listing of the northern spotted owl as a "threatened" species. If there could be said to be any evidence for an effect from environmental regulations in general, it would be that the era of environmental protection, dating back to the passage of the Wilderness Act in 1964, has been associated with a significantly improved outlook for logging and milling jobs, both in the Pacific Northwest and in the nation as a whole. In short, if indeed "those damned owls," as more than one logger has called them, are to be blamed for the decline in Pacific Northwest timber employment, then what will be required is a plausible argument as to how those birds could have started costing loggers their jobs more than 40 years before the protection of the owl became a focus of federal policy.

Notes

Freudenburg, William R., Lisa J. Wilson, and Daniel J. O'Leary. 1998. "Forty Years of Spotted Owls? A Longitudinal Analysis of Logging Industry Job Losses." *Sociological Perspectives* 41(1).

1. The most common metric of volume in the U.S. lumber industry involves the board foot, which measures 12 [inches] × 12 [inches] by 1 [inch], but the traditional practice has been to measure exports in terms of cubic feet (each of which of course measures 12 [inches] × 12 [inches] × 12 [inches]), or more recently, cubic meters. For the convenience of those who are accustomed to the traditional measures, we are using the same measures in the analyses being reported in this paper; in the interest of consistency, all metric figures have been converted to cubic-foot equivalents, as well.

References

Anderson, H. M., and J. T. Olson. 1991. *Federal Forests and the Economic Base of the Pacific Northwest*. Washington, D.C.: The Wilderness Society.

Batori, Stephen M. 1957. "Automation in Lumber Milling." *Forest Products Journal* 7:31A–32A.

Berger, Peter L., and Thomas Luckmann. 1967. *The Social Construction of Reality*. New York: Doubleday.

Beuter, John H., K. N. Johnson, and H. L. Scheurman. 1976. *Timber for Oregon's Tomorrow*. Corvallis: Oregon State Univ. School of Forestry, Res. Bull. 19.

Bland, J., and T. Blackman. 1990. "Forest Industry's Future: More Battling for Timber." *Forest Industries* 177:39–41.

Brown, Beverly A. 1995. *In Timber Country: Working People's Stories of Environmental Conflict and Urban Flight*. Philadelphia: Temple University Press.

Brunelle, A. 1990. "The Changing Structure of the Forest Industry in the Pacific Northwest." Pp. 107–24 in *Community and Forestry: Continuities in the Sociology of Natural Resources*, edited by Robert G. Lee, Donald R. Field, and William R. Burch, Jr. Boulder: Westview.

Carroll, Matthew S. 1995. *Community and the Northwestern Logger: Continuity and Change in the Era of the Spotted Owl*. Boulder: Westview.

Catton, William R., Jr. 1982. *Overshoot: The Ecological Basis of Revolutionary Change*. Urbana: University of Illinois Press.

Catton, William R., Jr., and Riley E. Dunlap. 1980. "A New Ecological Paradigm for Post Exuberant Sociology." *American Behavioral Scientist* 24:15–47.

Choldin, Harvey M. 1978. "Social Life and the Physical Environment." Pp. 352–84 in *Handbook of Contemporary Urban Life*, edited by D. Street. Chicago: University of Chicago Press.

Compton, K. C. 1956. "Increasing Sawmill Efficiency." *Forest Products Journal* 6:19–27.

Connor, Mary Roddis. 1978. "Logging in Northeastern Wisconsin." Pp. 31–38 in *Some Historic Events in Wisconsin's Logging Industry: Proceedings of the Third Annual Meeting of the Forest History Association*. Madison, WI: n.p., Sept. 9, 1978.

Conway, F. D. L. and G. E. Wells. 1993. *Timber in Oregon: History and Projected Trends*. Corvallis: Oregon State University Extension Service.

Dunlap, Riley E. 1990. "Trends in Public Opinion Toward Environmental Issues: 1965–1990." Pp. 89–116 in *American Environmentalism: The U.S. Environmental Movement, 1970–1990*, edited by Riley E. Dunlap and Angela G. Mertig. Philadelphia: Taylor and Francis.

Dunlap, Riley E., and William R. Catton, Jr. 1979. "Environmental Sociology." *Annual Review of Sociology* 5:243–73.

Egan, Timothy. 1994. "Oregon, Foiling Forecasters, Thrives as It Protects Owls." *New York Times* (Oct. 11): A1, C20.

Fitzgerald, Randall. 1992. "The Great Spotted Owl War." *Reader's Digest* 141 (Nov.):91–95.

Flynn, B. 1991. "Log Exports in Decline, Can't Offset Land Withdrawals." *Forest Industries* 118:13–15.

Force, J. E., Gary E. Machlis, L. Zhang, and A. Kearney. 1993. "The Relationship between Timber Production, Local Historical Events, and Community Social Change: A Quantitative Case Study." *Forest Science* 39:722–42.

Forest Ecosystem Management Assessment Team (FEMAT). 1993. *Forest Ecosystem Management: An Ecological, Economic, and Social Assessment.* Washington, D.C.: U.S. Government Printing Office.

Forest Industries. 1991. "Industry 'On Hold' as Haggling Over Owl Continues." 118:5–6.

Foster, John Bellamy. 1993. *The Limits of Environmentalism without Class.* New York: Monthly Review.

Freudenburg, William R. 1991. "A 'Good Business Climate' as Bad Economic News?" *Society and Natural Resources* 3:313–31.

Freudenburg, William R. 1992. "Addictive Economies: Extractive Industries and Vulnerable Localities in a Changing World Economy." *Rural Sociology* 57:305–332.

Freudenburg, William R. 1997. "The Crude and the Refined: Sociology, Obscurity, Language, and Oil." *Sociological Spectrum* 17:1–28.

Freudenburg, William R., Cynthia-Lou Coleman, James Gonzales, and Catherine Helgeland. 1996. "Media Coverage of Hazard Events: Analyzing the Assumptions." *Risk Analysis* 16:31–42.

Freudenburg, William R., Scott Frickel, and Robert Gramling. 1995. "Beyond the Society Nature Divide: Learning to Think about a Mountain." *Sociological Forum* 10:361–92.

Freudenburg, William R., and Robert Gramling. 1994a. *Oil in Troubled Waters: Perceptions, Politics, and the Battle over Offshore Oil.* Albany: State University of New York Press.

Freudenburg, William R., and Robert Gramling. 1994b. "Natural Resources and Rural Poverty: A Closer Look." *Society and Natural Resources* 7:5–22.

Glick, Daniel. 1995. "Saving Owls and Jobs Too: In the State Where Protection of the Northern Spotted Owl was Supposed to Destroy Jobs, a Booming Economy Debunks the 'Owls vs. Jobs' Premise." *National Wildlife* (Aug./Sept.):8–13.

Gramling, Robert, and William R. Freudenburg. 1996. "Environmental Sociology: Toward a Paradigm for the 21st Century." *Sociological Spectrum* 16 (#4, Oct.):347–70.

Greber, Brian J. 1993. "Impacts of Technological Change on Employment in the Timber Industries of the Pacific Northwest." *Western Journal of American Forestry* 8:34–37.

Greber, Brian J., K. Norman Johnson, and Gary Lettman. 1990. "Conservation Plans for the Northern Spotted Owl and Other Forest Management Proposals in Oregon: The Economics of Changing Timber Availability." Papers in Forest Policy, Vol. 1. Corvallis: Oregon State University, Forest Research Laboratory.

Gup, Ted. 1990. "Owl vs Man." *Time* 135 (June 25):56–62.

Hamilton, Lawrence C., and Carole L. Seyfrit. 1994. "Resources and Hopes in Newfoundland." *Society and Natural Resources* 7:561–78.

Heberlein, Thomas A. 1994. "This Train Goes to Los Angeles." Paper presented at Forestry and the Environment: Economic Perspectives II. Banff National Park, Canada, Oct. 14.

Hirt, Paul W. 1994. *A Conspiracy of Optimism: Management of the National Forests since World War Two.* Lincoln: University of Nebraska Press.

Humphrey, Craig R., Gigi Berardi, Matthew S. Carroll, Sally Fairfax, Louise Fortmann, Charles Geisler, Thomas G. Johnson, Jonathan Kusel, Robert G. Lee, Seth Macinko, Nancy L. Peluso, Michael D. Schulman, and Patrick C. West. 1993. "Theories in the Study of Natural Resource-Dependent Communities and Persistent Rural Poverty in the United States." Pp. 136–72 in *Rural Sociological Society Task Force on Persistent Rural Poverty: Persistent Poverty in Rural America.* Boulder, CO: Westview.

Kazis, R., and R. L. Grossman. 1982. *Fear at Work: Job Blackmail, Labor, and the Environment.* New York: Pilgrim Press.

Kendrick, John W. 1961. *Productivity Trends in the United States.* Princeton: Princeton University Press.

Krannich, Richard S., and A. E. Luloff. 1991. "Problems of Resource Dependency in U.S. Rural Communities." *Progress in Rural Policy and Planning* 1:5–18.

Lee, Robert G. 1993. "Effect of Federal Timber Sales Reduction on Workers, Families, Communities, and Social Services." Seattle: University of Washington (unpublished paper).

Lee, Robert G. 1994. *Broken Trust, Broken Land: Freeing Ourselves from the War Over the Environment.* Wilsonville, OR: Book Partners.

Levine, Jon B. 1989. "The Spotted Owl Could Wipe Us Out." *Business Week* (Sept. 18):94.

Liebier, Carol M., and Jacob Bendix. 1994. "Old-Growth Forests on the Evening News: News Sources and the Framing of Environmental Controversy." Paper Presented at the Media and Environmental Conference, Reno, Nevada, April.

Love, Ruth L. 1997. "The Sound of Crashing Timber: Moving to an Ecological Sociology." *Society and Natural Resources* 10:211–22.

Machlis, Gary E., Jo Ellen Force, and Randy Guy Balice. 1990. "Timber, Minerals, and Social Change: An Exploratory Test of Two Resource-Dependent Communities." *Rural Sociology* 55:411–24.

McClosky, Mike. 1992. "Twenty Years of Change in the Environmental Movement: An Insider's View." Pp. 77–88 in *American Environmentalism: The U.S. Environmental Movement, 1970–1990,* edited by Angela G. Mertig and Riley E. Dunlap. Philadelphia: Taylor and Francis.

McConnen, Richard J. 1967. "The Use and Development of American's Forest Resources." *Economic Botany* 21:2–14.

Mitchell, J. G. 1990. "War in the Woods: West Side Story." *Audubon* 92:82–99.

Mitchell, J. G. and D. Lamont. 1991. "Sour Times in Sweet Home: Frustration, Despair, Anger, and Political Manipulation Follow Layoffs in a Troubled Oregon Timber Town." *Audubon* 93:86–87.

Mitchell, Robert C., Angela G. Mertig, and Riley E. Dunlap. 1992. "Twenty Years of Environmental Mobilization: Trends among National Environmental Organizations." Pp. 11–26 in *American Environmentalism: The U.S. Environmental Movement, 1970–1990,* edited by Riley E. Dunlap and Angela G. Mertig. Philadelphia: Taylor and Francis.

Nash, Roderick. 1982. *Wilderness and the American Mind.* New Haven: Yale University Press.

O'Connor, James. 1988. "Capitalism, Nature, Socialism: A Theoretical Introduction." *Capitalism, Nature, Socialism* 1:8–38.

Peluso, Nancy L., Craig R. Humphrey, and Louise P. Fortmann. 1994. "The Rock, the Beach, and the Tidal Pool: People and Poverty in Natural Resource-Dependent Areas." *Society and Natural Resources* 7:23–38.

Rice, James Owen. 1992. "Where Many an Owl Is Spotted." *National Review* 44:41–43.

Runte, Alfred. 1987. *National Parks: The American Experience.* Lincoln: University of Nebraska Press.

Satchell, Michael. 1990. "The Endangered Logger." *U.S. News and World Report* 108 (June 25):27–29.

Schnaiberg, Allan. 1980. *The Environment: From Surplus to Scarcity*. New York: Oxford University Press.

Schnaiberg, Allan, and Kenneth A. Gould. 1994. *Environment and Society: The Enduring Conflict*. New York: St. Martin's.

Schurman, Rachel. 1993. "Economic Development and Class Formation in an Extractive Economy: The Fragile Nature of the Chilean Fishing Industry, 1973–1990." Madison: Dept. Sociology, Univ. of Wisconsin (unpublished Ph.D. dissertation).

Simmons, Fred C. 1947. "Mechanizing Forest Operations." *Journal of Forestry* 45:345–49.

Thernstrom, Stephan. 1965. "'Yankee City' Revisited: The Perils of Historical Naiveté." *American Sociological Review* 30:234–42.

Tisdale, Sallie. 1992. "Marks in the Game." *Sierra* 77:58–65.

United States Department of Labor. 1994. *Employment, Hours and Earnings—U.S., 1909–1994*. Washington, D.C.: U.S. Bureau of Labor Statistics.

Van Syckle, Edwin. 1980. *They Tried to Cut It All: Grays Harbor-Turbulent Years of Greed and Greatness*. Seattle: Pacific Search Press.

Wambach, Robert F. 1969. "Compatibility of Mechanization with Silviculture." *Journal of Forestry* 67:104–8.

Wilderness Society. 1996. "A Step Through Time." *Wilderness America* (Dec.):6.

Wood, Nancy. 1971. *Clearcut: The Deforestation of America*. San Francisco: Sierra Club.

Yoho, James G. 1965. "The Responsibility of Forestry in Depressed Areas." *Journal of Forestry* 63:508–12.

Young, J., and J. Newton. 1979. *Capitalism and Human Obsolescence: Corporate Control vs. Individual Survival in Rural America*. Montclair, NJ: Landmark Studies.

The Next Revolutionary Stage

Recycling Waste or Recycling History?

David N. Pellow

Some environmental sociologists have questioned whether recycling is unambiguously positive. This reading, excerpted from David Pellow's book Garbage Wars, *reveals that sometimes recycling advocates and other environmentalists have been guilty of caring more about narrowly defined "environmental" issues than about the daily lives of working people, in this case those who work at recycling centers. The materials recycling facilities (MRFs) described by Pellow are among the most dangerous and dirty kinds of facilities, but recycling is rarely a desirable job. This research begs the question "What about the 'environments' of recycling workers?"*

In Chicago] . . . non-profit recyclers have largely faded, and privatized, "dirty MRF" recycling has become dominant. The privatized model is characterized by the paradox of more public investment with less public control, by a stronger market orientation for recycling, rather than an ecological value base, and by control and degradation of labor.

. . . [Chicago adopted] the "Blue Bag" approach to recycling. Many curbside recycling programs are characterized by source separation, which entails putting recyclables into bins to be picked up by recycling (not garbage) trucks. This program is different. Through the Blue Bag program, residents place their recyclables in blue plastic bags, which are then collected along with household garbage in a single garbage truck. The trucks then dump their loads at materials recovery facilities operated by WMX [a corporate waste management company], where the bags are pulled out of the garbage and their contents separated. Recyclable materials not in bags are also pulled out of the garbage for processing.

In addition to meeting the requirements of the city's recycling ordinance (i.e., that 25 percent of the waste stream be recycled by 1996), the city promised that the new recycling system would create between 200 and 400 jobs. Like many industrial cities of the Northeast and the Midwest, Chicago had experienced continuous recessions and bouts with de-industrialization. Between 1967 and 1987 the city lost 326,000, or 60 percent, of its manufacturing jobs.[1] More and more, recycling was shaping up to be the next "win-win" urban policy for Chicago. It would solve the landfill problem, please the environmental community, and provide jobs in some of the city's depressed areas. The Blue Bag system seemed especially advantageous insofar as an infrastructure was in place, with the city already providing waste pickup service using a fleet of trucks and several transfer stations and landfills (owned by WMX). The Blue Bag program would fit right into this structure with no major changes. A cost-benefit analysis of a curbside recycling program versus the mixed-waste Blue Bag system revealed that the latter would cost millions of dollars less. Thus, ignoring ecological and social criteria, the *Solid Waste Management Newsletter* reported that "the primary reason given for adopting the commingled bag/MRF recycling program is its affordability."[2]

. . . With average annual revenue in the neighborhood of $11 billion, WMX has operations in Australia, Canada, Europe, and South Africa. Probably the corporation most vilified by the anti-toxics and environmental justice movements in recent years (owing to its ownership of scores of landfills, incinerators, and hazardous waste facilities), WMX is widely alleged to be an "environmental terrorist" and a perpetrator of flagrant corporate crimes.[3] What is curious is that several local community organizations that had been fighting WMX for years decided to support the Blue Bag program. For example, People for Community Recovery [PCR] had been fighting WMX since 1982. Specifically, PCR was up in arms about the location of WMX's landfill and chemical waste incinerator in the neighborhood and even held several public protests against the company. For years, PCR's diagnosis of the problem was "too much pollution and not enough good jobs" a mantra that was taken up by the rest of the environmental justice movement. When WMX announced that it was finally going to address the "jobs versus environment" issue by building a recycling plant that would hire local residents, PCR, the Mexican Community Committee, and several other organizations lent support. WMX made an arrangement with these organizations to recruit, interview, and pre-screen potential employees who later went to work for the company. What only WMX could know at that time was that the jobs at these recycling centers were terribly unsafe and unhealthy. In this way, many environmental justice movement organizations were unwittingly complicit in the imposition of environmental injustices on workers from their own communities. This is but a fraction of the complexity that is missing from much of the research on environmental racism. That is, the systems of institutional racism and classism produce arrangements that place certain people of color in the role of "the oppressed as oppressor," or stakeholders who receive small gains while more powerful actors benefit handsomely from both groups' oppression.[4]

... The City of Chicago provided taxpayer dollars and political will to WMX; neighborhood groups worked with WMX to recruit workers; residents participated in the Blue Bag program by providing their trash and recyclables to the waste haulers; and, while environmentalists were critical of WMX's recycling methods and its political power, they had earlier lent support to the campaign for a citywide recycling program, and they ignored the problems of occupational safety and environmental inequality inherent in the Blue Bag system. They objected mainly to the lack of non-profit involvement and the anticipated low quality of recovered recyclables. The stake-holders generally left out of the debate over the Blue Bag were the workers. This was in large part because the city and WMX had already determined that the jobs would be non-union.[5] And since residents were desperate for any type of work, and environmentalists were paying no attention to job quality or to wages, this exclusion went uncontested.

Green Business or Global Sweatshop? Management's Decisions and Environmental Hazards

... WMX created an unbearably hazardous work environment where people of color were concentrated.

The structure of the Blue Bag system and the MRFs in which they were processed was created by managers with a range of options available to them. These decisions had direct impacts on the nature of environmental inequality at the MRFs.

The Blue Bag program went into effect on December 4, 1995, serving single-family homes and low-density buildings in Chicago. Some 750,000 residents were expected to separate their paper and commingled products (plastic, metals, etc.) into blue bags that were to be mixed in with the regular garbage pickup and taken to four colossal MRFs in the city, where they would be processed for end markets. These "dirty MRFs"—MRFs processing both recyclables and garbage—were each the size of several football fields and were staffed by 100 African American and Chicano workers picking, sorting, lifting, and cleaning tons of trash and recyclables. WMX workers faced environmental inequalities in the plants based largely on managerial decision making. Two principal decisions contributed to environmental inequality at this MRF: to accept mixed waste and to maintain an oppressive work culture.

Decision 1: Accepting Certain Types of Waste

... [T]he Blue Bag system was touted as the most sophisticated and advanced waste management technology ever assembled under one roof. WMX's MRFs revealed that these buildings were packed with computer stations, assembly lines, magnetic sorters, air classifiers, balers, cameras, and other forms of machinery and technology. Although designed for efficient recycling, they seemed to neglect two important things: high-quality recycling (mixed-waste systems like the Blue Bag combine trash and recyclables, thus contaminating the latter) and a high-quality work environment. These two issues would haunt this recycling program from day one. The decision to use a capital-intensive production process to make recycling more efficient failed. This decision also created an occupational environment where workers were expected to adapt to machines and regularly increase production.

There are hundreds of MRFs across the United States that receive only source-separated materials and do not accept municipal solid waste (MSW). WMX departed from this model in an apparent effort to innovate and reap greater value from the entire waste stream. This managerial decision to build a "dirty MRF" had a major impact on the health and safety of workers at the MRF. MSW is material that a Danish inspection service ruled "presents a very high health hazard, and must not be sorted by hand."[6]

In a December 1995 episode of the television series *ER*, one emergency room patient was a man who had fallen from a 15-foot sorting line and impaled himself on a piece of machinery in a recycling center. The WMX recycling workers I interviewed faced hazards no less serious. And while these hazards are both physical and psychological in their impact, they originate in social structures.

WMX grossly miscalculated the environmental and safety issues, which is ironic since Blue Bag recycling was touted as an environmentally responsible initiative. I spoke to more than two dozen workers and managers who worked under the Blue Bag system. The stories they told resemble the experiences of laborers in the sweat shops, the mines, the steel and textile mills, and the slaughterhouses of the nineteenth-century United States and the contemporary Global South.

Recycling workers face a number of health and safety hazards. To the horror of most workers, the Blue Bag system was characterized by the routine manual handling of chemical toxins, hazardous waste, and infectious medical wastes, all of which are found in household garbage (whose contents are not regulated). Bleach, battery acid, paint, paint thinner, inks, dyes, razor blades, and homemade explosives are in the waste stream. Medical waste has recently emerged as a significant problem because, as hospital patients are more frequently sent home under managed care, so is their waste. Garbage and recycling workers are regularly exposed to these substances. One worker told me: "There are tons of medical wastes and construction wastes. Say, for instance, the red bags that have biohazards would drop down the chute. One time a bag went through marked 'asbestos' and I said 'Damn, that looks like asbestos, a cloud of asbestos dust just hanging there!'"[7]

Finger and arm pricks by syringes and hypodermic needles and battery acid sprays are becoming quite common in MRFs around the world.[8] Workers have died as a result of battery acid exposure. Needle pricks are particularly worrisome, as many employees fear exposure to HIV.

The newspapers covered the Blue Bag controversy from day one, but they generally focused on recycling rates and on whether recyclables were being contaminated when mixed with garbage. A few stories about working conditions appeared, but this issue never really concerned environmentalists. When city or WMX officials did address the question of labor, they pointed out that they had "created" 400 jobs in Chicago that were, according to Commissioner of the Department of Environment Henry Henderson, "clean and safe."[9] WMX management insisted that the working conditions were "excellent." A WMX MRF site-manager named Mitchell told me:

> The enclosures, we have nine of them, where the sorting will be done, are in a climate-controlled area, [which] will be heated and air conditioned. The system will produce six air changes an hour within the enclosure, for the sorters. . . . We've gone pretty far to make that environment safe.[10]

Another manager, Jake, added:

> The way the sorting is done, there are very different ways, some of them are "throw-across" into a backboard situation. And other sorts are pulling and dropping into a thing, where there's no lifting, so you don't have the back injuries. And what experts call "depth burden" (i.e. the length one has to reach) is greatly reduced. All of the exchanges of materials are outside of the enclosures, so any dust that's generated will stay out of the enclosure so the environment they're in is good.[11]

. . . On July 11, 1996, eight months after the Blue Bag program began, the U.S. Occupational Safety and Health Administration slapped WMX with a $10,500 fine for several labor violations. Although the penalty was not heavy, the fact that OSHA (by most accounts an immobile bureaucracy) actually came out and inspected the sites is notable. . . .These violations correspond strongly with the ethnographic data I gathered, but they fall short of capturing the full reality of the labor process WMX employees confronted every workday, including the oppressive work culture maintained at the plants. A year later, OSHA representatives made a follow-up visit to the WMX MRFs and found that the company had engaged in "willful violation" of previous orders to follow labor law and inoculate its employees against Hepatitis B. The penalty this time was $112,500. The charge regarding inoculations was verified by a WMX employee who told me that he and other workers had had "no physical examination, no inoculation":

> That's what made me kind of skeptical . . . because you were working around raw garbage and I was talking with a couple of guys that were *managers* for WMX, and before they were hired, they had to take physicals and they were inoculated. But the people from REM were given no physicals or inoculations.[12]

WMX hired its laborers from a temporary labor service called Remedial Environmental Manpower (REM). WMX therefore deliberately withheld legally mandated vaccinations from the REM "temps," all of whom were men and women of color. This withholding of medical care placed employees at great risk. WMX workers confronted other dangers too, many of which were seemingly unpreventable once the MRF was up and running. One employee told me: "WMX was built on top of a landfill. And for a while before they had the fire suppression systems working, we had to have people walking around with fire extinguishers because there was methane gas seeping through the floor. . . ."[13]

WMX's management made the decision to build this MRF on the methane-producing landfill. This decision contributed directly to a hazardous work environment.

In December of the Blue Bag's first year, a deceased infant was discovered on the recycling line. This was the fourth human body discovered at the city's recycling centers in a year.[14]

The workers I interviewed made it clear that dangerous incidents were routine. In two weeks' time, Ferris (a line sorter in the "primary" department, where the trucks dump the garbage) had witnessed "about six" accidents. He elaborated:

> I got cut on my finger with some glass, because the glass went through the glove. And then this other girl, she got stuck in the arm with needles, but they sent her to the company doctor. And they put her back to work the next day when her arm was still hurting. They gave her some medicine and she was still hurting, so I feel that wasn't right. Then this one guy fell down a chute and broke his arm. And then this one guy got burnt with some battery acid.[15]

These psychological and physical hazards intermingled as people desperate for gainful employment and job security continued working in the face of gross health and safety violations. In a city where the unemployment rate in many African American neighborhoods exceeds 50 percent, it is not difficult to understand why, as one worker explained, "you never turn down work when you're looking for it."[16] However, that worker also reasoned, "you also have to think of your safety because that *job* might be there next year, but if you contracted some disease, *you* might not be there next year."[17]

The sociological implications of hazardous jobs have always had effects beyond the workplace.[18] Often a worker exposed to hazards has unknowingly carried dangerous chemicals home on his or her clothing or body. Such transfers of toxins sometimes have direct impacts on the health of family members or friends. Other evidence of dangerous work is also often found in the home, in the form of the social and psychological wounds that risky jobs inflict upon employees and their families. WMX workers were no exception. Workers' families were stuck in a catch where either the lack of work or the presence of socially and physically oppressive work was a routine source of stress. . . . [The following] quote is typical of families facing the oppressive nature of the work. Jason's mother answered the phone one night when I called to schedule an interview with him. He was not home, but his mother had this to say about WMX: "They were just evil, and they treated him like the devil would treat somebody."[19]

Thus, management's decisions routinely produced an unsafe working environment for WMX employees and impacted their families. The way management "would treat somebody" is a part of what I would call "the work culture."

Decision 2: Maintaining an Oppressive Work Culture

Most secondary labor markets in which African Americans and Latinos are concentrated are characterized by low wages, a lack of mobility, a lack of health and safety guarantees, and a lack of hope. (Secondary labor markets are characterized by lower wages and lesser opportunities for mobility than are found in primary labor markets.) Thus we should not be surprised to find similar patterns at WMX. Most work in secondary labor markets is also characterized by the fact that workers are not encouraged or allowed to think creatively and independently on the job. Workers employed in hazardous occupations find that this problem is compounded by the job's dangers. "Workers in hazardous occupations are 50–90 percent more likely than workers in safe occupations to describe their jobs as uncreative, monotonous, meaningless, and as providing no worker control over work pace. They are 10–20 percent more likely to report no control over job duties, no control over work hours, and the pervasiveness of rules."[20]

The oppressive social environment at WMX contributes to the problem of dangerous and uncreative work. Richard Edwards's classic book *Contested Terrain* outlines three types of managerial control over labor: simple, technical, and bureaucratic.[21] WMX managers made use of all three. Simple control generally includes direct supervision of labor. WMX managers made the decision to use a disciplinarian management style. For example, workers regularly complained of being harassed by foremen and managers who rarely let them leave the sorting lines to use the bathrooms and arbitrarily instituted mandatory overtime. One whistle-blowing former manager put it as follows:

> John and Norm's [the general managers] philosophy was to "keep your foot in their ass." That was their verbal philosophy as communicated to us. That is bound to fail, nothing new about that. . . . Dan Karlson was an REM supervisor and he walked around the plant with a pistol in his holster. His philosophy was "whenever you get a disgruntled worker you have to slap them and shut them up."[22]

As evidence of this philosophy in action, one worker who contacted OSHA to file a complaint in early 1996 was fired within a week of doing so. After several workers spoke to journalists about the deplorable health and safety conditions in the plant, REM issued a memo to its employees "strictly prohibiting" any communication with the news media. Workers were explicitly instructed to respond with "no comment" to any inquiries about working conditions in the MRFs and were warned that "violation of this work rule may result in disciplinary action up to and including immediate termination of employment."[23]

These oppressive conditions had direct impacts on the state of occupational health at the WMX MRF. Collins, a worker at one facility, told of the following experience between the managers and workers who were determined to keep their jobs:

> They [management] was always threatening people. . . . They said that if you didn't work Saturdays then you might as well not come back Monday. . . . A couple of guys there had twisted ankles and everything. I looked for medical kits but didn't notice any of that. These guys with hurt ankles were frightened for their jobs because the way that these people were pushing them was like "Hey, I'm gon' tell you. We got more people standing in line waiting on this job." He wasn't lying because I noticed that they started bringing in Mexican Americans late at night who didn't speak English. They would bus them in late at night and then take them out in the morning.[24]

A female worker in the same plant told a similar story:

> One lady I know cut her foot. This was in where they pick the trash off the conveyor belts—it's upstairs. She had cut her foot and she was bleeding. And I was thinking you know, if you hadn't had a tetanus shot and you don't know what you stepped on, I thought the company should have took her right away to get her a tetanus shot. They didn't. They gave her a piece of toilet tissue, put it back in her shoes and sent her right back to work. When she got injured, that was part of her 15 minute break. She used her 15 minute break to take care of it.[25]

Workers and management battled over the control and the speed of assembly lines throughout the twentieth century.[26] In Richard Edwards's typology of forms of managerial control, the assembly line is the classic example of technical control. Like many workers on assembly lines, however, some recycling workers had the power to stop the process for emergencies and for certain quality-control purposes, although doing so was strongly discouraged. However, the *speed* of the line was inflexible. One female worker noted, "There was no way you could control the speed [of the line]. You could control it if you wanted to stop it, but not control how fast or how slow it went."[27]

Like major managerial decisions, the assembly line as technical control influenced occupational safety and health in the WMX MRF. Another worker recounted the following experience: "With the [conveyor] belt constantly running, you're reaching for one thing and, by the time you turn your head back, there's something in your face. . . . You couldn't see what was actually coming and that made it real dangerous for you."[28] For example, one female employee almost "got her neck slit on the line" when landscape clippings and branches poked her as she worked.[29]

The oppressive conditions also impacted workers' social psychological stability with regard to their status as "temps." Temporary status was difficult for many workers to contend with. Being a "temp" is alienating and marginalizing. Temporary status and the informal managerial practices that often accompany it were used as a form of bureaucratic control. That is, the presence or absence of company rules, regulations, and procedures can place many barriers to a worker's compensation, safety, status, power, and job advancement. Bureaucratic control: WMX included a structure of rules and procedures that guaranteed that African Americans and Latinos would remain temporary employees in the lowest-ranking, most dangerous jobs in the plant. . . .

As I mentioned above, workers at WMX were hired through a temporary job service called Remedial Environmental Manpower. In recent years, powerful lobbyists for temporary employment firms have fundamentally changed federal labor legislation. For example, workers at the WMX MRFs were technically not employees of WMX or of REM. Rather, as a result of recent changes in labor laws, they were "consumers" of REM's services.[30] They therefore had questionable legal rights as workers and an ambiguous legal relationship to WMX. This arrangement encouraged and rewarded abuse and exploitation by management.

All the workers at WMX were "temps," and therefore they could be fired arbitrarily. Two workers recounted the following:

When they did [suddenly] lay us off, they was telling us all that we did a great job . . . and that we'd be first for consideration when they started hiring back again. But I haven't heard anything else from 'em.[31]

They got our hopes up so high. They didn't tell us that we was experimental guinea pigs. They just had us psyched all up like we was gon' be there for a while and they just dropped us like that.[32]

... [M]any times workers worried about when they would be allowed to finish a shift and go home. Angela stated:

> I started work at 6 that morning and I got out by 7 that night. And that was during a time when some guys had been working from 6 in the morning until 2 the next morning straight. They had been threatened with immediate dismissal otherwise. The managers said that anybody that didn't want to work until they told them to go home should leave right then. Because they said they might work you 15–20 hours that day. And then it was more of a demand that you stay. It wasn't like eight hours and anybody that wanted to work overtime after eight could stay. It was like "You stay till we tell you to leave."[33]

Surprisingly, many laid-off workers expressed a desire to return. For example, after being laid off, rehired, then laid off, again, DJ told me: "I really liked that job and I wouldn't mind going back."[34] Leila, a temporary worker "on call" for WMX, told me she would like to become a permanent WMX worker because any job was better than no job: "It was just a job to get money, you know? Well I'm out of work now. Well, not really out of work because I'm working there, but I would like to do that, yes, for as long as possible. Being a temp is something that you never know what's going to happen."[35]

The decisions by management at the WMX MRF imposed a range of environmental injustices on workers of color. Despite the conditions workers face, WMX has insulated itself somewhat. Because the city was the client and because they recruited workers through a temporary agency and through local community organizations, they have been able to deflect some of the blame for the Blue Bag's shortcomings. WMX's public relations efforts have also helped. In early 1995, before the Blue Bag controversy erupted, WMX issued a "Good Neighbor Policy" indicating commitment to "environmental protection and compliance," "civic and charitable programs," "host community consultation," "communication and disclosure," and "local hiring and purchasing."[36] Relative to these promises, the resources workers sought were basic and direct. Most of them simply wanted better pay, job security, and more attention to occupational safety. Many of them drew on a range of strategies to express their opposition to the system.

Shaping Environmental Inequalities from Below: Worker Resistance to Occupational Hazards and Recycling Ideology

... Communities and workers are not always passive, and in fact they have measurable and powerful effects on government and industrial policies.

WMX workers responded to social subjugation and occupational hazards through various strategies. Resistance included responses to the ideology of recycling in the context of the dirty work it requires. It also included responses to management decisions that contributed to environmental inequalities.

Some workers gave voice to their concerns. One worker, in a letter to the press, pleaded for journalists to exercise "that constitutional freedom I do not [have]." He noted that he and his co-workers experienced "constant colds, flu, diarrhea and coughing"[37] as a result of working at WMX. He continued:

> Several people have been struck by discarded hypodermic needles. Air quality is bad. Others, including myself, have been injured by battery acid, muscle strains, lower back pain, pinched nerves, contusion, various types of trauma including emotional or psychological from witnessing dead bodies, parts of animal carcasses, live and dead rats, etc. And let's not forget the supervisors' bogus tactics. . . . [We have been] threatened to be fired by voicing your opinion. . . . Being talked to loud in front of other people. Not being able to take a day off even if you are sick or a family member dies.[38]

This worker was later fired for this action, after which he pressed for an OSHA investigation of unfair labor practices at his former workplace. Another worker, the aforementioned Collins, was very clear about his willingness to speak out and be heard in the media. When I said to him "I will use a false name for you in anything I write up from this interview, unless you would like it otherwise," he responded: "You can write my name anywhere you want to. I want people to know who I am and what I'm saying."[39]

Several workers articulated their circumstances and grievances through their experiences as men and women of color in an unequal society. One female worker referred to her manager as a "prejudiced, chauvinistic, racist."[40] Another female worker declared: ". . . it was mostly all black folks [working in the MRF], about four Mexicans and no white folks. Yes, they had us slaving, we was back in slavery."[41]

Critiquing the racially biased authority structure at one firm, one man said:

> I would love to see more people of color in positions of authority over there because all the bigwigs are white. Black folks are the peons over there. And then they shipped illegal immigrants in there at night under cover of darkness. . . . And they probably hired them so they could lower their wages to about half what we [blacks] is making. And if you ever go down to the facility you'll see what goes in and out of there . . . nothing but garbage, blacks, and Mexicans.[42]

Intraracial conflict was a theme in many interviews because the temporary agency (REM) that hired workers for WMX was owned by African Americans.

REM and WMX had promised that workers would be eligible to join a union after a 90-day probation period. Few if any workers ever made it through 90 consecutive days, however: One female recycler told me that she had confronted the manager of the agency and accused him of firing and rehiring workers to avoid paying them union wages:

> When people are getting their 90 days then you want to lay them off. He said "That ain't got nothing to do with it," but I said "Yes it do, it has a lot to do with it, because you don't want us in the union." And I told [the African American manager]: "You know what, Jack? Instead of you trying to help the next black man, you're a black person up there doing a little something for yourself and you're trying to kick the next black man in the back. You don't want them to get up there, and that ain't right."[43]

. . . Workers were well aware of the contradictions between recycling's public image and the personal trials they were subjected to. One worker said: "There are so many smells that you come across, they make your stomach queasy. Yet before we went to work, they showed us a safety film where all the stuff was really clean."[44]

A Chicana worker who later quit her job at the MRF explained:

> . . . they told us that it was going to be a clean work environment. They said that fresh air was going to be pumped through there every 15 minutes, so it wouldn't smell, and stuff like that, but it wasn't. It was a little different than they had described it. One time they had a dead dog . . . go through there. There was all garbage, you know [not just recyclables]. At first we thought they were only talking about plastic bottles and cans going through there. But that was plain garbage, everything, you know? Dirty diapers, cleaning products, and stuff like that.[45]

Many workers exercised their option to leave. The coercive work environment and the extreme occupational hazards were generally cited as the reasons for this type of response. Another Chicana worker told me: "The [heaters] upstairs would make you so hot sometimes we had to turn them off and that's how some people got sick. This one guy he got pneumonia. They told him he had to either work or go home, and that wasn't right. He left."[46]

Workers' responses to hazards and coercion included "everyday forms of resistance."[47] Some common forms this hidden resistance took were muted character assassinations directed at management and silent refusal to touch things on the line that were perceived as "too nasty" or "dangerous." The latter often angered managers because it had a negative impact on the plant's productivity. A worker named Seela explained: ". . . if [a pile of garbage] was too high, I just let it go. Plain and simple. That's what everybody else was doing. [The feeling was] 'Hey, if you can't see nothing, just back away from it and let it pass.'" I asked: "But would that end up messing up the process?" Seela replied: "Yeah 'cause then the manager would come up on the line and be all really bitchy" seeing that so much went down that wasn't supposed to go down. He would be talkin' all crazy."[48]

In one instance, a worker's response included an extreme form of resistance. A manager explained:

> Did you ever hear about the riot at WMX? You know, it was the week of Christmas and people had been working a lot of overtime. But management didn't keep track of their hours so people were getting checks for like $1.05 and $1.50 for two weeks of work. So the workers rioted. They tore the hell out of the dining room, knocking over tables, breaking out windows and all kinds of stuff.[49]

Management responded accordingly:

> Ken Carlson was the site supervisor for REM, and when things took a turn for the worse, when everybody started to riot at the plant . . . we had armed guards. I don't know if they were policemen or not, but they looked like street thugs. They were sitting around the dining room making sure that workers weren't going to bust any windows out or anything.[50]

Despite these efforts, the workers remained embedded in a larger social and political process that continued to support and ignore these environmental inequalities. In fact, most of WMX's employees were forced to choose between their safety and a job.

Workers at WMX resisted and shaped the nature of environmental inequalities at the MRFs. Through a variety of strategies, they responded to managerial decision making and environmentalist ideology, and this may have empowered them in significant ways. Even so, workers were never able to organize collectively in a formal sense, particularly since union eligibility was routinely denied them. In addition, the environmental injustices at WMX remain so entrenched and severe that, despite the range of resistance strategies workers drew on, most report that it remains a highly undesirable and unsafe job.

Notes

Pellow, David N. 2002. "The Next Evolutionary Stage: Recycling Waste or Recycling History?" In *Garbage Wars: The Struggle for Environmental Justice in Chicago*. Cambridge: MIT Press.

1. Wilson 1996.

2. *Solid Waste Management Newsletter* 1990....

3. WMX's Chem Waste subsidiary was the owner of a hazardous waste landfill whose expansion was challenged by El Pueblo para Agua y Aire Limpia and the California Rural Legal Assistance Foundation in what became a landmark environmental justice case....

4. Stockdill 1996.

5. There are conflicting reports about the union issue at the MRFs because, allegedly, some workers were allowed (or forced) to join a "union" that was clearly company run, while many others were fired before even being considered for union eligibility.

6. Ritter 1996. The RC's decision *not* to accept MSW had a positive impact on that work environment.

7. Interview by author, fall 1996.

8. Horowitz 1994; Powell 1992; Ritter 1996; van Eerd 1996.

9. Chicago Tonight 1996.

10. Interview by author, fall 1996.

11. Ibid.

12. Interview by author, fall 1996....

13. Interview by author, fall 1996....

14. Sullivan 1996....

15. Interview by author, fall 1996.

16. Ibid. The 50% unemployment figure is from Wilson 1996.

17. Interview by author, fall 1996.

18. Nelkin and Brown 1984.

19. Interview by author, summer 1996.

20. Robinson 1991, p. 84.

21. Edwards 1979.

22. Interview by author, summer 1996.

23. Memo to REM Employees 1996.

24. Interview by author, fall 1996.

25. Ibid.

26. Braverman 1974; Burawoy 1978; Edwards 1979; Thompson 1989.

27. Interview by author, summer 1996.

28. Interview by author, fall 1996.

29. Ibid.

30. Gonos 1997.

31. Interview by author, fall 1996.

32. Ibid.

33. Ibid.

34. Ibid.

35. Ibid.

36. WMX 1995.

37. REM Employees 1996.

38. Interview by author, fall 1996.

39. Ibid.

40. Ibid.

41. Ibid.

42. Ibid.
43. Ibid.
44. Ibid.
45. Ibid.
46. Scott 1990.
47. Interview by author, summer 1996.
48. Interview by author, fall 1996.
49. Ibid.
50. Bullard and Wright 1993; Johnson and Oliver 1989.

References

Braverman, Harry. 1974. *Labor and Monopoly Capital*. Monthly Review Press.
Bullard, Robert and Beverly Wright. 1993. "The Effects of Occupational Injury, Illness and Disease on the Health Status of Black Americans." In *Toxic Struggles: The Theory and Practice of Environmental Justice*, ed. R. Hofrichter. New Society.
Burawoy, Michael. 1978. *Manufacturing Consent: Changes in the Labor Process Under Monopoly Capitalism*. University of Chicago Press.
Chicago Tonight. 1996. PBS Television (WTTW).
Edwards, Richard. 1979. *Contested Terrain: The Transformation of the Workplace in the Twentieth Century*. Basic Books.
Gonos, George. 1997. "The Contest over 'Employer' Status in the Postwar United States: The Case of Temporary Help Firms." *Law and Society Review* 31: 81-110.
Johnson, John H., Jr. and Melvin Oliver. 1989. "Blacks, and the Toxics Crisis." *Western Journal of Black Studies* 13: 72-78.
Nelkin, Dorothy and Michael Brown. 1984. *Workers at Risk*. University of Chicago Press.
REM Employees. 1996. "Open Letter to the Press."
Ritter, Jack. 1996. "Recycling Plant Blues." *Chicago Sun Times*, March 10.
Robinson, James. 1991. *Toil and Toxics: Workplace Struggles and Political Strategies for Occupational Health*. University of California.
Scott, James. 1990. *Domination and the Arts of Resistance*. Yale University Press.
Solid Waste Management Newsletter (Office of Technology Transfer, University of Illinois Center for Solid Waste Management and Research). 1990. "Chicago Announces 1991 Recycling Plan." Vol. 4, no. 12.
Stockdill, Brett. 1996. Multiple Oppressions, and Their Influence on Collective Action: The Case of the AIDS Movement. Ph.D. thesis, Northwestern University.
Sullivan, Molly. 1996. "Baby Found Dead in Trash." *Daily Southtown,* December 27.
Thompson, Paul. 1989. *The Nature of Work: An Introduction to Debates on the Labor Process*. Macmillan.
Wilson, William J. 1996. *When Work Disappears*. Random House.
WMX. 1995. "Good Neighbor Policy." Oakbrook, Illinois.

PART

Corporate Responsibility

12

Silent Spill

The Organization of an Industrial Crisis

Thomas D. Beamish

Our ancestors were most likely to live and work in small, tightly knit communities of families and neighbors where many decisions were made informally. Much has changed in the last 100 to 200 years, and now most of us spend the majority of our time in larger impersonal organizations, like corporations, where most of the decisions are made according to a bureaucratic and impersonal set of rules and practices. The corporation, a uniquely modern entity that has its own legal rights and liabilities apart from that of its members, sits at the center of modern life. The two pieces in this section are sociological explorations into questions about the structure and nature of large corporations. One important question is the extent to which and under what conditions well-meaning individuals and groups are able to express concerns within powerful bureaucracies, such as a petrochemical company. Also, under what conditions do large corporations limit the public's control over its processes? Thomas Beamish examines one of the biggest oil spills in U.S. history—the spill at Guadalupe Dunes in California. While some environmental tragedies happen in a flash, like the Exxon Valdez oil spill described in the introduction, much environmental destruction happens slowly over many years. As Beamish shows, the cleanup at Guadalupe Dunes came late; the original problem had been allowed to mushroom into a total disaster over many years of inaction. Beamish, in his investigation of the systemic and institutional underpinnings of this long-term event, argues that the extent of this disaster was exacerbated by both the slowness in which it occurred and the style of decision making that takes place in particular kinds of organizations.

There's a strange phenomenon that biologists refer to as "the boiled frog syndrome." Put a frog in a pot of water and increase the temperature of the water gradually from 20°C to 30°C to 40°C . . . to 90°C and the frog just sits there. But suddenly, at 100°C . . . , something happens: The water boils and the frog dies. . . . Like the simmering frog, we face a future without precedent, and our senses are not attuned to warnings of imminent danger. The threats we face as the crisis builds global warming, acid rain, the ozone hole and increasing ultraviolet radiation, chemical toxins such as pesticides, dioxins, and polychlorinated biphenyls (PCBs) in our food and water—are undetected by the sensory system we have evolved.

—GORDON AND SUZUKI 1990

Underneath the Guadalupe Dunes—a windswept piece of wilderness[1] 170 miles north of Los Angeles and 250 miles south of San Francisco—sits the largest petroleum spill in US history. The spill emerged as a local issue in February 1990. Though not acknowledged, it was not unknown to oil workers at the field where it originated, to regulators that often visited the dunes, or to locals who frequented the beach. Until the mid-1980s, neither the oily sheen that often appeared on the beach, on the ocean, and the nearby Santa Maria River nor the strong petroleum odors that regularly emanated from the Unocal Corporation's oil-field operations raised much concern. Recognition, as in the frog parable, was slow to manifest. The result of leaks and spills that accumulated slowly and chronically over 38 years, the Guadalupe Dunes spill became troubling when local residents, government regulators, and a whistleblower who worked the field no longer viewed the periodic sight and smell of petroleum as normal. . . .

I first heard of the Guadalupe spill on local television news in August 1995. (My home was 65 miles from the spill site.) The scene included a sandy beach, enormous earth-moving machinery, a hard-hatted Unocal official, and a reporter, microphone in hand, asking the official how things were proceeding. The interplay of the news coverage and Unocal's official response caught my attention more than anything else. The representative asserted that Unocal had extracted 500,000 gallons of petroleum from a large excavated pit on the beach just in view of the camera. The newscaster ended the segment by saying (I paraphrase) "It's nice that Unocal is taking responsibility to get things under control." This offhand remark about responsibility set me to thinking about the long-term nature of the spill and about why it had not been stopped sooner, either by Unocal managers or by regulators.

A few months later, a colleague and I drove to the beach. My colleague, a geologist who was familiar with the area, had suggested that we visit the Guadalupe Dunes for their scenic beauty. We walked the beach and the dunes that border the oil field, alert for signs of the massive spill. The pit that Unocal had recently excavated had been filled in. The only hint of the project that remained was a small crew that was driving pilings into the sand to support a steel wall intended to stop hydrocarbon drift (movement of oil on top of groundwater) and the advancing Santa Maria River, which threatened to cut into an underground petroleum plume and send millions more gallons into the ocean.

Unocal security personnel followed along the beach, watching suspiciously as we took pictures. In fact, the spill was so difficult to perceive (only periodically does the beach smell of petroleum and the ocean have rainbow oil stains) that my impressions wavered. Was this really a calamitous event? The whole visit was imbued with the paradox of beauty and travesty.

Under my feet was the largest oil spill in California, and most likely the largest in US history. . . . Yet the "total amount spilled" continue to be, as one local resident noted in an interview, a matter of "political science." There is still controversy over just how big this spill really is. The smaller of the two estimates . . . (8.5 million gallons) comes from Unocal's consultants. State and local regulatory agencies do not endorse it (Arthur D. Little et al. 1997). The estimates quoted most often by government personnel put the spill at 20 million gallons or more, which would make it the largest petroleum spill ever recorded in the United States.

At first glance, it seems strange that so many individuals and organizations missed the spillage[2] for so many years; "passivity" seems to be the word that best characterizes the personal and institutional mechanisms of identification and amelioration. It is also clear that the Guadalupe spill is very different from the image of petroleum spills that dominates media and policy prescriptions and the public mind: the iconographic spill of crude oil, complete with oiled birds and dying sea creatures.

The Guadalupe Dunes spill is only the largest *discovered spill.* Representing an inestimable number of similar cases, it exemplifies a genre of environmental catastrophe that portends ecological collapse.

Describing his impression of the spill in a 1996 interview, a resident of Orcutt, California, explained why he remained unsurprised by frequent diluent seeps: "When you grow up around it—the smell, the burning eyes while surfing, the slicks on the water—I didn't realize it could be a risk. It was normal to us." In a 1997 interview, a local fish and game warden—one of those initially responsible for the spill's investigation—responded this way to the question "Why did it take so long for the spill to be noticed?": "It is out of sight, it's out of mind. I can't see it from my back yard. It is down there in Guadalupe, I never go to Guadalupe. You know, I may have walked the beach one time, but I never saw anything. It smelled down there. What do you expect when there is an oil field? You know, you drive by an oil production site; you are bound to smell something. You are bound to."

In the days and weeks after my initial visit to the dunes, I wondered why the spill had gained so little notoriety. Beginning my research in earnest, I visited important players, attended meetings, took official tours of the site, and followed the accounts in the media.

What makes the Guadalupe spill so relevant is that it represents a genre—indeed a pandemic—of environmental crises (Glantz 1999). Collectively, problems of this sort—both environmental and non-environmental—exemplify what I term *crescive troubles*. According to the *Oxford English Dictionary*, "crescive" literally means "in the growing stage" and comes from the Latin root "crescere," meaning "to grow." "Crescive" is used in the applied sciences to denote phenomena that accumulate gradually, becoming well established over time. In cases of such incremental and cumulative phenomena (particularly contamination events), identifying the "cause" of injuries sustained is often difficult if not impossible because of their long duration and the high number of intervening factors.[3] Applied to a more inclusive set of social problems, the idea of crescive troubles also conveys the human tendency to avoid dealing with problems as they accumulate. We often overlook slow-onset, long-term problems until they manifest as acute traumas and/or accidents (Hewitt 1983; Turner 1978).

There are also important political dimensions to the conception of crescive troubles. Molotch (1970), in his analysis of an earlier and more infamous oil spill on the central coast of California (the 1969 Santa Barbara spill), relates a set of points that resonate with my discussion. In that article, Molotch examines how the big oil companies and the Nixon administration "mobilized bias" to diffuse local opposition, disorient dissenters, and limit the political ramifications of the Santa Barbara spill. Two of his ideas have special relevance: that of the *creeping event* and that of the *routinization of evil*. A creeping event is one "arraigned to occur at an inconspicuously gradual and piecemeal pace" that in so doing diffuses consequences that would otherwise "follow from the event if it were to be perceived all at once" (ibid., p. 139). . . .

Our preoccupation with immediate cause and effect works against recognizing and remedying problems in many ways. It is mirrored in the way society addresses the origin of a problem and in the way powerful institutional actors seek to nullify resistance and diffuse responsibility. The courts and the news media, for instance, often disregard the underlying circumstances that led to many current industrial and environmental predicaments, focusing instead on individual operators who have erred and pinning the blame for accidents on their negligence (Perrow 1984; Vaughan 1996; Calhoun and Hiller 1988). Yet this ignores the systemic reasons why such problems emerge. In short, most if not all of our society's pressing social problems have long histories that predate their acknowledgment but are left to fester because they provide few of the signs that would predict response—for example, the drama associated with social disruption and immiseration. . . .

The inability of our current remedial systems, policy prescriptions, and personal orientations to address a host of pressing long-term environmental threats is frightening. There are, however, numerous examples of disconnected events—seemingly unrelated individual crises recognized after the fact—that have received widespread public attention. Through national media coverage, images of ruptured and rusting barrels of hazardous waste bearing the skull and crossbones have become icons that fill many Americans with dread (Szasz 1994; Erikson 1990, 1994). But these are only the end results of ongoing trends that have been repeated across the country with less dramatic consequences. In view of the startling deterioration of the biosphere, much of which is due to slow and cumulative processes, more attention should be devoted to how such scenarios unfold. . . .

My specific intent is to uncover how and why the Guadalupe spill went unrecognized and was not responded to even though it occurred under unexceptional circumstances. The industrial conditions were quite normal, and the regulatory oversight was typical. It would seem that there was nothing out of the ordinary, other than millions of gallons of spilled petroleum. This is, in part, why the spill is so instructive. It represents a perceptual lacuna—a blank spot in our organizational and personal attentions. . . .

Why didn't local managers report the seepage, as the law requires? How did field personnel understand their role? How could pollution of such an enormous magnitude be left so long before receiving official recognition and action? Why did the surrounding community take so long to react? . . .

The reality that surrounds crescive circumstances is characterized by polluters who are unlikely to report the pollution they cause, authorities who are unlikely to recognize that there is a problem to be remedied, uninterested media, and researchers who take interest only if (or when) an event holds dramatic consequence.[4] In short, all those who are in positions to address crescive circumstances are disinclined to do so. Forms of degradation that lack direct and immediate impact on humans, dramatic images of dying wildlife, or other archetypal images of disaster tend to be downplayed, overlooked, and even ignored.

The national print media certainly mirrored the propensity to ignore the Guadalupe spill (Hart 1995). Over the period 1990–1996, the national press devoted 504 stories to the *Exxon Valdez* accident and only nine to the Guadalupe spill.

In a 1996 interview, a reporter for the *Santa Barbara News Press* offered his opinion as to why the Guadalupe spill had received little public attention until 1993. His view resonates with three of the four social factors articulated above (social disruption, stakeholders, and media fit):

We didn't see black oily crude in the water and waves turning a churning brown. We didn't see dead fish and dead birds washing up. We didn't see boats in the harbor with disgusting black grimy hulls. This is largely an invisible spill. It took place underground. . . . Because it was not so visual, especially before Unocal began excavation for cleanup, I think that it just didn't capture the public. . . . But after Unocal began excavations, driving sheet pilings into the beach, scooping out massive quantities of sand, setting up bacteria eating machines, burning the sand. It began to dawn on people the magnitude of this thing, but again it wasn't in their back yards, Guadalupe is fairly remote. . . . And it's not a well-to-do city [the city of Guadalupe]—comparatively, anyway, with the rest of our area. . . . So I don't think it really sparked the public interest as much as it could have or would have if it was . . . a surface spill. . . .

Central to my research were field interviews with members of the local oil industry, government regulators, community members, and environmental activists. These interviews were tape recorded, transcribed, and systematically analyzed. In addition to the interviews, there were many spontaneous conversations— in hallways, in office waiting rooms, in the homes of those that were the intended interviewees—with individuals I had not originally contacted or planned to meet. Though not recorded, these conversations should not be seen as any less important than the others. I also pursued ethnographic context, recording scores of informal conversations concerning the spill. I accumulated and analyzed a substantial collection of archival materials, and I have followed media portrayals of the spill closely since 1989. . . .

In its early stages (from 1953 until 1978 or 1979), the leakage at the Guadalupe field was not troubling, nor was there anyone to whom to report it. Because it was part of routine fieldwork, it received little attention. According to those who read the meters that tracked the coming and going of the diluent, "many times there were little leaks; that was just normal" (field worker, telephone interview, 1996). A worker quoted in a local newspaper went so far as to say that "diluent loss was a way of life at the Guadalupe oil field" (Friesen 1993). Dumping hundreds of gallons of diluent into the dunes, as long as it was done a gallon at a time, was an ordinary part of production. This is not a great leap of reason; oil work obviously involves oil. Until the 1970s, Unocal sprayed the dunes with crude oil to keep them from shifting and thus to make field maintenance and transportation easier. If spraying crude oil over the dunes was unproblematic, why would diluent leaks, which were largely invisible as soon as they hit the sand, be unsettling? Although workers mention that they became alarmed in the 1980s when puddling diluent periodically appeared as small ponds on the surface of the dunes, the chronic leaks themselves evoked little attention. In brief, at Guadalupe the normalcy of spilling oil of all kinds (crude oil, lubricants, and diluent) worked to blunt perceptions of the leaks as problematic. The leaks were an expected part of a day in the life of an oil worker.

According to the *Telegram-Tribune* (Greene 1993b): "A backhoe [operator] at the field . . . for 12 years . . . cited 'an apparent lack of concern about the immediate repair of leaks or the detection of leaks.' Diluent lines would not be replaced unless they had leaked a number of times or were a 'serious maintenance problem. . . .' Although workers checked meters on the pipelines and looked for leaks if there was a discrepancy, often a problem wasn't detected until the stuff flowed to the surface' said . . . a field mechanic."

By both historical and contemporary accounts, oil spills have long been a common occurrence in oil-held operations.[5] This seems to have been especially the case at fields operated by Union Oil. But this does not help us understand why, once field personnel recognized the spillage as a significant problem, they denied it and failed to report it (as specified by state and federal law) for 10 years or more. A first step in understanding why workers failed to report their spill to the authorities once it had "tipped" toward becoming a grievous problem requires us to attend to the vocabulary, the structure, and the enactment of work and how these factors not only molded workers' perceptions of the leaks but also kept them from reporting outside their local work group.

The "Company Line," 1978–1993

Organizationally, oil work at the Guadalupe field was arranged, like work in many traditional industrial settings, around a hierarchical seniority system. Recruitment and promotion were internally derived, meaning the field workers relied on their immediate foremen and supervisors for instruction, guidance, and ultimately, future chances at success (promotion, salary increase, choice of shifts, and so forth). . . . This organizational structure helps to explain Unocal employees' silence about the Guadalupe spill after it was recognized as a threat. Even when the leaks began to look more like a bona fide spill, the rank-and-file workers were insulated from reporting it themselves by their position within the field's hierarchy and their immediate responsibilities. Reporting outside the work group was management's domain.

A Unocal field worker I interviewed in 1997 articulated his experience of the change from a normal to a problematic spill as follows: "You come up and you see a clamp [on a] diluent [pipe]line. It is leaking. You tighten it up, you change it, you . . . fix it and it . . . has made a puddle. That is not something you would turn in. When it went into the ocean . . . and you see the waves break and they weren't breaking white [but] brown water, there is a problem. [That happened] sometime in the 1980s. . . . We all knew right then . . . we had some kind of problem. Well, we all kind of estimated it could be rather large considering that this field had been here so long before we ever got there." Corroborating this worker's impressions, another worker quoted in the *Telegram-Tribune* (Greene 1993b) remembered finding large concentrations of diluent that were no longer the "leaks" that had created them but looked more like a typical "oil spill": "In 1980 a large puddle of diluent that had saturated the sand and bubbled up to fill a spot 5 to 10 feet wide. . . ." He told investigators that he and his co-workers realized at the time there were problems with the diluent system "even though management seemed to ignore the problems."

By this time, the problems brought on by "normal operating procedures" were obvious and destructive. This became especially apparent in the mid-1980s, when accelerated spillage periodically slowed oil production at the field (Greene 1992b, 1993a; Rice 1994). Yet, instead of self-reporting the spillage, the field workers turned to denial and secrecy. . . .

In view of the hierarchy in the field, they were not responsible; their managers were. The hierarchical insulation from responsibility thus helped to keep workers who watched diluent spill into the dunes from feeling obligated to do something about it. When relieved of making decisions, people tend to cede their personal responsibility to those who are in control (Milgram 1974; Asch 1951). . . .

The field's hierarchy had five major levels.[6] A new worker began as a utility man, then worked his way up to pumper and then to field mechanic. If able, with long enough tenure at the field he could become a foreman. Over the foremen were the field supervisors, who headed operations at specific fields; over them was a superintendent who oversaw Guadalupe and another oil operation in the area. . . .

The culpability of all those at the field, but especially the superintendent, supervisors, and field foremen, coupled with the field's organizational characteristics, meant that explicit knowledge concerning the scope and scale of the leaks stayed inside the local operation. "Each field is its separate own little field," said a field worker interviewed in 1997. "We were kind of out in the middle of nowhere. So once we reported to our superior [a field foreman] then he has to report it to the field supervisor, who has to report to the regional superintendent, who then reports it to Los Angeles. Some where along the line I think it stopped. I think that it stopped with the field supervisor." This field worker was describing a loosely *coupled* organizational arrangement—one with organizational units that are "somehow attached, but [whose] attachment may be circumscribed, infrequent, weak in its mutual affects, unimportant, and/or slow to respond"[7] (Weick 1976, p. 3). In this case, the slack that existed at the local field between workers and between workers and managers and the loose organizational coupling that existed between the local field and corporate offices (including environmental divisions) were reflected in the technical division of responsibilities, in the authorities of office, and in the expectations placed on each. A great deal of flexibility existed between these units as long as certain goals were met. In this case, petroleum continued to be produced and sent out at an acceptable rate. In view of the local field's autonomy and field personnel's collective interest in remaining a viable production unit, not telling outsiders about the spill made a great deal of sense. . . .

The long-term nature of the Guadalupe spill made it especially problematic for all those who worked the field for any length of time. Liability for it was diffuse—indeed, organization wide. For those in the lower echelons, going outside the proscribed line of command to report the spill created triple jeopardy: Not only would they risk being personally associated with an organizational offense; they also would have been informing on co-workers and endangering their careers by implicating their superiors. One does not succeed within a vertically organized work setting by "ratting out" one's superiors or co-workers.[8] Fear of social and organizational reprisal was evident in my discussions with field personnel, in California wardens' accounts of their interactions with subpoenaed field personnel, and in local newspapers' stories such as Greene 1993b: "Current employees contacted for this story were surprised and dismayed their names would become public because of what they told the state investigators. They worried about their superiors and co-workers at Unocal finding out."[9]

Workers at the Guadalupe field did not want to go over the head of their field foreman, their supervisor, or their superintendent. A field worker, interview by telephone in 1997, said: "There is somebody above you and someone above them and someone above them. One thing that you don't want to do is break the chain of command . . . that causes friction." Informing might have affected how many hours of work one received, one's chances of promotion, and ultimately whether or not one would keep a well-paying job. . . .

A Culture of Silence

Local managerial power and organizational routines did not wholly determine behavior at the field. The normative framework that prevailed there was also attributable to the subculture of oil-field work and to individual workers' agency. . . .

To understand more fully why workers kept quiet about a spill they knew was patently illegal while field managers covered it up and lied to authorities about its origins, we must look beyond matters of hierarchy and seniority. We must look at individual motivations and at the social glue that bound workers to their work group. In short, we must look at the dominant social milieu at the oil field in order to see how social relations between workers played into the initial normalization of the spill and how they reinforced the intra-organizational conditions that discouraged self-reporting. Taken separately, both structural and cultural explanations would predict that self-reporting was unlikely; together, they make self-reporting appear a dubious regulatory strategy. . . .

Social Ties and Field Secrets

Workers at Guadalupe inherited and developed a set of norms and beliefs about what were and were not appropriate in-group behaviors. This is a normal part of group unity. Moreover, that this unity led to the coverup of an ongoing petroleum spill becomes more understandable (even if socially inexcusable) when we address the threat it posed to each individual at the field and to the local outfit as a whole. . . .

. . . [P]ressure to keep the spill a secret, based in a de facto culture of silence was observable in how field workers reacted when they found out that one of them had called the authorities. (. . . the first admission came in an anonymous telephone call to state officials in February of 1990.) When interviewed in 1997, the field worker who initially blew the whistle related being overheard by the field office's secretary and described the secretary's reaction to his phone call as follows:

> I got on the phone in the office. I say [to the health department official], "Okay, I'll talk to you later," and I hear his click, and I'm still on the phone, and I hear another click. The secretary eavesdropped and heard my conversation. She came in, and she started yelling at me, "What are you doing! We will all lose our jobs!" And I said, "Not if we didn't do anything! If it isn't ours, why would we lose our jobs? We are not going to lose our jobs!" We knew [about the spill]. But I never thought it would come to the point where they would shut everything down. What I thought would happen is they would isolate the problem and go on producing.

[. . .]

. . . Individuals, in protecting themselves from association with the spill, also collectively shielded the organization from harm, at least in the short term. The threats of a shutdown of the field and a loss of jobs and the social pressure to remain silent kept workers from reporting the spill. (Once the spill was "discovered" by regulators, Unocal's corporate headquarters did shut the field down, and all the workers were either transferred or laid off.)

Moreover, breaking with one's peers and eliciting an out-of-group admission about what was (initially at least) a "normal" part of production was also unlikely for a set of more socially relevant reasons. Even once the spill had accumulated and became noticeable, reporting it would mean informing on coworkers and facing their opinions.[10] Once his identity became known at the site, the whistleblower was ostracized by many of his fellow workers. . . .

Inter-Organizational Location as Amplification

In conjunction with the organizational location of workers relative to one another and to management and with the culture of silence that characterized the field, the Guadalupe field's structural isolation from outside interference (both physically and organizationally from regulatory authorities and Unocal's head offices) and the corporate incentives worked against self reporting. Like many other corporations, Unocal was not a monolithic undifferentiated body with a single objective or universally shared knowledge. In organizational form, Unocal consisted of loosely coupled upstream corporate offices, production units, and downstream refinery and vending segments. Insulation from outside interference amplified the power that field routines and the local production culture had over individual perceptions and over field workers' choices.

Because the Guadalupe field was largely autonomous from its head offices, its day-to-day domestic affairs were largely internal. A report of an incident had to go to the top before making its way to outside authorities. Because the information stopped in the field's chain of command, it never made it out of the field, where action could be taken to stem it. This is not a claim that Unocal headquarters could not have known about the spill if they had wanted to investigate it. The argument forwarded here is more passive: Headquarters was interested only in specific information from Unocal's extraction divisions, and this information tended to consist of production quotas rather than of information as to whether environmental matters were being addressed. Again, Unocal, as a corporation, seemed to care little about how local operations performed their production as long as the fields continued to produce profits. . . .

Had efficiency included not wasting diluent, a case could be made that the loss of diluent into the dunes would have been a sign to those on the outside that something was amiss. In this instance, Unocal's head offices may have taken a more active interest if dollars were being lost. Had hundreds of thousands of gallons of refined petroleum product been purchased from an external source and subsequently lost, it would seem expensive and hard to cover up. But spilling was considered a part of producing oil at Unocal's operations, and it also was rather normal for others in the industry. Furthermore, it was considered largely an internal affair. The diluent used at the site beginning in the early 1950s originated at Unocal's refinery situated at the edge of the Guadalupe Dunes, literally a part of the Guadalupe field's production infrastructure. Oil extracted from the Guadalupe field was piped to the Nipomo refinery for initial separation. Diluent, as a by-product of this refining process, was then pumped back to Guadalupe for use. At Guadalupe, diluent was stored at a number of tank farms; from there it was transferred via pipeline to individual extraction wells. If production was consistent, lost diluent would not be missed, especially in view of the normality of spilling and the shoddy records that were being kept (because the price of refining was internal). Losing diluent cost the local operator little (at least, relative to getting caught or facing the prospects of personally reporting it), as long as crude oil was being produced at the expected rate. On the other hand, if the field supervisor reported the spills (which had "tipped" toward the obvious in the 1980s) he would have known that he had a big monetary and criminal problem on his hands. It would have tarnished his personal record, reflected badly on Unocal's image as a whole, potentially shut down the local operation, and presented the possibility of criminal prosecution. What is more, the potential fines for having not reported the spill are significant. . . .

Two examples of the penalties associated with pollution of this sort illustrate the predicament that field managers confronted when deciding whether to report the spill. The federal Clean Water Act specifies that violators can be fined between $5000 and $50,000 a day per violation for being "knowingly" negligent.[11] Estimating the potential fines involved for this single act would require starting with the date of the amendment's passage (in 1973) and calculating daily fines up to 1990 (when Unocal ceased using diluent at the field). The estimate ranges from $31,025,000 to $310,250,000. Likewise, under California's Proposition 65 (a citizen-sponsored "right to know" act passed in 1987) Unocal was also liable for not reporting its release of petroleum into local river and ocean waters frequented by recreationalists. Proposition 65 caps fines against violators at $2,500 per person per exposure day. These are but two examples.

Moreover, the field supervisor and superintendent personally stood to lose thousands of dollars in potential bonuses that were paid for meeting corporate expectations. Field supervisors received incentives in the form of commendations, quick advancement, and end-of-the-year bonuses for keeping production costs down and petroleum yields high. High production costs would have resulted from capital outlays for such items as Guadalupe's pipeline infrastructure. Much as in the system that prevailed in the Soviet Union into the 1980s, costs were "hidden" by a reward system that recognized only production goals and the accompanying steady income stream. Thus, the primary goal was keeping production high, not worrying about diluent costs that (at least on paper) were trivial, being locally internalized. According to a Unocal supervisor interviewed in 1997 for this research, why it took 38 years for the spill to be reported by field managers was rather easy to understand: "Unocal [did not report the spill] to the public because local managers received financial incentives to keep costs low. The corporate culture of the production outfits saw spills as a normal part of their routine." Although this was not the only reason that local Unocal managers would continue to spill, it certainly provided a strong incentive not to report it or stop the leakage at the field once it had become organizationally ominous. Only negative personal and organizational repercussions would have resulted if local managers reported the spill. As a latent product of the pressures articulated thus far, spilling and not reporting makes a great deal of sense from the production side of the equation. . . .

In brief, organization-sponsored complicity, the culture of silence, and the inter-organizational isolation of the field combined to make reporting of the Guadalupe spill improbable until the accumulation of diluent had gotten so bad that neither insiders (field personnel) nor outsiders (regulators) could fail to recognize it, a society of environmentalists was there to be concerned about it, and the insistent local media were eager to report on it. These are all factors that society can ill afford to either count on or wait for. . . .

. . . The predominantly social, cultural, and structural explanations I have put forth are powerful in part because of the nature of industrial regulation in the United States, where it has been left to corporate actors to report their own excesses. There is little interdiction, investigation, or active following up of problems by government authorities until a situation is so dire that a coverup is impossible to sustain. Thus, outside of personal motivation on the part of a worker, a foreman, or a supervisor to report a leak, there is little (aside from morals) that would impel anyone in a company to do so. And that is a slippery slope that takes us back inside the social dynamics that characterized the normative and cognitive institutions that characterized Unocal's local field operations. . . .

Notes

Beamish, Thomas D. 2002. *Silent Spill: The Organization of an Industrial Crisis.* Cambridge, MA: MIT Press.

1. The Guadalupe Dunes have been designated a National Natural Landmark. This designation, conferred by the US Secretary of the Interior, acknowledges the national significance of the dunes as an exceptional and rare ecosystem.

2. I use "spill" to refer to the total accumulation of diluent—the end product. I use "spillage," "leaks" and "leakage" to denote the process by which diluent was chronically lost over time, eventually becoming an enormous spill.

3. For analyses of cases of community contamination and the struggle to seek legal redress, see Calhoun and Hiller 1988; Brown and Mikkelsen 1990; Hawkins 1983.

4. I am as guilty of this as anyone. I did not become aware of the Guadalupe spill until the local news media reported that 500,000 gallons of petroleum had been recovered by Unocal.

5. My visits to oil fields, my interviews concerning other fields in the region, and historical accounts of oil work all indicate that petroleum production has been a messy business since its origins in the late nineteenth century. See also Pratt 1978, 1980; Dinno 1999.

6. There were also seniority-related grades within these five levels (field worker, telephone interview, 1997).

7. On "coupling" see Weick 1976, 1995; Perrow 1984.

8. On "whistleblowing" see Glazer 1987; Bensman and Gerver 1963; Elliston et al. 1985.

9. I encountered similar resistance to being interviewed on the part of oil-field personnel. One of the major constraints involved legal liability. Most potential respondents

who represented the legal side (state attorney general, Unocal lawyers) as well as both current and former Unocal employees were largely unwilling to speak to me, formally at least, about the spill because of potential liabilities.

10. See Garfinkel 1956.

11. On environmental law in the US, see Yeager 1991; Wolf 1988; Skillern 1981. When taken to court in 1992 and 1993, Unocal did not face a single charge, but 28 separate criminal and misdemeanor violations.

References

Arthur D. Little Inc. in association with Furgro West, Headley and Associates, Marine Research Specialists, and Science Applications International Corp. 1997. Guadalupe Oilfield Remediation and Abandonment Project.

Asch, S. 1951. "Effects of Group Pressure upon the Modification and Distortion of Judgments." In *Groups, Leadership, and Men*, ed. H. Guetzkow. Carnegie Press.

Bensman, J. and I. Gerver. 1963. "Crime and Punishment in the Factory: The Function of Deviancy in Maintaining the Social System." *American Journal of Sociology* 28: 588–598.

Brown, P., and E. Mikkelsen. 1990. *No Safe Place: Toxic Waste, Leukemia, and Community Action*. University of California Press.

Calhoun, G., and H. Hiller. 1988. "Coping with Insidious Injuries: The Case of Johns-Manville Corporation and Asbestos Exposure." *Social Problems* 35, no. 2: 162–181.

Dinno, R. 1999. *Protecting California's Drinking Water from Inland Oil Spills*. Planning and Conservation League, Sacramento.

Elliston, F., J. Keenan, P. Lockhart, and J. Van Schaick. 1985. *Whistleblowing: Managing Dissent in the Workplace*. Praeger.

Erikson, K. 1990. "Toxic Reckoning: Business Faces a New Kind of Fear." *Harvard Business Review* 90: 118–126.

Erikson, K. 1994. *A New Species of Trouble: The Human Experience of Modern Disasters.* Norton.

Friesen, T. 1993. "Criminal Charges May Be Eliminated against Unocal." *Five Cities Times-Press-Recorder*, December 7.

Garfinkel, H. 1956. "Conditions of Successful Degradation Ceremonies." *American Journal of Sociology* 61: 420–424.

Glantz, M., ed. 1999. *Creeping Environmental Problems and Sustainable Development in the Aral Sea Basin*. Cambridge University Press.

Glazer, M. 1987. "Whistleblowers." In *Corporate and Governmental Deviance*, ed. M. Ermann and R. Lundman. Third edition. Oxford University Press.

Gordon, A., and D. Suzuki. 1990. *It's a Matter of Survival*. Harvard University Press.

Greene, J. 1992b. "Unocal: A Leaky Environmental Record." *Telegram-Tribune*, August 5.

Greene, J. 1993a. "Unocal Spills May Have Gone Unreported." *Telegram-Tribune*, July 1.

Greene, J. 1993b. "Unocal Workers Confirm Leaks." *Telegram-Tribune*, June 17.

Hart, G. 1995. "How Unocal Covered Up a Record-Breaking California Oil Spill." In *The News That Didn't Make the News and Why*, ed. C. Jensen. Four Walls, Eight Windows.

Hawkins, K. 1983. "Bargain and Bluff: Compliance Strategy and Deterrence in the Enforcement of Regulation." *Law and Policy Quarterly* 5, no. 1: 35–73.

Hewitt, K., ed.1983. *Interpretations of Calamity: From the Viewpoint of Human Ecology.* Allen & Unwin.

Milgram, S. 1974. *Obedience to Authority.* Harper & Row.

Molotch, H. 1970. "Oil in Santa Barbara and Power in America." *Sociological Inquiry* 40 (Winter): 131–144.

Perrow, C. 1984. *Normal Accidents: Living with High Risk Technologies.* Basic Books.

Pratt, J. 1978. "Growth or a Clean Environment? Responses to Petroleum-related Pollution in the Gulf Coast Refining Region." *Business History Review* 52, no. 1: 1–29.

Pratt, J. 1980. "Letting the Grandchildren Do It: Environmental Planning during the Ascent of Oil as a Major Energy Source." *Public Historian* 2, no. 4: 28–61.

Rice, A. 1994. "Endless Bummer." *Santa Barbara Independent*, March 17.

Skillern, F. 1981. *Environmental Protection: The Legal Framework.* McGraw-Hill.

Szasz, A. 1994. *Ecopopulism: Toxic Waste and the Movement for Environmental Justice.* University of Minnesota Press.

Turner, B. 1978. *Man-Made Disasters.* Wykeham.

Vaughan, D. 1996. *The* Challenger *Launch Decision: Risky Technology, Culture, and Deviance at NASA.* University of Chicago Press.

Weick, K. 1976. "Educational Organizations as Loosely Coupled Systems." *Administrative Science Quarterly* 21, March: 1–19.

Weick, K. 1995. *Sensemaking in Organizations: The Mann Gulch Disaster.* Sage.

Wolf, S. 1988. *Pollution Law Handbook: A Guide to Federal Environmental Law.* Quorum Books.

Yeager, P. 1991. *The Limits of the Law: The Public Regulation of Private Pollution.* Cambridge University Press.

Corporate Responsibility for Toxins

Gerald Markowitz and David Rosner

The Beamish piece in this section focused on microlevel processes within corporations— the interactions between a corporation's decision-making structure and the members of that organization. Markowitz and Rossner focus on the macrolevel question of how corporations interact with other large social institutions such as the government and media. The authors examine the claim that industries will regulate themselves and voluntarily seek out the safest, least environmentally damaging production and disposal processes because they are under the scrutiny of the consumers of their products and are watched by a vigilant health-concerned public. The authors present three case studies that indicate the opposite. Their studies of the lead, silica-using, and plastics industries show that industry associations have sought to control information about their harmful production processes and products; in addition, industries have sought to challenge those regulations that do exist.

The protests around the World Trade Organization in Seattle, the Group of Eight Summit in Genoa, the World Economic Forum in New York, and the campaign to protect consumers and ordinary citizens raise important and difficult questions concerning the global economy in general and the problem of industrial pollution in particular. How can the physical environment be protected from the actions of huge multinational corporations whose activities have, until recently, gone virtually unchallenged and unregulated? Although this question of corporate responsibility sounds rather new, in fact it is the result of a century-long conflict over the rights of industry to control production and the obligation of industry to the broader society, whether global or national. How much should government regulate private companies to ensure that they act responsibly and in accord with the broader public interest? How can government and industry create incentives for responsible corporate behavior? From the beginning of its history, industry has responded to calls for government regulation by arguing that voluntary compliance was sufficient to ensure that it acted responsibly.

Here, we outline three cases that raise broad policy questions concerning the degree to which we can trust industry to control its own behavior with regard to industrial pollutants. First, we outline the experience of Americans with the lead industry, the producer of a well-known industrial toxin. Second, we look at the silica-using industries, whose central mineral caused innumerable deaths and disabilities to exposed workers in the 1930s. Finally, we trace the efforts of the plastics industry to keep knowledge about the carcinogenic potential of vinyl chloride secret from the government. As early as 1905, federal action was taken to protect consumers and the environment from irresponsible actions of industry. That year, Theodore Roosevelt and other conservationists established the principle of federal protection of national forests. In 1906, Congress passed the Pure Food and Drug Act that extended its authority to inspect and test for adulterated consumer products. In 1970, the federal government established the Environmental Protection Agency and the Occupational Safety and Health Administration to protect the environment and the workforce. Unfortunately, these measures have not always been adequate, as industry has used its enormous power and resources to challenge government's authority and to restrict information necessary for appropriate regulatory policies.

It is a tenet of democracy that citizens should have full access to information so that they can make informed decisions about risks that affect their lives. In the case of industrial toxins, such information has been regularly denied to workers and the public by the very corporations that have had the most up-to-date knowledge about the effects of exposure to chemicals, poisonous minerals, and dusts. As a result, factory workers have been assailed by noxious fumes and dangerous chemicals even while beseeching industry for information and protection. These toxins were all too often vented into the air, spilled into waterways, and dumped onto the land, both legally and illegally, making industrial pollution an issue of widespread public concern. It took catastrophes such as Love Canal in Niagara Falls, New York; Times Beach, Missouri; and Bhopal, India, to bring home to people the danger posed by industry to their lives and the environment (Fowlkes 1982).

Nonetheless, a great deal has happened outside of industry (often in spite of industry manipulation) to educate the public about the dangers of pollution and to begin to confront industry's negligence. In 1962, Rachel Carson published *Silent Spring*, which publicized the harm pesticides caused public health and the environment (Lear 1997). Ralph Nader began his crusade as a consumer advocate by exposing the willingness of General Motors to sacrifice human beings for profit, as exemplified in their promotion of the dangerously designed Corvair (Morris 1970). Paul Brodeur (1974) dramatized the duplicity of the asbestos industry's willingness to expose workers and entire communities to asbestos, despite the known risk of cancer and lung diseases. By the 1970s, questions were raised about the safety of a host of products: DES, red dye number 2, phosphates, Firestone radial tires, the Ford Pinto, tampons, Dalkon Shields, cyclamates, and saccharine (Gottlieb 1993). The Three Mile Island disaster led to widespread skepticism about the safety of the nuclear power industry (Houts, Cleary, and Hu ca. 1988; The Three Mile Island nuclear accident 1981). By the 1980s, civil rights groups developed the concept of environmental racism to describe the tendency of industry to situate polluting plants and toxic waste dumps primarily in poor and minority communities. Environmental activists made environmental justice a rallying cry when demanding that industry redress the race and class bias in many industry decisions (Bullard 1993, 1994a, 1994b).

In the 1990s, citizens became aware of perhaps the most serious breach of the social contract with corporations: major players in the tobacco industry, after decades of denial that cigarettes were addictive and carcinogenic, were finally forced to admit that they had manipulated the nicotine content of its products for the specific purpose of keeping smokers addicted and that they had falsified scientific research, thereby lying to the public about the deadly effects of tobacco smoking. Companies such as Johns-Manville, which mined and processed asbestos, and Phillip Morris, which grew and marketed tobacco products, were notorious for their willingness to hide information about the dangers of their products (Glantz et al. 1996). Although some have maintained that the activities of these companies were not typical of industry practices, that in fact these were rogue corporations acting outside the norms of industrial practice, our research points to a different conclusion. In the cases of lead, silica, and vinyl, entire industries actually have banded together to deny and suppress information about the toxic nature of their products and to call into question results by outside researchers that indicated their products posed a danger to the health of individuals (Markowitz and Rosner 2002).

In addition to withholding information, some industries have been able to assure the public that its products are benign by controlling research and manipulating science. Throughout much of the twentieth century, most scientific studies of the health effects of toxic substances have been done by researchers in the employ of industry or in universities with financial ties to industry. At times, their results were subject to review by industry, and if the results indicated a problem, the information was suppressed. At times, the independence of the academy has been undermined by industry's influence through grants and other support for research. . . .

Since the establishment of the Occupational Safety and Health Administration, the National Institute for Occupational Safety and Health, and the Environmental Protection Agency in 1970 and of independent foundations working with university researchers and public interest groups, a new generation of scientists not employed by industry are highlighting the risks and discounting industry's assurances about the benign nature of their products and production processes. They are providing research for the public and the public health community to consider. Newspaper articles, television specials, and other media bring home to people the personal toll industry practices take on people's lives. Increased knowledge has become a powerful weapon in the battle to hold corporations accountable for their impact on public health.

At the heart of the current struggle regarding the control of industrial toxins is the very difficult question of how industry or the government decides what is safe. Industry has always taken the position that there is no reason to hold up production of useful products if no danger has been proven. But the history of the twentieth century is riddled with disasters resulting from industry's moving forward with products whose danger became apparent only over time. Lead, asbestos, tobacco, and radioactive materials became widely used because scientific studies could not prove with certainty that these substances caused harm. In the realm of environmental health, it is extremely difficult to say that a particular substance causes a particular health problem; usually only after decades of observation can a scientifically valid correlation be made between exposure to a chemical and increased death and disease in a large population. Even then, it may not be possible to establish a connection conclusively and to the satisfaction of the entire scientific community.

As a result, the battle being waged today by public health advocates is to establish a different method for deciding how and when industry should proceed with the introduction of new substances or products. Many argue for the precautionary principle, according to which suspect substances must be held off the market until their potential dangers are more clearly understood and their safety is better established. Public health officials and some politicians are increasingly aware that the threats from dioxins, chlorinated hydrocarbons, and greenhouse gases in the environment are so high that social policy demands regulatory action—even before the existing data absolutely prove danger. Many argue that we should protect our citizens and not wait for objective studies to prove further danger. Industry has used scientific uncertainty to argue that its activities should not be interfered with (Markowitz and Rosner 2002).

The lead and plastics industries and industries in which toxic substances such as silica are used or produced have been central to the expansion of the American economy throughout the twentieth century. For the first half of the century, lead was critical to every industry involved in the building of the urban infrastructure, the modern suburb, and the expanded agricultural system. In the 1930s, 1940s, and beyond, silica was identified as a serious threat to workers in a host of industries critical to the American economy, such as mining, foundry work, and construction. After World War II, the plastics industry came to dominate American consumer society; plastics were used in vinyl siding, linoleum, tabletops, rugs, clothing, phonograph records, and computers. When, in the early 1970s, it was revealed that vinyl chloride monomer, the major constituent of polyvinyl chloride plastic, was a human carcinogen, the chemical industry in general and the plastics industry in particular were faced with serious questions about the safety of their products and how forthcoming it had been about its knowledge of these dangers. In many ways, the struggles over these three substances—lead, silica, and plastic—are paradigmatic of a broader struggle that continues to this day over the responsibilities of industry and government to protect public health.

Lead: A Strategy of Denial

As one of the central elements in the industrial revolution, lead was a known toxin throughout the nineteenth century. In the early twentieth century, reformers such as Dr. Alice Hamilton, often considered the founder of industrial hygiene in America, documented the extent of lead poisoning among the workforce and sought to clean up paint factories, battery manufacturing plants, potteries, and other industries where workers were being poisoned by lead. Despite this understanding of the toxic nature of lead, the automobile and gasoline industries decided in the 1920s to introduce tetraethyl lead into gasoline as a way to increase the power and efficiency of automobiles. Alarmed that five workers had died and dozens more had been poisoned producing tetraethyl lead, public health officials warned that in addition to the dangers to the workforce, there were possible long-range effects on consumers' health of depositing millions of pounds of inorganic lead dust onto the streets of cities all over the country from the exhaust pipes of cars. But the increasingly important automobile and petrochemical industries successfully argued that in the absence of absolute proof of tetraethyl lead's dangers to consumers, such a tremendously vital product to America's economy must not be banned or restricted. Frank Howard, first vice president of the Ethyl Gasoline Corporation, formed by General Motors and Dupont as a company devoted to the production and distribution of tetraethyl lead, stated the general position of industry "You have but one problem," he told a conference called by the surgeon general to consider the potential dangers to consumers and workers of leaded gasoline. "Is this a public health hazard? But," he answered, "unfortunately, our problem is not that simple." Rather, he argued, automobiles and oil were central to the industrial progress of the nation, if not the world. "Our continued development of motor fuels is essential in our civilization," he proclaimed, and the development of tetraethyl lead, after a decade of research, was an "apparent gift of God." The argument Howard developed would become a mainstay of industry's position regarding industrial toxins for decades to come: potential, or unproven, risks should not inhibit industry from being able to produce and market new products. It was the responsibility of government, consumers, and researchers to first prove danger before imposing controls on the private market. In the case of tetraethyl lead, this argument put the public health and labor communities on the defensive, making them appear to be reactionaries whose limited vision of the future could permanently retard human progress and stunt the nation's economic growth. . . . In the end, the Public Health Service concluded there was not enough evidence of long-term ill effects to justify restricting tetraethyl lead's sale and distribution. Industry learned a valuable lesson from the tetraethyl lead crisis in the early 1920s. It had to keep knowledge about harm out of the public eye or to find ways to argue that while these constituent materials might be toxic, the products produced from them were not.

The story of lead paint illustrates industry's efforts to downplay information about the dangers of its products. As children were identified as suffering from lead poisoning, industry sought to forestall a possible threat to its product's popularity. In many ways, the story of lead in paint is that of a guerrilla war fought by small groups of individuals—mostly doctors and a few public health officials—against the giant lead corporations. As evidence emerged, first of lead's dangers to factory workers and painters, then among children, then in the environment, the industry made it its business to frustrate the efforts of any who warned of the dangers of lead or called for lead regulation in consumer products. Throughout the 1920s, evidence of lead paint's toxic effect on children accrued in the medical literature. Articles in the *Journal of the American Medical Association* and a variety of pediatrics journals documented scores of children who had ingested lead paint from toys, cribs, windowsills, and woodwork and had developed palsies, seizures, brain damage, and listlessness, ultimately dying of classic symptoms of acute lead poisoning (Strong 1920; Holt 1923a, 1923b; Council of Queensland Branch 1922; McLean and McIntosh 1926; Weller 1925; Aub et al. 1926; Holt and Howland 1926; Hoffman 1927; Boston Health Department 1927). In 1928, partly in response to the growing attention to lead poisoning among workers and children, the industry organized the Lead Industries Association (LIA), a trade group aimed at promoting the various uses of lead and its products. Almost immediately, combating the negative image of lead within American society became a major part of the LIA's agenda. . . .

The most cynical response of the lead industry to reports of danger was a three-decade advertising campaign to convince people that lead was safe and, most insidiously, to target its marketing campaign specifically to children. From 1906 through the 1940s, for example, industry engaged in a massive effort to portray lead as a benign substance, essential to the economic, social, and cultural development of the country and of no threat to children. In one such series of promotions, the National Lead Company, manufacturers of Dutch Boy paints and lead pigments, sought to reach parents through their children by producing a series of "children's paint books," which carry "a message to the grownups, while its jingles and pictures amuse the little ones." . . .

One paint book, titled *The Dutch Boy's Lead Party*, extolled the advantages of white lead over nonleaded paints; its cover showed the Dutch Boy, bucket and brush in hand, looking at lead soldiers, light bulbs, and other members of what the paint book called the "lead family." Throughout the booklet, the Dutch Boy is featured carrying the bucket inscribed "Dutch Boy White Lead" (National Lead Company ca. 1923). The National Lead Company explained to parents that "the drawings afford [the child] pleasure," for the "story of lead, told in rollicking jingles," was meant to "capture his interest." But, again, the book was intended to do more than merely entertain the child; inside its pages there was a booklet for parents "so that a decisive paint message is placed in the hands of both parent and child." Through the marketing campaign to children, "business is built for the present and insured for the years ahead" (Insuring business 1924; National Lead Company 1925; *The Dutch Boy's hobby* 1926).

Not until the 1950s was there a significant challenge to the lead industry's dominance over lead research and the definition of lead poisoning. As a result of public health activities, municipalities restricted the use of lead as a pigment in paint, and in the 1960s, new attention was directed to the potential long-term damage caused by this mineral.

Finally, in the 1970s and 1980s, the federal government banned lead in paint and in gasoline, signaling a major victory for public health. Even so, the industry continued to seek ways of challenging researchers whose work pointed to lead as a continuing problem at lower and lower levels of lead in children's bodies. Herbert Needleman, a pediatrician whose research was critical in uncovering the effect of low-level lead exposure, was harassed by industry's consultants for nearly fifteen years (Cole 1977, 1980; Markowitz and Rosner 2002).

Silica: Controlling the Scientific Debate

While lead illustrates the levels to which industry will go to control the public perception of a product, silica presented a different dilemma. Unlike lead, silica in its most dangerous form, dust, was not widely used as a consumer product. Rather, it was a threat to a wide swath of workers in a host of critical American industries. These workers, largely out of the public eye, ultimately had to seek redress through the legal system, initiating in the 1930s millions of dollars of lawsuits that ultimately forced silica onto the national agenda. As a result, throughout the thirties, silicosis was considered the preeminent occupational disease, the subject of numerous conferences at the local and federal levels and the subject of major exposés in national periodicals. Industry's reaction to what was widely seen as a silicosis crisis was to seek ways to reassure the public, the workforce, and the government that industry itself had taken the lead in controlling exposures and that silicosis no longer posed any significant threat. Through the activities of the industry-sponsored Air Hygiene Foundation (later the Industrial Health Foundation), the various "dusty trades" banded together to set standards that they deemed acceptable. The issue of silicosis was being defined at the time of the foundation's initial meetings in terms that were extremely threatening to a business community already on the political defensive. In the midst of the Depression, silicosis was frequently framed by the government, industry, and workers as a labor and management problem, not solely a health issue. Industry recognized that it had to find a way of defusing the political conflict that a class-based model could ignite and sought to develop a viable, apolitical alternative. It wanted to return the discussion of silicosis to the realm of science and medicine where experts from the Public Health Service, industry, and other academic and research entities could reassert the dominance of industry-sponsored professionals over politicians and labor unions. Roger A. Hitchens (1935), chairman of the Temporary Organizing Committee, complained that "there is now no one place where there is available all information on all phases of the problem; and no *concerted* planning is being done either as to the present or future as to all many different phases." And Dr. E. R. Weidlein (1935), the director of the Mellon Institute of Industrial Research, argued that it was in the interests of industry to collect research studies as well as to initiate and direct research efforts: "It is necessary that there be some organization representing industries for the reason that no program of research investigation could be carried on without assurance of continuity and financial support." In the absence of government agencies able to carry out a long-term research program, industry could shape the research agenda to meet its political, scientific, and public relations objectives. The organization of industry behind a research effort was especially important when addressing chronic diseases, wherein it might be necessary to follow the long-term development of symptoms in humans and animal subjects. By organizing a research institute, it was possible to avoid the charge that the organization was a special interest that was pursuing a narrow agenda. The organization

> must have a broad outlook, a sympathetic understanding of the problem, and wide contacts with all cooperating agencies; and it must have the confidence of the industries, and of the physicians, engineers and all

institutions, groups or individuals who might cooperate to advantage. (Weidlein 1935)

Industry could achieve major benefits by providing the funding for such an organization. The new organization could determine which specialists "may not be properly qualified and whose results might, therefore, not be entirely acceptable . . . to be in position to make reasonable grants to qualified individuals and agencies to study special problems or phases of problems" (Weidlein 1935). Weidlein suggested that industry could retain some control over the dissemination of information if it were to support a permanent research organization. On a more practical level, such research could be "important from both medical and legal standpoints in the preparation of court cases" as well as assisting "in the preparation of safety codes and fair laws" (Weidlein 1935). . . .

. . . Faced with a crisis of overwhelming proportions, the affected industries were hardly defensive. Rather, they acted to gain control over dust research and public policy (Weidlein 1935).

Elsewhere, we argue that because silicosis was hidden from public view, a national tragedy developed in which workers were assured that they were safe from harm while many were still being exposed and ultimately killed by exposure to silica dust (Rosner and Markowitz 1991, 1995). Indeed, it is very probable that the epidemic of silicosis among shipyard workers and sandblasters in the 1970s and 1980s can be traced to the lethargy of industry and its active efforts to assure the government, the public, and the workforce that it had the situation well in hand (Markowitz and Rosner 1998).

Vinyl Chloride: Deceit as an Industrial Policy

During the post–World War II period, the production of new petrochemical synthetic materials gave rise to a new set of concerns. Unlike lead and silica, many of these chemicals and products were of unknown toxicity, and because they were so new, there was little history by which to judge the potential problems they posed for the broad community. But that was all to change in the late 1960s and early 1970s, when the chemical industry's own research indicated the systemic disease and carcinogenic properties of vinyl chloride monomer, the building block of polyvinyl chloride, the widely used plastic. It was then that the industry embarked on a serious effort to mislead the public and avoid federal regulation.

In 1970, just as the federal government vastly expanded the scope of its power in occupational and environmental health with the establishment of the Occupational Safety and Health Administration and the Environmental Protection Agency, the chemical industry learned that vinyl chloride at high dosages caused cancer in animals. Shortly after, the European chemical companies conducted secret animal studies that revealed that vinyl chloride monomer was an animal carcinogen in kidneys and livers at levels half the level that industry recommended as safe (Viola, Bigotti, and Caputo 1971; Knapp 1972; Markowitz and Rosner 2002). The American chemical industry was given access to these findings after signing a secrecy agreement with the Europeans that also committed the Europeans to not reveal the results of any future American experiments. When the newly established National Institute of Occupational Safety and Health requested, in January 1973, any information that industry or others might have about the potential hazards of vinyl chloride, the chemical industry trade association, Manufacturing Chemists Association, was faced with a serious dilemma (Request for information 1973). Since the American companies had signed the secrecy agreement with the Europeans, the Manufacturing Chemists Association had to choose between what it acknowledged was its "moral obligation not to withhold from the government significant information having occupational and environmental relevance" and its self-interest in not disclosing information that might lead the government to impose regulations on the industry (Best 1973). The chemical industry decided to act in its own, rather than the public's, interests and not tell the National Institute for Occupational Safety and Health what it knew about the carcinogenic potential of vinyl chloride for humans. In an elaborately planned and scripted event in July of 1973, representatives of the industry met with National Institute for Occupational Safety and Health officials at their headquarters in Rockville, Maryland. While appearing to be forthcoming by providing information to the government, at the same time, the industry studiously withheld the key findings of the secret European studies that indicated vinyl chloride's carcinogenic nature (Notes on meeting 1973). Even at the time, industry officials recognized that their actions could be interpreted as "evidence of an illegal conspiracy by industry" (Wheeler 1973), but they gloated that the meeting had accomplished its purpose, forestalling any "precipitous action" by the government (Kusnetz 1973).

But in 1974, physicians hired by the Goodrich Company discovered that during several years, four workers in its Louisville, Kentucky, polyvinyl chloride plant had died from angiosarcoma of the liver, an extremely rare cancer, leading the chemical industry to inform state and federal officials. In the case of lead, no federal agencies existed to oversee the regulation of environmental and work-related diseases for the first seven decades of the twentieth century. With vinyl, however, the existence of the Environmental Protection Agency and Occupational Safety and Health Administration, along with a strong environmental movement and a labor movement that paid more attention to occupational disease, led to a fierce battle over the regulation of vinyl chloride. Although strict controls of the production process were achieved, in the years following the vinyl crisis, the business community mounted a fierce public relations and political offensive that caused the Occupational Safety and Health

Administration to be more wary of future confrontations with industry (Markowitz and Rosner 2002).

Conclusion

The struggle over environmental exposures continues to this day with uneven results. Certainly, there have been successes. Lead, identified as a major danger to children in the 1920s, was brought under greater control as a threat to children in the 1980s and 1990s, despite continuing efforts by the industry to control research and information about its dangers. Standards regulating exposure of workers and community residents to vinyl are considered models of strong and effective government regulation. Responsible CEOs of major corporations must now reconcile their fiduciary responsibilities to their stockholders with their environmental responsibilities to the public. They must, for example, reduce toxic air and water emissions from their plants to satisfy government regulations. In recent decades, industry, to protect its interests, has escalated its efforts to oppose the work of environmental groups. Organizations such as the Business Roundtable, made up of the CEOs of 200 of the largest corporations in the country, have intensified their lobbying efforts among government officials and established well-funded and large offices in Washington. Through political contributions, message ads, support for proindustry legislators, and direct contact with members of the executive branch—at the very highest levels—industry attempts to gain protection for its interests.

The effect of environmental toxins does not end simply because regulators have done their job. The impact of lead and vinyl will continue to be felt for generations. Recent studies show that all Americans carry in their bodies materials not normally found in human tissue and whose health effects may not be understood for many years to come (Centers for Disease Control and Prevention 2001). Because of their developing physiology, children especially are at risk. The walls of millions of homes are still coated with lead paint, which continues to pose a serious threat to children. The landfills of our country are absorbing millions upon millions of pounds of polyvinyl chloride that will deteriorate, releasing vinyl chloride monomer, a known carcinogen, into the air and groundwater. Computers, a commonplace of contemporary American life, pose a problem as their casings and components end up as refuse in landfills, where they deteriorate, producing vinyl chloride monomer. Computer monitors, on average, contain four pounds of lead, and millions of them are crowding landfills and leaching into drinking water. Even new methods of waste disposal pose new problems—for example, the burning of plastics, particularly vinyl, produces dioxins in all but the most efficient and up-to-date incinerators.

The activities of the lead, silica, and vinyl industries with regard to the known dangers of their product were not exceptional. Lying and obfuscation were rampant in the tobacco, automobile, asbestos, and nuclear power industries as well. In this era of privatization, deregulation, and globalization, the threat from unregulated industry is even greater. In fact, a deeper schism than ever separates the broader population's concerns about industrial pollution and the current political establishment's infatuation with market mechanisms and voluntary compliance. For this reason, it is imperative for future policy decisions that those concerned with responsibility for the public's health be aware of industry's response to environmental danger.

Note

Markowitz, Gerald, and David Rosner. 2002. "Corporate Responsibility for Toxins." *Annals, AAPSS* 584: 159–74.

References

Angell, Marcia. 2000. Is academic medicine for sale? *New England Journal of Medicine* 342:1516–18.

Aub, Joseph, Lawrence Fairhall, A. S. Minot, Paul Reznikoff, and Alice Hamilton. 1926. *Lead poisoning.* Baltimore: Williams and Wilkins.

Best, George. 1973. Letter to John D. Bryan. Manufacturing Chemists Association Papers, 26 March.

Blackman, S. S., Jr. 1936. Intranuclear inclusion bodies in the kidney and liver caused by lead poisoning. *Bulletin of the Johns Hopkins Hospital* 58:384–97.

———. 1937. The lesions of lead encephalitis in children. *Bulletin of the Johns Hopkins Hospital* 61:40.

Boston Health Department. 1927. Lead poisoning in early childhood. *Monthly Bulletin* 16:266.

Brodeur, Paul. 1974. *Expendable Americans.* New York: Viking.

Bullard, Robert. 1994a. *Dumping in Dixie: Race, class, and environmental quality.* Boulder, CO: Westview.

———, ed. 1993. *Confronting environmental racism: Voices from the grassroots.* Boston: South End.

———. 1994b. *Unequal protection: Environmental justice and communities of color.* San Francisco: Sierra Club Books.

Carson, Rachel. 1962. *Silent spring.* Cambridge, MA: Riverside Press.

Centers for Disease Control and Prevention, National Center for Environmental Health. 2001. National report on human exposure to environmental chemicals. NCEH Publication no. 010164. Atlanta, GA: Centers for Disease Control.

Cole, Jerome. 1980. Letter. *New York Times,* 3 June, C5.

Cole, Jerome F. 1977. Letter to Gershon W. Fishbein. 21 July.

Council of Queensland Branch. 1922. An historical account of the occurrence and causation of lead poisoning among Queensland children. *Medical Journal of Australia* 1:148–52.

Dublin, Louis. 1933. Letter to Ella Oppenheimer. 14 September. Courtesy Christian Warren.

The Dutch Boy's hobby: A paint book for girls and boys. 1926. New York: National Lead Company.

Dutch Boy's jingle paint book. 1922. New York: National Lead Company.

Fowlkes, Martha. 1982. *Love Canal: The social construction of disaster.* Washington, DC: FENIA.

Glantz, Stanton A., John Slade, Lisa A. Bero, Peter Hanauer, and Deborah F. Barnes. 1996. *The cigarette papers.* Berkeley: University of California Press.

Gottlieb, Robert. 1993. *Forcing the spring: The transformation of the American environmental movement.* Washington, DC: Island Press.

Hitchens, Roger A. 1935. The industrial dust problem. In W. G. Hazard, letter to L. R. Thompson, 21 March. National Archives, Record Group 90 (Records of the Public Health Service), File 087596-49. Pittsburgh, PA: State Boards of Health.

Hoffman, Frederick L. 1927. Deaths from lead poisoning. U.S. Department of Labor bulletin no. 426. Washington, DC: Bureau of Labor Statistics, Government Printing Office.

Holt, L. Emmett. 1923a. General function and nervous diseases. In *The diseases infancy and childhood for the case of students and practitioners of medicine*, 8th ed., 645–94. New York: D. Appleton.

———. 1923b. Lead poisoning in infancy. *American Journal of Diseases of Children* 25: 229–33.

Holt, L. Emmett, and John Howland. 1926. *The diseases of infancy and childhood.* New York: D. Appleton.

Houts, Peter S., Paul D. Cleary, and TeWei Hu. ca. 1988. *The Three Mile Island crisis: Psychological, social, and economic impacts on the surrounding population.* University Park: Pennsylvania State University Press.

Insuring business for the years ahead. 1924. *Dutch Boy Painter,* July, p. 139.

Knapp, W. A. 1972. Allied Chemical Corporation memorandum to J. C. Fedoruk and A. P. McGuire. Manufacturing Chemists Association Papers; Law Offices of Baggett, McCall, Burgess & Watson; Lake Charles, Louisiana, 20 November.

Kusnetz, Shell. 1973. Letter to files. Manufacturing Chemists Association Papers, 17 July.

Lear, Linda. 1997. *Rachel Carson: Witness for nature.* New York: Henry Holt.

Lead Industries Association (LIA) annual meeting. 1935. Lead Industries Association Papers, 13 June.

Markowitz, Gerald, and David Rosner. 1998. The reawakening of national concern about silicosis. *Public health Reports* 113:302–11.

———. 2002. *Deceit and denial: The deadly politics of industrial disease.* Berkeley: University of California Press/Milbank Memorial Fund.

McLean, Stafford, and Ruston McIntosh. 1926. Studies of the cerebral spinal fluid in infants and young children. In *The human cerebrospinal fluid*, 299300. New York: P. B. Hoeber.

Morris, J. D. 1970. Nader says G.M. suppresses data. *New York Times,* 5 September.

National Lead Company. ca. 1923. *The Dutch Boy's lead party.* New York: National Lead Company.

———. 1925. *Dutch Boy in story land.* New York: National Lead Company.

Notes on meeting between representatives of Manufacturing Chemists Association Technical Task Force on Vinyl Chloride Research and NIOSH. 1973. Manufacturing Chemists Association Papers, 17 July.

Painting the house that Jack built: Do not forget the children—Someday they may be your customers. 1920. *Dutch Boy Painter* August: 126.

Request for information. 1973. *Federal Register* 30 January: 2782.

Rosenstock, Linda. 1999. Global threats to science: Policy, politics, and special interests. In *Contributions to the history of occupational and environmental prevention*, edited by A. Grieco, S. Iavicoli, and G. Berlinguer, 111–13. London: Elsevier Science.

Rosner, David, and Gerald Markowitz. 1991. *Deadly dust: Silicosis and the politics of occupational disease in twentieth century America.* Princeton, NJ: Princeton University Press.

———. 1995. Workers, industry, and the control of information: Silicosis and the industrial hygiene foundation. *Journal of Public Health Policy* 16:29–58.

Strong, Robert. 1920. Meningitis caused by lead poisoning in a child of nineteen months. *Archives of Pediatrics* 37:532–37.

The Three Mile Island nuclear accident: Lessons and implications. 1981. *Annals of the New York Academy of Sciences* 365.

U.S. Public Health Service. 1925. Proceedings of a conference to determine whether or not there is a public health question in the manufacture, distribution or use of tetraethyl lead gasoline. *Public Health Bulletin* 158:4, 69, 105–7.

Viola, P., A. Bigotti, and A. Caputo. 1971. Oncogenic response of rats, skin, lungs, and bones to vinyl chloride. *Cancer Research* 31:516–32.

Weidlein, E. R. 1935. Plan for study of dust problems. In W. G. Hazard, letter to L. R. Thompson, 21 March. National Archives, Record Group 90 (Records of the Public Health Service), File 087596-49. Pittsburgh, PA: State Boards of Health.

Weller, Carl Vernon. 1925. Some clinical aspects of lead meningo-encephalopathy. *Annals of Clinical Medicine* 3:604–13.

Wheeler, R. N. 1973. Letter to Eisenhour et al. Manufacturing Chemists Association Papers, 31 May.

PART

V

Globalization

Tangled Routes
Women, Work, and Globalization on the Tomato Trail
Deborah Barndt

Globalization is a complex and multifaceted process that involves, among other things, connecting individuals, groups of people, and institutions in different parts of the world. In this piece, we see that globalization has entailed the privatization of that which was once owned communally, technological developments that allow for rapid exchanges of information and goods, export-oriented production, and trade agreements such as the North American Free Trade Agreement (NAFTA). NAFTA removed many barriers to investment and trade among the countries of North America. In this excerpt from her book Tangled Routes: Women, Work, and Globalization on the Tomato Trail, *Deborah Barndt, who teaches environmental studies at York University in Toronto, argues that NAFTA's provisions primarily benefit large agribusiness but do little to help small farmers. In making this argument, she uses a commodity chain perspective that highlights the numerous nodes of production and consumption in the life cycle of a product, tomatoes.*

Across Space and through Time:
Tomatl Meets the Corporate Tomato

The history of the tomato can reveal the unfolding global food system and the shifting role of women workers within it. In this chapter, we . . . follow the tangled routes of the tomato through the intertwining stories of Tomatl and the corporate tomato.

... Two main characters introduce the contrasting approaches to growing food: Tomatl is the homegrown tomato, named with the Indigenous name it was given in Aztec times; the corporate tomato is the fruit in its more familiar commodified form, produced in large quantities through multiple technological interventions. While the focus will be on the journey of the corporate tomato, we will periodically refer to the contrasting and shorter journey of Tomatl, from precolonial to postcolonial times.

In tracing the trail from Mexican field to Canadian fast-food restaurant, we move through the three NAFTA countries, from south to north, on a journey that is clearly not a straight line. To simplify the story, I am dividing the process into three major stages following the trip north:

- the production of tomatoes in Mexico;
- their transport, trade, and distribution into the United States and Canada; and
- their commercialization and consumption in Canada.

Each stage will be graphically summarized, providing the traveler with a kind of road map. While the linear south-north trajectory suggested here reflects a predominant dynamic of the south producing for the north,[1] all three phases—production, distribution, and consumption of tomatoes—take place in each of these three countries, as well as in others around the planet. ...

I am building on the tradition of global commodity chain (GCC) analysis, an approach developed by Gary Gereffi and others to understand the current forms of capitalism in which production and consumption not only have crossed national boundaries but have been reorganized under a "structure of dense networked firms or enterprises."[2] While my framing of the tomato story does not follow a classic commodity chain analysis,[3] it does try to link the particular and general, the local and global aspects of tomato production and consumption.

Gereffi distinguishes between producer-driven commodity chains, in which transnational corporations control production networks, and buyer-driven commodity chains, in which large retailers and brand-named merchandisers shape and coordinate decentralized production networks while controlling design and marketing themselves. Because the corporate tomato moves from globalizing Mexican agribusiness and processing plants to Canadian supermarkets and restaurants that are also globalized, we will see both types of chains in their overlapping complexity. ...

... In my particular telling of the tomato tale, the overall impression is of a long and twisty trail, a many-staged journey that no one understands in its entirety. We were most struck by this fact as we interviewed campesino and company vice president, cashier and chief buyer alike: no one person has the whole picture; each actor in this complex chain perhaps has some sense of the steps that come just before and after his or hers, but sometimes not even that much. . . .

I am arbitrarily dividing the journey into twenty-one steps (for the corporate tomato's journey) and five stages (for Tomatl and alternative practices). . . . The tangled journeys make us realize we cannot separate our survival in the north from the survival of people in the south, nor the fate of human beings from the fate of the earth. The corporate journey is described in steps (numbers 1, 2, 3), while we can follow Tomatl's story (designated as stages and marked by Roman numerals—I, II, III) from the margins: the survival of more sustainable locally controlled growing practices.

Tomatl's Story from the Margins

Stage I: Tomato's Beginnings in Pre-Hispanic America The tomato originated as a wild plant in the Andean region (what is now northwest Peru), its seeds then probably carried north by birds to what is now Mexico, centuries before the time of Christ. First domesticated by the Mayans and the Aztecs, the fruit was named *tomatl* which in Nahuatl, the language of the Aztecs, means "something round and plump." For centuries the tomato was a native crop grown by Indigenous peoples in Mexico to feed their families. Using traditional agricultural practices . . . , they grew tomatoes in great variety, interplanted them with other crops, and rotated crops from year to year, in the context of complex local ecosystems. Wild tomato species, for example, supplied other varieties of tomatoes with resistance to nineteen major plant diseases.

The Journey of the Corporate Tomato

Step 1: Colonial Conquest of the "Love Apple" In the sixteenth century, the Spanish conquistadores received tomatoes as part of tributes from Indigenous peoples in the Americas and eventually took the plant back to Europe along with other natural riches they had "discovered." There it was initially feared as poisonous and primarily considered decorative as a "love apple," until Italians began to embrace it in their cuisine. French settlers carried tomatoes to Quebec and Louisiana in the eighteenth century, and it was soon proclaimed medicinal and promoted by agricultural innovators such as Thomas Jefferson. Since then the tomato has been central to diets in the Americas and considered rich in vitamins (A and C) and minerals (calcium and potassium), especially when ripe. It has been bred into hundreds of hybrid forms; the most common big round red version, *Solanum lycopersicon* in Latin, is known in Mexico as *jitomate*.[4] The tomato is now the most widely grown fruit in the Americas as well as the most heavily traded.

Step 2: The Struggle for Land (campo) In recent decades, many Mexican *campesinos* (which means literally "of the land," or *campo*) have lost access to lands for cultivating the plant, either individually or collectively in peasant communities. Indigenous peoples have struggled for land for centuries, especially after the Spaniards arrived and sent them to work as peons in the mines and plantations. Mestizo and Indigenous campesinos gained greater access to land through the Mexican Revolution (whose battle cry was "Land and liberty!") and through agrarian reforms under President Lazaro Cardenas in the 1930s.... In the 1980s, Mexican neoliberal policies privatized *ejidos* (communal lands) and encouraged foreign investment, and in the 1990s, NAFTA increased agroexports. Since then, more and more campesinos from the southern states of Mexico have migrated to richer northern states to work as salaried labor for large agribusinesses....

Land, or the *campo*, is thus central to the story of the corporate tomato, particularly as it has become viewed as a natural resource and as private property by Western science and industrial capitalist interests, both national and international....

Step 3: Monocultures Led by U.S. Industrial Agriculture Tomatoes were the first fruit produced for export in Mexico, beginning in the late 1880s, but their production intensified with the development of capitalist production in Sinaloa in the 1920s. Often financed by U.S. capital and inputs, Mexican companies adopted American industrial practices such as Taylorization, the assembly line production and standardization developed after World War I. The work was divided into small manageable units, and technology was introduced that didn't depend on physical force, opening up jobs for women. In the late 1920s, U.S. surplus and protectionist policies forced Mexican producers to standardize packing tomatoes in wooden crates[5] to compete with U.S. producers. In the 1950s, two technologies revolutionized tomato cultivation: the use of plastic covering that kept the plants from direct contact with the earth[6] and the growth of seedlings in greenhouses.[7] By 1994, tomatoes accounted for 22.6 percent of the fruit and vegetable production in Mexico, even though they took up only 3.5 percent of the arable land.[8]

Monocultural and cash crop production is a central feature of the global food system today. It has, however, eliminated many types of tomatoes; 80 percent of the varieties have been lost in this century alone.[9] Now Indigenous and mestizo campesinos tend tomatoes as salaried workers in agribusinesses built on a Western scientific logic and rationalism. Each worker is relegated to a specific routinized task, in large monocrop fields or more recently in greenhouses (called "factories in the fields"), where the goal is to harvest thousands of tomatoes at the same time and in identical form....

Stage II: Combining Salaried Work with Subsistence Agriculture While large monocultural agribusinesses dominate tomato production in Mexico, the campesinos who work seasonally for them cannot survive without also culti-vating their own staple crops. As the case study of Empaque Santa Rosa, the Mexican agribusiness, shows, the low wages of industrial agriculture are based on the assumption that workers will combine salaried work with subsistence agriculture. For the poorer Indigenous migrant farmworkers, this is becoming less possible as they must migrate to more and more harvests to survive and as they lose access to arable land in their home states. But many peasants. . . main-tain their subsistence knowledge and more environmentally sustainable prac-tices by growing basic foods in plots on hillsides outside their village, working in their *milpa* (cornfield) after returning from picking tomatoes in large plan-tations. This double day not only assures their survival but keeps traditional knowledges alive alongside more industrialized practices. The interplanting of corn, squash, and beans (called the "three sisters" by North American Aboriginal people) uses the advantages of each crop to improve the growth of the others while maintaining the fertility of the soil.[10]

Step 4: Multinationals Control the Technological Package Even though many tomato seeds originated in Mexico, they have now become the "intellectual property" of multinational companies, which claim patents on genetically mod-ified forms of the seeds. . . . They have been recreated in thousands of varieties, hybridized and more recently genetically engineered by multinational agribusi-nesses such as the U.S.-based Calgene and Monsanto and their counterparts such as Western Seed of Mexico. In 1996, Western Seed created, for example, a seed that is immune to the whitefly that destroyed thousands of tons of tomato production in Autlán, Jalisco, in the early 1990s.[11] These seeds, selling for $20,000 a kilogram and geared entirely to the export market, have also been altered with genes that make the tomatoes last much longer before ripening ("long shelf life" tomatoes), so they can make the journey from Mexico to Canada without rotting en route.

For many Indigenous peoples and campesinos, this has meant not only a loss of ownership and control of the seeds but also a loss of their own knowl-edge about how to grow tomatoes in endless varieties. Ironically, Mexican pro-ducers such as Empaque Santa Rosa must now buy tomato seeds from foreign companies in the United States, Israel, and France; they also hire French and Israeli engineers who bring a whole technological package that must be used with the seeds, as well as an entire production process adopting European and North American management and work practices.[12]

. . . Long before tomato seedlings are planted in the ground, for example, the soil has been treated with fertilizers to enrich the soil for growth. As the tomatoes grow, there is a constant barrage of a variety of agrochemicals—pesticides, herbicides, and fungicides—aimed at killing pests, bacteria, and fungi. Under the mantra of efficiency and productivity, they are heralded as making the plants grow faster, stronger, more uniform, and in greater quantity; they are also critical to the production of the blemish-free tomatoes demanded by the export market.[13] The agrochemicals themselves are primarily imported from U.S. multinationals: Bayer, Dupont, Monsanto, Cargill. There is neither training in their use, however, nor protective gear provided for workers in fields where pesticides are sprayed by hand, combine, or small plane. Every year an estimated three million people are poisoned by pesticides.[14]

Stage III: Zapatistas, NAFTA, and Food It is no coincidence that the poorest field-workers are Indigenous families from the south, forced away from their land for the myriad of reasons named earlier. Nor was it an accident that the Zapatistas chose 1 January 1994, the inaugural day of the North American Free Trade Agreement, as the moment for an uprising of Indigenous communities who have lost their land and livelihoods through colonial practices and neoliberal policies. The Zapatista struggle, for bread and dignity, has been transformed into an international movement that is reclaiming Indigenous rights and knowledges as critical not only for the survival of poor campesino communities but also for the survival of the planet. Food is a political centerpiece of this initiative, reflecting the continuing struggle for the land (campo) as well as for cultural identity of campesinos and Indigenous peoples.

Step 5: Gendered Fields: Women Workers Plant and Pick Primarily young women plant the seeds in Empaque Santa Rosas large greenhouses in Sinaloa and nurture them into seedlings, ready to be distributed to production sites in other parts of the country. Once shipped to Sirena, they are transplanted in the surrounding fields by the few full-time workers hired by Santa Rosa from neighboring villages. The young plants are watched carefully over the first few weeks, pruned by campesino women who pluck off the shoots so the stems will grow thicker, faster, and straighter. If tomatoes grow from a main stalk, they take up less space, are less vulnerable to pests on the ground, and are easier to pick. When the plants reach a certain height, women workers tie the vines to strings that hold them up, so they can grow without being crushed on the ground.[15]

As one of the most labor-intensive crops, tomato picking requires many more person hours and careful work than does picking bananas, for example. While most agribusinesses in the United States now have mechanical harvesters that pick tomatoes very fast and in massive amounts, in most Mexican monocultural plantations, tomatoes are still handpicked by campesinos. Hired by the companies, many of them are Indigenous families who have been brought on a one- to two-day journey from the poorer southern states for the harvest season, and they live precariously in migrant labor camps near the fields.

At Empaque Santa Rosa, the tomato workers usually start picking tomatoes at 7:30 a.m., stop for a lunch at 10:30, and are finished by 2:30 p.m., by which time the sun has become unbearably hot. They pluck them fast, too, so that they can fill the quota of forty pails a day to earn their twenty-eight pesos (approximately U.S. $5 in 1997).[16] Both men and women (as well as children) pick tomatoes, but women pickers are considered more gentle, so there is less damage to the crop. Men, on the other hand, are the ones who stack crates on flatbed trailers that they pull by tractor from the field to the packing plant. This gender dynamic needs to be understood in the context of a *machista* culture perpetuated by an international sexual division of labor. . . .

Step 6: Selecting and Packing the Perfect Tomato Men unload the tomatoes in crates from the trucks and dump them into chutes that send them sailing into an agitated sea of 90 percent chlorinated water, a bath to remove the dirt, bacteria, and pesticide residue from their oversprayed skins. They are dried by blasts of warm air, then moved along on conveyor belts through another chute that coats them with wax. It keeps the moisture in and the bacteria out, protecting the tomatoes from further breakdown during the long journey, but it also gives them a special shine that makes them more attractive to wholesalers and shoppers in the north.[17]

Not all tomatoes will make the longer trip north, as only the "best" are selected for export. To be chosen, they must be large, well-shaped, firm, and free of any cracks, scars, or blemishes. The "nimble fingers" that decide which tomato goes where belong to young women, many of them brought by Santa Rosa from its larger production site in Sinaloa to handle this delicate task. They sort the fruit according to grades and destinations but also by size (determined by how many fit into a box— e.g., 5 × 5s or 6 × 7s) and by color (from shades of green to red), because this is how the importers order them.[18] In Santa Rosa's packing plants, tomatoes are sorted by hand, while in the greenhouses, they are sorted partially by a computerized system that weighs and scans them by laser, then sends them down specific chutes for packing by size and color.

As the tomatoes move along the conveyor belt, primarily women sorters determine their destiny. If they are perfect by international standards, they are deemed "export quality" and divided into second and first grades.[19] If they are regular sized, they go to belts for national consumption and are again categorized as second and first grade. The domestic tomatoes are sent to the big food terminals in Guadalajara and Mexico City, where they may be sold at one-third the price that they will draw internationally.

Women packers have even more responsibility with the tomatoes. They pick them up from depositories that have divided them by color but often have to re-sort them, checking on the sorters' work. Then they put them gently but quickly into boxes. It's a contradictory tension for these women because they are paid by the box and not by the day (as the sorters are); so they try to put several tomatoes into boxes at the same time, while also being careful not to damage the fruit. The contents are inspected before being closed. In the past few years, as Empaque Santa Rosa has more fully entered the global export market, little round stickers are pasted on the skin of the tomatoes before they are packed up and sent off. Also delicately applied by women, these stickers indicate the particular variety of tomato, according to an international numbering system (e.g., Roma tomatoes are #4064, while cherry tomatoes are #4796).[20]

Step 7: Tomatoes, Trade, and Agroexports   It is easy to tell the difference between those destined for local or export markets: if they're going north, they're packed in cardboard boxes with "Mexican tomatoes" written in English on the outside, often with Styrofoam or plastic dividers that hold each tomato in place; those chosen for domestic consumption are packed, without separators, in wooden crates marked with the company's Mexican label, Empaque Santa Rosa. The real rejects are dropped unceremoniously through a big chute into a truck outside the packing plant and sold to local farmers as animal feed.

Once packed and stickered, the boxes that will carry the tomatoes north are sealed, stacked, wrapped, and moved by men working in the packing plant. They are stacked into skids of 108 boxes and wrapped with a plastic netting that keeps them intact en route. Bar codes are also stuck on the skids by ticketers (usually men); when scanned, the lines on the bar code identify the company, tomato variety, the field they were grown in, the day they were packed, and so forth, allowing inventory to be recorded and problems to be traced.[21] An additional sticker bears a number identifying the worker who packed and inspected the boxes at the point of origin. Men driving motorized forklifts deposit most skids directly on to big trailer trucks, while leaving others in temporary storage.

Structural adjustment programs and neoliberal policies in Mexico in the 1980s encouraged agroexports, and NAFTA in the 1990s opened the doors for competition with northern producers. . . . [T]omatoes are one of the few Mexican crops to really "win" with NAFTA, because Mexico maintains the comparative advantage with more intense and consistent sun, easier access to land, and cheaper labor than the United States and Canada. Empaque Santa Rosa, for example, used to produce tomatoes as much for domestic production as for export, but it now sends 85 percent of its harvest north across the border; an ever-increasing number of greenhouse operations produce cherry tomatoes entirely for export. Mexico ships seven hundred thousand tons of tomatoes annually to the United States and Canada.[22] Prices are better in the north, and with the asymmetry of currencies and wages, companies like Santa Rosa can make much more money in the export market.

Tomatoes are ordered by international brokers who request them not only in specific sizes, but also in different shades, from green to red (1 = green, 6 = red).[23] Their journey north may be delayed while the company owners wait for the prices in the United States to rise so they can be sold for more profit. Thus, they might be stored away in refrigerated rooms at the packing plants or near the food terminals, at a temperature that keeps them from ripening too fast, remaining there for a few days up to a week, until the market is more favorable. When the producers decide to fill an order, then, depending on the color requested as well as the destination, the tomatoes may be gassed with ethylene, the same substance that naturally causes ripening, so that the ripening process, temporarily slowed down, is now speeded up. The doors of the storage rooms are closed for twenty-four hours, while the tomatoes are gassed, as the ethylene is dangerous for humans to inhale.

Step 8: Erratic Weathers: El Niño or Global Warming? Besides being sprayed incessantly with chemicals, tomatoes have been subjected recently to intense rains and even freak snowstorms. If a premature freeze occurs in the fields, the juice and pulp of the tomato freeze like ice, as though they had been put in a refrigerator. The journey for some tomatoes dead-ends here, causing the company economic losses and ending the work season prematurely for thousands of poor campesinos.

These erratic weather conditions are often blamed on El Niño, which originated in Peru and is caused by the clashing of hot and cold currents off the Pacific coast. But many contend that human intervention is also affecting global weather patterns, and crops have suffered from their erratic nature in recent years. Global warming is particularly accelerated by the emission of greenhouse gases into the atmosphere, slowly depleting the ozone layer. Among the greatest culprits of this process are the large trucks that transport food long distances, the focus of step 10.

Step 9: Detour to Del Monte Processing Adds "Value" Second-rate tomatoes are sent in wooden crates to the major food terminals (in Guadalajara, Mexico City, and Monterrey), to local markets, and sometimes to food-processing plants. Santa Rosa, for example, supplies Del Monte with tomatoes for processing into canned tomatoes, ketchup, or salsa at its plant in Irapuato, Guanajuato. Tomatoes received at Del Monte are dumped into an assembly line production that moves them along to be weighed and washed, sorted and mashed, then processed through cooking tanks, evaporating tanks, and pasteurizing tanks. Again, primarily women workers fill the bottles through tubes, and the bottles are capped, cooled, labeled, and packed into boxes.[24]

While one might think Del Monte would prefer overripe tomatoes for processing, they actually prefer firmer varieties, so that the tomatoes won't get caught in the automated conveyor systems and mess up the technology for transporting them into the plant.[25] More and more, however, ketchup producers like Del Monte are buying tomato paste rather than whole tomatoes, because the paste-making business draws on cheap labor and facilitates the process for the manufacturer. In bottled form, tomatoes join many other processed and frozen foods that are increasingly replacing fresh food in North America; they are sometimes called "value-added" products, although the real added value is reflected mainly in the price.

Step 10: Trucking: A Nonstop Dash North with Perishable Goods   Empaque Santa Rosa owns a few of its own trailer trucks to transport tomatoes to both domestic and northern markets; they guzzle fossil fuel and also contribute to the depletion of the ozone layer. More often, however, Santa Rosa contracts independent truckers to deliver tomatoes to the Mexican-U.S. border at Nogales. It often hires UTTSA, for example, a trucking company whose refrigerated units can carry fifty thousand-pound shipments of fresh produce. The tomatoes are sometimes precooled in a hydrocooling machine that brings their core temperature from 75 degrees down to 34 degrees, because if the temperature drops from 75 to 34 during the two-day journey north, the fruit might deteriorate.

Trucking is a male job. . . . Truckers often work in pairs, so that one can sleep in the back of the cab, while the other takes over the driving. The trip to Nogales from Sirena may take thirty to forty hours, depending how many drivers there are; time is of the essence, because tomatoes are highly perishable and preferred at a certain ripeness, but not overripe. Their average life span, in fact, is 4.7 days, so the faster the drive, the quicker they arrive, and the more market days remain for the critical activity of selling them.

We now enter the second phase of the journey of the corporate tomato from Mexico to Canada, highlighting issues of trade and transport, inspection and distribution. While it involves processes in all three NAFTA countries, this phase is clearly controlled by U.S. regulatory agencies, political interests, and multinational corporate needs. Contending political economic, and legal interests converge in activities around the borders, especially the highly charged U.S.-Mexican line.

Step 11: Controlling the Gates: Dumping, Drugs, and Deportees   To better control and facilitate the border inspection process, the U. S. Department of Agriculture (USDA) has installed its own inspectors within many Mexican agroexport plants to check the tomatoes before they're even loaded into the trailer trucks. Mexican environmental laws are not as strict as those in the United States and Canada, though NAFTA has provided some pressure to "harmonize." U.S.-based companies, however, sometimes "dump" pesticides in Mexico after they have been banned in their own country. The problem comes back to haunt them when tomatoes are exported back to the United

States, carrying higher concentrations of agrochemicals and threatening the health of U.S. consumers.

The USDA hopes to eventually complete all inspections at the point of origin, in the Mexican plants where the tomatoes are packed. Nonetheless, loads of tomatoes are inspected again and again along the route to the border, and the trucks carrying them are stopped regularly by inspectors at four checkpoints. Usually it is not the tomatoes that interest them as much as other possible cargo that could be smuggled within the trucks, such as narcotics or Mexicans seeking illegal entry into the United States. Narcotraffic is actually a much more lucrative (and volatile) enterprise than tomato production, and a lot of the border activity centers on attempts to control or eradicate it.

The border patrol complex is located in a sandy ravine with desert brush competing with large-armed spotlights and police cruisers on the hillside, a veritable militarized zone.[26] U.S. Customs officials, guns bulging at their hips, check for truck fraud and narcotics; the work of sniffing dogs has recently been complemented by high-tech X-ray equipment which can scan entire truckloads for suspicious objects. . . .

Second to drugs is concern for the growing number of desperate Mexicans who try to escape poverty and unemployment, by illegally crossing the border, seeking work in the United States where they earn in one hour what they would make in a day at home. Horror stories abound about the ways they try to smuggle themselves in, under truck cabs, amid produce, or across rivers at night, and about how they are often captured, mistreated, and sent back to Mexico. It's ironic that tomatoes, as well as capital, are so welcome in the north, while Mexican workers are not, except when they are wanted for menial tasks, at specific times, and under limited conditions. . . .

Tomatoes account for 56 percent of the cargo of the nine hundred to thirteen hundred big trailer trucks that cross the Nogales border daily. Truck traffic has been increasing at such a dramatic pace since NAFTA (in peak season in 1998, over twenty-seven thousand trucks crossed here in one month) that new lanes are being added to the highway to ease the congestion.

Step 12: Checking for Quality: Appearance Matters   Most food inspection actually takes place on the Mexican side of the border.

At the complex of the Confederation of Agricultural Associations (CAADES), in Nogales, Sonora, six kilometers south of the U.S. border,[27] tomatoes are run through a series of checks by the USDA officials. First they weigh the trucks, to be sure they don't surpass the total limit of eighty-eight thousand pounds; if the loaded trucks are overweight, they must unload and reload the tomatoes in smaller trucks. Then a USDA inspector goes through a truckload and randomly stamps boxes of tomatoes at the top, middle, and bottom of a skid. Ten boxes are opened and inspected at a time. . . . Some tomatoes get their temperature taken to be sure that the refrigeration of the truck has not failed; if they were packed pink and register higher than 50 degrees, they may be deteriorating too fast and are turned back. Of the long list of potential "quality defects" and "condition defects" used to check the tomatoes, most (such as "smoothness" and "color") relate primarily to the appearance of the fruit.[28] To be deemed suitable as a U.S. No.1 grade, no more than 10 percent of a load can have either quality defects or condition defects.

Step 13: The Line Is Drawn: Border of Inequalities There is a stark contrast at the border between the huts dotting the hillsides on the Mexican side and the more elegant homes on the U.S. side; just as the price of tomatoes rises the minute they cross the line, the wages and standard of living also rise. The way business is organized on both sides of the border area also reflects this asymmetry between nations. A growing number of maquiladora plants, set up since the 1960s in the northern Mexican border by multinational companies, employ thousands of young women in assembling electronics, in piecing together garments, and, in lesser quantity, in food processing. On the U.S. side, on the other hand, an immense infrastructure of administrative offices and warehouses has been established to facilitate the speedy movement of tomatoes beyond the border to northern consumers. The border thus also separates the workers in the south (Mexico) from the managers in the north (the United States).

Step 14: Keeping Pests and Pesticides at Bay Truckers that have passed the inspection in Nogales, Sonora, on the Mexican side, and are transporting tomatoes from a reputable agribusiness, can pass through the rapid transit lane, merely handing in the paperwork and moving quickly north. Others, however, may be directed into the U.S. Customs complex on the Arizona side of the border, for further inspections by the FDA and USDA. The Food and Drug Administration officials randomly select a box from a truck and cut a chunk out of a sample tomato to send to an FDA lab in Phoenix, Arizona. About 1 percent of the produce are tested for pesticide residues. This is one way officials can check to see whether Mexican tomato producers are following the standards regarding the acceptable levels of pesticide residue permitted in the United States. The lab testing may take a few weeks, by which time the chemically suspect tomatoes may have already been unwittingly digested by U.S. or Canadian consumers. Growers whose produce is proven to have certain chemicals[29] above the legal limit are warned that enforcement action might be taken if the problem continues.

What can be detected more immediately, however, are the pests or plant life that may be carried inadvertently in the trucks or boxes in which the tomatoes are packed. USDA botanists don rubber gloves and check the fruit for microbes or markings (a hard scar may be evidence of a pest). If found defective, they may be sent back to Mexico for domestic consumption, sent on to Canada "in bond" (quarantined and wrapped with unbreakable metal straps), or sprayed by a Nogales fumigation company, with USDA officials monitoring the process.[30] If, on the other hand, all goes well in the inspection, the border-crossing process will be complete within three to four hours, and the tomatoes are given official entry into the United States.

Step 15: Exporting/Importing: Brokers and Wholesalers When a Mexican trucker is not certified to cross the border, he will pay an American trucker $20 to drive the truck through customs and to a warehouse a few miles north of the border. The warehouses are owned by exporters as well as brokers; Empaque Santa Rosa, for example, has its own office on the U.S. side to manage international sales and distribution within the United States and into Canada. The skids are unloaded in thirty to sixty minutes and stored temporarily in the warehouse. Throughout the day, brokers arrange sales by phone, fax, and increasingly by e-mail. This is clearly a man's world, and tomatoes are constantly repacked and reloaded on the trucks of brokers or distributors for U.S. and Canadian wholesalers and retailers.

The Blue Book lists hundreds of wholesalers and retailers in the United States and Canada who purchase tomatoes, especially during peak season. Loblaws supermarkets in Ontario, for example, brings up three truckloads of tomatoes daily from the Nogales border. Like other wholesalers and retailers in Canada, they deal with brokers or shippers in Nogales who receive their orders and seek out the best deal from warehouses in the area.

It takes about three days in refrigerated trucks (kept at 48 degrees Fahrenheit) for the tomatoes to reach Ontario from the Mexican border; if coming from Florida it's only two days, while from California it may be four....

Stage IV: Challenging Globalized Production: Ecological Footprint Activists and academics concerned about the often hidden ecological costs of production and distribution in a global food system that depends on moving tomatoes long distances have developed tools for measuring the impact of such practices. Neither transportation, which is heavily subsidized by government, nor environmental degradation (exacerbated both by the burning of fossil fuels and by the hydroflurocarbons in refrigerated units of trucks) appears either in the balance sheet of the companies or in the price we pay as consumers. One such tool, the ecological footprint, developed by William Rees,[31] calculates both primary energy consumption and carbon dioxide emissions.

... Of the tomatoes imported annually into Ontario, 74 percent were from the United States, 22 percent were from Mexico, and 4 percent were from other countries.[32] In 1997, Ontario's forty thousand tons of tomato imports (from North America) traveled over ninety-one million kilometers (i.e., 2,320 kilometers/ton).[33] A recent study[34] estimates that most tomatoes enter Toronto by truck but that North American imports emit 221 tons of carbon dioxide into the atmosphere while the transportation of Ontario greenhouse tomatoes only emitted 67 tons. Air transport is even more damaging to the environment; according to a 1994 SAFE Alliance study, tomatoes arriving by air contributed 1,206 grams/ton · kilometer, compared to 207 for road travel, 30 for water, and 41 for rail.

Step 16: A More Permeable Border: Slipping into Canada   It's difficult to know how many Mexican tomatoes actually make it into Canada. One-quarter of the tomatoes sold in Ontario come directly from Mexico, but this doesn't include those that are shipped from Mexico to border states, then repacked under new U.S. trademarks and sent on to Canada.[35] While the journey from Nogales to Sarnia may take as long as the Sirena-Nogales trip within Mexico, tomatoes have a much easier time at the Canadian border. Fortunately, the elongated inspection at the Mexican-U.S. border is not repeated, because the standards in the United States and Canada are pretty much the same. . . . Truckers who, since the deregulation of transportation in the late 1980s and NAFTA in the early 1990s, cross the border more regularly merely present a "confirmation of sale," which has often been previously faxed or sent electronically to both Canadian Customs and the Canadian Food Inspection Agency (CFIA). A small number (about 4 percent) of the shipments are inspected by customs officials (and dogs), usually initiated as a search for contraband (drugs, weapons, liquor, tobacco), and secondarily a check on the quality of the tomatoes. If the fresh produce smells or appears spoiled, a CFIA inspector will be called in.[36]

Tomatoes are the subject of intense communications between the brokers (shippers) at the U.S.-Mexican border, and U.S. and Canadian buyers (wholesalers), and then again between the brokers at the Canadian border and the buyers awaiting the arrival of fresh tomatoes. Ontario Produce, one of the key wholesalers at the Ontario Food Terminal, for example, has its own brokers negotiating the crossing of tomato shipments at Sarnia, Ontario. Ontario Produce faxes its record of the load, and its broker helps shepherd it across the Sarnia border. Customs officials check the shipper's manifest and the buyer's manifest, and if there are no problems, the tomatoes are allowed to enter Canada.[37]

Finally, we move on to the third phase of the corporate tomato's journey north, as it is received, inspected, and distributed in Canada, to terminals and then to supermarkets and fast-food restaurants.

Step 17: The Morning Zoo: Food Terminals Work While We Sleep   Tomatoes are delivered (by truck via Sarnia) to the Ontario Food Terminal, often in the middle of the night, to be ready for sale when wholesalers and retailers arrive from 4 a.m. on.[38] Ontario Produce,[39] one of largest of the twenty companies in the terminal, has eight buyers who order tomatoes from all over the world (Belgium, Spain, Italy, Mexico); while most are beefsteak tomatoes from Florida, they also buy tear drops, cherry, hothouse, and Roma (demanded by the ethnic market), and sell eleven truckloads a day. Ontario Produce sends its own three trucks out to pick up orders as well as to receive deliveries. If they arrive too early, the trucks may have to wait for hours before unloading, while wholesalers close down for a couple of hours to clean up and prepare displays of the best samples for the following day.

Once unloaded at the terminal, the tomatoes may be returned to refrigerated storage units, similar to the ones in Sirena, but with computerized temperature control, where they're kept at a temperature of 36 to 40 degrees. These units are equipped with catalytic generators to produce ethylene, a liquid that when released creates vapor that accelerates the ripening of the fruit. Signs around the heavily locked door warn of its highly flammable nature, indicating that smoking around it could cause an explosion. Whether or not the tomatoes are gassed depends on demand and price.

Wholesalers (or jobbers) like Joyce Foods arrive in the early morning to buy for fast-food restaurants, the primary customers for tomatoes. They prefer the firmer Florida tomatoes (without any markings) because they are more sliceable (e.g., for McDonald's Arch Deluxe or for pizzas). While importers and wholesalers have noted an increase in Mexican tomatoes since NAFTA, they are often too watery for fast-food use. Tomatoes that have been traveling several days from Mexico ripen at different times and, to different degrees, suffer from stem puncture, or deteriorate. Importers can claim for damages, but this involves lengthy court procedures, so they may just send them to be repacked. Women workers at Bell City packers near Toronto, for example, eliminate the decayed tomatoes, wash them again, and re-sort them into six different colors through a computerized laser system and a mechanized assembly line similar to the one in the Santa Rosa greenhouse in Mexico.

Stage V: Local Tomatoes Picked by Mexican Hands In the summertime, fewer tomatoes are imported from Mexico, since Canadians can get them fresh from local farmers. Even the sales manager at Ontario Produce recognizes the difference: "Anything that is grown locally has a better taste than the imported merchandise. If you compare it with the local stuff, it is totally different, just like night and day."[40]

There is also a growing greenhouse production of hothouse tomatoes in Ontario year-round. The climate can be carefully controlled in these sophisticated glass greenhouses, and the production is more predictable. Many people also prefer them, because they can ripen on the vine and thus are tastier than tomatoes picked green and sent on the long journey north from Mexico. With a growing demand for organic tomatoes, biological methods of pest control are being used in greenhouses as well.

National Grocers buys tomatoes for Loblaws from Mennonite farmers in western and southern Ontario. Ironically, many of the locally produced tomatoes are harvested by Mexican migrant farmworkers, men and women, who come north every summer as part of a government program called FARMS. Irena Gonzalez, for example, has been coming for thirteen years from July to October. While she is still only paid at minimum wage, she makes in an hour in Canada what she would make in a day in Mexico picking tomatoes. . . .

Step 18: Designer Supermarkets and Multicultural Labels A mixture of corporate tomatoes and (in season) locally grown tomatoes is delivered to Loblaws supermarkets, where they become part of a simulated village market within megastores, which now combine selling groceries with gourmet takeout, dry cleaning, pharmacy, photo processing, plant nurseries, art galleries, and even banking . . . The produce section has been moved to the front of stores to create the illusion of being closer to the source of food. The walls surrounding the tomatoes are brightly colored and well lit. With their waxed shiny surfaces,[41] they are arranged artfully on carts, under umbrellas and a sign that says "Fresh from the Fields."

As part of a global retail market, Loblaws now proclaims that "Food Means the World to Us." Having come from Mexico, tomatoes are part of its global reach, either in fresh form or processed into one of Loblaws' corporate brand President's Choice sauces, Italian style plum tomatoes, or salsas and tacos being promoted in colorful Latino-style aisles, introducing Canadians and its multicultural population not only to new tastes but to new ways of being. The seduction of consumers into lifestyle foods and an illusion of diversity is a key theme in the story of Loblaws. . . .

Step 19: High-Tech Tomatoes and Computerized Cashiers Corporate tomatoes can be purchased in fresh or processed form, and they are either punched in or scanned at the supermarket checkout lane. Fresh tomatoes are given PLU (product look-up) codes[42] either on tickets or on stickers.[43] Because they are of variable weight, they must first be weighed and their PLU numbers punched in to calculate their price. If a cashier is not sure of the type of tomato (nine different varieties are sold by Loblaws), she may check the visual inventory on her computer screen. Canned tomatoes or bottled salsa, on the other hand, can be quickly swiped through the scanner; their bar codes are read by a laser beam, and the type and price appear immediately on a printed receipt.

Global food production has become highly technologized in recent decades, and work practices have been transformed by the information revolution. The high-tech corporate tomato mediates a complex relationship between the worker and the technology; the electronic devices that control pricing and inventory, for example, can also monitor the productivity of cashiers. . . .

Step 20: Fast Food: Homogenized Tomatoes and Toys McDonald's, which traditionally has targeted a young market, doesn't include tomatoes in the Big Mac, since many children don't like them.[44] In 1996, however, when the fast-food giant's domestic sales were slipping, it created the Arch Deluxe[45] for grown-ups, adding tomatoes; in 1999, it was replaced by the Big Xtra. McDonald's prefers to buy Florida beefsteak tomatoes that are pulpier, firmer, and easier to slice for a hamburger bun, while the tastier Mexican produce are juicier and more likely to fall apart.[46] It is clearly a question of appearance and not of taste. The draw of McDonald's is often more the lifestyle, reflected in the glossy ads, billboards, TV commercials, toys, and videos that promote dominant popular culture. At the final destination of the corporate tomato, it becomes clear that tomatoes and hamburgers are not just food, nor even mere commodities, but symbols of a way of life. . . .

Perfectly sliced tomatoes on a cookie cutter hamburger and bun are part of a global trend toward homogenized diets. In fact, the term *McDonaldization* is now equated with this rationalizing and homogenizing process, which is built on principles of efficiency, calculability, predictability, and control; other businesses and social institutions are increasingly modeled on practices similar to the fast-food restaurant. . . .

The standardization of meals also fits a frantic lifestyle that devalues the preparation or savoring of food such as tomatoes, and the experience of commensality, or enjoying sharing a meal as an intimate social act. Homogenized consumption patterns in Canada parallel the monocultural production of tomatoes in the Mexican fields, both representing dominant practices at either end of the current global food system.

Step 21: Waste or Surplus: Compost or Charity? There is a "paradox of hunger"[47] reflected by a deepening poverty in the context of a relatively affluent Canada. Food retail giants such as Loblaws are by far the largest donors to a burgeoning network of food banks. Besides getting a tax write-off for its contribution, Loblaws also invites customers to buy goods, such as canned tomatoes, and add them to a donation box in the store. There are critics of this practice, and food bank organizers complain that the kind of items donated aren't always the most needed. Organic produce such as tomatoes, however, are disposed of in a different way; in collaboration with Organic Resources, Loblaws uses a system of underground holding tanks where wasted tomatoes are stored until they can be recycled as compost on experimental farms outside of Toronto. . . .

Stage VI: Full Circle: Seeds in the Multicultural City Tomatl is well and alive and living in Toronto. One sunny May day, I made my way to Field to Table, the warehouse where FoodShare Metro Toronto is promoting a variety of food alternatives for low-income communities. . . . Lauren Baker, who helped me trace the corporate tomato journey for two years, today sells me heritage tomato seedlings that she has grown on the first certified organic rooftop garden in Canada. There she has transformed organic waste into rich composting soil, which is then recycled in diverse urban agricultural projects.

Part of a growing food security movement, FoodShare recently launched a new project, entitled "Seeds of Our City." It draws on the rich knowledges of ethnic communities that are now part of Toronto's multicultural population and that have brought their own growing practices and food traditions to community gardens sprouting all over the city. They are vibrant examples of the survival and recovery of more ecologically sustainable growing traditions built on a closer relationship between production and consumption. . . .

The two intertwining journeys [of Tomatl and the corporate tomato] provide stark contrasts between producing tomatoes for local consumption, a relatively short trail, and the corporate production of tomatoes in the south for massive consumption in the north, a much longer and more convoluted trail. . . .

Notes

Barndt, Deborah. 2002. *Tangled Routes: Women, Work, and Globalization on the Tomato Trail.* Lanham, Md.: Rowman & Littlefield.
1. Mexico predominates in tomato production, compared to the United States, Canada, and Holland. While it is the only country that can produce all year round, it is in close competition with Florida, where intense production occurs for nine months. Source: The Blue Book for exporters and importers of fruit and vegetables, Carol Stream, Ill.: Producer Reporter Company Blue Book Services, 1999.

2. Gary Gereffi, Miguel Korzeniewicz, and Roberto Korzeniewicz, "Introduction: Global Commodity Chains," in *Commodity Chain and Global Capitalism,* ed. Gary Gereffi and Miguel Korzeniewicz (Westport, Conn.: Praeger, 1994), 1.

3. Gereffi suggests that global commodity chains (GCCs) have three main dimensions: an input-output structure (set of products and services), a territoriality (production/distribution networks), and a governance structure (authority and power relationships).

4. The historical information on the tomato is drawn from several sources: Sophie D. Coe, *Americas First Cuisines* (Austin: University of Texas Press, 1994), 46–50; Jennifer Bennett, ed., *The Harrowsmith Tomato Handbook* (Camden East, Ontario: Camden House, n.d.), 6–13; Philip Hardgrave, *Growing Tomatoes* (New York: Avon, 1992), 7–9; World Resources Institute, "Food Crops and Biodiversity" (Washington, D.C.: World Resources Institute, 1989), also on its website: www.wri.og/wri/biodiv/foodcrop.html.

5. The production of wooden crates, in fact, upset the ecosystem balance in rural Mexico and contributed to a plague of *la roña* (a whitefly), which destroyed tomato harvests in Atlàn in the early 1990s. The forests surrounding the plantations were cut down

to make crates, and the white fly that had lived from the leaves of the trees was forced into the tomato fields for sustenance. Personal communication with Antonieta Barrón, Miami, Florida, March 2000.

6. The plastic sheets have several functions: they keep the moisture in and the weeds out, they maintain uniformity among the plants, and the shine on their surface repels pests.

7. Sara Lara, "Feminizacion de los procesos de trabajo del sector fruti-horticola en el estado de Sinaloa," *Cuicuilco* 21 (April–June 1988): 29–36.

8. Yolanda C. Massieu, "Comercio bilateral Mexico–Estados Unidos y logros del TLC: 'La guerra del tomato,'" *El Cotidiano* 79: *Revista de la Realidad Mexicana Actual*, 13 October 1996, 114.

9. A study by the Rural Advancement Foundation International (RAFI) of seventy-five types of vegetables found that 97 percent of the varieties on the old USDA lists are now extinct. Of the 408 varieties of the common tomato, *Lycopersicon exculentum*, existing in 1903, only 79 varieties are now held by the U.S. National Seed Storage Laboratory, perhaps the major seed bank in the world. Cary Fowler and Pat Mooney, *Shattering: Food, Politics, and the Loss of Genetic Diversity* (Tucson: University of Arizona Press, 1996), 51, 62–63, 67.

10. Sometimes called *polycropping*, this approach has multiple advantages. "The beans 'fix' organic nitrogen, thereby enhancing soil fertility and improving corn growth. The corn in turn provides a trellis for the bean vines, and the squash plants, with their wide shady leaves, help keep the weeds down." Scientists have proven that total yields of these three crops grown together are higher than if the same area were sown in monocultures. John Tuxill, "The Biodiversity That People Made," *World Watch* (May/June 2000): 27.

11. "Jalisco produce un jitomate libre del virus de la mosca blanca," *Economia*, 5 December 1996, and "Western Seed colocará 400 kilos de semilla de jitomate hibrido," 6 December 1996.

12. Interview by author and Sara San Martin with Yves Gomes, San Isidro Mazatepec, Jalisco, 24 July 1996.

13. The cosmetic standards that Mexican agroexport companies feel pressured to adhere to have been written under pressure from U.S. growers as one tactic to keep the competition down. Mexican agronomists admit that much higher amounts of pesticides are used to avoid blemishes and irregularities caused by pests. Entomologist Mayra Aviles Gonzales suggests that anxious growers who use an irrationally high quantity of pesticides could get the same production and cosmetic results with 50 percent the amount used. Angus Wright, *The Death of Ramón González: The Modern Agricultural Dilemma* (Austin: University of Texas Press, 1990), 33–35.

14. Tuxill, "The Biodiversity That People Made," 32.

15. Interview by Maria de Jesus Aguilar with Milagros Baltazar, Santa Rosa field-worker, Sayula, Jalisco, 23 August 1997.

16. Interview by author and Sara San Martin with Santa Rosa field-worker, Gomez Farias, Jalisco, 26 April 1997.

17. Interview by author and Lauren Baker with Cesar Gil, plant manager, Santa Rosa packing plant, 7 December 1996. See also Wright, *The Death of Ramon Gonzalez*, 34.

18. Interview by author with Angelo Vento, Ontario Food Terminal, Toronto, January 1999.

19. Lara, "Feminizacion de los procesos de trabajo," 29–36.

20. Produce list used by Loblaws supermarkets, for the week of 28 July–3 August 1996. It identifies almost five hundred items by names and PLU number.

21. Interview by author and Lauren Baker with Enrique Padilla, greenhouse worker, San Isidro Mazatepec, Jalisco, 10 December 1996.

22. Linda Tons, *The Toronto Star*, 16 March 1997, 5F.

23. The Blue Book used by exporters and importers of tomatoes uses the following categories to classify the range of colors: Green—completely green, Breakers—not more than 10 percent turning, Turning—10 to 30 percent turning, Pink—30 to 60 percent pink or red, Light red—60 to 90 percent pink or red, and Red—more than 90 percent red. Tomatoes may be ordered in any of these categories. Those harvested for distant transport are often picked in the mature green state, are gassed, and ripen during transport if kept at temperatures between 55 and 70 degrees Fahrenheit.

24. Interview by author and Maria Dolores Villagomez with Alfredo Badajoz Navarro, Irapuato, Guanajuato, 28 April 1997. See also Lara, *Cuicuilco*, 35.

25. According to California producers, juicy and tasty tomatoes are not ideal for processing; rather, they prefer those that are "bred for thick walls and lots of 'meat' per tomato." While dry and flavorless when raw, such tomatoes apparently "provide just the right color and texture for prepared sauces, salsas, and paste." The major concern, however, seems to be for the transport, not the taste: "The thick walls are what allows a pretty red tomato to survive at the bottom of one of those big truck bins. That durability means it goes bounce, instead of splat, when it becomes vegetarian road kill" on the highway. Carlos Alcala, "California Really Is the Big Tomato," *The Sacramento Bee*, 10 August 1997, B6.

26. Somewhat naively, I crossed this border by foot with three graduate students, armed with video cameras and tape recorders. It wasn't until we had passed by several checkpoints that one of the customs officials stopped us. "Do you realize you are in a highly sensitive area?" he queried and then admonished us: "You really shouldn't be here; we would be liable if there was a shoot out or if something happened to you."

27. By February 1999, a new CAADES complex had opened at a location thirteen kilometers south of the border and equipped so that all of the inspections could be done on the Mexican side of the border, facilitating quick passage through customs and easing the growing bottlenecks.

28. The "Quality Defects" listed in the Blue Book include maturity, cleanness, shape, smoothness, development, bacterial spot, bacterial speck, catfaces, puffiness, growth cracks, field scars, hail injury, insect injury, cuts or broken skins, and sun scald, while "Condition Defects" include color, sunken and discolored, sunburn, internal discoloration, freezing injury, chilling injury, alternaria rot, gray mold rot, and bacterial soft rot.

29. One problem with the FDA testing procedure is that it only tests for a limited number of chemicals and for single chemicals, while the impact of pesticides on humans comes from a cumulative or synergistic effect that results from being exposed to a variety of pesticides. Wright, *The Death of Ramon Gonzalez*, 196.

30. Personal interview with Jonathon Barnes, USDA inspector, Customs Complex, Nogales, Ariwna, 17 February 1999.

31. William Rees and Mathis Wackernage, *Our Ecological Footprint: Reducing Human Impact on the Earth* (Gabriola Island, Canada: New Society, 1996).

32. StatsCan, 1998.

33. OMAFRA, 1998.

34. Alex Murray (with Eric Krause), "The Ecological Footprint of Food Transportation," *Proceedings form Moving the Economy, An International Conference* (Toronto: Detour Publications, 1999), 84.

35. Linda Tons, *The Toronto Star,* 16 March 1997, 5F.

36. Interview by author with Greg P. Hummell, customs superintendent, Blue Water Bridge, Sarnia, Ontario, 2 January 2000.

37. Personal interview with Angelo Vento, manager, Ontario Produce, Ontario Food Terminal, 26 January 1999.

38. *The Morning Zoo,* a documentary by Daisy Lee, takes an intimate look at the early morning operations of the Ontario Food Terminal, where the Chinese Canadian filmmaker went frequently as a child with her farm family. A Grindstone Films Inc. Production, 1989.

39. Most of the information in this section is drawn from an interview by author and Anuja Mendiratta with Angelo Vento, sales manager, Ontario Produce, and Dominique Stillo, Joyce Foods, Toronto, 26 January 1999.

40. Interview with Angelo Vento.

41. According to Gary Lloyd, chief buyer for National Grocers, importers sometimes won't accept the tomatoes if there is too much wax on them. "It's really just like body builders; they put oil on and they look bigger and better. But customers go crazy when they pick it up and their hands are all covered." Interview by author and Stephanie Conway with Gary Lloyd, Erin Mills, 7 May 1997.

42. According to Larry Tenelia, sales manager for Pacific Produce, Canadian importers of fresh fruit and vegetables in Vancouver, the PLU code is for products that have to be sold by weight and are not scannable, while the UPC bar code is for commodities that are fixed weight and can be scanned. An importer must first get a code for any new item from the Produce Electronic Identification Board, located in the United States, having proven that they are moving a significant number of items. Personal interview by author and Stephanie Conway with Larry Tenelia, Vancouver, 8 August 1997.

43. Stickers on other fresh produce, such as Chiquita bananas, have also become microadvertisements for other products such as computers.

44. While many North American children have an aversion to tomatoes, they are generally drawn to ketchup, a highly sweetened tomato-based sauce, often added to McDonald's sandwiches or french fries. Ketchup recipes, however, vary by country and cultural tastes; a Del Monte executive in the Mexican plant revealed that Mexicans prefer acidic to sweet ketchup, so a different formula is used there. Interview by author with Jorge Betherton, vice president, Del Monte, Irapuato, Guanajuato, Mexico, 19 July 1996.

45. The Arch Deluxe, however, lasted for less than five years, recently replaced by the Big Xtra. See chapter 3, note 15.

46. Interview by author and Lauren Baker with Dominique Stillo, Joyce Foods, 1996; interview by author and Stephanie Conway with Larry Tenelia, sales manager for Pacific Produce, 8 August 1997.

47. "The coincidence of hungry mouths with overflowing grain silos may seem to be a paradox, but it is a paradox not of our analysis, but of capitalist agribusiness itself." Fred Magdoff, Frederick H. Buttel, and John Bellamy Foster, "Hungry for Profit: Agriculture, Food, and Ecology," *Monthly Review* 50, no. 3 July/August 1998): 3.

Driving South
The Globalization of Auto Consumption and Its Social Organization of Space
Peter Freund and George Martin

The development of a global economy is among the most important trends of the last several decades. Aided by heavy governmental support for transportation infrastructures, such as roads, wealthy countries became completely dependent on auto transport by the middle of the twentieth century. Now the automobile culture is spreading throughout the world. Freund and Martin explore the effects that the automobile, the most mass-traded durable good, is having on the organization of social space in poorer countries. The authors show that poorer countries face challenges in their attempts to adopt the automobile culture, including a lack of automobile infrastructure and the high costs of importing oil. Because of these challenges, pollution, health problems, dangers to pedestrians and bicyclists, and the loss of green space are exacerbated.

1. Introduction

Since its introduction over a century ago, the automobile has become an icon of freedom, progress, and modernity throughout the world. Auto ownership is used as an indicator of economic development; it is the leading mass-produced durable good in the world.[1] In developed countries, it has become the dominant transport modality; in parts of the U.S. it is the *only* viable means of every-day mobility. Most significantly, the attainment of individualized auto con-sumption for the great majority of populations in places like the U.S. has structured general perceptions of what is the most desirable and the most practi-cal means of mobility.[2] This attainment is made material by the now-embedded public infrastructures necessary for private auto use, from roads to regulating bureaucracies.

Countries of the South[3] are rapidly adopting Western modes of consump-tion such as mass auto ownership as a path to economic development. Between 1993 and 1997, vehicle registrations rose by 40 percent in the North, compared to 61 percent in the South. But what kind of development is it?

> But what, then, should we mean by development? And how do we mea-sure higher living standards? Are the living standards of a family in Bangkok raised, for example, when their cash income rises enough so that they can purchase an automobile—when the pleasure of car ownership is offset by the loss of free time caused by the need to work more hours, or even take a second job, to pay for the car and all its attendant costs, by the longer commuting time caused by other motorists exercising their free-dom, by lung cancer, by the generalized urban blight of jammed roads, car parks, petrol stations, used-car lots, drive-in fast-food franchises, brown skies and howling car alarms? Is this a higher standard of living?[4]

It is the replication of auto transport, etched in concrete and asphalt, that is the embodiment of a dream to which many in the world aspire. However, the auto represents more than transport. Its mass consumption has transformed the physical structure of landscapes, contributed to atmospheric, soil, and water degradation, and shaped, often in negative ways, the social organization of communities.

It is important, when considering these changes, not to see them as a prod-uct of "the automobile" but rather as a consequence of the way its use is pro-moted and organized, as part of an auto-centered transport *system*.[5] This system includes a vast material infrastructure of roadways, repair facilities, auto supply shops, gas stations and service facilities, motels and tourist destinations, storage spaces, and an extensive social infrastructure of bureaucracies for the control of traffic, the education of drivers, and the regulation of drivers, vehicles, and fuels (among other things).

In addition to their dependence on elaborate infrastructures and to their increasing scale of use, autos have qualitative features that maximize their use of space; they require multiple, dedicated spaces. Spaces have to be allocated not only in driveways, but also on roads, in parking lots, and at work sites. All of these spaces are difficult to use for any purpose other than temporary car storage; at other times they are vacant. Fundamentally, in auto-centered transport systems (ACTS) the auto is not only the dominant mode of transport; its use commands the disposition of much public space, which it wastes much more than it employs.

The emergence of auto-centered systems and their hegemony over public space in many parts of the world is not only a testament to the success of a particular technological form and a system for organizing its use, including the social life around it, but also to the extreme way in which capitalism has come to structure consumption. As a *mode of consumption,* auto-centered transport is a highly individualized, privately-owned form which is heavily subsidized by the state. The relationships among the auto, the environment, and capitalist economy revolve around the use of a technology which is the central durable commodity in the mature capitalist economy. This mode of consumption—not just the auto itself—is being diffused to countries in the South, a diffusion that has accelerated in the global, neoliberal 1990s.

2. The Globalization of Auto Hegemony

While auto production and consumption became worldwide phenomena at the end of the 20th century, great variation in the concentration of ownership remains. . . . [J]ust three nations in the world—the U.S., Germany, and Japan—account for 53 percent of world vehicle production and 49 percent of consumption, while they have only eight percent of world population. Their domination of auto production and consumption is a major factor in their supremacy over the world economy.[6] Despite the ongoing economic restructuring based on technological advances in computers, electronics, and telecommunications, the auto-oil industrial complex retains a leading position in the global economy; in 1997, it accounted for six of the world's 10 largest business firms.[7]

Auto densities in the world range from a high of 1.3 persons per vehicle in the U.S. to a low of 12,913 persons per vehicle in Afghanistan. For purposes of analysis, the nations of the world can be sorted into three groups by their degree of auto-centered transport systems (ACTS):[8]

1. The counties with the most developed ACTS feature three or less persons per vehicle, in a total of 38 places. The most populous among these places are the U.S., Japan, Germany, Italy, the UK, and France. Africa and Central and South America have no developed ACTS, but several oil-rich nations in Asia—Brunei, Kuwait, and the United Arab Emirates—do have developed ACTS. In places with the most developed ACTS, autos dominate the cities, auto-induced urban sprawl is considerable, and the countrysides have become auto-dependent.

In the nation with the most developed ACTS, the U.S., we can speak of the emergence of *hyperautomobility*—characterized by very high auto density (highlighted by the specialization of vehicles within households) and by very high auto use (indicated by increasing trips and distances, but with decreasing vehicle occupancy).[9] The hyperautomobile U.S. is different from the other nations with the most developed ACTS. While U.S. cities average 54 percent more autos per person than European cities, auto use is even more pronounced. Auto kilometers per year per person average 143 percent more in U.S. cities than in European cities.[10]

2. The mid-range ACTS countries have 4–50 persons per vehicle, in a total of 70 places. The most populous mid-range ACTS nations are Russia, Brazil, Mexico, Turkey, and Thailand. Mid-range ACTS countries include all of the former centrally planned economies of Eastern Europe and most of South America, as well as Israel, South Korea, and Taiwan. In these places with mid-range ACTS, the cities are also auto-dominated, but auto-induced urban sprawl is not yet pronounced and cities maintain a moderate level of transport diversification. The countrysides retain a mix of motorized and non-motorized transport.

3. Finally, the least developed ACTS countries have more than 50 persons per vehicle, in a total of 50 places. All these are in Africa or Asia, with the exception of the following nations in the Americas: Bolivia, Cuba, El Salvador, Guatemala, Haiti, and Honduras. The most populous nations are China, India, Indonesia, Pakistan, Bangladesh, and Nigeria. In these countries with the least developed ACTS, cities have been penetrated by autos but retain considerable diversity of transport modalities, while the countrysides are virtually auto-free. There is little or no auto-induced urban sprawl. . . .

In many respects, car troubles are an urban phenomenon, and this is starkly illustrated in the growing transport problems of the megacities of the South. Autos have been increasing at relatively high rates in many Asian, African, and Latin American cities, so that these cities at first glance resemble those of the North—featuring super highways and super traffic congestion. However, there are several significant differences in the current status of auto transport in cities of the North and the South:

1. In the South, auto transport has not become a mass phenomenon; it is largely restricted to the relatively small elite and middle class sectors, which constitute a small minority of the population.

2. Because auto transport is the privilege of a minority in Southern cities, these cities retain a more diversified transport modal split relative to the North. The working and poor classes of Southern cities depend on public transport (where available) and upon non-motorized transport, especially cycling, walking, and animal power.

3. Because many of the nations of the South are relatively poor and debt-laden, the public costs of auto transport come at a greater social price than they

do in Northern cities. Principal among these costs is the provision of an adequate sociomaterial infrastructure, including roadways and regulating bureaucracies. The development of infrastructure lags behind the introduction of the car in Southern cities. Infrastructures for walking and other non-motorized forms of transport are even less developed than auto infrastructures. For instance, over 70 percent of Jakarta's roads have no sidewalks.[11] Mass transit is similarly neglected as scarce resources are diverted to subsidize ACTS. One of the more dramatic illustrations of this uneven development is the relatively high toll in roadway-related human carnage in Southern nations.

4. Related to this infrastructure issue is a special problem for many Southern nations: They have to import oil to operate autos. Since autos can run without an adequate material and social infrastructure but not without oil, the oil-poor nations of the South are further disadvantaged, as their meager resources can be sapped by oil import costs.

... [S]ince auto markets in the developed nations have matured, the great potential for new markets exists in the South. The greatest increases in new car sales from 1997 to 2005 are forecast to be in China, India, and the Philippines.[12] For auto makers such a trend is hopeful, yet the globalization of auto-centered transport represents an immeasurable ecological threat and badly aggravates already existing social inequities in access to transport. In short, ACTS are neither sustainable nor equitable.

The growth of auto-centered transport in the South thus brings into sharp relief issues of social inequality, the inappropriate use of transport technology, and the limits of globalizing such an energy-and-resource intensive and environmentally unfriendly system. The ideal of a motorized world built in the image of Southern California is simply not physically and economically feasible on a global scale.

The continuing diffusion of ACTS into the South is a major contributor to the South's growing ecological problems. The environmental problems of auto-centric transport in the South, especially air, soil, and water pollution, have been well documented.[13] Here, our focus is on the social organization of space promoted by auto-centered transport.

3. Modes of Consumption and the Social Organization of Space

In considering the diffusion of auto-centered transport systems, it is important to emphasize the sociomaterial aspects of consumption, of which two are central. First the resource (including especially land) and energy-intensive nature of auto-centered transport systems is at the core of the social and ecological problems that these systems cause. In the 1990s, the environmental problems caused by such systems have become global issues (e.g., global warming). Second, most central for our analysis is the impact of auto-centered transport as a mode of consumption on the *social organization of space.*

While the negative aspects of the diffusion of the auto are obvious enough in countries of the North, in the poorer and less developed South they are gravely aggravated. Surface transport (in this case the auto) has been described by some as an "engineering industry" that is not carried out inside a factory but outside in public space.[14] In the South, just as the economic and technological development of the means of production is uneven, so is the development of the modes of consumption. Uneven development here means that the spatial and other contradictions of auto-centered transport are accentuated in countries of the South.

What are being exported to the South are not only technological forms that originated in the North, but consumerism, mainly modeled on that of the U.S. In this evolving model of consumerism, the auto is the transport commodity for local elites and is rapidly becoming the "privileged means of urban transportation."[15] These newly emerging auto consumption patterns have influenced urbanization and drained scarce public resources in the South.

One looming consequence of the diffusion of such a mode of consumption to the South revolves around issues of sustainability or what some have called the "China factor."[16] What happens to space in the form of arable land if, for instance, auto transport was to be adopted by all 1.2 billion Chinese? Total vehicle production increased by 152 percent between 1991 and 1998 in China and its mix changed as well. While the proportion of passenger cars was only six percent in 1991, it rose to 31 percent in 1998.[17]

While arable land is not in scarce supply in many Northern countries, in many nations of the South it is.

> China, which has almost exactly the same land in area as the U.S., has four times as many people living on it. Since such a large proportion of China is desert or mountains, its population is crammed into dense concentrations around the great river valleys. As a result, the country must feed more than one-fifth of the world's population on less than one-fifteenth of its farmland.[18]

Does it make sense for China to pave over arable land or land usable for dwelling spaces? Other countries of the South face similar problems, especially Egypt, Bangladesh, and Indonesia.[19] For example, each year in Indonesia, 250 square kilometers of agricultural land, forest, and wetland become roads and urban spaces, displacing large numbers of people.[20]

The use of land by autos reflects the great social inequality that exists in the South.

> In developing countries in general—and in Brazil in particular—transport and traffic policies, coupled to economic and social policies, have crystallized remarkable differences between those with and without access to private transport. Most decisions have a common objective: to adapt space to the use of the automobile for selected social groups.[21]

In the North, the disenfranchisement of poor people, people displaced by freeways, people with disabilities and older people and children is one harsh feature of auto-centered systems. In the South, such disenfranchisement is amplified, particularly in African and Latin American cities in which alternative modalities are underdeveloped and human-powered vehicles are not as widespread as in many Asian cities.

In the less developed auto-centered systems of the South, spatial contradictions and their impact on safety are glaring, since the possibilities of technological fixes (e.g., resources for more benign organizations of traffic) are not available, and the technical potentials of mass automobility have not been realized. While road deaths in the developed countries are down to less than five per 10,000 vehicles per year, in the developing countries the picture is quite different: 40 deaths per year per 10,000 vehicles in India, 77 in Bangladesh, and 192 in Ethiopia.[22] Transport space is in poor condition and it disenfranchises and is unsafe for the great majority who are not auto users. Traffic control is poor, as are mass transit alternatives. . . .

In South Africa, extremes of wealth and poverty contribute to one of the world's worst safety records. Luxury cars mix with overcrowded trucks (used as buses), donkey carts, cows, and pedestrians to produce a deadly combination.[23] Black townships do not have sidewalks, adequate lighting, or pedestrian overpasses on the roads through which affluent-owned high-powered vehicles race. For most blacks, transport (even now, in post-apartheid society) in cities is confined to public transport. Whites travel mostly by auto: one out of two white South Africans owns a car; only one of 100 blacks do.[24] South Africa, unlike many other countries of the South, has a developed transport infrastructure—but one which excludes modalities other than the automobile.

The growing space demands of auto owners are also degrading public places, the urban commons, in cities of the South; for example, the principal plaza in Mexico City. What used to be a popular and pleasant place to walk and socialize is now "filled with deafening traffic and the air is blue with car exhaust—trying to walk is more dangerous than driving."[25] Another example is Bangkok. Once called the "Venice of Asia," the city has paved over its canals; still, gridlock and pollution grow.[26] . . .

Autocentric transport demands massive public investment of land and resources for its infrastructures. The public purses of most nations of the South are not big enough to make these investments. In many cities of the South, the result is traffic congestion on an unprecedented level, congestion that carries great inefficiency costs in the transport of goods as well as workers. For example, in Sao Paulo, the lack of an adequate subway system and of well-developed auto infrastructures has produced perhaps more traffic noise, air pollution, and congestion than in any other city in the world. However, consistent with their positions of power and their orientations to technologies of the North, wealthy Paulistanos are buying themselves out of this morass—by purchasing helicopters. At over 400, the fleet of private helicopters is the biggest of any city in the developing world, trailing only the fleets of New York City and Tokyo. The social class contradictions of such uneven development is highlighted by the fact that "it is easier for a wealthy person to buy a helicopter than it is for a working class person to buy a car."[27] . . .

. . . Poor people, particularly poor rural women, have an unequal transport burden. In effect, they are invisible to transport planners mainly concerned with large scale economic activity and with motorized traffic.[28]

In many countries of the South where other modalities are in pervasive and intensive use (e.g., bicycles in Chinese cities), these modalities are being pushed to the side and increasingly marginalized. Human-powered vehicles such as rickshaws and bicycles are seen by the middle classes and elites as archaic and as impeding the smooth flow of motorized traffic. In Bombay and Jakarta, such vehicles have been banned and in Manila they have been removed from the main roads.[29] Transport planners dislike large amounts of mixed traffic and their bias is to eliminate any obstacles to motorized traffic. Calcutta, despite a shortage of revenues, is investing in a huge new road infrastructure complete with flyovers, motor ways, and the rest. These structures will exclude the cycle rickshaws which are a source of jobs for many of Calcutta's poor.[30] In Bangladesh approximately 1.25 million people are directly involved with driving and maintaining rickshaws, while five million people depend on them for subsistence. Rickshaws are not part of government planning and in a government report, they were described as "slow moving" vehicles that should eventually be eliminated and replaced by automobiles and trucks.[31]

4. Transport Policy and Change

Governments in Southern countries are eagerly adopting the auto as the primary means of transport. Automobility is viewed as a sign and a means of economic development. Malaysia's experience with its government's "National Car Project" in the 1980s is an example of the huge costs involved in developing an auto industry. The auto is an up-market product that costs more than the average Malaysian house. In addition to sinking funds into production of the car, the government built new auto infrastructures, while cutting expenditure for rail and bus transport. Despite this public investment, the Malaysian car faces immense challenges.[32] Local critics argue that the experience demonstrates the folly of such industrial mega-projects for developing nations.

While most cities in the South have allowed free rein to the auto, some have restricted its use. One successful example of diversifying transport in the South is Curitiba, Brazil, a city of 1.6 million. Beginning in the 1970s, the city adopted a series of transport-related measures, including improved bus transit, cycle ways and pedestrian ways, and integration of transport and zoning policies, in which higher densities were encouraged along major arterials and a mix of jobs, homes, and services were included in local areas. Additional policies have included traffic calming schemes and in-fill, in which new development is sited in abandoned land in the existing city rather than sprawling outward. Several improvements have been attributed to these policies, including the facts that Curitiba's rate of accidents per vehicle is now the lowest of Brazilian cities and its fuel consumption per vehicle is 30 percent less than in other Brazilian cities of its size. Finally, residents of Curitiba spend about 10 percent of their incomes on transport, one of the lowest such rates in Brazil. This is despite the fact that the city's auto ownership rate is high by Brazilian standards, second only to Brasilia's.[33] Various grassroots projects, such as those sponsored by the Institute for Transportation and Development Policy, are trying to develop and sustain transport diversity, especially those modalities that are less energy-and-resource intensive and are useful to poor people, women, and people with disabilities. "Afri Bike," for instance, promotes bicycles as development tools, sponsoring an urban-based center where vendors can lease load-carrying bikes.[34] . . .

Mobility is a growing problem for poorer people in the South. There, the increasing cutbacks in the public sector because of debt repayment, coupled with population growth in urban areas that outdistances the availability of public transport, deprives many people of any form of mobility except walking. . . .

Not only do many Southern governments subsidize auto transport and the auto industry at the expense of alternatives, so, too, do banks. Non-motorized forms of travel such as improved bicycles (e.g., using light metal) and carts have received virtually no subsidies. Alternative transport, which local elites view as "backward," is sacrificed to the auto. Yet, in the South, bicycles are a source of jobs and foreign exchange, and generate small-scale entrepreneurial activities such as vending, scrap collecting, and delivery services. Moreover, their manufacture and maintenance are labor-intensive enterprises.

Because of government support for the wholesale introduction of auto technology many Southern cities have become caricatures of the most auto-centered cities in the U.S. Cities such as Caracas have almost unimaginable journeys to and from work (up to seven hours per day), elaborate freeways without adequate feeder arterials that result in massive congestion, and highways that stop abruptly at the edges of old city centers because of lack of space. These cities could hugely benefit from improved bus service, bicycle and jitney use, and other less energy-and-resource intensive modes of transport.

5. Conclusion

The automobile and its socioenvironmental consequences revolve around its dual aspects as a transport *system* and a *mode of consumption*. When speaking of development, one is not simply talking about technological diffusion, but also the diffusion of consumption patterns. In countries of the South, the automobile has spread to local elites and is beginning to dominate (especially urban) space and to drive other modalities out. Just as with the diffusion and globalization of a mature capitalist mode of production, so too the diffusion of a mature capitalist mode of consumption takes place in an uneven fashion.

A consumption mode of transport characterized not only by auto hegemony but by auto dependence is increasingly proving itself to be a form of unsustainable development and a socially destructive mode of consumption. Its diffusion to countries of the South aggravates its problematic social and environmental consequences in the short term. Even in the long term, if countries of the South could fully develop an auto-centered transport system such as that which exists in North America, the results would be disastrous in terms of the impact on the ambient environment and, above all, on social space. Deconstructing the taken-for-granted notion that autocentric transport is "progress" can lead us to reconsider such general questions as what is the "good life" and what constitutes material prosperity, and what kind of transport investments and planning need to be made—in the *North*, not only the South.

What distinguishes the auto from other consumer goods (including other durable goods) is that while the latter may also be energy-and-resource intensive, the car's pervasive and intensive use requires a great deal of *space*. The bulk of this space, furthermore, is the most desirable *public* space (e.g., in urban centers). The auto appropriates valuable public space and, particularly in countries of the South, makes this space virtually unusable for other non-motorized modalities (including walking). In this way, the adoption of autocentric transport can spearhead a more general development of capitalist consumption, i.e., the car provides a material inlay for Western consumerism. For example, in Venezuelan cities, auto-centered consumption patterns have been the catalyst in privatizing public space and in individualizing consumption.[35]

The auto is a private consumer good that is (in the North as well as South) heavily subsidized by the state. Scarce public resources go into highways and the infrastructures that automobiles need in order to be intensively and pervasively used by a minority of citizens in the South. Yet little is invested in ameliorating the impact of emerging automobility on social and physical environments. Technical fixes (e.g., systems of traffic control) and educational campaigns for safety are not nearly as developed as in countries of the North. There, the systemic limits to emission control and accident prevention are being reached, with further significant reductions only possible through social changes—changes in travel patterns and modal splits, both of which lead to a consideration of the reorganization of space.[36]

As a means of empowering the mass of the population and contributing to a more sustainable form of transport, a *diversity* of modalities should be developed and subsidized. Non-motorized vehicles can be appropriate technologies, not only for countries of the South, but also in the North (e.g., countries like Denmark and the Netherlands). Mass transit and non-motorized transport need to be valorized and taken seriously in technological development (e.g., new metal alloys, solar power). Developing such modalities can represent a move "back to the future." Asian cities, particularly those of China, could benefit a great deal from such a shift in development policy.

Contemporary auto technologies do not co-exist well with other uses of their spaces. Highways built for autos are not hospitable for walkers and cyclists. Even motorized public transit like buses find it difficult to use auto-centered space. For example, effective bus transit requires frequent stopping for on- and off-loading passengers. For this reason, local buses are unable to make full use of auto highways. In the North, where auto-centered transport has matured, other modalities have been pushed aside, especially in the U.S. It will be difficult to re-engineer social space in the U.S., so that other modalities can be effectively used—even if the political will were present. In cities of the South, diversification in transport still remains, but it is gradually being pushed aside as auto consumption rises. While the South could profit from the negative lesson of the North and preserve (and modernize) their transport diversification, it is not happening. So far the neoliberal global development model has been too powerful.

... Transnational corporations based in Western Europe, North America, and Japan view the South—especially the so-called emerging market countries—as fast-growing, potentially vast markets for their commodities, as their own markets become saturated with goods and tend to stagnate. These corporations and the nation-states of the North are trying to globalize the way of life that evolves from automobility in general and to help develop the material infrastructure for the use of automobiles in the South in particular. The auto industry, petrochemicals, and other sectors dependent in part or whole on mass automobility are clamoring for a greater play in the South. Mass motorization in the South, it is thought, is the answer to excess productive capacity and profit shortfalls in the North.

While it is easy for countries of the North to tout sustainable development (having reaped the dubious benefits of unsustainable development), the countries of the South suffer the most. It is thus the North that needs to take the initiative in "greening" both production and consumption. This is because the

North has done the most damage to the environment (especially as imperialists and neo-imperialists in exploiting resources in the South); has the material resources to make needed changes (some have termed the countries of the North post-scarcity societies); and as a model for development (particularly the U.S.) has the moral obligation to provide an example of a greener and socially less destructive way of life.[37]

Whether or not there can be a greening of consumption in the North *or* the South depends on planning the global market place, shifting production to less energy-and-resource intensive consumer goods, encouraging more sustainable and socially democratic modes of consumption—in all countries of the world.

Notes

Freund, Peter, and George Martin. 2000. "Driving South: The Globalization of Auto Consumption and Its Social Organization of Space." *CNS*, 11, no. 4: 51–71.

1. Unless otherwise indicated, data sources were: American Automobile Manufacturers Association, *Motor Vehicles Facts & Figures*, 1997 (Detroit); United Nations, *Human Development Report*, 1996 (New York); United Nations, *Statistics of Road Traffic Accidents*, 1994 (New York); U.S. Department of Commerce, *Statistical Abstract of the United States*, 1998 (Washington); Ward's Communications, *Motor Vehicles Facts & Figures*, 1999 (South field, MI); World Health Organization, *The Global Burden of Disease*, 1996 (Geneva).

2. Peter Freund and George Martin, "The Commodity That Is Eating the World: The Automobile, the Environment, and Capitalism," *CNS*, 4, 1996.

3. We would agree with Wolfgang Sachs, et al. (*Greening the North*, London: Zed Books, 1998) that with the dissolution of the Second World, the Third World is no longer (if it ever was) a homogenous category, with its countries dividing into underdeveloped and "emerging market economies." Following Sachs and others, we will use "North" and "South," accepting all of their imprecision. We also separate the South into two parts, based on levels of auto-centeredness.

4. Richard Smith, "Creative Destruction: Capitalist Development and China's Environment," *New Left Review*, 222, 1997, p. 28. The debate over measuring standards of living in quantitative versus qualitative terms first began between those looking at growth statistics during the English Industrial Revolution versus those whose focus was the loss of community, the decline of social cohesion, environmental destruction, and the like.

5. While we speak of autos, our data refer to motor vehicles, which includes passenger cars, light trucks, commercial trucks, and buses. However, passenger cars are numerically dominant. In the world in 1997, they represented 71 percent of all motor vehicles. It is also valid to include trucks as autos because in the U.S. about one-half of current passenger car sales are classified as light trucks, a category which includes SUVs, vans, and pickup trucks. The SUV is of course a new phenomenon that deserves special treatment.

6. See Peter Dicken, *Global Shift: Transforming the World Economy* (New York: Guilford Press, 1998), Chapter 10.

7. *Forbes*, July 28, 1997, p. 180.

8. The long-standing assumption of a close link between automobility and economic development (wealth) may be no longer valid. See Peter Newman and Jeffrey Kenworthy, *Sustainability and Cities: Overcoming Automobile Dependence* (Washington, DC: Island Press, 1999), pp. 111–114.

9. George Martin, "Hyperautomobility and Its Sociomaterial Impacts," Working Paper, Centre for Environmental Strategy, University of Surrey, Guildford, February, 1999.

10. Newman and Kenworthy, op. cit., p. 80.

11. ITDP, "Jakarta's Non-motorized Modes, 'Living Dangerously,'" *Sustainable Transport*, 6, 1996, p. 8.

12. *The Economist,* June 13, 1998, p. 142.

13. See Kenneth Button and Werner Rothengatter, "Global Environmental Degradation: The Role of Transport," in David Banister and Kenneth Button, eds., *Transport, the Environment and Sustainable Development* (London: E. & A.N. Spon, 1993); Asif Fail, "Motor Vehicle Emissions in Developing Countries: Relative Implications for Urban Air Quality," in Alcira Kreimer and Mohan Munasinghe, eds., Discussion Papers, No. 168 (Washington, DC: World Bank, 1992); Asif Fail et al., *Automotive Air Pollution—Issues and Options for Developing Countries* (Washington, DC: World Bank, 1990); World Bank, *Sustainable Transport: Priorities for Policy Reform* (Washington, DC: World Bank, 1996).

14. David Banister, *Transport Planning in the UK, USA, and Europe* (London: E & AN Spon, 1994), p. 20.

15. Tom Angotti, "The Political Economy of Oil, Autos and the Urban Environment in Venezuela," *Review of Radical Political Economics*, 30, 1998, p. 101.

16. John Whitelegg, *Critical Mass* (London: Pluto Press, 1997).

17. Today, Volkswagen has three car factories in China; Suzuki has four factories; Daihatsu, Citröen, General Motors, Honda, Audi, and Daimler-Chrysler have one each (*New York Times*, August 3, 2000).

18. Smith, op. cit., p. 33.

19. Lester Brown, "The Future of Automobiles," *Society*, 21, 1984, p. 65.

20. Walter Hook, "Jakarta: A City in Crisis," *Sustainable Transport*, 8, 1998, p. 14.

21. Eduardo A. Vasconcellos, "Urban Transport and Equity: The Case of Sao Paulo," *World Transport Policy and Practice*, 4, 1998, p. 16.

22. Oliver Tickell, "Death Duties," *The Guardian*, June 24, 1998, pp. 4–5.

23. Donald G. McNeil, "South Africa's Well-Kept Roads, a Grim Harvest of Traffic Deaths," *The New York Times*, December 25, 1997, p. 5.

24. John Griffin, "South Africa: Mobility in the Post Apartheid City," *Sustainable Transport*, 3, 1994, p. 8.

25. Wayne Ellwood, "Car Chaos," *New Internationalist*, 195, 1989, p. 6.

26. Smith, 1997, op. cit., pp. 29–30.

27. Simon Romero, "Rich Brazilians Rise Above Rush-Hour Jams," *The New York Times*, February 15, 2000, p. A4.

28. Priyanthi Fernando, "Gender and Transport," in Saskia Everts, ed., *Gender and Technology* (London: Zed Books, 1998).

29. Whitelegg, op. cit., p. 47.

30. Paul Brown, "Road Accidents Set to be the World's Biggest Killer," *The Guardian*, June 24, 1998, p. 14.

31. Whitelegg, op. cit., p. 47.

32. Halinah Todd, "The Proton Saga Saga," *New Internationalist*, 195, 1989, pp. 14–15.

33. Marcia D. Lowe, "Shaping Cities," in Lester Brown, ed., *State of the World* (New York: W. W. Norton, 1992); Jonas Rabinovitch and Josef Leitman, "Urban Planning in Curitiba," *Scientific American*, March, 1996.

34. Karen Overton, "Women Take Back the Streets," *Sustainable Transport*, 3, 1994, p. 6.

35. Angotti, op. cit.

36. Deaths from roadway accidents (relative to the number of motor vehicles for vehicle miles driven) in the U.S., for example, have been falling. However, the absolute number of deaths remains stubbornly high. Deaths slowly declined from a high of 51,093 in 1979 to a low of 39,250 in 1992. Since 1992, they have inched up again, to 41,480 in 1998, despite the introduction of improved safety technology (such as airbags) and the tightening of social controls (over drunk drivers, for instance).

37. Sachs, et al., op. cit.

The Environmental Justice Movement

Equitable Allocation of the Costs and Benefits of Environmental Management Outcomes

David N. Pellow, Adam Weinberg, and Allan Schnaiberg

In the final piece of this section, the authors present a case study of an incinerator siting in Illinois within the context of an analysis of the international trade in hazardous waste. The authors argue that successful opposition to an environmental threat in one community can lead to another, less powerful community becoming exposed. The solution proposed in this essay is for local, regional, national, and international policymakers to include environmental justice principles and research into their rule making.

The majority of EJ [environmental justice] research has focused on the distribution of hazardous facilities in vulnerable communities and on local responses to these policies. The latter studies generally centered on the importance of organized community-based resistance to facility siting—the environmental justice movement. Recently, researchers have begun to explore several other areas of environmental justice concern, including the workplace, housing, and transportation. These studies reveal that, paralleling LULU [locally unwanted land use] conflicts, the poor and people of color are disproportionately impacted by pollution on the job, in their homes, and via transportation systems (Bullard and Johnson, 1997; Robinson, 1991).

Scholars have typically argued that environmental racism occurs when the poor or people of color are "dumped on" or exposed to hazards, because they are less powerful than corporations and the state. . . .

An Alternative View: Environmental Inequality as an Unfolding Process

We argue that environmental inequality unfolds in ways that are more complex than most written accounts of EJ struggles reveal. Not only are industry and government often guilty of perpetrating these acts of injustice, but many times community leaders, neighbors, and even environmentalists are also deeply implicated in creating these problems. In order to understand how and why environmental inequality disrupts communities, we must extend our research further into the history and the roles of the many people and organizations involved. We need to study the impacts of social inequality, and the power of the disenfranchised to shape the outcomes of these conflicts. . . .

. . . Environmental inequality is a social process involving and impacting many actors, institutions, and organizations. These actors, or "stakeholders," often include social movement organizations, private sector firms, the state, residents, and workers. Each group's interests are complex and often involve crosscutting allegiances. Thus it is often difficult to distinguish between perpetrators and targets. Our environmental justice framework allows us to move beyond a view of environmental racism where hazards are unilaterally and uniformly imposed upon victims, who then react. Instead, we offer a scenario wherein many actors are viewed in their full complexity. Putative victims often become active agents in resisting and shaping environmental inequalities before, during, and after they emerge. Within this framework, we also move beyond a view wherein outcomes are defined simply as the presence or absence of hazards. In contrast, we can account for variations in *patterns* of environmental inequalities. In its present state of development, the environmental justice literature does not provide an adequate accounting for why such variations occur.

Environmental inequality occurs as different stakeholders struggle for access to valuable resources within the political economy, and the benefits and costs of those resources become distributed unevenly. Stakeholders who are unable to effectively mobilize resources are most likely to suffer from environmental inequality. Conversely, stakeholders with the greatest access to valuable resources are able to deprive other stakeholders of that same access. This perspective captures the dynamic nature of environmental inequality. Valuable resources can include clean living, recreational, and working environments. They can also include power, wealth, and status. Thus, the inability to access these resources often means living and working under dangerous conditions, with very little power, wealth, or status. Conversely, those stakeholders with the ability to access these resources live and work under safer, healthier conditions with more power, wealth, and status. In communities where environmental inequality is evident, immigrants, people of color, low-income populations, and politically marginal groups tend to bear the brunt of the pollution, the toxins, and the risk. Environmental inequality has a negative impact on people's physical and psychological well-being, and on the health of entire communities.

Since power is often correlated with race, over and above class in U.S. society, we argue that environmental inequality also has a racial dimension. People of color are likely to lack economic power because institutional racism is embedded in every facet of the labor market process (Wilson, 1996). Communities of color also continue to lack equal access to political power. They have a harder time getting elected to political office and experience barriers to raising the massive amounts of funds needed to exert influence through lobbying organizations. Thus, many communities lack power for racial reasons. This is not a surprising sociological argument (Bullard, 2000; Hurley, 1995; Marbury, 1995; Rodney, 1982).

Our framework stresses four major points. . . . First is the importance of process and history; second is the role of multiple stakeholder relationships; third is the impact of social stratification such as institutional racism and classism; and fourth is the ability of those actors with the least access to resources to resist toxics and other hazards. It should be evident from these four factors that environmental inequalities are not always simply imposed unilaterally by one class or race of people on another. Rather, like all forms of stratification, environmental inequalities are relationships that are formed and often change through negotiation and conflict among multiple stakeholders.

To illustrate this framework, we present the recent and well-publicized battle over the Robbins Incinerator in Chicago, IL. As this historical case study makes clear, a focus on the sociopolitical processes by which the Robbins Incinerator emerged reveals a more complex process than can be captured by a simple "perpetrator–victim" analysis.

Divide and Conquer: The Fight for and against the Robbins Incinerator

Historical Factors

In 1992, Robbins, an all–African American suburb of Chicago, was literally begging a waste company to locate an incinerator in its borders. The typical responses by EJ activists in Chicago were "How in the world did that happen?" and "Is that environmental racism or not?" These are two of the many perplexing questions this case raises. Without addressing them, we will never successfully move toward a coherent theoretical framework for understanding EJ struggles.

In 1987, EJ groups around the nation were attacking the waste industry, blocking landfills and incinerators in scores of conflicts. Fearing a sudden exhaustion of landfill capacity, the Illinois state legislature responded by passing the Retail Rate Law. This law provided significant tax subsidies to incinerator companies that chose to locate in the state. This was the state of Illinois' way of saying "we're open for business and we'll pay you to burn trash here." Not surprisingly, this legislation sparked an enormous growth in incinerator proposals in Illinois.

This is an example where the EJ movement was in fact influencing (if not driving) policy. The shift from landfills to incinerators was both an unintended consequence of the anti-LULU movement. In 1987 there was only one incinerator in the state of Illinois. In the next several years, nine incinerators were either proposed or built in the state, seven of which were in African American communities, while the other two were in working class and/or white ethnic neighborhoods (Portney, 1991, 138ff.). African Americans comprise only a fraction of the state's population, so any observer of Illinois' incinerator siting practices would likely conclude that this pattern did not emerge by chance. Both class and race intersected to position communities of color, working class, and low-income and ethnic European populations as the most attractive candidates for LULU sitings.

Robbins is an historic African American town. Founded in 1917, it is the oldest all-black governed town in the northern United States. Today, unfortunately, this village of 7000 people is also one of the poorest communities in the United States. It faces mounting financial debts, including a $1.3 million unpaid water bill owed to the City of Chicago. Since the 1980s, the city has held charity drives to raise funds to pay for the city's operations. This is a ghost town as far as most businesses are concerned, a barren landscape with few prospects for economic development. In fact, the major economic activity in Robbins consists of a nursing home, liquor stores that open at 8 a.m., mom-and-pop convenience stores, and the illicit trade in drugs and sex. Like many cities, Robbins experienced a mass exodus of both businesses and residents during the period between 1970 and 1990. Since that time, each mayoral administration has courted all manner of businesses, to no avail—until the early 1990s.

Robbins had been working for some time to attract any sort of business, including waste management facilities. In the 1970s, the administration developed a plan for an "energy park," where a waste-to-energy incinerator and several ancillary industries would be sited. This idea languished for a while because landfilling was cheap and even waste firms had reservations about the level of poverty and destitution in the village. Finally, in 1986, the Illinois Senate approved a measure to acquire land for an incinerator in the village of Robbins. Robbins struck a deal with the Reading Energy Company (and later the Foster Wheeler company). With the taxpayer subsidy from the Retail Rate Law, the company could boost annual profits from the burner to $23 million, and it would pay the village nearly $2 million in rent each year, thus doubling the city's revenue.

Resistance against Inequality: Stakeholder Battles

This proposal immediately spawned a major campaign against the incinerator. A regional EJ coalition emerged to oppose the "burner." Most of the organizations in this coalition were located outside Robbins. Soon thereafter, there emerged several strong-willed residents and leaders within Robbins who stood up against the incinerator as well. These groups quickly joined the EJ coalition. In a memo written by the coalition, these organizations outlined the hazards of incineration and the environmental injustices of locating this burner in Robbins:

> Minority and low-income communities have long been the favorite dumping ground for society's pollution. Recently, there has been a growing movement demanding environmental justice. To that end, we should oppose facilities such as Robbins that wish to add pollution to our already overburdened populations. Although our communities need additional jobs, this cannot be justification enough for exposing our families to a considerable health risk. Ironically, it has been certain African American politicians who seem to have blocked an effort to repeal an Illinois law enacted in 1987 (the Retail Rate Law) that subsidizes the construction and development of incineration facilities like Robbins, using Illinois taxpayers money.

This was a case of environmental racism, they argued. While the company had not targeted the community, the community was in such dire economic straits they felt they had no choice but to court such an incinerator (Portney, 1991, 138ff.). In flyers, speeches, and letters to newspapers, the EJ coalition sent a message to the public that an incinerator burning 1600 tons of refuse per day would release dangerous levels of mercury, dioxin, and other carcinogens into the air. This would not only impact the health of people in Robbins, but also put people in adjacent communities at risk. Also, many of the 80 jobs Foster Wheeler was promising to Robbins residents would be hazardous and low-wage, while any white collar jobs would go to non-Robbins residents. The price tag that taxpayers would pay for this facility was $730 million and that figure alone catalyzed a lot of support for the anti-incinerator cause. The coalition also claimed that incinerators are hazardous operations because they routinely experience a high rate of accidents. The movement argued that recycling initiatives would be much more economically and environmentally sound, and could produce many more jobs with less funding.

At a hearing in 1992, scores of incinerator supporters (outnumbering opponents) attended, wearing painters' caps that read: "Yes. In My Backyard," "Don't Trash Jobs," and "Right 4 Robbins." Incineration opponents wore shirts that read: "Ban the Burn" and "Over Our Dead Bodies." The *Chicago Tribune* was staunchly in favor of approving the Robbins incinerator, printing editorials decrying antiburner protests and the regulatory hurdles placed before the facility's developers. In 1993 the incinerator received the green light from the Illinois EPA and was given a permit. But the movement against the facility moved forward.

The EJ coalition drew on a two-pronged strategy. First, they put the spotlight on the Robbins facility and framed it as environmental racism. Second, they sought to repeal the Retail Rate Law. Without this law, Foster Wheeler and other incinerator companies would lose profits and eventually leave the state. In 1993, nearly a dozen south suburban municipalities passed resolutions opposing the Robbins incinerator. This meant that the potential regional sources of trash to fuel this burner were dwindling rapidly.

Robbins' Mayor Irene Brodie, the person leading the campaign to court the burner, cleverly framed this conflict as one stirred up by "outside agitators." In doing so, she challenged the EJ coalition's legitimacy. The mayor also used the observation that vocal white citizens in the region were universally against the incinerator as "evidence" that whites did not want to see an all-black town "pull itself up and out of poverty." Furthermore, Brody claimed that if activists prevented Robbins from building the incinerator, that such an outcome would constitute environmental racism! In what some activists have called a "twisted logic," Brodie proudly proclaimed, "There's always been environmental racism. We're just making it work for us for once."

The Foster Wheeler company's reaction to the EJ coalition was predictable: similar to incineration proponents a century before, they claimed in billboard and newspaper ads and in trade journals that they planned on constructing a state-of-the-art facility that would be clean and safe: "Pollution emissions will be low," they argued. These promises were dashed when, after its construction in 1997, the Robbins incinerator was plagued by several accidents, Clean Air Act violations, and continued negative campaigns by environmentalists. Environmentalists helped repeal the Retail Rate Law that same year, halting all subsidies to Foster Wheeler, and in the year 2000, the decision was finally made to shut down the facility.

Because Robbins was actually recruiting an incinerator—rather than being targeted by the waste company—this case raises the questions (a) is this environmental racism? and (b) how did this happen? As Jim Schwab of the American Planning Association told a group of environmental scholars and activists at a conference at the University of Chicago, "asking the question 'is this environmental racism?' begs the question as to why Robbins would seek an incinerator in the first place. When you answer that second question, you have to confront the fact that racism had everything to do with the current position the village finds itself in." In short, Schwab would answer "yes" to the question as to whether the Robbins struggle was an example of environmental racism. Those analysts who might view this struggle as rooted only in one town's efforts to climb out of poverty would be ignoring history and the legacy of racism.

The sort of history that unfolded in Robbins is generalizable more globally. To make this clear, we now turn to the transnational trade in hazardous waste.

The Transnational Trade in Hazardous Waste: Global Environmental Racism

Historical Forces: Nation, Race, Class, and Unintended Consequences

Since the end of World War II, industrialized nations have generated increasing volumes of hazardous chemical waste—the result of several technological "advances" in the manufacturing, transportation, and military sectors, among others. The amount of toxins produced around the globe has risen exponentially in the last five decades. Today, it is estimated that nearly 3 million tons of hazardous waste from the United States and other industrialized nations cross international borders each year. Of the total volume of hazardous waste produced worldwide, 90% of it originates in industrialized nations. Much of this waste is being shipped from Europe and the United States to nations in South America, Southeast Asia, and Africa—the Third World (Critharis, 1990; Marbury, 1995; Tiemann, 1998).

There are two principal reasons for this shifting of the burden to southern or Third World nations. First is the emergence of more stringent environmental regulations in nations in the North. These changes have driven up the costs of waste treatment and disposal, which are magnitudes greater than those found in "developing" nations, which allow for dumping at a fraction of the cost. Similarly, the typical legal apparatus found in industrialized nations is much more complex when compared to the lax regulatory regimes in many Third World nations. This is in part due to a comparatively more influential environmental movement sector in industrialized nations. This sector has successfully produced a regulatory structure that provides at least a minimal level of oversight over polluting firms. The unintended consequence of this "success" in the North is to provide an incentive for the worst polluters to seek disposal sites beyond national borders (Marbury, 1995).

The second factor driving the waste trade is the widespread need for fiscal relief among Third World nations. This need—rooted in a long history of colonialism and contemporary loan/debt arrangements between Third World and industrialized nations—often leads government officials in Africa, Asia, and South America to accept financial compensation (i.e., bribes or officially sanctioned payment) in exchange for permission to dump chemical wastes in their borders (Porterfield and Weir, 1987). Many observers (economists and business leaders in industrial countries) have described these transactions as "economically efficient," while others (African leaders and environmentalists) prefer the terms "toxic colonialism" and "garbage imperialism" (Marbury, 1995).

Stakeholder Struggles and Resistance to the Toxic Trade

Despite the existence of several international treaties and domestic legislation in many nations intended to regulate and even prohibit this trade, toxic dumping in the Third World continues to grow. In response, environmental and social justice advocates have continued to monitor and resist these practices. A sampling of notable cases will provide a more substantive context. Some years ago, the African nation of Guinea-Bissau was on the verge of permitting the dumping of several tons of toxic waste in its borders, in return for payment equal to four times that country's Gross National Product. Several environmental and social justice groups in Africa and around the world protested this arrangement and persuaded Guinea-Bissau officials to reject the shipment. The fiscal incentive was very tempting, given the dire poverty in which most of Guinea-Bissau's citizens find themselves.

In another example, Formosa Plastics Company attempted to dump a load of the toxic polyvinyl chloride (PVC) in the Philippines in 1997. Local activists in the Philippines and Taiwan teamed up with Greenpeace and several other labor, human rights, and environmental justice organizations around the world to repel the shipment from the Philippines and several other nations over a period of 2 years. The company was eventually forced to return the waste to the original site for disposal.

In Koko, Nigeria during the 1980s, several thousand tons of highly poisonous wastes were illegally dumped by an Italian chemical firm. The result was the release of dioxin, lead, mercury, and other chemicals into the local environment, which produced elevated rates of cancer, lead and mercury poisoning, birth defects, miscarriages, kidney disease, and mortality among the local population. Nigerian officials responded by arresting Italian diplomats and ambassadors and threatened to execute any individuals responsible for future waste dumping.

Nigeria, the Philippines, and Guinea-Bissau have each been colonized by other nations. Their own natural and human resources were extracted for the benefit of foreign stakeholders, leaving these states destitute. "Development" loans provided by international banking institutions have generated few benefits, and left these states mired in debt. Their workforces and citizenry are desperate for economic development to meet subsistence needs and their political leaders are often willing to do whatever it takes to ensure that this happens. And bribes, or "ground rent," provided an extra incentive for officials and leaders to cooperate.

Scholars researching the transnational waste trade have focused mainly on the legislation and treaties that have been enacted to regulate these activities. This literature has centered on one major pressing question: to what extent can domestic regulations and international agreements control or minimize the waste trade? The majority of the existing research emphasizes the legal aspects of this global form of environmental racism without paying attention to the driving forces behind the waste trade. However, if one only takes a cursory look at the nations importing waste (legally or illegally) into their borders, it immediately becomes clear that they are nations on the geopolitical and economic periphery, they are nations that have endured colonization during the last two centuries, and they are often nations populated by a majority of people of color. This parallel with environmental inequalities within industrial nation-states like the United States might indicate that similar forces are causing global environmental inequalities. For example, the African nation of Benin was colonized by France. After independence, Benin is now deeply in debt to France and several financial institutions. French waste traders have recently offered to pay Benin large sums of money, as compensation for accepting toxins. Benin's motivation to accept such payment stems largely from its desire to repay its loans to France—hence, the term "toxic colonialism" and a brief explanation for one of the causes of global environmental racism.

Shaping Waste Trade Policy

The movement against toxic dumping in poor and people of color communities in the United States emerged during the 1980s, just as the movement against the global waste trade was taking shape. Our reading of this history is that these two parallel events were related. Shortly after the movement for environmental justice in the United States made headlines in the early 1980s, activists and policymakers began to take notice of similar patterns of environmental inequality around the globe. The Basel Convention, a transnational accord regulating the hazardous waste trade among nations, was first signed in 1989, during the height of the EJ movement's visibility in the United States. It is also probably not coincidental that Greenpeace, an organization that has been intimately involved in struggles against environmental racism across the United States, has been the principal advocate for a ban on the transnational trade in hazardous waste. As a result of the pressure applied by movement organizations, politicians, and the media in the late 1980s and early 1990s, significant segments of the toxic waste trade are shifting away from Africa, and toward Eastern Europe and Russia. Thus, like the flow of solid waste eventually being burned in Robbins, IL, the waste was simply shifted to a new location, rather than being reduced at the source.

Domestic versus Global Environmental Racism

There are major similarities and differences between environmental inequalities in the domestic United States and around the globe. We will begin with the similarities. The domestic U.S. case of the Robbins incinerator resembles the transnational waste trade in three ways. First, Robbins and all Third World nations have a history of externally imposed oppression and subordination (Rodney, 1982). All-black towns like Robbins may instill pride in African Americans, but they are also peripheral to society. They receive fewer federal funds, experience sparse business development, and enjoy little-to-no political influence in state politics. Similarly, nations in the Third World, by definition, are on the periphery of the "global village" economically and politically. While inner-city African American communities have some of the highest rates of unemployment and mortality within the United States, nations in West Africa experience some of the highest levels of poverty and infant mortality in the world.

Second, both types of communities shoulder a disproportionate burden of toxic waste. African American communities like Robbins, have been described as environmental "sacrifice zones" (Bullard, 2000) while entire nations and regions in the Third World have been referred to as a "global dumping ground" (Moyers, 1990) and the "outhouse for industrialized nations."

Third and finally, local leaders and citizens are willing to accept compensation for having their communities shoulder the toxic burdens, whether they are illicit "payoffs" or legally sanctioned remuneration. The Robbins incinerator represented "blood money" and "environmental blackmail" to many activists. They felt the mayor could have demonstrated greater integrity if she had rejected the notion that the only jobs poor black towns can attract are hazardous jobs. Other observers view Mayor Brodie's decision to court the incinerator as pragmatic planning in a difficult situation. Many African American leaders have scorned those who accept dirty industries in the name of economic development because of the greater long-term environmental and human health costs the citizens will pay. Similarly, leaders of many Third World nations have been exposed for accepting bribes for—or simply permitting—toxic dumping in their borders as well.

The differences between domestic EJ struggles in the United States and those in the Third World have mainly to do with regulatory structures, governmental apparatus, and movement infrastructures. In the first case, the regulatory structures in the United States versus those in Third World nations are vastly different. Hazardous waste dumping in the United States can cost up to $4,000 per ton while the charge in many African nations is as little as $5 per ton

(Marbury, 1995). The existence of a relatively strong domestic regulatory regime provides U.S.-based communities of color with a body of law with which to fight environmental injustices. Communities in many Third World nations have few such resources.

The second major difference between United States versus international EJ struggles is less apparent. Communities of color—in fact *all* communities—in the United States have no formal autonomy with regard to the waste trade. In the 1990s, the U.S. Supreme Court handed down a ruling that "flow control" was unconstitutional—thus denying municipalities the right to decide where they would import or export their solid waste for landfilling, incineration, or recycling. Furthermore, state and U.S. federal laws take precedent over municipal laws if the latter are found to be in violation of the former. Third World nations, on the other hand, are not bound by the U.S. Supreme Court, and can control the export and import of their waste as they please (as long as it does not violate treaties they have signed with other nations, or agreements with the World Trade Organization). In this way, Third World nations can theoretically exert much more control over the waste trade than can domestic U.S. communities.

Third, the social movement infrastructure in many U.S. communities of color is often much more influential (and less endangered by severe state repression) than advocacy groups in most Third World nations. This has allowed movements in the United States to have a stronger influence on state and industrial policy making than movements in Third World nations. One manifestation of this power differential is the successful development of an environmental policy infrastructure in the United States that has—although unintentionally—produced an increase in waste exports away from domestic communities to the Third World. So taken as a whole, Third World nations are in a much more tenuous position than communities of color in the United States.

Conclusion

Returning to the EJ framework suggested earlier, we emphasize the following key points for understanding these conflicts:

- the importance of the history of environmental inequalities and the processes by which they unfold;
- the role of multiple stakeholders in these conflicts;
- the role of social stratification by race and class; and
- the ability of those least powerful segments of society to shape the contours of environmental justice struggles.

In the cases discussed in this [essay], we find that examining the *historical bases* of many environmental justice conflicts can present new data and a deeper, more accurate understanding of the problem. We also find that the poor and people of color are generally the most vulnerable to environmental inequalities. We have sought to suggest some of the ways that racism, classism, and our changing knowledge of environmental hazards interact to shift the burdens of environmental risks to different populations over time. Within and across communities, *stakeholders* are constantly jockeying for quality living and working environments.

Departing from conventional accounts of environmental racism, this study critically examines the role of *social inequality* and finds that class and political privilege often place certain people of color in a position to benefit from (and perhaps to perpetrate) acts of environmental racism. When EJ struggles take this direction, the very claim that environmental racism is at work becomes problematic. These dynamics add an ingredient long absent from much of the literature—the way that class and political power often divide communities and racial groups, creating intraracial, intracommunity, and class conflicts. These tensions reveal the importance of analyzing the motivations and actions of *multiple stakeholders* in environmental conflicts. Unfortunately, these internal community dynamics often serve to divert attention from the larger political economic structures where more affluent classes in the Global North (and usually white middle and upper classes in the United States) still remain in the cleanest, best-protected living and working environments (Portney, 1991, 138ff.).

Finally, the framework we have introduced in this paper emphasizes that "it's not over 'til it's over." In other words, just because a population is exposed to environmental hazards does not mean the struggle is finished or that the battle has been lost. Environmental injustices are a work in progress, they are constantly in process because people are continually resisting them. Most observers might have predicted that after the Robbins incinerator was constructed the battle was over. But EJ activists continued to mobilize until they succeeded in shutting the burner down—3 years after it was built. This story should instill hope for environmental justice and a deeper understanding of the processes by which environmental racism unfolds and evolves. The shifting nature of the transnational waste trade is another example of the power of continuous movement resistance. The U.S. environmental movement must share some of the accountability for producing a shift in waste disposal from domestic to Third World locations. Similarly, the transnational network of environmental organizations resisting the waste trade in certain Third World nations must bear some responsibility for shifting toxic exports to Eastern Europe, Russia, and parts of Asia. The good news is that the EJ movement does have a measurable impact on policy. The bad news is that the movement is often faced with limited and inappropriate choices, or is simply excluded from the implementation process.

Our hope is that local, regional, national, and international legislators would incorporate these findings into policy making around environmental and social justice issues. . . .

Note

Pellow, David N., Adam Weinberg, and Allan Schnaiberg. 2001. "The Environmental Justice Movement: Equitable Allocation of the Costs and Benefits of Environmental Management Outcomes." *Social Justice Research* 14, no. 4: 423–39.

References

Bullard, R. (2000). *Dumping in Dixie: Race, Class and Environmental Quality*. Westview Press, Boulder.

Bullard, R., and Johnson, G. (eds.). (1997). *Just Transportation: Dismantling Race and Class Barriers to Mobility*. New Society Publishers, Philadelphia.

Critharis, M. (1990). Third World nations are down in the dumps: The exportation of hazardous waste. *Brooklyn J. Int. Law* 16: 311.

Hurley, A. (1995). *Environmental Inequalities: Class, Race and Industrial Pollution in Gary, Indiana, 1945–1980*. University of North Carolina Press, Chapel Hill.

Marbury, H. (1995). Hazardous waste exportation: The global manifestation of environmental racism. *Vanderbilt J. Transnatl. Law* 28: 1225–1237.

Moyers, B. (1990). *Global Dumping Ground: The International Traffic in Hazardous Waste*. Seven Locks Press, Washington.

Porterfield, A., and Weir, D. (1987). The export of U.S. toxic wastes. *The Nation* 137: 245–325.

Portney, K. (1991). *Siting Hazardous Waste Facilities: The NIMBY Syndrome*. Auburn House, New York.

Robinson, J. (1991). *Toil and Toxics*. University of California Press, Berkeley.

Rodney, W. (1982). *How Europe Underdeveloped Africa*. Howard University Press, Washington, DC.

Tiemann, M. (1998). Waste trade and the Basel convention: Background and update, Congressional Research Service Report for Congress. Committee for the National Institute for the Environment, Washington, DC.

Wilson, W. J. (1996). *When Work Disappears: The World of the New Urban Poor*. Vintage Books, New York.

PART

Media and Popular Culture

17

Touch the Magic

Susan G. Davis

When social scientists talk about "culture," they are referring to our values, beliefs, and human-made material objects. Over time and across geographic spaces, different cultures have generated very different belief systems about nature. For instance, some cultures worship cows, while others see them mainly as a source of food. Media and popular culture shape our values and attitudes toward nature. One phenomenon of our modern leisure culture that offers a unique, and potentially problematic, experience of the environment/human relationship is the nature-oriented theme park. This reading by Susan Davis, a folklorist, shows how corporations, both in theme parks and through the media, influence the way we think about nature. Specifically, Davis shows how Sea World constructs nature as highly visual, adventure filled, and appealing to humans' emotions and desires for closeness. She also reveals how Sea World's particular presentation of nature appeals largely to educated, middle-class white people. Davis argues that corporations such as Anheuser-Busch, owner of Sea World, use theme parks, educational programs, and advertising to send a message that they are good environmental stewards—a phenomenon often referred to as "greenwashing" (though Davis does not use that term).

The television commercial opens on a little boy, his face pressed to glass, "in absolute amazement," as a gigantic black-and-white killer whale swims past. The scene dissolves to a seal jumping, and then to an old man hugging a small girl. On the audio track a male singer asks, "Do you remember . . . the feeling of wonder? . . . Bring back the smile to the child inside of you. . . ."[1]

The scene dissolves back to the tank of the black-and-white whale, where an athletic, wet-suited man is thrust into the air on the whale's nose. As the stadium crowd cries out at the sight, the singer intones "Touch the magic. . . . Touch the world that you once knew."

Now the visual and aural pace of the commercial turns quiet and gentle. A little girl dips her hand into a tide pool, grasps a starfish, and turns it over. Another little girl and her grandfather lean over a wall to pet a dolphin as the chorus fades away, singing "From the heart, Sea World touches you. . . ."

The images on the television screen alternate between shots of human groups and shots of animals. Waddling penguins dissolve into shots of a little girl waddling in imitation. A flock of flamingos is intercut with a woman in a wheelchair and a little boy watching dolphins at a glass wall. A mother duck swims by with her brood, and a small boy swings on a playground rope. The male singer returns to tell us that we will have good friends forever in this place.

The chorus swells in song as a pair of killer whales performs a synchronized jump: "A world apart. . . . That brings our world together. . . ." People in the stadium crowd throw back their heads and laugh. A father takes a picture of his wife and children, a mother tends her baby, a courting man and woman snuggle.

At the commercial's conclusion, the camera returns to the little boy and girl at the edge of the killer-whale tank. They approach the clear wall, and the camera lingers a few seconds in close-up as, with faces pressed close to the glass, they try to touch the black-and-white whale. The whale's head is close to the children; it appears to gaze back. Stringed instruments surge, and the chorus sings, "From the heart . . . Sea World touches you." Then silence for a few reverent seconds, as the television screen displays a blue, white, and black logo and the words "Sea World: Make Contact." And, at the very bottom of the screen, "An Anheuser-Busch Theme Park."

"Touch the Magic," as the television commercial is titled, is the centerpiece of a recent advertising campaign for the Busch Entertainment Company's four Sea World theme parks.[2] From the spring of 1993 through the spring of 1994, Busch Entertainment urged millions of prime-time viewers to "make contact."[3] At the simplest level the powerful "Touch the Magic" images sum up the product. A visit to a Sea World park, the commercial argues, is not just a day out of the house; it's a "magical," "touching" experience that "brings our world together."

In this essay I want to explore Sea World's nature magic as it appears in this commercial and as it is constructed and delivered at the theme parks. "Touch the Magic" is a rich text, presenting many ideas about the meaning of nature in contemporary American culture. It summarizes, as the Sea World parks elaborate, twentieth-century mass culture's dominant arguments about human relations to nature. Carefully thought out and expensively produced, this piece of publicity is a tiny but typical portion of a much larger field of images and arguments emanating from the corporate media. As such, it is part of a media world that helps shape how Americans understand nature and the environment. "Touch the Magic" has much to tell us about how Anheuser-Busch in particular and large businesses in general represent nature to vast audiences.

A look at Anheuser-Busch and its theme parks will help contextualize "Touch the Magic." Anheuser-Busch is the planet's largest brewer. The company controls 43 percent of the U.S. beer market and in 1993 saw about $11.6 billion in sales worldwide.[4] While Busch Entertainment does not report theme park attendance figures to the public, this subsidiary is very successful. With the second-largest total attendance of the five U.S. theme park chains, Busch Entertainment is expanding abroad. (Attendance at the Disney parks far outranks that of all other contenders.) At a conservative guess, the four Sea Worlds together entertain about 11.35 million paying visitors annually, most of them North Americans. These visits alone evidence a wide exposure to and interest in Anheuser-Busch's powerful and colorful representations of nature.[5]

Nature exhibits and marine animals are the central themes of the Sea Worlds. Superficially, the ways the parks present nature seem varied. At the San Diego park, for example, some settings mimic the new environmental zoo. The ARCO Penguin Encounter is a sort of living diorama that simulates parts of the Antarctic environment. At Orlando and San Diego the shark exhibits carry customers on a moving walkway and through a Plexiglas tube, to encounter the "Terrors of the Deep." "Rocky Point Preserve" re-creates a bit of the Northwest coast in San Diego and houses dolphins and sea otters, albeit in separate tanks.

Despite all the variety within individual parks, however, Sea World's nature is not only highly artificial but also standardized. Busch Entertainment has overarching supervision of the parks, so each new exhibit or entertainment is carefully vetted in St. Louis. Successful displays from one park are exported to the others, as are the animal shows. And the killer whale in "Touch the Magic" is more than just a striking creature; it is a registered trademark. "Shamu" appears at all four Sea Worlds, in shows that emphasize similar themes of loving, caring, and closeness between whales and people. Among the most popular attractions at all the parks are the pools holding dolphins, sea stars, rays, and skates that visitors can handle and feed. The themes of "touch" and "contact" in the television commercial connect directly to one of the most appealing facets of a visit to any Sea World: the opportunity to encounter and pet wild animals.

Sea World's reconstructed nature is, in the 1990s, a highly commercial production. Like all theme parks, each Sea World draws customers and keeps them spending as long as possible by offering a diverse, even clashing, array of activities and diversions. Ice skating extravaganzas, power boat shows, game arcades, and sky rides coexist with aquariums and tide pools. Lush landscaping and gardens provide a backdrop for country music concerts, company picnics, and disco dancing. The animal shows, though, are the major draw. But compared with an old-fashioned zoo, a publicly supported museum, or a free park, Sea World is very expensive. In 1994 a child's admission to Orlando's Sea World was $29.95, the highest in the theme park industry.[6] Just as important as admissions for each park's profits are the almost endless concession stands, boutiques, and gift shops offering refreshments and souvenirs of many sorts. As a student in one of my undergraduate classes aptly put it, "Sea World is like a mall with fish."[7]

Anheuser-Busch is not the only company profiting from theme park nature. The Sea Worlds subsidize some of their displays with corporate sponsorships. Although "Touch the Magic" ignores this by presenting Sea World's environment as if it were noncommercial, a good number of the displays inside the parks have outside funding. In San Diego, ARCO helps present the Penguin Encounter, Home Federal Bank sponsors Rocky Point Preserve, and am-pm mini-markets support the shark reef. But many other businesses and manufacturers, from Kodak film to Pepsi-Cola and Southwest Airlines, take part in joint promotional and advertising ventures inside and outside the parks.[8] Perhaps more important, the parks provide a kind of advertising *and* public relations for the parent company. For example, besides selling Anheuser-Busch's Budweiser beers and Eagle snack foods, the Sea Worlds' "hospitality centers" feature "microbreweries" and free beer tasting. The theme parks house and display the company's trademark Clydesdale horses, converting the registered trademark of one corporate division into an attraction at a wholly owned subsidiary. And, of course, associating Anheuser-Busch and its products with animals, nature, education, and families positions the world's largest brewer as environmentally and socially concerned.[9] At the Sea World theme parks, advertising, marketing, and public relations are so thoroughly collapsed into entertainment and recreation that it is very hard to tell what is publicity and what is "just fun." Inside the park the advertising–promotional–public relations mix is an intricate maze.

Given this commercial and promotional environment, it is striking that the nature theme park styles itself a public facility.[10] To do this the Sea Worlds draw on today's concern with environmental issues and animals, and they emphasize education. Busch Entertainment and Anheuser-Busch make much of their involvement with research on whales and dolphins, their efforts at fostering the reproduction of species (including a few endangered species), and their part in marine animal rescue and rehabilitation efforts.[11] Sea World's rescue and research activities are also heralded in educational programs offered to public school systems, courses taught at the parks, and credentialing programs for teachers. Anheuser-Busch estimates that 650,000 children annually participate in education programs inside the Sea World parks. Direct satellite broadcasts now reach an estimated 16 million viewers via schools throughout the United States.[12]

This push by a private entertainment corporation into the public school classroom is not unique. "Channel One" of Whittle Communications is only the most famous recent example. On the environmental education front, the Walt Disney Company is at work on a curriculum for use in the California public schools. Procter & Gamble, one of the world's largest consumer goods producers and the world's biggest advertiser, has recently developed and distributed a free packet of "Decision Earth" materials for elementary school use.[13] However, Busch's expansion into public education suggests that the Sea World theme park is a new kind of institution. Unlike the older amusement park, the nature theme park combines the search for private profits through entertainment with an attempt to occupy the cultural space and functions of the nonprofit, publicly funded zoo, natural history museum, or aquarium. Courses are taught, whales are bred, and the nesting patterns of the least tern are studied at the theme park. But these activities are inseparable from Sea World's marketing and Anheuser-Busch's public relations. Education, corporate image, and luxuriant profits go hand in hand.

"Touch the Magic" is just one small segment of the mass of pictures and print that Busch Entertainment commissions from ad agencies to support not just its theme parks but all its marine-themed products. Like all theme parks, the Sea Worlds are integral to a media culture that extends far beyond their physical boundaries.[14] For example, the Sea Worlds create whale- and dolphin-based entertainment specials. These network television shows, like the satellite broadcasts mentioned above, serve to advertise and promote the park as a tourist destination, while the theme park helps build an audience for the television programs.[15] Aggressive public relations departments make sure that the parks' animal-saving activities are regularly featured and highlighted in network entertainment and news programming. At present, Anheuser-Busch lacks only a syndicated television show to deliver to the widest possible audience filmed and animated media products based on its performing celebrity animals.[16] Many other Sea World-themed commodities result from licensing agreements: stuffed animals, postcards, T-shirts, story books, nature study books, and video tapes only begin the list of commodities marketed through the parks.[17] Again, the products support the parks, while these support the goods far outside the theme park gate. In creating and distributing its own imagery so extensively, Sea World has gone far beyond the traditional educational functions of the zoo or natural history museum, even while it claims to be providing those services.

This sprawling commercial operation is the context for "Touch the Magic." Let's return to the world of the nature theme park as proposed by the ad. In a very real sense, the expansive corporation is as much the author of the television commercial as any copywriter or video producer. What stories does Anheuser-Busch use "Touch the Magic" to tell? How do the commercial and the company ask the audience to think about relationships between humans and the natural world? What happens if we touch the magic?

"Touch the Magic" presents a condensed, more perfect world. The advertisement does a good job of delivering Sea World's visual richness. As other scholars have stressed, theme parks and television are intimately connected media. Not only does the advertisement aim to reproduce the qualities of the park, but the park in many ways tries to be "televisual." The advertisement compresses a richness of experience into the visual mode, and so does the park, which pays careful attention to color, detail, landscape, and sight lines. Sea World's leafy foliage and expensive plantings, its brightly colored birds and massing fish, the contrast of sun and shade, wet and dry, and the variety of its surfaces are all emphasized in the ad. Here is a wealth of things to see and touch.

Sea World defines nature as an overwhelmingly visual experience, but its way of seeing extends beyond kaleidoscopic abundance. In the commercial and the theme park, we see animals, greenery, and performers with an easy immediacy. As in other heavily photographic media, such as *National Geographic* or the television program "Nature," our approach to nature is simple and unobstructed. Our sight is unimpeded—we see nature in a way that we rarely could out in the "real world," where brush and trees block a view or murky water hides the fish. Like television, the parks are full of tricks of visual purification. The many carefully realistic aquariums, for example, are designed to make the unseen visible. Sea World assures us that its way of seeing nature will be natural.[18]

At the same time that we are tempted by perfect seeing, great distances are collapsed and we are promised adventure. As "Touch the Magic" offers to take viewers to spectacularly inaccessible, invisible, or little-known places, so Sea World designs a mass version of nature tourism implicitly built on an older tradition of exploration. A trip to the park is a condensed voyage that circles and surveys the world, without requiring us to go very far from home. Again, this is similar to the experience offered by other mass media, most famously *National Geographic* magazine.[19] Sea World is full of exhibits structured like a journey: the customer consults a map of the "World," navigates a place in line, and finally, slowly draws near. The supposedly strange sight is nevertheless thoroughly familiar from popular culture and the literature of travel and tourism: a penguin-packed Antarctic ice floe, a Polynesian atoll seething with sharks, a "forbidden reef" infested with moray eels, the rugged stretch of the Pacific Northwest coast, a ghostly sea bottom in the "Bermuda Triangle." On the one hand, this selective tour of the globe defines nature as exotic and remote. On the other, without ever calling up a precise history, the theme park offers its customers a chance to stand in the shoes of the European discoverers as they mapped the non-European world.

While Sea World locates this highly visual nature in an implied historical narrative of exploration, it paradoxically decontextualizes nature. Indeed, what customers don't see is as significant as what they do. At Sea World crowds of people watch a natural world seemingly uncrowded, unpeopled. Any human cultures or histories of the environments shown in the dioramas linger only as background information. At the "Shark Encounter" we see no Pacific Islanders, but the mysterious sound of drums enhances the carefully painted and sculpted set. Ersatz totem poles frame the park's centerpiece show, "Shamu New Visions," which is introduced with a (genuine?) Northwest Coast Indian story about a boy and a whale.[20] Otherwise, long human connections to animals and environments vanish. In Sea World's presentation nature and animals have been discovered and are being protected in a pristine state by white North Americans. This pristine state is physically produced within the context of intense consumerism.

The most extreme examples of decontextualization are also the most popular: the trained animal performances. The theme park exhibits rebuild nature and bring the faraway close. But in the whale stadium nature is isolated and held up to the collective gaze. The surprise, awe, and wonder of crowds focus on individual animals. In these shows performers seem to push the possibilities of animal bodies and animal-human interactions to their limits, Shamu's launching the trainer off its snout being only the most famous example.

There is one exception to the decontextualization of nature: Sea World's technicians and scientists are ever present in the theme park frame. In its stories of exploration, Sea World always identifies itself with pioneering experts. Again, the trained killer whales are the famous and telling examples. Long a part of Northwest Coast Native American knowledge and mythology, orcas were first kept in captivity and developed as an entertainment resource by Sea World's founders in the mid-1960s. Anheuser-Busch makes much of this "pioneering" role, casting Sea World's success in inducing killer whales to reproduce successfully in captivity as an important scientific accomplishment.[21] In the Sea World version of natural history, the theme park itself brings exciting unknown things to public awareness, and the audience travels vicariously with the experts. Similarly, the Sea World Penguin Encounter in San Diego claims the capacity to support the penguin reproductive cycle. While visitors to a natural history museum's dioramas might view a scene that looks like part of Antarctica, at the ARCO-sponsored exhibit they are treated to something new in mass culture— the proliferating Antarctic. It appears that the skill of the theme park enterprise itself has made life multiply.[22]

Besides creating a larger-than-life, decontextualized nature, as the ad argues, Sea World has another magical effect. Having decontextualized nature, Sea World makes it a powerful carrier of human emotions. In "Touch the Magic" intense emotion is signaled by physical contact (the hugs and kisses in the crowd), power and speed (the trainer rocketing off the nose of the killer whale), and sound (the sentimental music track). Visuals and music underscore the ad's verbal references to happy feelings: "Bring back the smile to the child inside of you."

Here nature is not only about seeing, exploring, and collecting; it is about relationships, feelings, and families. "Touch the Magic" shows people close to each other, in groups and couples. Similarly, animals appear in flocks or pairs. We see two whales, two dolphins, and a mother duck leading her brood. Animal groupings parallel human groupings of families and potential families (boy- and girlfriends).

As the ad's theme underscores, relationships at Sea World are intimate, expressed through touch. Grandparents hug children, parents watch their kids cavort and clown, a young woman cuddles her boyfriend. Even that ambivalent image of human community—the crowd—seems not to consist of strangers. Laughing, close together, the people enjoy themselves. The advertisement's argument is that the theme park as a cultural space can help bring people closer together—across boundaries of gender, races, handicaps, and generations, and perhaps even across the barriers of anonymity that separate people in crowds and audiences. As the song has it, Sea World is "another world . . . that brings our world together."[23]

But "Touch the Magic" contains a tension it sets up, resolves temporarily, and then calls up again. Gentle physical contact takes place not just among people but also between people and animals. The ad's opening and closing shots feature one of Sea World's most famous advertising images: the children and Shamu pressed nose to nose, reaching out to one another. But their intimacy is incomplete: the transparent barrier of Plexiglas allows humans and whales to come close while keeping them separated. Since the nose-to-nose shot precedes and follows so many images of direct contact between people, and between people and animals, it seems to express a fantastic wish for a total merging with wild nature.

The stark print text "Make contact with another world" accompanying the nose-to-nose image emphasizes Sea World's claim to create a kind of communication that is otherwise impossible. Making contact with "another world" may mean several things. Certainly the video shot of children and whale separated but trying to touch each other implies that nature and wild animals constitute a world distinct from humans, one that humans should wish to approach more closely. The notion of far-away worlds trying to contact us might be here, too—in a reference to spirits and unseen forces or, perhaps, as in science fiction, to other planets and parallel universes.[24] Given the surrounding emphasis on relationships, however, it seems more likely that the other world of nature is also an interior world, one of emotions and feelings. We've already been exhorted to remember feelings, to "bring back the smile to the child inside of you." "Touch the Magic" suggests that the theme park offers customers access not only to nature and exotic animals but to themselves. Asking us to "remember the feeling of wonder" and "bring back the smile" suggests that we need to return to authentic feelings.

In short, communication and contact with nature promise to remake people. This is familiar: most contemporary advertising presents the product as magically transforming the consumer herself from an alienated and isolated state to a meaningful, whole identity.[25] Consumer products also promise to alter social identities, and perhaps a visit to Sea World is no different. To make contact with nature is to have real feelings and to become someone different and more desirable. But how does this transformation work?

[N]ature is a vast complex of ideas as well as a biological world. This is part of what makes the idea of a more intimate contact with it so evocative, powerful, and magical. Sea World offers its customers transformation through contact with a long historical tradition of nature's social meanings, a tradition the theme park version of nature, with its emphasis on visual realism and scientific expertise, paradoxically obscures.[26]

For at least the last two hundred years, Europeans and Euro-Americans have made nature a visual, touchable, "out there" object. Making nature literally "another world," Euro-Americans endowed it with cultural and spiritual properties. At the same time, art and literature emphasizing labor and the complex interactions between humans and the biological world have been marginalized in the Euro-American aesthetic tradition.[27] Appreciation of this aestheticized, separate version of nature has been used to distinguish people from each other and to normalize the differences between them. For example, in the eighteenth century, as Raymond Williams argues, the gentry justified its expanding property rights and dominance over the rural poor through aesthetic practices. Sculpting nature into country estates, celebrating it in pastoral poetry, manipulating it in the form of lovingly landscaped gardens, the gentry literally naturalized its vast social and economic power.[28] In the nineteenth century appreciating nature by viewing it, painting it, or hiking through it helped factory owners and businessmen define themselves as rational and sensitive men, especially in comparison with rural and urban workers.[29] Toward the end of this period, zoos and museums did more than popularize the global scientific world view and integrate nature into a Euro-centered colonial map. Their boards of directors also hoped museums would teach respect for law and "natural order" to the urban, immigrant working classes, whom they saw as having anarchic, un-American ideas.[30]

For at least two centuries the propertied have used nature as a material and a symbolic resource and as a favored tool of improvement aimed down the social scale at class and racial others. And while nature education in classrooms and summer camps has surely had many democratic and socially progressive uses, it has also been closely tied to efforts to model a hierarchical social order. The right sort of person, as advocated by nature educators, was an English-speaking, self-controlled, property-respecting, refined middle-class citizen. Sea World's nature magic partakes of this uplifting tradition. As "Touch the Magic" argues, contact with nature creates or affirms a customer's identity as a caring, sensitive person. A visit to Sea World offers nature as a source of rational pleasure albeit in a context of irrational prices and endless throw-away souvenirs. There are no gut-wrenching, mind-blowing roller coasters here. Perhaps a visit to Sea World also helps customers distance themselves, however little they think about it, from people defined as uneducated, insensitive, and irrational.

In "Touch the Magic" Sea World's visitors dress casually but well. They are mostly white adults with children, respectfully excited by the sights they behold. The ad reflects Sea World's marketers' understanding of their audience as defined by age, income, ethnicity, and education. Although Sea World's managers, like most white Americans, are reluctant to speak in terms of social class or to acknowledge a racial pattern among their customers, market research for the San Diego park reveals that their audience is heavily southern Californian and white, with a very high level of income and education.[31] The San Diego park describes its customers as consisting heavily of "parents" who are "usually college educated and . . . interested in learning about ocean life."[32] It is not that the people who run Sea World do anything active to keep ethnic minorities out of the park.[33] Indeed, the education programs recruit minority children via public school field trips. But considering the multiethnic demographics of the five southern California counties, the whiteness of Sea World's paying audience seems to be an example of extreme self-selection.[34] Perhaps the high admission prices alone tend to keep poor people away; however, other expensive theme parks in southern California, most notably Disneyland and Six Flags–Magic Mountain, have strong followings among people of color. Is it possible that the version of nature marketed by Sea World appeals positively to white people as part of being appropriately middle-class? Conversely, does Sea World's discrete, aestheticized nature seem unfriendly or irrelevant to working-class people and nonwhites? Whatever the answer, Sea World's marketers clearly craft their nature product for the affluent.

The children in the ad give another clue that Sea World's nature magic is in part about social class. In white middle-class culture, the positive association between children and animals, children and nature, reaches far back.[35] Children are supposed to have special things to gain from contact with nature, but perhaps these special things are related as much to social ideals as to children's practical growth.[36] According to long-standing theories of education, nature and the outdoors teach the child about the inner self.[37] And, at a more mundane level, contact with nature is thought to lay the groundwork for the child's future success in biology or some other important science. It is not that any of these ideas are entirely false. But in this ideal of childhood, nature, social mobility, and the sense of self all run together. The nature-children-class connection is expressed in the century-old middle-class emphasis on suburban yards, summer homes, summer camps, and nature study in the classroom.[38] That "Touch the Magic" and all Sea World's advertisements and publicity feature children learning points to the theme park's claim that it helps produce the right sort of person. That children are shown learning in the context of the family implies that the park also helps reproduce social position.

So perhaps "Touch the Magic" and Sea World itself show us that mass-mediated nature is constructed to appeal to its consumers as much in terms of who they want to be, as individuals and members of families and communities, as in terms of the aesthetics of clarity and purity. In any case, when Sea World claims to create feelings of awe, wonder, and joy, we might understand this as an argument that awe, wonder, and joy—as opposed to fear, boredom, hostility, or exhaustion—are feelings the "right sort" of person should have in the presence of nature.

While the desires and identities of Sea World's customers have a long past, something new is happening in the theme park, too. Touching the magic of nature is a way of making contact with a world of possibilities as well as a way of finding one's feelings and confirming a social identity. All of Sea World's and Anheuser-Busch's important publicity materials appeal strongly to the environmental interests of the American public, carefully positioning the corporation as at once a good environmental citizen and a responsible producer of goods. It is in the context of wishes and worries about the future, I think, that Anheuser-Busch and Sea World try to redirect popular environmental concerns. In the process of presenting nature as something for people to contact and care about, the company is arguing for nature's reinvention.

Once again, the shot of the children pressed nose to nose with Shamu is suggestive. This image sums up all Sea World's urgings about "making contact," but specifically making contact with another form of higher life, not just scenery and science. The promise of mutuality between whale and people is what is different. Traditional zoos and aquariums, the aesthetic theorist John Berger writes, reveal people's distance from the world of nature by bringing them close to animals. Because animals are turned into spectacle, and because their boredom and passivity is inescapable, zoos underline the extreme marginalization of nature from human life in industrial society.[39] The new zoo and the nature theme park actively try to override this disappointing perception by showing animals in more natural-looking environments, where they seem to have privacy, autonomy, and the ability to avoid the human gaze.[40] The aesthetic of the new zoo does not reverse the long separation of humans and nature, but it makes nature seem less dominated, less captive.

"Touch the Magic," however, makes a different argument for Sea World. If we look closely at the way the killer whales are represented in Sea World's commercials, we see that the whale appears to be meeting the children's gaze. It is looking back, and this is exactly what Berger argues zoo animals can never do. Why should the whale seem to be trying to make contact? Certainly the audience is being asked to have feelings with and about whales, but is it possible that by seeming to reach out, the whales ask people to join with them? Sea World constructs itself as "another world," a parallel world in a watery realm, and the parks surely argue that orcas parallel humanity, or at least that segment of humanity defined as Sea World customers. The largest, most popular mammals on display, the orcas are always discussed in terms of social organization, intelligence, and especially reproduction. Much of the pleasure of watching their performances comes from seeing them humanized. Orcas are made to seem so like us—caring for their babies, working hard for their rewards, and getting the better of their foolish trainers—that a subtle identification takes place. Perhaps without intending it, Sea World asks its audience to form a relationship with nature—under the theme park's auspices. But which relationship? What prospect for humanity's relationship to nature is being promised to the audience as it makes contact?

The relationship Sea World proposes between humans and nature emerges more clearly when we recall that all the park's publicity contains claims for the social and scientific responsibility of the company and Anheuser-Busch. The nature images and animal performances flowing through the overlapping channels of entertainment, advertising, and promotion all tell the same story: Sea World and Anheuser-Busch conserve, protect, study, and foster nature. In this context, contact between whale and children invokes the beginning of a journey into the future, in which the protected, biologically reproducing animal and the learning, socially reproducing children travel together under the same, benign auspices. The right sort of person not only expresses interest in animals and science; she trusts and knows that companies like Anheuser-Busch are taking good care of nature. This caretaking goes beyond paying taxes, making donations to appropriate philanthrophic organizations, or establishing foundations. Corporate America is taking care of nature right down to the structure of DNA, and in the process, as Donna Haraway argues, it is transforming definitions of humanity, community, and nature. At the theme park customers collaborate in this corporate assumption of responsibility as audiences and consumers. As a neighbor of mine said to me, "Even the high admission fees are worth it, because they do such good things for the oceans there." And as a line from one of Sea World's shows has it, "Just by being here, you're showing that you care."

For all its seductive imagery, "Touch the Magic" is just one commercial, a tiny piece of media culture. But it is a good example of how much corporate culture hopes its audiences will understand nature and environmental issues. It is not that Sea World and Anheuser-Busch made up the idea of nature as pure, separate from humans, and under the benign care of experts and multinational capital. Rather, Anheuser-Busch, its parks division, and the collaborating sponsors have very intelligently recast and reworked some much older ways of seeing nature, in part for purposes of direct commercial appeal and profit, in part as a more general strategy for creating a positive public view of a very large corporation. In the process "Touch the Magic," Sea World, and many similar mass media products advance a vision of nature's future that is consonant with the interests of corporate America. The green public relations version of nature not only obscures a long history of relationships between humans and nature; it makes democratic pressures for environmental preservation, safety, and health invisible. Although parks like Sea World and ads like "Touch the Magic" appeal to popular environmental concern, neither the problems of pollution and resource exhaustion nor solutions from outside the corporate sphere have a place in the Sea World scenario. Rather, nature theme parks show corporations like Anheuser-Busch rising to the conservation occasion with spontaneous good will.

Anheuser-Busch holds no monopoly on this corporate-friendly version of nature. The same story is told, with variations, in many streams of the mass media. The arguments of Sea World's nature magic resemble those conjured in corporate-image campaigns such as Chevron's "People Do" ads or the Du Pont "Ode to joy" commercials.[41] The nature magic is also familiar from other kinds of television, film, and print. Environmentalists, students of environmental history, and communication scholars would do well to take mass-media representations of nature and environmental issues seriously. Environmental public relations, "green advertising," and nature programming occupy a significant proportion of network and cable television broadcast hours.[42] Nature magic is summoned in much, though certainly not all, of what one can see on the corporately sponsored Public Broadcasting System or the commercial Discovery Channel.[43] Its core ideas radiate into the larger culture from many different sources. But arguably the "Walkthrough TV" environment of the theme park—the synthesis of entertainment and advertisement that reaches into the day-care center, school, library, bookstore, and home—is more thoroughgoing than any earlier medium. Because it is at work on so many levels, Sea World's privately produced version of nature takes up a huge amount of space. This space is physical and cultural, imaginative and psychic. As Candace Slater argues . . . , such a spectacular but limited way of seeing nature necessarily displaces or hides other kinds of connections and contacts that need to be made. In the late twentieth century what are the alternatives to theme park nature? Where can missing connections be picked up?

In asking these questions about nature theme parks, we must recognize that most Americans—scholars included—live in the vast environment of mass-mediated culture, as surely as we drive cars and eat cheeseburgers. And we must confront the fact that many older forms of popular contact with nature and science that once enjoyed public support in the form of tax-based funding have been extensively reorganized. Zoos and natural history museums always had an elite bias; they often communicated the social vision of their philanthropic funders and directors. The establishment of national parks may have expressed, in part, an elitist wilderness aesthetic. But these institutions were not, in theory or practice, private property. Operating with some governmental funding and oversight, they were open to pressure for popular use and had to respond to often conflicting concerns.[44] As basically nonprofit institutions, zoos and museums were spared the pressure of having to turn every square foot of exhibit space into profits. But today aquariums, museums, zoos, and state and national parks are increasingly tied into the same tourist economies that shape theme parks; at the same time, declining tax support means that they are forced to rely heavily on corporate funding. The new funds come with new strings attached—for example, the need to show that the zoo or museum has the appropriate audience demographics. All these institutions rely more and more on blockbuster exhibitions or special events that can be promoted to garner paying audiences.[45] Because funds are tighter, and corporate support unreliable, at many museums admission is no longer free or even modestly priced. The gift shop has become crucial to the institution's budget. As zoos and museums compete with theme parks, shopping malls, and television, they resemble these commercial forms ever more closely.

Nevertheless, what remains of the public educational sphere deserves the support of environmental activists just as much as the corporate version of nature demands critical analysis. San Diego's small and underfunded Museum of Natural History offers a good example. Competing directly with Sea World of California for visitor dollars, and often drawing on Sea World staff for resources, the museum has been forced to promote itself as an attraction for regional tourists, even while it struggles to provide a local educational resource. Despite its reliance on local corporate sponsorship, the museum has been able to mount some small but sophisticated exhibitions on serious environmental issues. These have covered topics ranging from the relationship between development, land use, and habitat destruction in San Diego County to the *Exxon Valdez* oil spill and to scientific controversies over global warming. If they are small and limited, spaces such as this one have great potential for enlivening informed debate about what nature is and how we might struggle to rethink its problems. Try to imagine a theme park mounting a thoughtful exhibit on ocean pollution—a problem of grievous immediacy in southern California.[46] While we criticize commercial culture's nature, we also need to create and support sites for less magical images and information. Defending older quasi-public spaces from the tyranny of the bottom line is one way to do this. To argue for public spaces for environmental discourse is not to argue for doing away with entertainment. Neither do I disdain people who enjoy the remarkable qualities of animals. But not all entertainment should be commercial, and education should not be collapsed into public relations and a ruthless drive for corporate profits.

Definitions of nature and the solutions to its problems are now massively authored by the private sphere of conglomerate, corporate culture, at the same time that corporations claim to further the public good. Sea World's theme park nature is only one example, but it is striking that it asks America's most affluent, educated, and influential citizens to trust nature's future, and their own, to the corporate matrix. Finding a new environmental ground depends on contentious debate, not on easy consensus that corporations and citizens are each doing their part. To conduct this debate, it is vitally important that all Americans have a wide range of ideas, information, and images to draw on.

Notes

Davis, Susan G. 1995. "Touch the Magic." In *Uncommon Ground: Toward Reinventing Nature*, ed. William Cronon. New York: Norton, 204–17. Special thanks for research assistance to Kay Mary Avila.

1. Descriptions and direct quotations refer to ad copy written by D'Arcy Masius, Benton and Bowles, Inc. USA, and the commercial as aired.

2. Anheuser-Busch owns ten theme parks: Sea Worlds in San Diego, Calif., Aurora, Ohio, San Antonio, Tex., and Orlando, Fla.; Busch Gardens parks in Williamsburg, Va., and Tampa, Fla.; Adventure Island in Tampa; Cypress Gardens in Winter Haven, Fla.; Sesame Place in Langhorne, Penn.; and Water Country USA in Williamsburg, Va. An Anheuser-Busch publicity brochure estimates that "over the past 30 years, more than 160 million people have visited Sea World parks." Anheuser-Busch, public relations brochure on environmental responsibility (untitled), "Item No. 001-584," St. Louis, 1993. The first Sea World was launched in San Diego in the early 1960s by a small group of private investors. Anheuser-Busch purchased the Sea World from the publishing, real estate, and insurance conglomerate Harcourt Brace Jovanovich in 1989.

3. Thirty- and sixty-second versions of "Touch the Magic" ran in major television markets nationally.

4. Richard Melcher, "Anheuser-Busch Says *Skoal, Salud, Prosit*," *Business Week*, Sept. 20, 1993, 76–77.

5. I estimate total attendance for all the Busch theme parks as about 18 million visits for 1993. The American Disney parks totaled about 41 million visits in the same year. *Amusement Business*, Year End Report, chart, Dec. 20, 1993–Jan. 2, 1994, 68–69. The Busch Entertainment Division brought $55 million in profits to the corporation in 1992, up 22 percent from the previous year (but still weaker than in 1990). Melcher, "Anheuser-Busch," 76.

6. Tim O'Brien, "Theme Park Admission Prices Reach Record High," *Amusement Business*, April 18–24, 1994, 1,35. Adult admission is $34.95.

7. According to their management, Sea World of San Diego earns roughly 50 percent of its profits from admissions and 50 percent from concession sales.

8. Through sponsorship arrangements the Sea World parks reduce their advertising costs; sponsors gain cross-promotion advantages, exclusive merchandising rights (for example, Kodak and Pepsi are "official suppliers"), and association of their name in connection with animals, the environment, and family entertainment. Sponsorship may have been more important when the Sea Worlds were owned by Harcourt Brace Jovanovich, especially as HBJ had to fight off takeover attempts. Anheuser-Busch's pockets are much deeper.

9. This may be especially important in an era of public concern over advertising alcoholic beverages to youth, and over commercial connections between beer, rock music,

and sports. Beer companies' extensive sponsorship of rock concerts and band tours has been criticized by health professionals and advocates of drunk-driving prevention. Paul Grein, "Suds 'n' Bucks 'n' Rock 'n' Roll: Beer Companies Rock Sponsorships Stir Controversy," *Los Angeles Times*, Sunday Calendar, July 30, 1989, 8, 85, 86. Budweiser has been the major sponsor for U.S. concerts by Mick Jagger and the Rolling Stones, most recently providing millions of dollars for the 1994 "Voodoo Lounge" tour.

10. This commercial confusion of retail space with public space is increasingly common, as shopping malls, for example, begin to house playgrounds, museums, libraries, city halls, even public schools and community colleges. See Margaret Crawford, "The World in a Mall," in Michael Sorkin, ed. *Variations on a Theme Park: The New American City and the End of Public Space* (New York: Hill and Wang, 1992), 3–30; Leah Brumer, "Discovery Zone" (unpublished paper, summer 1994).

11. This argument is made over and over again throughout the parks and their public relations materials. Anheuser-Busch also emphasizes the parks' membership in the American Association of Zoological Parks and Aquariums. Anheuser-Busch, public relations brochure on environmental responsibility (untitled), "Item no. 001-584," St. Louis, 1993.

12. Busch Entertainment allows cable systems, school districts, and teachers to tape these broadcasts for repeated use; thus their reach is expanded.

13. The maker of many disposable paper and plastic items, Procter & Gamble has been criticized for distributing factually inaccurate and self-serving materials disguised as science study resources. Michael Parrish, "Environmentalists Criticize Firm's Educational Material," *Los Angeles Times*, Dec. 17, 1993, D2.

14. With the recent acquisition of parks chains by Time-Warner (Six Flags parks), MCA (Universal Studios), and Viacom (Paramount Parks), and with the merger of Viacom with Blockbuster, the theme park industry is dominated by mega-media corporations. Most of the large theme parks in the United States are now sites for the integrated marketing of diverse media products, following the Disney model. The five biggest companies in the industry are Disney, Anheuser-Busch, Time-Warner, MCA, and Viacom. Anheuser-Busch uniquely specializes in "nature."

15. Walt Disney pioneered this technique. Richard Schickel, *The Disney Version: The Life, Times, Art and Commerce of Walt Disney,* rev. ed. (New York: Simon and Schuster, Touchstone Books, 1985), 295–338. George Lipsitz, "Discursive Space and Social Space: Television, Highways, and Cognitive Mapping in the 1950s City" (paper presented at the American Studies Association Meeting, Nov. 1989).

16. One such model might be Discovery Channel's "Those Amazing Animals," filmed at another nature theme park, Marine World Africa USA. A feature-length film and an animated television show featuring "Shamu" are reported to be in development. Kim Kowsky, "Busch to Buy Rights to Films about Shamu," *Los Angeles Times* (San Diego County ed.), Jan. 5, 1990, D1.

17. In 1994 Third Story Books published twelve children's books based on the Sea World parks and characters. These were promoted in the Barnes and Noble bookstores, among others. "New Firm to Do Sea World Books," *Publisher's Weekly*, March 14, 1994, 11. LaPorta and Company produces the "Shamu and You" educational video series. "Video Treasures of the Deep and Wild," *Billboard*, Feb. 8, 1992, 47.

18. At the same time, we are always aware of the artifice of Sea World's realism. For example, the blue-painted pool holding Shamu is constructed to appear bottomless. The seeming depth of the pool is a kind of perceptual support system, letting us think the whale has boundless room to move way from us, even while we are aware that it cannot get away.

19. On the history and ideological structure of travel literature, see Mary Louise Pratt, *Imperial Eyes: Travel Writing and Transculturation* (London: Routledge, 1992); on nature, knowledge, and empire, Harriet Ritvo, *The Animal Estate: The English and Other Creatures in the Victorian Age* (Cambridge: Harvard Univ. Press, 1987), 1–44; on *National Geographic*, Herbert I. Schiller, *The Mind Managers* (Boston: Beacon, 1974), 79–103, and Catherine A. Lutz and Jane L. Collins, *Reading National Geographic* (Chicago: Univ. of Chicago Press, 1993).

20. The public relations department at Sea World was unable to tell me the provenance of the myth referred to in "Shamu New Visions."

21. Anheuser-Busch, untitled public relations brochure, "Item no. 001-584."

22. Between 1983 and 1988, hundreds of eggs were "imported" from Antarctica, incubated, and the hatchlings hand-raised at Sea World of California. At least some of the penguins on display are descendants of these chicks.

23. Sea World's publicity materials and entertainments make much of teaching "naturally antagonistic" species "how to live together," although it's not clear what "naturally antagonistic" means, or that the parks really do this. (For example, killer whales and sea lions do not share tanks in the park.) *Amusement Business*, Jan. 14, 1984. In 1993–94 San Diego's Sea World featured a multispecies whale and dolphin show called "One World."

24. Sea World's exhibits do make use of such references, especially to "Star Trek."

25. Stuart Ewen, *All Consuming Images: The Politics of Style in Contemporary Culture* (New York: Basic Books, 1988). Very little work has been done on nature as an ingredient in contemporary consumer advertising. Judith Williamson's *Decoding Advertisements: Ideology and Meaning in Advertising* (New York: Marion Boyars, 1984) remains helpful.

26. Raymond Williams, "Ideas of Nature," in *Problems in Materialism and Culture* (London: Verso, 1980), 67–85; idem, *The Country and the City* (New York: Oxford Univ. Press, 1980).

27. See, for example, Williams, "Ideas of Nature."

28. Ibid. and Williams, *Country and the City*, 87–107.

29. Jonas Frykman and Orvar Löfgren, *Culture Builders: A Historical Anthropology of Middle-Class Life*, trans. Alan Crozier (New Brunswick: Rutgers Univ. Press, 1987), 42–87.

30. Pratt, *Imperial Eyes*, 15–37; John Michael Kennedy, "Philanthropy and Science in New York City: The American Museum of Natural History, 1868–1968" (Ph.D. diss., Yale Univ., 1968); Donna J. Haraway, *Primate Visions: Gender, Race, and Nature in the World of Modern Science* (New York: Routledge, 1989), 1–58; Peter J. Schmitt, *Back to Nature: The Arcadian Myth in Urban America* (Baltimore: Johns Hopkins Univ. Press, 1990), 77–95.

31. Each of the Sea Worlds commissions extensive market and "psycho-graphic" (lifestyle) research on its customers, interviewing as many as five hundred visitors in person per month and distributing numerous take-home questionnaires to others. In 1992 Sea World of California's interview research showed that only 15 percent of the customers reported family annual income under $30,000. (During this same period median family

income in San Diego County was about $35,000.) Some 51 percent of the customers claimed more than $40,000 annual income; 33 percent stated that their family earned more than $50,000. Even allowing for the many problems inherent in self-reporting, this is obviously an affluent audience. Fully 43 percent of those interviewed claimed a college or higher degree, while 22 percent had a high school diploma or less. Market research for the same period reports that 89 percent of the customers interviewed are" Anglo" and 11 percent "non-Anglo" (only these two categories were used). It is unclear whether this identification is based on observation or self-description, whether it refers to color, historical identity, or mother tongue. Nevertheless, the figure is sharply divergent from the general ethnic makeup of the southern California counties. The average audience member was a baby boomer, about thirty-eight years old, and nearly 65 percent of the audience was between the ages of twenty-five and fifty. (In other words, the audience closely resembled the people who wrote this book.)

32. *Amusement Business*, April 4, 1987, 4.

33. Other theme parks in California use dress codes to discourage the presence of "gang members." The content of youth music concerts at theme parks is also carefully vetted.

34. The five counties are Los Angeles, Riverside, Orange, San Bernardino, and San Diego.

35. Frykman and Löfgren, *Culture Builders,* passim; Yi-Fu Tuan, *Dominance and Affection: The Making of Pets* (New Haven: Yale Univ. Press, 1984), esp. 115–31; Schmitt, *Back to Nature,* 77–124.

36. See, for example, Richard Louv, *Childhood's Future* (Boston: Houghton Mifflin, 1990).

37. Frykman and Löfgren, *Culture Builders.* When thinking about what children "need," we should distinguish contact with nature from unstructured play or autonomous activities. There is little unstructured or autonomous about the "nature" children encounter at Sea World.

38. Schmitt, *Back to Nature.*

39. John Berger, "Why Look at Animals?" *On Looking* (New York: Pantheon, 1980), 1–26.

40. The rationale behind the architecture of the new zoo is summarized by Melissa Greene, "No Rms, Jungle View," *Atlantic Monthly,* Dec. 1987, 62–78.

41. The Du Pont television commercial shows marine mammals applauding the chemical manufacturer's environmental record, to the strains of Beethoven's "Ode to Joy." Chip Berlet and William K. Burke, "The Anti-environmental Movement," *Democracy Watch,* July 1992, 3, 11. See also *Hold the Applause* (Washington, D.C.: Friends of the Earth, 1992).

42. Yet nature on TV has received almost no attention from media scholars. For example, while there has been extensive study of network news, soap opera, and crime drama, we have no broad and accurate picture of how much nature TV is broadcast, who produces it, and whom it reaches. Presumably a media genre this prolific has some important effects on how its television audience understands environmental problems and their social and political context.

43. On the general structure and corporate funding of PBS, see William Haynes, *Public Television for Sale: Media, the Market and the Public Sphere* (Boulder, Colo.: Westview Press, 1994), esp. 89–114. Haynes notes that "corporations such as BASF, DuPont,

W. F. Grace and Waste Management, Inc., sponsor such programs as *Adventure, Discoveries Underwater, Victory Garden* and *Conserving America* (102–3). For criticism of the "public" nature of public television, see also David Choteau, William Haynes, and Kevin M. Carrageen, "Public Television and the Missing Public: A Study of Sources and Programming," *Extra*, Sept./Oct. 1993, 6–14; Janine Jackson, "When Is a Commercial Not a Commercial? When It's on Noncommercial TV," ibid., 17–18.

44. For example, Roy Rosenstein and Elizabeth Blackmar emphasize that the social uses of parks, landscapes, and museums are contested, and shaped by their diverse and often antagonistic users. *The Park and the People: A History of Central Park* (Ithaca: Cornell Univ. Press, 1992).

45. Although institutions such as Chicago's Museum of Science and Industry have a long history of connections to corporate sponsors, the interpenetration of the for-profit and nonprofit spheres has accelerated in the last decade. As zoos and museums have found their public funding and philanthropic support dwindling, they have come to depend heavily on gift shop and souvenir sales, and in this and other ways they have become more like the pay-to-enter, concession-driven theme park. Famous nonprofit science centers, such as San Francisco's Exploratorium, seek development funds through licensing and coproduction agreements with media giants like Time-Warner. For a critique of related developments in the world of art museums, see Debora Silverman, *Selling Culture: Bloomingdale's, Diana Vreeland, and the New Aristocracy of Taste in Reagan's America* (New York: Pantheon, 1986.) A more general and thoroughgoing critique is offered by Herbert I. Schiller, *Culture, Inc.* (New York: Oxford Univ. Press, 1991). On national parks, see Dean MacCannell, "Nature Incorporated," in *Empty Meeting Grounds: The Tourist Papers* (New York: Routledge, 1993), 114–20.

46. Interestingly, Sea World's references to the *Exxon Valdez* oil spill are limited to its display of rehabilitated Alaskan otters. Wildlife biologists, writing in *Science,* have criticized the media focus on otter and bird rescue after the spill. They argue that media coverage of the rehabilitation of a small number of animals has diverted attention from the extent of environmental damage, and from the need for the absolute prevention of oil spills, and given the public a false sense that catastrophic environmental damage can be mitigated. James A. Estes, "Catastrophes and Conservation: Lessons from Sea Otters and the *Exxon Valdez*," *Science* 254 (1991): 1596.

The Nature of Future Myths

Environmental Discourse in Science Fiction Film, 1950–1999

Christopher W. Podeschi

Christopher Podeschi conducts a content analysis on a selection of popular science fiction films. Based on this analysis, Podeschi argues that science fiction films, such as Alien, have largely devalued nature and portrayed it as hostile and inferior to human culture. These films both foster and reflect a modern technocratic culture that sees the human-controlled domain of science and technology as superior to the "wild" domain of nature. Podeschi argues that cultural analyses of popular media are important not just for illuminating our taken-for-granted belief systems but also for potentially pointing the way to a better future. As Podeschi shows, though the devaluation of nature is the dominant trend, alternative, pro-environment ideas do exist within films, and recognizing those is the first step toward nurturing their growth and development.

Around the world, it is clear that human societies have damaged, and are damaging in profound ways, the environments upon which all life depend. Proximate causes, like technology or population can be indicted, but it is also clear that the global and destructive structural level tread-mill of production (see Gould, Schnaiberg, and Weinberg 1996; Schnaiberg and Gould 1994; Schnaiberg 1980) has a momentum that cannot be ignored. This is no alarmism. Finite and fragile Earth cannot withstand the continually expanding barrage of withdrawals, additions, and transformation which global capitalism and industrialism demand.

. . . Fifty years ago there was perhaps less awareness of the universality or the severity of the problem, but ecologist Aldo Leopold (1949) knew redirection was needed. He called for the extension of ethics to the nonhuman, an opening of community beyond town, city, region, nation, indeed, beyond *humanity.* Leopold argued for a "land ethic" to change humanity, "from conqueror of the land-community to plain member and citizen of it" (p. 204). Beyond his recognition that the domination by human society over the rest of nature needed to be stopped, we can read in his now-famous lines two other important principles. First, that nature is in large part a cultural construction and therefore the nature/culture dichotomy is a false one. Nature is cultured. As Fine (1998) puts it, "'Nature' does not exist. . . . The lumping of diverse objects together within a category (nature) is a human creation" (p. 4). This is not to stake out some naive idealism; material reality and therefore things like environmental crisis are all too real. The point is that material reality is also unavoidably cultural, under-stood through discourses and language, though material reality can clearly have some influence over our cultural constructions of it (see Freudenberg, Frickel, and Gramling 1995).[1] The second point is that if we learn to think in a new way about the land and its inhabitants, if we culture nature differently, the contra-dictory human societal presence can perhaps be resolved. In short, nature and the societal relationship with nature are cultural constructs, and they are con-structs with material consequences. The form they take is critical to preventing crisis from becoming catastrophe. "The culture of nature" can generate social acquiescence and reproduction of destructive relations.[2] But, it can also fuel struggle to remedy the relationship with nature.

. . . Environmental concerns are a topic in the news media and while there have been periods of decline in public opinion toward the environment since the 1960s, surveys show that people think of themselves largely as environ-mentalists. More attention, however, needs to be given to key cultural narra-tives in order to explore the meaning of nature and the environment (see Shanahan and McComas 1999). Popular science fiction films from the latter half of the twentieth century are examined toward this end. Science fiction films are a good choice not only for their popularity, but because they con-ceptualize and problematize the voyage into the near and distant future. These projections of social reality are our future myths, and whether intended or not, these productions comment on nature and the social relationship with nature. . . .

... The question at hand is of the way in which nature and the relationship with nature will be represented in Hollywood science fiction films. ...

To get a broad sense of the status of environmental discourse in science fiction film, films were selected from the 1950s to the 1990s. Within this period, choice of films was guided by popularity to maximize the sociological relevance of the project, i.e., these films reached and potentially impacted the most people. ...

Because films are complex, nuanced, and often internally contradictory texts, and because the films sampled here need to be read as in dialogue with one another and with various forms of environmental discourse, qualitative methods seemed best suited for the analysis. ...

Analysis

Themes of environment and nature are indeed evident in science fiction film from the latter half of the twentieth century. Much of this treatment is of course latent and/or marginal to the central narrative of these films, though a handful of films directly address nature, environment, or other relevant issues. The analysis is presented here in terms of the specific ways in which the film texts resonated with resistant and/or reproductive discourse. Technology, the material relationship with nature, and the value of nature emerged as the themes key to a reading of science fiction films via environmental discourse. Each of these are in turn divided into a number of inductively derived subcategories centered on more specific clusters of meaning. Technology is considered in terms of (1) portrayals of the potential power of technology and (2) debated technologies. The material relationship with nature is considered in terms of (1) portrayals of colonization of space and (2) portrayals of environmental consequences. Finally, the value of nature is considered in terms of (1) the portrayal of nature relative to society, and (2) discourse that resists devaluation *of* nature.[3]

Technology

As *the* driving force behind the genre, technology gets ample attention in these films. The all too familiar technology texture of unexplained and mostly unused flashing lights, computer screens, and other assorted gadgetry attests to this. This presence is important here because technology is fundamental to the relationship with nature and to discourse about nature and the environment. In general, resistant perspectives protest technology as risky and as a root of environmental crisis while reproductive thinking focuses on benefits it has brought or promises to bring to humanity. Not surprisingly, most films resonate with the latter perspective, oriented toward a future in which technology continually increases in power, scope, and safety.

The Power of Technology

If we take the word of those who produce science fiction, there are nearly no bounds to the power of future technology. Twenty-one out of the 27 films analyzed show us powerful and almost magical technology."[4]

The main vehicle for glorifying future technology is simply the repeated presentation of amazing technology. There are abundant examples of powerful technologies: intelligent and feeling robots, enormous space stations, space travel at the speed of light, energy shields protecting space ships, and handheld laser guns. . . . Other technologies go beyond these types by having control over nature at a seemingly fundamental level. . . . In *Forbidden Planet*, Robbie the Robot can actually create matter in great abundance given a moment to analyze the type of material to be produced. . . .

Finally, many of the films also fit with reproductive discourse by arguing that technology is innocent and risk free. In these films, technology is quite dependable as it does not, on its own, malfunction or create hazards. Environmental consequences are scarce as well. . . . *Destination Moon* is about the launch of a nuclear-powered rocket being delayed by government and public concern over the safety of the technology. Despite the opposition, the project's backers launch anyway and without a test even though this is the first rocket of its kind. The trip is a success. Not only is concern about the technology contradicted, but the protagonists make it clear that fear of and interference with technology is unfounded. In the end, we are not only told to trust technology and its makers, but that powerful technologies are not a threat to us or the environment.

This virtually risk free technological apparatus is found in one-third of the films sampled. . . .

Debated Technology

The march of technology does not go completely unchecked. Technology is questioned at times by science fiction, a tension which undoubtedly transcodes the emergence of resistant environmental thought and fears about technology's effects. It must be emphasized beforehand that in holistic context, interrogations of technology are difficult to "hear." They are limited in number and contradicted by the abundance of positive portrayals; critiques are generally found within the hi-tech futures described above. Furthermore, when technology is questioned, it is almost always focused on one of the three particular types discussed below, technologies that transform nature, technologies that unite with humanity, and nuclear technology. There are only a few exceptions to this narrow form of critique. The *Star Wars* films and *Waterworld* have subtle subtexts linking antagonists to technology and destruction, while the protagonists are more rooted in nature or rely on simpler technology. *Jurassic Park* is more explicit. Though the sentiment is fleeting, one character calls science and technology "rape of the natural world."

Transforming Nature

Transforming nature here refers to productive control of nature, i.e., control which changes nature in a fundamental way. This sort of instrument is present in ten films, the vast majority of which come from the period between the late 1970s and the 1990s. Eight of these fit with reproductive discourse by portraying such technology favorably or without critique.[5] There are, for example, technologies which control, alter, or create atmosphere in the films *Back to the Future, Part 2, Aliens*, and *Total Recall*. In *Back to the Future, Part 2*, we are told that the weather service controls the weather more efficiently than the postal service delivers mail. . . .

The two films that are critical of technologies which transform nature are *Star Trek* and *Jurassic Park*. The conflict in *Star Trek* stems from the return of Voyager, a U.S. exploration satellite. In the film, Voyager has been changed by a race of living machines it encountered on its journey. It is now able to do more than collect information about objects in the universe that it encounters; it actually sucks these objects (e.g., planets) into itself. This of course could have devastating results as it approaches Earth, and so this conflict functions as a critical comment on such powerful technology. In *Jurassic Park*, genetic engineering is the criticized technology. This film visits a doomed theme park populated with ancient plants and dinosaurs cloned from preserved DNA. Bad luck and greed play a part in the park's failure, but the power of the artificial creatures incompatible with present social and ecological reality is more significant. Foreshadowing an ecological nightmare in which dinosaurs reclaim Earth, the supposedly sterile dinosaurs end up reproducing. Critique is expressed as well, as the parks inventor is told, "Genetic power is the most awesome force the planet's ever seen, but you wield it like a kid that's found his dad's gun." The entire film resists genetic engineering in this way, and overall, nature is defined as too powerfully chaotic to be controlled at this fundamental level.

Uniting with Humanity

Technology that unites with *humanity* is another form of technology which gets attention. Present in 13 of the 27 films analyzed, again primarily from the period of the 1970s to 1990s, this "wedding" is portrayed in two predominant ways, either as simulation of humanity (i.e., artificial intelligence or robots) or as actual combination with the human body.[6] Of the three types of debated technology discussed, this form is actually most resisted. All but three of the films that include it provide critique and in many cases quite strong critiques, though some also validate the technology at the same time. This level of concern is important because, as noted, science fiction provides very little critical vision of technology. It must be emphasized, however, that while this discourse asserts that this sort of technology is quite threatening, the resistance is brought to bear not on technology's environmental effects, but on technology's impact on humanity alone. As mentioned above, technology is rarely resisted in a general sense in these films.

There are numerous examples of technology that combines with humanity. In almost all cases it is repulsed. *Total Recall*, *Demolition Man*, and *The Matrix* all resist technology implanted in the body for surveillance purposes. Other films focus on robotic attachments to the human body. Perhaps the most well known example is the embodiment of evil from the *Star Wars* films, Darth Vader. With a body so abused that numerous parts are artificial and he must continually wear a life-support system, this wedding of body and technology is explicitly denigrated: "He's more machine now than man; twisted and evil."[7] . . .

Computer technology that has direct access to the brain is another combination with humanity given attention in these films. . . . In *Total Recall* and *The Matrix*, near perfect illusions of reality are created this way. *Total Recall* is fully critical here, as the realism can cause psychological trauma or "schizoid embolisms" and death. *The Matrix* provides mixed messages. On the one hand, the actual physical connection of computer to the body is treated negatively, but more importantly, the virtual reality it creates is used for total social control. Artificial intelligence has taken over the world and uses human beings as an energy source, keeping them asleep to this fact by immersing them in a virtual world. Despite this, the technology is also depicted as powerful and liberating. When used voluntarily it provides entertaining illusions and human enhancement. Expertise in any human practice, for example, can be downloaded into the brain and learned in moments.

Considering simulated humanity, or artificial intelligence, this form of union with humanity is resisted powerfully in *Alien*, *2001*, and *the Matrix*. In *Alien*, Ash is an artificial person who functions as a traitor without care for human life. Emphasizing the rigidity of technology in opposition to human reflexivity, Ash fuels the plot by following his programmed orders and subversively seeking to ensure that a hyperviolent alien makes it back to Earth despite the danger to his crewmates. He also mentions that he admires the alien for its similarity to him: like a machine, it lacks conscience, remorse, or morality. . . .

Despite this resistance, the novelty and power of these simulations must be remembered. Ash from *Alien* is amazing regardless of his flaws; perhaps, he can be perfected. In fact, in *Aliens*, the follow-up to *Alien*, another artificial person is depicted, but as a flawless contrast to Ash. Fitting reproductive discourse closely, the second film redefines Ash as technology waiting to be perfected. Perfect robots are found in other films as well. Throughout the *Star Wars* trilogy, we get to know the lovable, humanlike robots R2D2 and C3PO. Other loved robots are found in *Forbidden Planet* and *The Black Hole*. . . .

Nuclear Technology

The final technological subtheme is the most abundant: nuclear technology is a subject in 18 of the 27 films sampled and is present throughout the sampling period. Like technologies that transform or unite with nature, it is also supported and resisted. Transcoding the Cold War and fears about radiation, films primarily from the 1950s and 1960s contest nuclear technology sharply, some presenting it as a useful savior and some condemning it for potential, if fantastic, problems. Then the futuristic films of the 1980s and 1990s drop this debate and present nuclear power simply as a part of future social reality.[8]

Starting with the debate, three films, *The Beast from 20,000 Fathoms*, *Destination Moon*, and *Voyage to the Bottom of the* Sea each provide strong support for nuclear technology. . . . In *Voyage*, a nuclear scientist captains a nuclear submarine and plans to fire a nuclear warhead into the Van Allen belt, which has somehow caught fire and is cooking the planet. This plan is globally opposed and the captain's sanity is even questioned. Despite the opposition and even attempts to stop him with force, he tires his warhead and saves the planet. In the end, the film tells us quite simply that nuclear technology is of great value and all who fear or oppose it will look foolish in the end. . . .

The five films which form the other side of this contest about nuclear weapons and war do provide strong critiques. . . . In The *Incredible Shrinking Man* exposure to radiation in the Pacific causes a man to shrink away to nothingness. Similarly irradiated ants grow into enormous monstrosities in *Them!* . . .

As mentioned, after 1970, this contest about nuclear weapons and war ends in science fiction film.[9] Fitting well with reproductive discourse, futuristic science fiction from the 1980s and 1990s simply naturalizes nuclear power into the imagined social landscape. For example, though they are defined as dangerous if damaged, large nuclear reactors are the unquestioned power source at a deep space colony in *Aliens*. *Star Wars*, *The Black Hole*, *Star Trek II*, *Star Trek IV*, and *Demolition Man* treat nuclear technology in a similar manner. *Back to the Future, Part 2* even predicts nuclear power will be so safe that it can be integrated into everyday life: there is a Cuisinart-like nuclear generator for personal use. The naturalization in *Total Recall* and *Forbidden Planet* is quite different. These films give us ancient, abandoned, an enormous, yet still perfectly functioning, nuclear reactors. These two films implicitly define nuclear technology as powerful and absurdly dependable.

The Material Relationship with Nature

Technologies certainly demonstrate a relationship with nature, but it is also useful to examine the films in terms of more general environmental relationships. Included here are issues of the human presence in nature and the treatment of nature. . . .

Colonizing the Cosmos

An important treatment of the human presence in nature in the films analyzed
is of society filling the wild frontier of space. Spread across the sampling period,
all but two of the futuristic films sampled present futures in which civilization
has expanded its boundaries well beyond the terrestrial, often deep into the cosmos to some grand scale." . . .[10]

Many futuristic films present this colonization of the cosmos simply with
space travel and space stations. Films like *2001* and *Total Recall* only take us as
far as this solar system, but *Planet of the Apes*, *Forbidden Planet*, and the *Star
Wars* and *Star Trek* series all show civilization bounding about the cosmos. The
full colonization of other planets is also a powerful manifestation of the drive
to colonize. *2001* and *Total Recall* portray colonization of the moon and planets of our solar system. . . . In *Alien* and *Aliens*, space is divided between "frontier" and "core" systems and in the *Star Trek* films, there are galactic political
boundaries, implying control over large amounts of space. . . .

Environmental Consequences

. . . A number of the films, primarily those which engage nuclear weapons and
war, bring a notion of environmental consequences to the fore apart from the
portrayal of technology. Some of these do not allow environmental concern to
survive the plot. *Destination Moon* and *The Beast from 20,000 Fathoms* raise and
then strongly undermine concern about the effects of nuclear technology while
Star Trek II raises and then undermines concern about the effects of the Genesis
device. Other films examined remain critical about the environmental consequences of technology. *Them!*, *The Incredible Shrinking Man*, *On the Beach*,
Planet of the Apes, and *Beneath the Planet of the Apes*, from the 1950s to 1970
each center on the potential effects of nuclear weapons. Both from the 1990s,
Jurassic Park predicts catastrophe from genetic engineering and *Waterworld*
indicts global warming-inducing technologies. A notion of negative environmental consequences, and environmental protection, appear in a number of
other films as well. In addition to those listed here, eight other films, primarily
from the late 1970s to the 1990s, include a notion of environmental consequences and/or protection.[11]

The vast majority of the additional material is, however, fleeting, minor,
or indirectly environmental. Only one of these eight explicitly foregrounds an
environmental problem in the way *On the Beach* or *Jurassic Park* do: *Star Trek
IV* functions as a film-length critique of activities which have caused the extinction of humpback whales. As examples of the minor material, *Demolition Man*
provides images of a future with an apparently cleaner environment, while *Back
to the Future, Part 2*, projects a generalization of recycling practices into the
future. . . .

In opposition to this protectionist sentiment, other content in the sample latently validates environmental damage. *Aliens* and *The Empire Strikes Back* both define space as an appropriate repository for society's wastes. In *Aliens*, despite environmental protection laws, we learn that there is so much rubbish in space that salvage teams search the vastness of space for valuable trash. Violence also needs to be considered as a latent validation of environmental damage. In seven films there is an exclusion of consideration of the consequences of violence for the nonhuman.[12] Resistant discourse would make such consequences central, but in these films, the business of civilization is paramount and the results for nature are not just subordinate to this, they're absent. For example, the destruction of technology, usually large spaceships, occurs with no regard for the impact this will have on the surrounding environment. From the *Star Wars* films, the planet-sized and nuclear-powered Death Star is twice blown up, once right above a forested planet. In *Aliens* the means to resolve the conflict with the aliens is to destroy with nuclear warheads a large colony on the surface of a distant planet that could simply be avoided. . . .

The Value of Nature, the Value of Civilization

Technology and the more general relationship with nature discussed above most certainly speak to the value of nature. In these areas, we see some minor environmental concern and some concern about technology voiced, but overall, there is a predominance of fit with reproductive discourse. These futures of continual domination of nature through things like violence, colonization, and technology can be understood to connote the lesser value of the nonhuman relative to civilization. A more direct look at the value of nature is needed as well. Basically, the question is of the status of the nonhuman relative to humanity or society in the projected futures. . . .

Hostile Nature

Both inanimate nature and environments as well as animate nature are primarily defined unfavorably relative to civilization. Considering environments and inanimate nature first, the pattern is one in which place is not only nonsalient, functioning as a backdrop for the exploits of civilization, but when foregrounded, portrayals tend to be negative. With the exception of moments in which landscapes are valued for their beauty, foregrounded environments in films from the 1970s and 1980s are hostile, harsh, or even disgusting, fitting well with the idea that nature is wild, dangerous, and in opposition to civilization.[13] In the *Star Wars* films, despite the fact that the environments appear pristine and often beautiful, they are also harsh and ecologically monolithic. In *Star Wars* and *Return of the Jedi*, we go to an all-desert planet and in *The Empire*

Strikes Back, we go first to an ice planet and then to a swamp planet. The swamp planet teems with life, but in forms which are meant to spark fear and disgust, like monsters lurking in murky waters. . . .

The other side of the coin is animate life forms, and again there is resonance with reproductive discourse. . . .

The *Star Wars* trilogy and *Alien* and *Aliens* have particularly strong anti-creature stances. The protagonists from *Star Wars* repeatedly confront foul creatures ready to consume them. While these films give us plenty of disgusting aliens and even an abominable snowman, it is snakes, worms, and tentacled things which are more common. In *Star Wars* a tentacled thing lurks in and attacks from murky water in a garbage dump. In *The Empire Strikes Back*, the protagonists encounter snakes and another tentacled hunter hiding in swamp waters. Later in the same film, a cavern on an asteroid turns out to be the belly of a giant worm, out of which the protagonists barely escape. Similarly, in *Return of the Jedi*, a giant desert worm attempts to grab victims with long tongues in much the same fashion as the other tentacled hunters. . . .

The juxtaposition of wild and domesticated animals is also telling, widening the rupture between nature and civilization. In three films, domesticated animals are juxtaposed and contrasted with wild creatures, reinforcing the notion that domesticated nature is good and wild nature is bad. In *Jurassic Park*, huge and gentle herbivorous dinosaurs are compared to cows and contrasted with carnivorous dinosaurs that are coded as evil, ruthless, and even intentional killers (as opposed to being simply, carnivores). . . .

Nature Devalued

Beyond the duality of hostile nature and "civilization," also present in the films analyzed is commentary that simply devalues nature as inferior to or of lesser worth than people. This tendency is again present in films primarily from the 1970s and 1980s. In *Star Trek II*, for example, a planet is described as "a great rock in space," with "unremarkable ores," the connotation being that nature is only valuable through human development. In *Star Wars* the Death Star is involved in a direct assault on nature, destroying an Earthlike planet. This is defined as reprehensible, but only in terms of loss of human life: the nature-priest figure, Obi Wan Kenobi, senses the destruction of the planet because he hears millions of voices crying out. Similarly, *Forbidden Planet* and *Aliens* destroy planets with no regard for anything but human interests. Also, in *Forbidden Planet*, the inventor of Robbie the Robot demonstrates the safety of his creation by showing that even when ordered, the robot will not attack a "rational being." By contrast, Robbie most willingly destroys nature, irrational and hence expendable things.

Portrayals of civilized nonhumans (i.e., aliens) in the *Star Wars* and *Star Trek* films support this tendency as well. Nonhuman protagonists are not spared, but it is primarily nonhuman antagonists that are portrayed as animal-like and closer to nature, inherently hostile, or disgusting. What is natural, or closer to nature, is lesser. One of the best examples is the Klingon group from the *Star Trek* series of films. This people has animalistic and primitive connotations revealed in their appearance and their aggressive demeanor and culture. The *Star Wars* trilogy is ripe with this sort of content as well. The best example is the wealthy criminal Jabba the Hut from *Return of the Jedi*. Disgusting and like a giant slug, he is already a denigration of the nonhuman, but beyond this, he is a sadist who eats live animals and surrounds himself with pig-like body guards and a "court" of other foul creatures.

Resistant Strains: Nature Valued and Civilization Critiqued

Alluded to already, positive valuations of nature are overall quite uncommon. There is one exception. In well over half the films and from throughout the sampling period there is clear aesthetic valuation of nature, primarily in the form of long shots of landscapes, planets, or starfields.[14] This valuation must be put into context. Not only are these scenes peripheral to the plot, but when set in opposition to actual encounters with nature discussed above, the films tell us that nature can be beautiful from the comfortable distance of civilization, but that contact is another story.

Considering the handful of other positive valuations of nature, two films provide moments of empathy with pest nature, forms denigrated in *Alien* and *Aliens*. In *Fantastic Voyage*, a group of people is shrunk to microscopic size in order to enter an injured man's body and perform surgery. One of the men overseeing the operation resists his instinct to kill an ant, the connotation being that he now empathizes with small, vulnerable things because of the danger his shrunken surgeons are experiencing. Similarly, in *The Incredible Shrinking Man*, as the protagonist shrinks away to nothing, he begins to feel "closer to nature." And in contrast to the imbalanced portrayal of nature as hostile discussed above, he comes to understand a spider that hunted him in ecological terms: "I no longer felt hatred for the spider, [who] like myself, had struggled blindly for the means to survive."

Resonant with "save the whales" discourse in recent decades, *Star Trek IV* arguably goes further than other films in its valuation of nature. A key protagonist, Mr. Spock, provides an ecocentric voice a number of times in the film. Mentioned above, in this film, a group of whale aliens comes to Earth searching for comrades left in the waters long ago. When it is assumed that the space whales are trying to communicate with humanity Mr. Spock elevates the nonhuman, saying, "There are other forms of intelligent life on Earth. . . . Only human arrogance would assume the message must be meant for man." . . .

In this context it is also appropriate to consider not just valuation of nature, but devaluation of civilization, the normally privileged term in this opposition.

Key here are films which provide general visions of civilization's future. For the most part, the sampled films do not provide a clear vision or the vision is optimistic and uncritical about society's direction. If considered in terms of overall future technological prowess, a key dimension of these films, most of the futures in these films can be considered optimistic. Six are opposed to this tendency and provide negative or dystopian views of the future that are warnings about society's direction. Most relevant here, however, four of these grim futures focus on the environment, i.e., a key facet *of* our grim future is environmental trouble.[15]

. . . *Waterworld* envisions a future marred by global warming. This film provides a clear warning about societal dependency on fossil fuels. *On the Beach, Planet of the Apes,* and *Beneath the Planet of the Apes* all predict that we will wipe ourselves out with nuclear war. The latter two films deserve further attention because beyond the implicit condemnation of nuclear technology, there is more explicit critique of civilization in these films than in any others sampled. Considering just *Planet of the Apes* makes the stance of these films evident. It opens with a scathing monologue about society from the protagonist spaceship captain named Taylor. Dislocated into the future by his space traveling, Taylor is glad to leave the twentieth century and societies marred by greed and war and is hoping to find a better breed when he returns to the Earth of the future. When he and his crew land on the Earth of the future, his hopes are dashed, but his critiques are proven correct—the societies he despised have destroyed themselves. Earth is now part wasteland from the nuclear war, but it is also controlled by civilized apes, with humans living like animals in the wilderness. This is meant to be a shocking reversal of fortune: the nature that humanity once dominated has risen to dominate humanity. Furthermore, the apes have learned to fear and hate humans for their civilization's failings, environmental and otherwise.[16] A bit of their mythology reinforces Taylor's critique and summarizes the warning provided by these films:

> Beware the beast man, for he is the devil's pawn. Alone among god's primates, he kills for sport or lust or greed. Yea he will murder his brother to possess his brother's land. Let him not breed in great numbers for he will make a desert of his home and yours. Shun him. Drive him back into his jungle lair for he is the harbinger of death.

Discussion and Trends over Time

The analysis thus far has focused on key areas of environmental meaning that emerged in the readings of these films. Portrayals of technology, the broader material relationship with nature, and the value of nature have been examined as both resistant and reproductive environmental discourse. A fuller understanding of the ways in which environmental struggle has or has not been "discursively transcoded" (Ryan and Kellner 1988) requires a more complete overview of trends in time than has been provided. This analysis also serves well to summarize and contextualize the key findings. Throughout the period examined technology is strongly valorized by science fiction film, reproducing notions of continual technological prowess and progress. More than seventy-five percent of the films sampled fit this profile. Concern about technology is not wholly absent. Undoubtedly rearticulating environmentalist critiques and other fears of technological society, these films do debate technology to some extent. Overall, however, the debates are subordinate to the support given to technological futures. With a few exceptions, technology is not given the broad critique found in resistant environmental discourse. Focused concerns commonly play out in contexts of technological splendor. This valorization and biased debate are the trends despite the fact that in recent decades various forms of resistant environmental discourse and analyses have faulted science and technology for playing a significant role in environmental problems. From Commoner's well-known indictments of technology to feminist and ecofeminist critiques of Western patriarchal science and technology, the questioning of modern technology has been sustained and strong from environmentalists in these years.

Despite the strong support for technology, there are nonetheless a number of apparent trends in the portrayal of debated technology worth mentioning. Technology that transforms nature and technology that unites with humanity both became subjects primarily after 1970, perhaps because more futuristic films, films more apt to imagine such powerful technologies, were popular in this time. Of these two, technology that transforms nature is given very little critical attention. Technology united with humanity is, however, given numerous negative portrayals. Key in the films analyzed here are technological augmentation of the body and the imagined pinnacles of information technology, artificial intelligence and perfect virtual reality technology. These forms have been abundant since Hal from *2001* in 1968 until the last film analyzed, *The Matrix* from 1999. This presence is clearly a product of the rise of the computer age and the postmodern information society (see Lyotard 1984). Haraway's (1994) translation of this shift is appropriate here, emphasizing that "By the late twentieth century ... we are all chimeras, theorized and fabricated hybrids of machine and organism; in short, we are cyborgs" (p. 83). While computers in general are well naturalized into these films, these "cyborg" technologies are resisted more strongly than any other type. There are films and content that valorize artificial intelligence and virtual reality: Hal from *2001* may be our nightmare, but C3P0 from *Star Wars* is our friend. Apart from these supports, it is clear that the centrality and increasing power of computer technology have produced significant anxiety. From an environmental perspective, however, it must be again emphasized that this theme focuses on consequences for *humanity* and not on technologies' effects on the broader environment.

There is also clear historical patterning in the debate over nuclear technology. According to Worster (1994), the nuclear era, begun in 1945 with the detonation of the first atomic bomb, gave birth to modern environmentalism. Clearly manifestations of this birth, eight of all eleven films sampled from 1950 to 1970 transcode Cold War fears of nuclear war and fears about the effects of radiation from nuclear weapons. In this period, there is strong support for the technology in films like *Voyage to the Bottom of the Sea*, but concern outweighs validation, with films like *The Incredible Shrinking Man, On the Beach*, and *Planet of the Apes* providing nightmarish scenarios that are products of nuclear weaponry. The nuclear test ban treaty in 1963 likely eased concerns to some extent, but Cold War tensions escalated and the antinuclear movement saw significant strength through the 1980s. Nevertheless, this subject is dropped by popular science fiction films after 1970, perhaps again an artifact of the switch to futuristic science fiction and imagined times in which fears of present-day nuclear war fade.

Though nuclear weapons disappear from the sample, popular science fiction films after 1970 pick up the subject of nuclear power. After the 1970s, films overall naturalize this side of nuclear technology in future society. Such support is worth noting not only because the period after 1970 is the height of the environmental movement, but because it is punctuated by nuclear disasters at Three Mile Island in Pennsylvania and Chernobyl in the Soviet Union. These decades also saw nuclear power severely hampered by the public and the antinuclear movement (Andrews 1999). Unlike the film *The China Syndrome* (1979), which transcodes fears about nuclear power, science fiction films are comfortable with the technology.

Concern about nuclear weapons and concern about technology that unites with humanity are the strongest threads of resistance in the entire sample. Taken together, they cover the entire period, one largely prior to 1970 and one largely after 1970. Combined with the naturalization of nuclear power that is also evident after 1970, these films may be articulating a shift in societal anxiety, from the tangible fear of radiation and nuclear war to explorations of futures in which computer technology is central and powerful in our lives. If this switch is more than coincidence, and more than an artifact of the prevalence of futuristic science fiction after 1970, it is surprisingly out of step with Beck's (1991) risk society thesis which could be interpreted as predicting continued focus in culture on powerful and insidious hazards like toxic chemicals and radiation. It is out of step with environmental and antinuclear struggle as well, which in these decades has maintained focus on nuclear weapons, nuclear power, and radioactive waste disposal (Rothman 2000). This is not meant as an argument that people no longer fear nuclear technology, only that the story these films are telling imply that "cyborg" technologies have taken on special significance. It is of course also possible that this shift is to some extent "greenwashing" or ideological intervention.

The general material relationship with nature, like the overall treatment of technology, resonates strongly with reproductive environmental discourse as well. In films covering the entire period, one can point to a notion of environmental consequences of human actions or to the need for environmental protection. Much of the strength in this trend, however, comes from the films articulating concern about nuclear war and weapons tests. After the period of nuclear radiation fear, a notion of environmental consequences or protection is largely absent or fleeting. Only three films in this entire thirty-year period (1970–1999), *Star Trek IV*, *Jurassic Park*, and *Waterworld*, focus specifically on an environmental issue in the way pre-1970s films focused on nuclear weapons. Along with this weak presence, portrayals of violence and waste disposal undermine notions of environmental consequences or protection. Again, the period of greatest strength for environmental struggle brings science fiction films with minimal environmental concern.

The strongest resonance with reproductive discourse in this area involves society colonizing space. The colonization of the cosmos is present in almost all of the futuristic films, and from throughout the sampling period. America's frontier was officially closed over a century ago (see Nash 1982), but science fiction's futures find new space to colonize, showing future society conquering the solar system and beyond. This is important itself, but the connotations are also of leaving earthly limits, and therefore concerns about scarcity and degradation, far behind. As North America once seemed a vast and infinite source of material wealth, space appears so conceived by science fiction, and in a way which counters environmentalist concerns about sustainably inhabiting this planet. Also on an Earthly scale, one is reminded not just of colonialism, but of neo-colonialism, underdevelopment, poverty, and environmental destruction in the global economic order. The connections Shiva (1988) draws are apt, revealing how the Western Enlightenment-rooted model of technological and economic progress is a force for exploitation of nature, women, and marginalized peoples in underdeveloped nations. Science fiction films latently validate and reproduce this model.

The value of nature relative to society is also a crucial question. Films from throughout the sampling period tell us that nature can be beautiful and a handful of films elevate the value of the nonhuman. A few more provide critical visions of society's environmental future. The primary finding, however, is that numerous films, though primarily from the 1970s and 1980s, provide markedly negative portrayals of both animate and inanimate nature. Some films simply tell us that civilized humans are more worthy than other life forms. Other films provide stronger contrasts. Environments are harsh and unpleasant obstacles, something Shanahan and McComas (1999) find with regard to televisions' narrative as well. Likewise, the animals encountered are hostile and dangerous, a sense that is further reinforced by contrasts with domesticated animals. Overall, these films reproduce Arluke and Sanders' (1996) "socio-zoologic scale." Good animals are docile pets or tools, e.g., the Ton-tons from *The Empire Strikes Back*. Bad animals are freaks, vermin that contaminate or disgust, or demons seeking to kill and eat people. Notably, the alien from *Alien* and *Aliens* and the variety of snakes and tentacled things from the *Star Wars* films are a mix of both vermin and demon.

While in reality nature can certainly be hostile, wild, powerful, and deadly, when foregrounded by these films, this is quite nearly the only sense of wild places and creatures we are offered. Devaluation of nature like this is clearly not something new to human storytelling; enhancing the drama of these stories may be a reason for this emphasis. Despite the possible causes, because we are not given a more balanced picture of nature, this tendency gels all too well with the destructive nature/civilization dualism and the age-old demonization of wilderness (see Nash 1982). That this would be the case, especially in the 1970s and 1980s, arguably runs counter to the efforts of the preservationist and other branches of the environmental movement from throughout the twentieth century. This dimension of environmental struggle had by these decades been long supporting and defending the wild in national parks and monuments, had succeeded in getting a wilderness system created in the 1960s, and had worked to convince Americans of the value of the wild. Nash (1982) in fact finds evidence that this work was a success in changing the meaning of wild for Americans. Science fiction film does not support his argument.

Conclusion

. . . The period in which these films were produced witnessed the ascendance of environmental struggle to a key position on the social and political landscape. But consonant with other research on the media and the environment, this emergence has gone largely unreflected. Resistance is present and undoubtedly transcodes environmentalist discourse and struggle. This orientation is simply less common and less powerful than its opponent. The future distilled from these films is not one of technological caution, of sustainable living on Earth, or of harmony with wild places and creatures. Fitting well with reproductive environmental discourse and a manifest destiny perspective on nature and society, we are instead given a future of ubiquitous and powerful technologies, of colonization of space and other planets, and of conflict with hostile environments and creatures found there. Popular future myths are not providing a vision of change in the direction of society with regard to the environment. . . .

Notes

Podeschi, Christopher, W. 2002. "The Nature of Future Myths: Environmental Discourse in Science Fiction Film, 1950–1999." *Sociological Spectrum* 22: 251–271.

1. Rather than continually and awkwardly referring to the constructedness of nature with quotation marks or other means, especially since this paper explores this very topic, this problematic is assumed throughout.

2. This phrase is from the title of Wilson's (1992) book of the same name.

3. The analysis may appear to privilege certain films over others. This is simply due to the fact that, while every film had relevant content, some films were simply richer than others.

4. See *The Beast from 20,000 Fathoms, Forbidden Planet, Destination Moon, Voyage to the Bottom of the Sea, Planet of the Apes, 2001, Fantastic Voyage, Beneath the Planet of the Apes, Star Wars, Alien, Star Trek, Black Hole, The Empire Strikes Back, Star Trek II, Return of the Jedi, Star Trek IV, Back to the Future, Part 2, Aliens, Total Recall, The Matrix, and Demolition Man.* Note that some of these films contain critique of technology, but they also provide images of futures in which technology is valorized.

5. See *Forbidden Planet, Fantastic Voyage, Star Trek, The Black Hole, Star Trek II, Star Trek IV, Back to the Future, Part 2, Aliens, Total Recall,* and *Jurassic Park.*

6. See *Forbidden Planet, 2001, Star Wars, Alien, Star Trek, The Black Hole, The Empire Strikes Back, Return of the Jedi, Back to the Future, Part 2, Aliens, Total Recall, Demolition Man,* and *The Matrix.*

7. A comic version of Darth Vader appears in *Back to the Future, Part 2.*

8. Films reflecting concern about nuclear weapons and war include *The Beast from 20,000 Fathoms, Them!, The Incredible Shrinking Man, Destination Moon, On the Beach, Voyage to the Bottom of the Sea, Planet of the Apes,* and *Beneath the Planet of the Apes.* Those that support nuclear power by its unquestioned presence include *Forbidden Planet, Fantastic Voyage, The Black Hole, Star Wars, Star Trek II, Star Trek IV, Aliens, Back to the Future, Part 2, Demolition Man,* and *Total Recall.*

9. *Forbidden Planet* and *Fantastic Voyage,* not from this later period, also naturalize nuclear technology.

10. See *Forbidden Planet, Planet of the Apes, 2001, Beneath the Planet of the Apes, Alien, Star Wars, The Black Hole, The Empire Strikes Back, Star Trek II, Return of the Jedi, Aliens, and Total Recall.*

11. See *Them!, The Incredible Shrinking Man, On the Beach, Planet of the Apes, Beneath the Planet of the Apes, A Clockwork Orange, Alien, Star Trek, Star Trek IV, Aliens, Back to the Future, Part 2, Jurassic Park, Total Recall, Demolition Man,* and *Waterworld.*

12. See *The Beast from 20,000 Fathoms, Forbidden Planet, Voyage to the Bottom of the Sea, Star Wars, Return of the Jedi, Star Trek II,* and *Aliens.*

13. See *Star Wars, The Empire Strikes Back, Return of the Jedi, Alien, Aliens,* and *Star Trek II.*

14. See *Forbidden Planet, Voyage to the Bottom of the Sea, Planet of the Apes, 2001, Star Wars, Alien, Star Trek, The Black Hole, The Empire Strikes Back, The Return of the Jedi, Star Trek II, Star Trek IV, Aliens, Jurassic Park, Total Recall,* and *Waterworld.*

15. Bright futures are provided by *Destination Moon, The Beast from 20,000 Fathoms, Forbidden Planet, Voyage to the Bottom of the Sea, Fantastic Voyage, 2001, Star Trek, Star Trek II, Star Trek IV,* and *Back to the Future, Part 2.* Added to this list are films with bright technological futures, including *Star Wars, Alien, The Black Hole, The Empire Strikes Back, Return of the Jedi, Aliens, Total Recall,* and *Demolition Man.* Dystopias or other bad futures include *On the Beach, Planet of the Apes, Beneath the Planet of the Apes, A Clockwork Orange, Waterworld,* and *The Matrix. Demolition Man* is arguably an "anti-utopia," something much different than the usually critical vision of a dystopia (see Moylan 2000). The environmental dystopias include *On the Beach, Planet of the Apes, Beneath the Planet of the Apes,* and *Waterworld.*

16. *Beneath the Planet of the Apes* is even more cynical for condemning civilization in a general sense by showing ape society following in human society's footsteps and becoming racist and aggressive.

References

Andrews, Richard. 1999. *Managing the Environment, Managing Ourselves: A History of American Environmental Policy.* New Haven, CT: Yale University Press.

Arluke, Arnold, and Clinton R. Sanders. 1996. *Regarding Animals.* Philadelphia: Temple University Press.

Beck, Ulrich. 1991. *Risk Society: Towards a New Modernity.* London: Sage.

Commoner, Barry. 1971. *The Closing Circle.* New York: Knopf.

Fine, Gary. 1998. *Morel Tales: The Culture of Mushrooming.* Cambridge, MA: Harvard University Press.

Frank, David John. 1997. "Science, Nature, and the Globalization of the Environment." *Social Forces.* 76, 2:409–37.

Freudenberg, William R., Scott Frickel, and Robert Gramling. 1995. "Beyond the Nature/Society Divide: Learning to Think About Mountain." Sociological Forum. 10,3:361-92.

Gould, Kenneth A., Allan Schnaiberg, and Adam S. Weinberg. 1996. *Local Environmental Struggles: Citizen Activism in the Treadmill of Production.* New York: Cambridge University Press.

Haraway, Donna. 1994 (1985). "A Manifesto for Cyborgs: Science, Technology, and Socialist Feminism in the 1980s." Pp. 82–115 in Steven Seidman, ed. *The Postmodern Turn: New Perspectives on Social Theory.* New York: Cambridge University Press.

Harding, Sandra. 1986. *The Science Question in Feminism.* Ithaca, NY: Cornell University Press.

Leopold, Aldo. 1949/1987. *A Sand County Almanac and Sketches Here and There.* New York: Oxford.

Lerner, Jennifer, and Linda Kalof. 1999. "The Animal Text: Message and Meaning in Television Advertisements." *Sociological Quarterly.* 40:565–86.

Liebler, Carol M., and Jacob Bendix. 1996. "Old Growth Forests on Network News: News Sources and the Framing of an Environmental Controversy." *Journalism and Mass Communications Quarterly.* 73,1:53–65.

Lyotard, Jean Francois. 1984. *The Postmodern Condition: A Report on Knowledge.* Minneapolis: University of Minnesota Press.

Moylan, Tom. 2000. *Scraps of the Untainted Sky: Science Fiction, Utopia, Dystopia.* Boulder CO: Westview.

Nash, Roderick F. 1982. *Wilderness and the American Mind.* New Haven, CT: Yale University Press.

Rothman, Hal. 2000. *Saving the Planet: The American Response to the Environment in the Twentieth Century.* Chicago: Ivan R. Dee.

Ryan, Michael, and Douglas Kellner. 1988. *Camera Politica: The Politics and Ideology of Contemporary Hollywood Film.* Bloomington: Indiana University Press.

Schnaiberg, Allan. 1980. *The Environment: From Surplus to Scarcity.* New York: Oxford.

Schnaiberg, Allan, and Kenneth Gould. 1994. *Environment and Society: The Enduring Conflict.* New York: St. Martin's Press.

Shanahan, James, and Katherine McComas. 1999. *Nature Stories: Depictions of the Environment and their Effects.* Cresskill, NJ: Hampton Press.

Shiva, Vandana. 1988. *Staying Alive.* London: Zed Books.

Wilson, Alexander. 1992. *The Culture of Nature: North American Landscape from Disney to the Exxon Valdez.* Cambridge, MA: Blackwell.

Worster, Donald. 1994. *Nature's Economy: A History of Ecological Ideas.* New York: Cambridge University Press.

Selling "Mother Earth"
Advertising and the Myth of the Natural
Robin Andersen

As Podeschi suggests in the previous article, film has the potential to serve as an outlet for expressions of "green" ideals. However, what we find most often is that wealth determines who controls the media, and, as Robin Andersen shows in this selection, corporations use their resources to very successfully express their own version of environmental concern in their advertisements. Greenwashing *refers to a corporate strategy that directs resources to public relations—especially to the presentation of images of their organization's "greenness"—rather than to effecting substantial change toward environmental sustainability. This reading examines how corporations use images of nature to sell their products, forging symbolic connections between pristine nature and consumer goods. At the same time that these ads appear to revere nature, the production processes associated with many consumer products destroy the environment.*

A television advertisement for KitchenAid opens with spectacular vistas of purple mountains framing a desert landscape. A foregrounded spire juts heavenward, rising from scarred hills cut by centuries of water runoff. Mist hangs between the barren ranges. This image fades to an oven framed by kitchen counters decorated in the same sun-washed shades as a voice intones, "KitchenAid ranges mirror nature in surprising ways." The camera moves across another exquisite desert view. This time the camera is closer, the hills are smaller, and sparse desert flora cover the dry earth. The camera moves over ancient hills rounded by centuries of wind erosion; superimposed are the words "KitchenAid."

These desert contours morph into the visually similar curved forms of twisted bread baking inside the oven. The voice adds, "with even warmth." Next, as the camera glides high above the protruding formations, it moves toward us and passes over a sheer, free-standing rock face. The image transposes from the sun-washed rock wall to the off-white oven door, moving upward to close. The voice assures, "strength that endures." Next, "a mystery that unfolds," reveals, through time-lapse photography, cactus flowers opening. With the same pacing, the image turns into a rising soufflé. "The fire of creation" fills the screen with the desert sun's warm golden glow. The golden circle fades into the lit stovetop burner, then into a circular skillet sautéing vegetables, "all someplace a little closer to home. KitchenAid freestanding ranges, built-in ovens and cook tops," the ad continues.

Inside the kitchen now, a woman moves around the appliance-laden, picture-perfect room. She wears a long earth-mother, printed gauze skirt, and as we hear, "because we took a cooking lesson from Mother Nature," the image moves back to a panorama of the desert with "KitchenAid" slashed across the hot, dry landscape. Throughout the ad, the female voice, soft and low, mixes with continuous new-age spiritual humming that rises to an inevitable crescendo as the commercial ends.

There is no essential, authentic, or inherent connection between the desert and the oven. A pristine ecosystem bears little logical connection to a manufactured appliance. They are both hot, though not comparable in degree, but the sensuous analogy establishes a connection. More important, however, stunning visual juxtapositions and editing forge a powerful and compelling association that unites the natural world with the commodity. They become indistinguishable; our social constructions and commonsense conceptions of "untouched wilderness" are now compatible with the highly materialistic lifestyles of first-world consumer culture.

This chapter seeks to demonstrate how corporate consumer culture creates enchanting and persuasive advertising messages, using both visual and textual strategies, that celebrate the environment while their own business practices continue on a path of ecological destruction. Left unexamined, such influential messages help facilitate corporate environmental destruction, especially in the absence of information and public debate about the destructive consequences of consumer culture.

The KitchenAid ad inserts the oven into the untarnished landscape through the use of a highly persuasive mode of visual language. Visual persuasion uses symbolic, associative terms that lack "propositional syntax," that is, they do not explicitly indicate causality or other logical connections.¹ This syntactical indeterminacy is one important aspect of the persuasive uses of images of nature in advertising.

The symbolic landscapes of consumer culture, including visually enhanced images of the natural world, allow advertisers to make insinuations about their products without explicitly claiming anything. To state openly that an oven is like nature because it is hot, and because a rising soufflé can be made to look like the opening of a cactus flower, would sound ridiculous. However, because the visual implications have not been stated plainly, they are not logically rejected. Such stunning associations make powerful connections between the natural world and the world of products; pleasing aesthetic representations and feelings about awe inspiring natural landscapes are united with the product. The sense of psychic pleasure and inspiration is linked to the KitchenAid freestanding ranges. KitchenAid used these techniques in a series of ads for their appliances.

Perhaps most significant is their refrigerator advertisement. The aerial camera glides though clear, blue skies over a high-mountain forest covered in pure, white snow. As we hear the words "crisp, freshness," the image cuts to a refrigerator vegetable compartment. And with "brilliant light," another snow-covered peak and a dazzling bright-white mountain meadow are transfigured into the refrigerator's interior. The voice proclaims, "The refrigerator designed with a blueprint from Mother Nature, by KitchenAid." The last shot is a high-mountain lake luminous with the reflection of snow-covered peaks, as a woman's voice whispers, "For the way it's made." But what *about* the way it's made?

Advertising's symbolic culture inserts the product into the natural world, removing it from the context of its production. The ad never explains how it is made, who made it, or the effect on the environment, either during its production or during its use. The remarkable juxtapositions silence those realities and hide the social and environmental relations of the production and use of the commodity.

However, there is much to say about the "way it's made." The primary industrial cause of ozone depletion is the gases associated with refrigeration. Chemicals such as chlorofluorocarbons and hydrochlorofluorocarbons (CFCs and HCFCs), widely used as refrigerants and insulating foams, destroy the ozone layer by releasing chlorine in the upper atmosphere. Refrigeration, however, no longer requires ozone-depleting gases. Over one million "greenfreezes" have been sold in Europe, but such environmentally friendly refrigerators are not available in the United States. A campaign by Greenpeace encourages American manufacturers to offer nonozone-depleting refrigerators to the American public. The campaign targeted Whirlpool, the parent company that manufactures KitchenAid, because it is a huge company that produces parts for the greenfreeze sold in Europe. Whirlpool actually sells compressors to some leading manufacturers of the greenfreeze. However, the attempts have failed. U.S. companies will not make them here.

Instead, Whirlpool chose to respond with a slick advertising campaign. Even as Whirlpool capitalized on persuasive portraits of natural beauty, it refused to invest in the new technology available for making environmentally safe refrigerators, creating a symbolic culture that seemingly reveres the environment as it helps destroy it. This public relations strategy has been referred to as "greenwashing"; it has become a cornerstone of American corporate promotional culture.

Instead of creating popular support by moving toward sustainable resource management, less toxic ingredients and manufacturing, waste reduction and reusable packaging, corporations rely on perception management through advertising as a major feature of their strategies. Perception management creates symbolic associations that evoke a sense of well-being. Since the advertising taps directly into psychic associations using the language of art, poetry, and the unconscious, the logical mind does not reject absurd representations about "taking a cooking lesson from Mother Nature," because as psychoanalysts Haineault and Roy argue, such impressions are psychologically pleasing.[2] Rejecting them would require a degree of distress the psyche seeks to avoid. The destructive qualities of the product and its negative environmental effects are successfully hidden. In an age when the public is bombarded with 1,500 advertising messages a day, and when corporate ownership and advertising influence successfully block information about environmental destruction by corporations, it is difficult to nourish citizen awareness and action.

Such false connections exist in a media environment that offers little information to counter advertisers' representations. With a few notable exceptions, major advertising clients have been successful in putting direct pressure on editors and producers not to contradict the messages of their advertising campaigns in programming content.[3] The only way the impressions created by advertising culture can persist is in an atmosphere devoid of contradiction.

There is another aspect of advertising's appropriation of nature that also makes it a powerful tool for greenwashing and, more broadly, an enchanting and influential celebration of capitalism as a social system. In addition to the visual persuasions and the lack of information are the underlying belief systems and commonsense conceptions of nature that much advertising successfully mines. How can we accept the assertion that Whirlpool "took a cooking lesson from Mother Nature," or that ideas about the industrial design of appliance manufacturing "flowed from Mother Nature"? Why do these and other absurd presentations, even without direct information on ozone depletion, sound acceptable to a public willing to revere nature but at the same time participate fully in consumer culture?

Media Representations and Cultural Conceptions of Nature

In large measure, the visual constructions of nature in advertising can be linked to, and have been appropriated from, some key conceptions of the environmental movement. The advertisements are effective because they do not contradict the "commonsense" understandings of the natural world shared by many. These compatible portrayals, rooted in ideological assumptions about nature, must be examined and redrawn in order to foster public awareness and citizen action and achieve sustainable social and environmental ecosystems.

Many environmental writers regard the space missions as key cultural markers in our contemporary conception of nature, the Earth, and the need for conservation: "It is almost commonplace to note how the first pictures of Earth from Space in 1966 made evident the frailty of the planet and sparked a global ecological consciousness."[4] Ecofeminist Chaia Heller (1993) writes, "Awareness of the ecological crisis peaked in 1972 when the astronauts first photographed the planet, showing thick furrows of smog scattered over the beautiful green ball. 'The planet is dying' became the common cry. Suddenly the planet, personified as Mother Earth, captured national, sentimental attention."[5] The mass media presented pictures of the Earth as well as representations of the significant environmental issues of the last 30 years, from protecting endangered species around the globe to preventing the destruction of the rain forests and saving the whales.

Indeed, even though most people have not been to the moon, or the faraway wild places they become concerned with, television creates critical impressions. As DeLuca observes, "in particular PBS documentaries, I learn of and become concerned about the Amazon rainforests."[6] However, television and other media representations are just that, (re)presentations. They are stories of nature, not the Earth made real. We know Mother Nature through the stories we tell about her. Our cultural narratives help shape our perceptions.[7] Ecosystems, wilderness areas, and wild things are most often unknown to us without the framing of such cultural texts. And those texts will affect not only our perceptions of nature but also the way we act in and upon the natural world.

Heller argues that we tell a particular type of cultural narrative about Mother Nature. This narrative features characters and relationships borrowed from the medieval romanticization of women as depicted in passionate love poetry: "The metaphors and myths of this eco-drama are plagiarized from volumes of romantic literature written about women, now recycled into metaphors used to idealize nature."[8] The medieval narrative drama of romantic love plays out in the wistful longings of a man for an idealized, pure woman whom he vows to protect, for his love can never be consummated. Rooted in Platonic dualism, the ideal love is unpolluted by physical contact. She is powerless and in need of protecting. The lover realizes his romantic fantasy only through noble self-control and heroic acts of protection and sexual self-restraint. The beloved is deserving because she is pure and chaste.

Heller argues that the imagery of the contemporary ecology movement finds its ideological roots in this cult of the romantic. Mother Nature takes the role of the helpless beloved, and those in the ecology movement vow not to defile her. "In our modern iconography, nature became rendered as a victimized woman, a Madonna-like angel to be idealized, protected, and saved from society's inability to constrain itself."[9]

Procter & Gamble's advertising campaigns display similar iconography and the same sentimentalized message. In soft focus, a little girl drinks water from a beautiful indoor sink. In the next picture a deer drinks from a pond in an idealized outdoor setting. The caption reads, "everybody deserves a clean home." This series also includes images of a baby girl playing with a little yellow duck, while fuzzy little yellow ducks swim in a placid soft-focus lake in the picture next to her. Helpless and idealized, Mother Nature must be protected from all those who would destroy her.

The sentimentalized renderings of romantic ecology raise many questions. For example, whom does Mother Nature need to be protected from? This category is always vague and sweeping, but the implied formulation is often "human nature," as with Devall and Session's formulation in *Deep Ecology*:

> Excessive human intervention in natural processes has led other species to near-extinction. For deep ecologists the balance has long been tipped in favor of humans. Now we must shift the balance back to protect the habitat of other species.[10]

Condemning all of humanity as equally responsible for environmental degradation ignores oppressive social and economic relations, as well as the unequal distribution and use of natural resources and wealth. Certainly "failing to expose the social hierarchies within the category of human erases the dignity and struggle of those reduced to and degraded along with nature."[11] The exploited laborers of the third world who toil in extreme conditions in sweatshops making commodities for first world consumers are not equivalent to the ravages of global capitalism.

Overpopulation must certainly be a concern, not only for environmental organizations, but for all of the globe's peoples. Yet some would put the central responsibility for the problem squarely on the shoulders of women—human mothers defiling an idealized Earth Mother. While women own and control only about 1 percent of the Earth's wealth, they account for 80 percent of human labor power. United Nations population studies demonstrate that in countries where women have become less economically exploited and more educated and socially empowered, the birth rate has declined.

The first world consumes almost four times more of the Earth's resources than its third world counterparts. Those who are victimized by and struggle against the ravages of global capitalism are not to blame for the ecological crisis, corporate global practices are. Habitually emphasizing birthrates out of context only facilitates capitalism's ravages of the Earth.

The idea that every person is equally culpable of defiling the Earth conceals the role of corporations in ecological degradation and distracts attention from the production, use, and disposal of resources. Public service announcements and advertising campaigns often suggest that solutions to environmental problems are the private responsibility of individual consumers. "We recycle" and recycling logos on many products are omnipresent, but such campaigns help manufacturers prevent mandatory deposit and recycling regulations. As long as manufacturing does not make recycling, especially of plastics, more effective, recycling will have little positive impact. The romantic narrative focus on individual responsibility and such socially constructed stories place corporate responsibility outside their discourse.

In this cultural ecodrama that we present, and read ourselves into, how are the characters drawn? What is the nature of this relationship between the beloved Mother Nature, and the protective ecoknight? As Heller points out, the romanticization of nature is "based on the lover's desires, rather than on the identity and desires of the beloved." The role of nature as the beloved is to offer emotional well-being to dedicated environmentalists and consumers alike. Some "daily affirmations" of new-age environmentalism begin to sound very similar to advertising messages. "I hold in my mind a picture of perfection for Mother Earth. I know this perfect picture creates positive energy from my thought, which allows my vision to be manifest in the world."[12]

An advertisement for Evian water mirrors these sentiments. Evian's campaign features healthy people delighting in pristine natural settings. On the left half of a two-page ad, a man runs across a blurred, cold, dark-blue background. Across his chest are the words, "In me lives a wildcat who chases the moon and races the wind and who has never measured his life in quarterly earnings." The opposite page is an image of the French Alps, rendered in soft beige tints. The perspective places us on one of the mountains, peering over, but within, the snow-covered, jagged peaks. The caption claims, "In me lives the heart of the French Alps. Pure, natural spring water perfected by nature. Untouched by man, it is perfect for a wildcat." On the left page the word "wildcat" stands out in boldface, and on the right page "heart" appears in boldface. This graphic design unites the two pages; at a glance the viewer sees wildcat heart.

As with many ecologically posed ads that feature the natural world, this one constructs an alternative sense of place, a pure, untouched environment offered to the ecoknight turned consumer as an escape from the unpleasant realities of modern industrial life. The goodness of Mother Nature is generally depicted in advertising by what she can do for us and most often by the way she can make us feel. Turned into a commodity, nature promises a return to a state of emotional well-being from the strain and stress of urban and social life.[13]

In addition, the absence of a social presence in the renderings of wild places reinforces a solitary, individual relationship to nature that negates collective action. In the Evian ad, the heart of Mother Nature nurtures the wildcat, the human spirit supernaturally transformed and empowered. This is what Mother Nature can do for us. Such ideas are compatible with a new-age environmentalism based in the romantic tradition. However, once turned into a commodity through advertising, romantic attainment can only be realized through the possession of the product tied to those sentiments. With all those wildcat spirits consuming water sold in plastic bottles, what happens to the environment?

The Costs of Consumption

Over the last decade the marketing of bottled water has vastly increased sales of the product. Instead of relying on municipal water supplies, especially when away from home, it has become popular to buy and carry individual water bottles made of plastic. Plastic is made from nonrenewable fossil fuels and manufactured from fractions of crude oil or natural gas changed into solid form through the use of different chemicals. Many of these chemicals are highly toxic, such as benzene, cadmium and lead compounds, carbon tetrachloride, and chromium oxide. They are used as solvents, for coloring, and as catalysts in chemical manufacturing. Over the past two decades, the production of synthetic chemicals has almost tripled, and much of this increase is attributed to the manufacture of plastics. Synthetic chemicals released into the environment create dioxins and other toxins.

Thermoplastics, the soft type of plastic, such as polyethylene, are used for everything from milk jugs and margarine tubs to pipes and tubing. Polystyrene is used to make styrofoam, other packaging material, and tiles. The most toxic plastic is polyvinyl chloride (PVC), and it is used for pipes, shower curtains, and insulation, and as Greenpeace points out, many children's toys and accessories. Polyethylene tereplithalate (PET) is used to make plastic bottles.

According to the U.S. Environmental Protection Agency's (EPA) home pages on the World Wide Web, in 1994 only 4.7 percent of all plastic was recycled in the United States. Recycling is difficult because of the mix of plastics that cannot be degraded together. Incineration reduces the plastic headed for the landfill, but it also releases toxic pollutants, including heavy metals, into the air and through the remaining ash. These pollutants also enter the groundwa-ter and contaminate the water supply. The EPA recommends what is called "source reduction" for plastics, to reduce the amount of plastic purchased and thrown away.

Instead of reducing the use of plastic, however, the marketing and distrib-ution of bottled water has led to a dramatic increase in its use. The development of polyethylene terephthalate (PET), the lightweight, durable yet malleable plas-tic, allowed the manufacture and distribution of smaller, individual containers which continue to proliferate. Portable plastic, huge advertising budgets, and a profit margin of 40 to 45 percent (compared with 30 to 35 percent on soda) have led to an increase in the sales of bottled water by an amazing 144 percent over the past decade.[14]

Thus we have come full circle. The concept Mother Nature helps associate bottled water with purity and inspiration and compels people to express their love for nature by buying individual plastic containers, a substance that degrades the environment, presents hazards to public health, and ultimately has the potential to contaminate our drinking water.

Such uses of mythic nature help cloud judgment, as well as the knowledge and understanding of how consumer culture actually affects the environment. Nature is perfect purity, undefiled, "untouched by man," waiting there, after cen-turies, just to quench the spiritual thirst of the running man in the Evian ad. As with many advertisements, nature exists to provide solace and satisfy the spiritual longing of those who love her. Like all myths used to reinforce ideologies, this ide-alization of pure, untouched nature has been drained of any essential meaning. Nature exists only as it is reflected in the needs and desires of the Evian runner.

Loss of any true knowledge of the beloved is another consequence of grounding our love of nature in romantic mythology. "The romantic's love depends on his fantasy of his beloved as inherently powerless and good as he defines good."[15] This type of "knowing" nature is unidimensional and is "wed-ded to ignorance. Certainly the romantic does not know his lady to be a woman capable of self-determination and resistance."[16] Just as the true qualities of the beloved are never known to the lover in romantic poetry, when romanticized, nature itself remains a mystery. As long as nature is made myth, is docile and willing to give, she is deserving of love and protection.

However, this is a selfish love indeed. For once Mother Nature challenges the ego, refuses the needs, or engages in self-expression and determination, any-thing can be done to her. These points are borne out in the most common rep-resentations of nature in advertising today, those promoting sport utility vehicles (SUVs).[17] While the male would-be drivers are invited to experience the wild world of jungles, forests, deserts, and even underwater ecosystems to gain a sense of inspiration and adventure (again, nature making the hero feel good), nature can also be unruly and anything but docile.

Nature is often depicted acting up, creating bad weather and rough roads. She will burn you, she will freeze you, she will try to blow you away, as one 4×4 ad threatens. This invites the worst kind of treatment, and nature must be tamed. SUVs are most often depicted off-roading; Mother Earth flies out from under the tires that are tearing up wildlands, disturbing wildlife, and breaking down river banks as the 5,000-pound vehicles cleave through them. Domination and mastery over nature is the result when she steps out of line.

However, simply portraying nature as female is enough to invite a certain behavior. In a Wrangler ad for jeans, a man is given a pink cake by his wife. In the next scene, he is shown smashing through the now huge pink cake in a forest, expressing a distinctive adolescent male rebellion.

The romantic gendering of nature has resulted in complicated and contradictory advertising messages that are used for very destructive purposes. Originally, the concept of Mother Nature was borrowed from Native American philosophy and other indigenous cultures. While it may have had a noble purpose, when this concept is lifted from its social and cultural contexts and reinserted within patriarchal capitalist culture, it is turned into a commodity and commercialized in ways that ironically come to support the economic practices that misappropriate the Earth's resources. This new hybrid of Mother Earth is now a woman destined to be dominated in the same ways that women are subjugated by the legacy of patriarchal capitalist culture.[18] In such a culture, protective paternalism quickly loses its tolerance when those subjugated—women, minorities, or those outside the dominant social order—express rage at their unequal treatment or make demands for self-determination.

This is how we come to such contradictory uses and depictions of nature in consumer culture. For example, studies show that those who purchase SUVs think of themselves as environmentally conscious. They want to enjoy the great outdoors and have adopted a vehicular consumer style that expresses that. At the same time, the ads invite them to dominate nature. Because nature has been portrayed as an unruly female, these two messages are not seen as contradictory.

In addition, a mythicized conception of nature, existing only to please those who admire her, does not invite knowledge and understanding of how humans might best interact with nature. Because of the ways we formulate our love for nature, environmental consciousness can sit comfortably in the cab of an SUV, a machine that uses far more gas than a smaller car. Extracting the greater amounts of fossil fuel needed to run SUVs helps destroy the wilderness areas and animal habitats, such as the last Alaskan wilderness, so lovingly depicted in the persuasive ads for these vehicles. SUVs increase greenhouse gases and therefore global warming because of their lower emission standards. Because of lobbying by the automobile industry, this situation is not likely to change. The same industry that professes a love of nature fights fuel efficiency, emission standards, and public spending on mass transportation while it continues to make as much as $10,000 gross profit on the biggest and most destructive SUVs.[19] This is greenwashing at its worst.

The Beauty of Nature and the Cosmetics Industry

If we turn to another major aspect of consumer culture, that of the beauty and cosmetics industry, we find that the advertising of such products often features the globe, the concept or language of Mother Earth, and the natural world. It is common for beauty and hygiene product advertising to extol "natural" ingredients, proclaim the benefits of nature, and in general, associate the Earth's goodness with a wide range of cosmetics and their ingredients. In cologne advertisements, women sit in fields of sun-drenched flowers, and in hygiene ads they are compelled to douche with fresh, flower-scented liquids. Advertisements feature the goodness of Mother Nature in striking contrast to the depiction of real women's experience of nature.

In a comical essay reprinted in the *Utne Reader* titled, "Forever Fresh: Lost in the Land of Feminine Hygiene," journalist Alison Walsh writes that she was "astonished to discover that most daytime TV commercials have one clear message: women leak, dribble, and smell. . . . Apparently women must buff, douche, diet, gargle, and primp constantly if they are to overcome their basic vileness." Walsh, noting a certain gender imbalance observes that if we are to judge from television commercials aimed at men, "Evidently men are just fine the way they are. They have a small problem with weight gain and graying hair, but mainly they are handsome, playful and successful." She ends her piece with a query as to why there is no masculine hygiene aisle in the drugstore.[20]

However, women have been told, "how not to offend," by the cosmetics industry for a long time. In an essay with the same title, Marshal McLuhan noticed an anxiety-producing message from Lysol in the 1940s. The magazine advertisement features a woman waist deep in swirling water as the words "doubt, inhibitions, ignorance, and misgivings" seem to be pulling her down. Distressed, with her arms raised helplessly in the air, the ad rebukes, "Too late to cry out in anguish. Beware of the one intimate neglect that can engulf you in marital grief. For complete Feminine Hygiene rely on Lysol."[21] Advertising, as McLuhan comments, continues to remind us of "the terrible penalties . . . that life hands out to those who are neglectful . . . left in sordid isolation because they 'offend.'" When a lovely woman "stoops to BO, she is a Medusa freezing every male within sniff. On the other hand, when scrubbed, deloused, germ-free, and depilatorized, when doused with synthetic odors and chemicals, then she is lovely to love."[22] For McLuhan, "the cult of hygiene and the puritan mechanisms of modern applied science" were all part of technological progress. "Fear of the human touch and hatred of the human smell are part of this landscape of clinical white-coated officials and germ and odor proof laboratories."[23] Feminist writers ... have long noted that the domination of women's bodies coincides historically with the domination of the natural world. Today, the white-coated hygienists of the 1940s and 1950s have been replaced with images of Mother Earth, but the message is the same.

Just as the domination of nature often results in its destruction, the advertising imperative that the female body be sanitized, tamed, powdered, and redolent only of perfumes has led to dire health consequences. In "Taking a Powder," journalist Joel Bleifuss documents epidemiological evidence connecting the use of talcum powder with ovarian cancer, the fourth leading cause of death in American women.[24] While talc and other toxins found in cosmetic products pose a serious health risk to American consumers, especially women, the industry is not regulated. A U.S. Food and Drug Administration (FDA) document found on their WWW home pages states, "a cosmetic manufacturer may use any ingredient or raw material and market the final product without government approval." The FDA does prohibit seven known toxins such as hexachlorophene, chloroform, and mercury, but for the remaining 8,000 ingredients used in the manufacture of cosmetics, the industry regulates itself. While the FDA has the power to pull dangerous products off the shelf, it rarely does so, "despite mounting evidence that some of the most common cosmetic ingredients may double as deadly carcinogens."[25]

It is impossible to assess advertising claims of "natural" ingredients of cosmetic products most of the time, but cosmetic products "are often contaminated with carcinogenic byproducts, or contain substances that regularly react to form potent carcinogens during storage and use."[26]

Identified as one of the most dangerous toxins by FDA doctors and cancer researchers are nitrosamines, a group of carcinogens found in a variety of products from shampoos to sunscreens. One of these nitrosamines, N-nitrosodiethanotamine (NDELA), forms when some common cosmetic ingredients, such as triethanotamine (TEA) and Cocamide diethanol amine (DEA), interact with the nitrites that are used to preserve many products. Vidal Sassoon shampoo, for example, contains the toxin, Cocamide DEA.

Joel Bleifuss also uncovered a number of research reports documenting the toxins contained in cosmetic products.[27] In 1992, tests conducted by the FDA revealed a product that contained NDELA at a concentration of 2,960 parts per billion, but the agency will not publish the brand name. The European Union does not allow more than 50 parts per billion of nitrosamine-producing chemicals in cosmetic products. In 1992, all 14 products tested that year by the FDA were contaminated with the nitrosamine carcinogens. Individual FDA scientists are speaking out. Drs. Harvey and Chou assert that with the "Information and technology currently available to cosmetic manufacturers, N-nitrosamine levels can and should be further reduced in cosmetic products." A social goal should be to keep "human exposure to nitrosamines to the lowest level technologically feasible by reducing levels in all personal care products."[28]

Even as Redkin hair product features a picture of the globe, contrasting heaven (beautiful hair) and Earth (an image of the globe), plastic bottles appear in the lower right corner of the ad. The pervasive plastic packaging used for cosmetic products (with the exception of those made by the Body Shop) usually end up in the landfill (as noted above). In addition, one common general-purpose plasticizer, adipate, or DEHA, used in processing polyvinyl and other polymers, is also used as a solvent or plasticizer in such cosmetics as bath oils, eye shadow, cologne, foundations, rouge, blusher, nail polish remover, and moisturizers. It can contaminate foods wrapped in plastic films. This common plasticizer contaminates groundwater through fly ash from municipal waste incineration and wastewater effluents from treatment and manufacturing plants.

The $20 billion plus a year cosmetics industry relies on a lack of media scrutiny, which was made apparent in 1998 when epidemiological studies and the carcinogenic contents of some cosmetics were identified as important censored news stories for that year.[29] "Few publications put effort into investigating the cosmetics industry, which is not surprising since the industry is a major magazine and newspaper advertiser. This is especially true of the women's magazines. Consequently, there is almost no coverage of the industry."[30]

What is necessary to unveil the toxic substances and environmental destruction hidden behind advertising's compelling symbolic culture? Corporate cultural conceptions of nature must be recognized and transformed. Romantic protection of a mythic version of nature must become obsolete because of the ease with which its appropriation serves the interests of economic exploitation and environmental destruction. Instead, we should strive "to know and care for the resistance of all living things that dwell in poisoned eco-communities, offering ourselves as allies in resistance to social and ecological degradation."[31]

Authentic love is based on knowledge not myth, and "allied resistance" enjoins citizens to offer their support for the struggles against the global corporate domination that sustains the production and use of toxic substances. Given the power of global corporations, we need to identify and support innovative forms of resistance in our homes around the planet with a unified strategy against industries that pollute the environment, oppress their workers, and promote toxic substances. Such solidarity would demonstrate an authentic love for nature, rejecting the image of drinking water bottled in plastic, or smelling "naturally" fresh with a toxic cosmetic product, or driving an SUV as having any relation to a healthy life or preservation of the ecosystem.

Notes

Andersen, Robin. 2000 "Selling 'Mother Earth': Advertising and the Myth of the Natural." In *Reclaiming the Environmental Debate: The Politics of Health in a Toxic Culture*, ed. Richard Hofrichter. Cambridge: MIT Press.

1. Paul Messaris, *Visual Persuasion: The Role of Images in Advertising* (Thousand Oaks, Calif.: Sage Publications, 1997).

2. Doris-Louis Haineault and Jean-Yves Roy, *Unconscious for Sale: Advertising, Psychoanalysis and the Public* (Minneapolis: Univ. of Minnesota Press, 1993).

3. Robin Andersen, *Consumer Culture and TV Programming* (Boulder, Colo.: Westview Press, 1995); Michael Jacobson and Laurie Ann Mazur, *Marketing Madness: A Survival Guide for a Consumer Society* (Boulder, Colo.: Westview Press, 1995); Matthew McAllister, *The Commercialization of American Culture: New Advertising, Control and Democracy* (Thousand Oaks, Calif.: Sage Publications, 1996).

4. Kevin DeLuca, "Constituting Nature Anew through Judgment: The Possibilities of Media," in *Earthtalk: Communication Empowerment for Environmental Action*, Star A. Muir and Thomas Veenendall, eds. (Westport, Conn.: Praeger, 1996, p. 60).

5. Chaia Heller, "For the Love of Nature: Ecology and the Cult of the Romantic," in *Ecofeminism: Women, Animals and Nature*, Greta Gaard, ed. (Philadelphia: Temple Univ. Press, 1993, p. 219).

6. DeLuca, "Constituting Nature."

7. Ibid.

8. Heller, "Love of Nature."

9. Ibid.

10. Ibid., p. 223.

11. Heller, "Love of Nature," p. 226.

12. Ibid., p. 223.

13. See Paul Messaris, "Pristine, Damaged, and Nightmare Landscapes: Visual Aesthetics of American Environmentalist Imagery," unpublished paper, presented at the Seventh Annual Visual Communication Conference, Jackson, Wyo. (June, 1993).

14. See the cover story by Corby Kummer, "What's in the Water?" in *New York Times Magazine*, August 30, 1998:41. Ironically, "standards set for municipal drinking water supplies are mandatory and are monitored and tested more often than for bottled water," while the bottled water industry has been self-regulated for years (p. 41). In addition, a good portion of the bottled water sold in the United States is simply filtered or deionized.

15. Heller, "Love of Nature," p. 222.

16. Ibid.

17. Robin Andersen, "Road to Ruin! The Cultural Mythology of SUVs," in *Critical Studies in Media Commercialism*, Robin Andersen and Lance Strate, eds. (London: Oxford Univ. Press, 1999).

18. For a discussion of a related position on "cultural essentialism," see Laura Pulido, "Ecological Legitimacy and Cultural Essentialism: Hispano Grazing in the Southwest," in *The Struggle for Ecological Democracy: Environmental Justice Movements in the United States*, Daniel Faber, ed. (New York: Guilford Press, 1998).

19. See Andersen, 1999.

20. Alison Walsh, "Forever Fresh: Lost in the Land of Feminine Hygiene," *Utne Reader* (September/October, 1996):32.

21. Marshall McLuhan, *The Mechanical Bride: Folklore of Industrial Man* (New York: Basic Books, 1967, p. 61).

22. Ibid.

23. Ibid., p. 62.

24. Peter Philips (ed.), *Censored 1998: The News That Didn't Make the News* (New York: Seven Stories Press, 1998).

25. Ibid., p. 30.

26. Ibid.

27. Ibid.

28. Ibid., p. 324.

29. Ibid.

30. Ibid., p. 31.

31. Heller, "Love of Nature," p. 235.

PART

Science and Health

The Social Construction of Cancer
A Walk Upstream
Sandra Steingraber

One of the core concepts of sociology is the "social construction of reality"—the idea that we shape our understanding of reality through our interactions with each other. Sandra Steingraber, a biologist, traces her own personal history with cancer to explore the links between cancer and environmental pollutants. In this excerpt from Steingraber's book, Living Downstream, *we see how, despite the growing body of scientific evidence that links cancer to environmental causes, popular and public education still promote a model of cancer that places responsibility on individual behaviors and genes. The development of a new cancer model depends on not only the generation of more scientific evidence but also the construction of a social reality that "sees" the connections between cancer and pollution. In addition, Steingraber calls for a new model of policymaking based on the precautionary principle, whereby toxins would be reduced at the point of production and the burden of proof for showing that the product is safe would be placed on the manufacturer. Under the current system, producers of potentially hazardous substances are not called on to prove their safety; instead, the burden of proof is placed on citizens.*

In 1979, in between my sophomore and junior year in college, I was diagnosed with bladder cancer. Four years later, while a doctoral student in biology, I took the train from Ann Arbor, Michigan, to central Illinois for an appointment with my original urologist. This particular appointment was destined to turn out fine: there were no recurrences. What I remember most clearly is my journey there by train.

Something about the landscape changes abruptly between northern and central Illinois. I am not sure what it is exactly, but it happens right around the little towns of Wilmington and Dwight. The horizon recedes, and the sky becomes larger. Distances increase, as though all objects are slowly moving away from each other. Lines become more sharply drawn. These changes always make me restless and when driving, I drive faster. But since I am in a train, I close the book I am reading and begin impatiently straightening the pages of a newspaper strewn over the adjacent seat.

That is when my eye catches the headline of a back-page article: Scientists Identify Gene Responsible for Human Bladder Cancer. Pulling the newspaper onto my lap, I stare out the window and become very still. It is only early evening, but the fields are already dark, a patchwork of lights quilted over and across them. They have always soothed me. I look for signs of snow. There are none. Finally, I read the article.

Researchers at the Massachusetts Institute of Technology, it seems, had extracted DNA from the cells of a human bladder tumor and used it to transform normal mouse cells into cancerous ones.[1] Through this process, they located the segment of DNA responsible for the transformation. By comparing this segment with its unmutated form in noncancerous human cells, they were able to pinpoint the exact alteration that had caused a respectable gene to go bad.

In this case, the mutation turned out to be a substitution of one unit of genetic material for another in a single rung of the DNA ladder. Namely, at some point during DNA replication, a double-ringed base called guanine was swapped for the single-ringed thymine. Like a typographical error in which one letter replaces another—*snow* instead of *show, block* instead of *black*—the message sent out by this gene was utterly changed. Instead of instructing the cell to manufacture the amino acid glycine, the altered gene now specified valine. (Nine years later, other researchers would determine that this substitution alters the structure of proteins involved in signal transduction—the crucial line of communication between the cell membrane and the nucleus that helps coordinate cell division.)

Guanine instead of thymine. Valine instead of glycine. I look away again—this time at my face superimposed over the landscape by the window's mirror. If, in fact, this mutation was involved in my cancer, when did it happen? Where was I? Why had it escaped repair? I had been betrayed. But by what?

Thirteen years later, I possess a bulging file of scientific articles documenting an array of genetic changes involved in bladder cancer.[2] Besides the oncogene just described, two tumor suppressor genes, p15 and p16, have also been discovered to play a role. Their deletion is a common event in transitional cell carcinoma, the kind of cancer I had. Mutations of the famous p53 tumor suppressor gene, with guest-star appearances in so many different cancers, have been detected in more than half of invasive bladder tumors. Also associated with transitional cell carcinomas are surplus numbers of growth factor receptors. Their overexpression has been linked to the kinds of gross genetic injuries that appear near the end of the malignant process.

The nature of the transaction between these various genes and certain bladder carcinogens has likewise been worked out in the years since a newspaper article introduced me to the then-new concept of oncogenes. Consider, for example, that redoubtable class of bladder carcinogens called "aromatic amines"—present as contaminants in cigarette smoke; added to rubber during vulcanization; formulated as dyes for cloth, leather, and paper; used in printing and color photography; and featured in the manufacture of certain pharmaceuticals and pesticides.[3] Aniline, benzidine, naphthylamine, and o-toluidine are all members of this group. The first reports of excessive bladder cancers among workers in the aniline dye industry were published in 1895. More than a century later, we now know that anilines and other aromatic amines ply their wickedness by forming DNA adducts in the cells of the tissues lining the bladder, where they arrive as contaminants of urine.

We also now know that aromatic amines are gradually detoxified by the body through a process called "acetylation." Like all such processes, it is carried out by a special group of detoxifying enzymes whose actions are controlled and modified by a number of genes. People who are slow acetylators have low levels of these enzymes and are at greater risk of bladder cancer from exposure to aromatic amines. Members of this population can be readily identified because they bear significantly higher burdens of adducts than fast acetylators at the same exposure levels.[4] These genetically susceptible individuals hardly constitute a tiny minority: more than half of Americans and Europeans are estimated to be slow acetylators.

Very likely, I am one. You may be one, too.

We know a lot about bladder cancer.[5] Bladder carcinogens were among the earliest human carcinogens ever identified, and one of the first human oncogenes ever decoded was isolated from some unlucky fellow's bladder tumor. Sadly, all of our knowledge about genetic mutations, inherited risk factors, and enzymatic mechanisms has not been translated into an effective campaign to prevent the disease. The fact remains that the overall incidence rate of bladder cancer increased 10 percent between 1973 and 1991. Increases are especially dramatic among African Americans: among black men, bladder cancer incidence has risen 28 percent since 1973, and among black women, 34 percent. Somewhat less than half of all bladder cancers among men and one-third of all cases among women are thought to be attributable to cigarette smoking, which is the single largest known risk factor for this disease.[6] The question thus still remains: What is causing bladder cancer in the rest of us, the majority of bladder cancer patients, for whom tobacco is not a factor?

I also possess another bulging file of scientific articles. These concern the continuing presence of known and suspected bladder carcinogens in rivers, groundwater, dump sites, and indoor air. For example, industries reporting to the Toxics Release Inventory disclosed environmental releases of the aromatic amine o-toluidine that totaled 14,625 pounds in 1992 alone.[7] Detected also in effluent from refineries and other manufacturing plants, o-toluidine exists as residues in the dyes of commercial textiles, which may, according to the *Seventh Annual Report on Carcinogens,* published by the U.S. Department of Health and Human Services, expose members of the general public who are consumers of these goods: "The presence of o-toluidine, even as a trace contaminant, would be a cause for concern."[8] A 1996 study investigated a sixfold excess of bladder cancer among workers exposed years earlier to o-toluidine and aniline in the rubber chemicals department of a manufacturing plant in upstate New York.[9] Levels of these contaminants are now well within their legal workplace limits, and yet blood and urine collected from current employees were found to contain substantial numbers of DNA adducts and detectable levels of o-toluidine and aniline. Another recent investigation revealed an eightfold excess of bladder cancer among workers employed in a Connecticut pharmaceuticals plant that manufactured a variety of aromatic amines.[10]

What my various file folders do not contain is a considered evaluation of all known and suspected bladder carcinogens—their sources, their possible interactions with each other, and our various routes of exposure to them. Trihalomethanes—common contaminants of chlorinated tap water—have been linked to bladder cancer, as has the dry-cleaning solvent and sometime-contaminant of drinking-water pipes, tetrachloroethylene. I possess individual reports on each of these topics. What I do not have is a comprehensive description of how all these substances behave in combination. What are the risks of multiple trace exposures? What happens when we drink trihalomethanes, absorb aromatic amines, and inhale tetrachloroethylene? Furthermore, what is the ecological fate of these substances once they are released into the environment? What happens when dyed cloth, colored paper, and leather goods are laundered, landfilled, or incinerated? And why, almost a century after some of them were so identified, do powerful bladder carcinogens such as amine dyes continue to be manufactured, imported, used, and released into the environment? However improved the record of effort to regulate them, why have they all not been replaced by safer substitutes? To my knowledge, these questions remain largely unaddressed by the cancer research community.

Biased Focus

Genes

Several obstacles, I believe, prevent us from addressing cancer's environmental roots. An obsession with genes and heredity is one.

Cancer research currently directs considerable attention to the study of inherited cancers.[11] Most immediately, this approach facilitates the development of genetic testing, which attempts to predict an individual's risk of succumbing to cancer, based on the presence or absence of certain genetic alterations.

Hereditary cancers, however, are the rare exception. Collectively, fewer than 10 percent of all malignancies are thought to involve inherited mutations.[12] Between 1 and 5 percent of colon cancers, for example, are of the hereditary variety, and only about 15 percent exhibit any sort of familial component.[13] The remaining 85 percent of colon cancers are officially classified as "sporadic," which essentially means that we don't know what causes them.[14] Breast cancer also shows little connection to heredity (probably between 5 and 10 percent).[15] Finding "cancer genes" is not going to prevent the great majority of cancers that develop.

Moreover, even when inherited mutations do play a role in the development of a particular cancer, environmental influences are inescapably involved as well. Genetic risks are not exclusive of environmental risks. Indeed, the direct consequence of some of these damaging mutations is that people become even more sensitive to environmental carcinogens. In the case of hereditary colon cancer, for example, what is passed down the generations is a faulty DNA repair gene.[16] Its human heirs are thereby rendered less capable of coping with environmental assaults on their genes or repairing the spontaneous mistakes that occur during normal cell division. These individuals thus become more likely to accumulate the series of acquired mutations needed for the formation of a colon tumor.

Cancer incidence rates are not rising because we are suddenly sprouting new cancer genes. Rare, heritable genes that predispose their hosts to cancer by creating special susceptibilities to the effects of carcinogens have undoubtedly been with us for a long time. The ill effects of some of these genes might well be diminished by lowering the burden of environmental carcinogens to which we are all exposed. In a world free of aromatic amines, for example, being born a slow acetylator would be a trivial issue, not a matter of grave consequence. The inheritance of a defective carcinogen-detoxifying gene would matter less in a culture that did not tolerate carcinogens in air, food, and water. By contrast, we cannot change our ancestors. Shining the spotlight on inheritance focuses us on the one piece of the puzzle we can do absolutely nothing about.

Lifestyle

Risks of lifestyle are also not independent of environmental risks. Yet, public education campaigns about cancer consistently accent the former and ignore the latter. I collect the colorful pamphlets on cancer that are made available in hospitals, clinics, and waiting rooms. When I was teaching introductory biology and also spending many hours in doctors' offices, I began to compare the descriptions of cancer in the tracts displayed in the racks above the magazines with the chapter on cancer provided in my students' textbook. Here are some of my findings.

On the topic of how many people get cancer, a pink and blue brochure published by the U.S. Department of Health and Human Services offers the following:

> Good News: Everyone does not get cancer. 2 out of 3 Americans never will get it.[17]

Whereas, according to *Human Genetics: A Modern Synthesis*:

> One of three Americans will develop some form of cancer in his or her lifetime, and one in five will die from it.[18]

(Since these materials were published, the proportion of Americans contracting cancer has risen from 30 to 40 percent.) On the topic of what causes cancer, the brochure states:

> In the past few years, scientists have identified many causes of cancer. Today it is known that about 80% of cancer cases are tied to the way people live their lives.

Whereas the textbook contends:

> As much as 90 percent of all forms of cancer is attributable to specific environmental factors.

In regard to prevention, the brochure emphasizes individual choice and responsibility:

> You can control many of the factors that cause cancer. This means you can help protect yourself from the possibility of getting cancer. You can decide how you're going to live your life—which habits you will keep and which ones you will change.

The genetics book presents a somewhat different vision:

> Because exposure to these environmental factors can, in principle, be controlled, most cancers could be prevented. . . . Reducing or eliminating exposures to environmental carcinogens would dramatically reduce the prevalence of cancer in the United States.

The textbook identifies some of these carcinogens, the routes of exposure, and the types of cancer that result. In contrast, the brochure emphasizes the importance of personal habits, such as sunbathing, that raise one's risk of contracting cancer. Thus, in my students' textbook, vinyl chloride is identified as a carcinogen to which workers making polyvinyl chloride (PVC) are exposed, whereas in the brochure, occupations that involve working with certain chemicals are called a risk factor. The textbook declares that "radiation is a carcinogen." The brochure advises us to "avoid unnecessary X-rays." Both emphasize the role of diet and tobacco.

In its ardent focus on lifestyle, the Good News brochure is typical of the educational pamphlets in my collection. By emphasizing personal habits rather than carcinogens, they present the cause of the disease as a problem of *behavior* rather than one of *exposure* to disease-causing agents. At its best, this perspective can offer us practical guidance and the reassurance that there are actions we as individuals can take to protect ourselves. (Not smoking, rightfully so, tops this list.) At its worst, the lifestyle approach to cancer is dismissive of hazards that lie beyond personal choice. A narrow focus on lifestyle—like a narrow focus on genetic mechanisms—obscures cancer's environmental roots. It presumes that the continuing contamination of our air, food, and water is an immutable fact of the human condition to which we must accommodate ourselves. When we are urged to "avoid carcinogens in the environment and workplace," this advice begs the question. Why must there be known carcinogens in our environment and at our job sites?

Cancer is certainly not the first disease to inspire this kind of message. In 1832, at the height of an epidemic, the New York City medical council announced that cholera's usual victims were those who were imprudent, intemperate, or prone to injury by the consumption of improper medicines.[19] Lists of cholera prevention tips were posted publicly. Their advice ranged from avoiding drafts and raw vegetables to abstaining from alcohol. Maintaining "regular" habits was also said to be protective. Decades later, improvements in public sanitation would bring cholera under control, and the pathogen responsible for the disease would finally be isolated by the bacteriologist Robert Koch in 1883. Of course, the behavioral changes urged by the 1832 handbills were not all without merit: uncooked produce, as it turned out, was an important route of exposure, but it was a fecal-borne bacteria—and not a salad-eating lifestyle—that was the cause.

The orthodoxy of lifestyle today finds its full expression in the public educational literature on breast cancer. Scores of cheerful pamphlets exhort women to exercise, lower the fat in their diets, perform breast self-examinations, ponder their family history, and receive regular mammograms. "Delayed childbirth" (after age twenty) is frequently mentioned as a risk factor. (I have never seen "prompt childbirth" in the accompanying list of cancer prevention tips—undoubtedly because such advice would be tantamount to advocating teenage pregnancy.)[20]

By itself, a lifestyle approach to preventing breast cancer is inadequate.[21] First, the majority of breast cancers cannot be explained by lifestyle factors, including reproductive history. We need to look elsewhere for the causes of these cancers. Second, mammography and breast self-examinations are tools of cancer detection, not acts of prevention. The popular refrain "Early detection is your best prevention!" is a non sequitur: Detecting cancer, no matter how early, negates the possibility of preventing cancer. At best, early detection may make cancer less fatal.

Finally, the adage that high-fat Western diets are the cause of breast cancer has not yet been supported by data.[22] Dietary fat has long been a centerpiece of study in the investigation of breast cancer risk. Yet, several long-term studies have indicated that dietary fat is unlikely to play a major role by itself.[23] Rather than continuing to focus single-mindedly on the absolute quantity of fat consumed, several researchers have called for a more refined, ecological approach to diet.[24] Two obvious starting points would be to assess the link between breast cancer and diets high in animal fat and to launch a definitive investigation into the extent to which various kinds of fats are contaminated by carcinogens. We already know with certainty that animal-based foods are our main route of exposure to organochlorine pesticides and dioxins.[25]

Even reproductive choices have environmental implications. Breasts, for example, do not complete their development until the last months of a woman's first full-term pregnancy. During this time, the latticework of mammary ducts and lobules differentiates into fully functioning secretory cells. This process of specialization permanently slows the rate of mitosis, dampens the response to growth-promoting estrogens, and renders DNA less vulnerable to damage. According to the leading hypothesis, a full-term pregnancy early in life protects against breast cancer precisely because it reduces a woman's vulnerability to carcinogens and other cancer promoters, such as estrogens.

One of the principal proponents of this hypothesis, the Harvard epidemiologist Nancy Krieger, has urged its further testing. She has also urged a redirection of breast cancer research toward environmental questions.[26] Investigators have repeatedly confirmed that reproductive history contributes to breast cancer risk. We need to know now, Krieger argues, whether women with similar reproductive histories but divergent exposure to carcinogens have marked differences in breast cancer incidence. This need is made urgent by the results of animal studies showing that exposure to certain organochlorines hastens the onset of puberty.[27] Early first menstruation—along with late parenthood—is considered a risk factor for breast cancer in women.

Within the scientific community, grand arguments have ensued from the attempt to classify and quantify cancer deaths due to specific causes.[28] Traditionally, the final result of this task takes the visual form of a great cancer pie sliced to depict the relative importance of different risk factors. "Smoking" is always a big wedge, monopolizing about 30 percent of the circle. "Diet" is also a sizable helping. Depending on who's doing the apportioning, an array of other lifestyle factors—"alcohol," "reproductive and sexual behavior," and "sedentary way of life"—make up the remainder, along with "occupation" and "pollution."

The quarreling begins immediately. How do we account for malignancies, such as certain liver cancers, to which both drinking and job hazards contribute? Or lung and bladder cancers where both job hazards and smoking conspire? Should the effects of pesticides be tallied under "pollution" or under "diet"? What about pollution's indirect effects—such as hormonal disruption, inhibition of apoptosis (programmed death of damaged cells), and immune system suppression that act to augment the dangers of risk factors across the board? What about formaldehyde, which seems to bind with DNA in such a way that it prevents repair of damage induced by ionizing radiation, possibly raising the cancer risk from medical X-rays? Interactions between risk factors aside, how can the death toll from environmental factors be calculated at all when the vast majority of industrial chemicals in commerce have never been tested for their ability to cause cancer?

The futility of what the cancer historian Robert Proctor calls "the percentages game"[29] has not deterred public health agencies from using this kind of simplistic accounting to formulate cancer control policies and educational programs. Lifestyle is the bull's-eye of cancer prevention efforts, while targeting environmental factors, perceived as making a small contribution to the cancer problem, is seen as inefficient.[30] Moreover, the rationale continues, not enough is known about environmental risks to make specific recommendations. (On the other hand, incomplete and inconsistent evidence about the role of dietary fat in contributing to breast cancer does not appear to be an obstacle to advising women to change their diets.)

In my own home state, a recent county-by-county cancer report reproduced an old cancer pie chart published in 1981 that relegated environmental factors to a single, tiny slice and depicted tobacco and diet as major risk factors. The report concluded, "Many persons could reduce their chances of developing or dying from cancer by adopting healthier lifestyles and by visiting their physicians regularly for cancer-related checkups."[31] It never mentions or considers that Illinois is a leading producer of hazardous waste, a heavy user of pesticides, and home to an above-average number of Superfund sites. Nor does this report correlate cancer statistics with Toxics Release Inventory data or attempt to determine whether cancer might follow industrial river valleys, rise in areas of high pesticide use, or cluster around contaminated wells.

Lifestyle and the environment are *not* independent categories that can be untwisted from each other: To talk about one is to talk about the other. A discussion about dietary habits is necessarily also a discussion about the food chain. To converse about childbirth and breast cancer is also to converse about changing the susceptibility to carcinogens in the breast. And to advise those of us at risk for bladder cancer to "void frequently" is to acknowledge the presence of carcinogens in the fluids passing through our bodies.

The Right to Know

During the last year of her life, Rachel Carson discussed before a U.S. Senate subcommittee her emerging ideas about the relationship between environmental contamination and human rights.[32] She urged recognition of an individual's right to know about poisons introduced into one's environment by others and the right to protection against them. These ideas are Carson's final legacy.[33]

The process of exploration that results from asserting our right to know about carcinogens in our environment is a different journey for every person. For all of us, however, I believe it necessarily entails a three-part inquiry. Like the Dickens character Ebenezer Scrooge, we must first look back at our past, then reassess our present situation, and finally summon the courage to imagine an alternative future.

We must begin retrospectively for two reasons. First, we carry in our bodies many carcinogens that are no longer produced and used domestically, but which linger in the environment and in human tissue. Appreciating how even today we remain in contact with banned chemicals such as polychlorinated biphenyls (PCBs) and DDT requires a historical understanding. Second, because cancer is a multicausal disease that unfolds over a period of decades, exposures during young adulthood, adolescence, childhood—and even prior to birth—are relevant to our present cancer risks. We need to discover what pesticides were sprayed in our neighborhoods and what sorts of household chemicals our parents stored under the kitchen sink. Reminiscing with neighbors, family members, and elders in the community where one grew up can be an eye-opening first step.

This part of the journey is, in essence, a search for our ecological roots. Just as awareness of our genealogical roots offers us a sense of heritage and cultural identity, our ecological roots provide a particular appreciation of who we are biologically. It means asking questions about the physical environment we have grown up in and the molecules of which are woven together with the strands of

DNA inherited from our genetic ancestors. After all, except for the original blueprint of our chromosomes, all the material that is us—from bone to blood to breast tissue—has come to us from the environment.

Going in search of our ecological roots has both intimate and far-flung dimensions. It means learning about the sources of our drinking water (past and present), about the prevailing winds that blow through our communities, and about the agricultural system that provides us food. It involves visiting grain fields, as well as cattle lots, orchards, pastures, and dairy farms. It demands curiosity about how pests in our apartment buildings are exterminated, how our clothing is cleaned, and how golf courses are maintained. It means asserting our right to know about any and all toxic ingredients in such products as household cleaners, paints, and cosmetics. It requires a determination to discover the location of underground storage tanks, how the land was used before a subdivision was built over it, what is being sprayed along the roadsides and rights-of-way, and what exactly goes on behind that barbed-wire fence at the end of the street. Acquiring a copy of the Toxics Release Inventory for one's home county, as well as a list of local hazardous waste sites is a simple place to begin.

In full possession of our ecological roots, we can begin to survey our present situation. This requires a human rights approach. Such an approach recognizes that the current system of regulating the use, release, and disposal of known and suspected carcinogens—rather than preventing their generation in the first place—is intolerable. So is the decision to allow untested chemicals free access to our bodies until they are finally assessed for carcinogenic properties. Both practices show reckless disregard for human life.

A human rights approach would also recognize that we do not all bear equal risks when carcinogens are allowed to circulate within our environment.[34] Workers who manufacture carcinogens are exposed to higher levels, as are those who live near the chemical graveyards that serve as their final resting place. Moreover, people are not uniformly vulnerable to the effects of environmental carcinogens. Individuals with genetic predispositions, infants whose detoxifying mechanisms are not yet fully developed, and those with significant prior exposures may all be affected more profoundly. Cancer may be a lottery, but each of us does not hold equal chances of "winning." When carcinogens are deliberately or accidentally introduced into the environment, some number of vulnerable persons are consigned to death. The impossibility of tabulating an exact body count does not alter this fact. A human rights approach to cancer strives, nonetheless, to make these deaths visible.

Suppose we assume for a moment that the most conservative estimate concerning the proportion of cancer deaths due to environmental causes is absolutely accurate. This estimate, put forth by those who dismiss environmental carcinogens as negligible, is 2 percent.[35] Though others have placed this number far higher,[36] let's assume for the sake of argument that this lowest value is absolutely correct. Two percent means that 10,940 people in the United States die each year from environmentally caused cancers.[37] This is more than the number of women who die each year from hereditary breast cancer—an issue that has launched multimillion dollar research initiatives. This is more than the number of children and teenagers killed each year by firearms—an issue that is considered a matter of national shame. It is more than three times the number of nonsmokers estimated to die each year of lung cancer caused by exposure to secondhand smoke—a problem so serious it warranted sweeping changes in laws governing air quality in public spaces. It is the annual equivalent of wiping out a small city. It is thirty funerals every day.

None of these 10,940 Americans will die quick, painless deaths. They will be amputated, irradiated, and dosed with chemotherapy. They will expire privately in hospitals and hospices and be buried quietly. Photographs of their bodies will not appear in newspapers. We will not know who most of them are. Their anonymity, however, does not moderate this violence. These deaths are a form of homicide.[38]

According to the most recent tally, forty possible carcinogens appear in drinking water, sixty are released by industry into ambient air, and sixty-six are routinely sprayed on food crops as pesticides.[39] Whatever our past exposures, this is our current situation.

Guiding Principles for Reducing Toxics

After having carefully appraised the risks and losses that we have endured by tolerating this situation, we can begin to imagine a future in which our right to an environment free of such substances is respected. It is unlikely that we will ever rid our environment of all chemical carcinogens. However, as Rachel Carson herself observed, the elimination of a great number of them would reduce the carcinogenic burden we all bear and thus would prevent considerable suffering and loss of human life.[40] Three key principles can assist us in this effort.

One is the idea that public and private interests should act to prevent harm before it occurs. This is known as the *precautionary principle*, and it dictates that *indication* of harm, rather than *proof* of harm, should be the trigger for action—especially if delay might cause irreparable damage.[41] Central to the precautionary principle is the recognition that we have an obligation to protect human life. Our current methods of regulation, by contrast, appear governed by what some frustrated policymakers have called "the dead body approach": Wait until damage is proven before taking action.[42] It is a system tantamount to running an uncontrolled experiment using human subjects.

Closely related to the precautionary principle is the *principle of reverse onus*.[43] According to this edict, it is safety, rather than harm, that should necessitate demonstration. This reversal essentially shifts the burden of proof from the public to those who produce, import, or use the substance in question. The principle of reverse onus requires that those who seek to introduce chemicals into our environment first show that what they propose to do is almost certainly not going to hurt anyone. This is already the standard we uphold for pharmaceuticals and yet for most industrial chemicals, no firm requirement for advance demonstration of safety exists. Chemicals are not citizens. They should not be presumed innocent unless proven guilty, especially when a verdict of guilt requires some of us to sicken and die in order to provide the necessary evidence.

Finally, all activities with potential public health consequences should be guided by the *principle of the least toxic alternative*, which presumes that toxic substances will not be used as long as there is another way of accomplishing the task.[44] This means choosing the least harmful way of solving problems— whether it be ridding fields of weeds, school cafeterias of cockroaches, dogs of fleas, woolens of stains, or drinking water of pathogens. Biologist Mary O'Brien advocates a system of assessment of alternatives in which facilities regularly evaluate the availability of alternatives to the use and release of toxic chemicals. Any departure from zero should be preceded by a finding of necessity. These efforts, in turn, should be coordinated with active attempts to develop and make available affordable, nontoxic alternatives for currently toxic processes and with systems of support for those making the transition—whether farmer, corner dry cleaner, hospital, or machine shop. The highest priority for transformation should be assigned to all processes that generate dioxin or require the use or release of any known human carcinogen such as benzene and vinyl chloride.

The principle of the least toxic alternative would move us away from protracted, unwinnable debates over how to quantify the cancer risks from each carcinogen released into the environment and where to set legal maximum limits for their presence in air, food, water, the workplace, and consumer goods. As O'Brien observed, "Our society proceeds on the assumption that toxic substances will be used and the only question is how much. Under the current system, toxic chemicals are used, discharged, incinerated, and buried without ever requiring a finding that these activities are necessary" (personal communication, M. O'Brien, 1997). The principle of the least toxic alternative looks toward the day when the availability of safer choices makes the deliberate and routine release of chemical carcinogens into the environment as reprehensible as the practice of slavery.

Notes

Steingraber, Sandra. 2000. "The Social Production of Cancer: A Walk Upstream." In *Reclaiming the Environmental Debate: The Politics of Health in a Toxic Culture*, ed. Richard Hofrichter. Cambridge: MIT Press.

1. R. A. Weinberg, "A Molecular Basis of Cancer," *Scientific American* (November 1983):126–42.

2. I. Orlow et al., "Deletion of the p16 and p15 Genes in Human Bladder Tumors," *Journal of the National Cancer Institute* 87 (1995):1524–29; S. H. Kroft and R. Oyasu, "Urinary Bladder Cancer: Mechanisms of Development and Progression," *Laboratory Investigation* 71 (1994):158–74; P. Lipponen and M. Eskelinen, "Expression of Epidermal Growth Factor Receptor in Bladder Cancer as Related to Established Prognostic Factors, Oncoprotein Expression and Long-Term Prognosis," *British Journal of Cancer* 69 (1994):1120–25.

3. D. Lin et al., "Analysis of 4-Aminobiphenyl-DNA Adducts in Human Urinary Bladder and Lung by Alkaline Hydrolysis and Negative Ion Gas Chromatography-Mass Spectrometry," *Environmental Health Perspectives* 102 (Suppl. 6) (1994):11–16; P. L. Skipper and S. R. Tannenbaum, "Molecular Dosimetry of Aromatic Amines in Human Populations," *Environmental Health Perspectives* 102 (Suppl. 6) (1994):17–21; S. M. Cohen and L. B. Ellwein, *Environmental Health Perspectives* 101 (Suppl. 5) (1994):111–14.

4. P. Vine and G. Ronco, "Interindividual Variation in Carcinogen Metabolism and Bladder Cancer Risk, "*Environmental Health Perspective* 98 (1992):95–99.

5. One researcher offers the following reflection on the bladder cancer situation in England: "The continued use of known carcinogenic substances in British industry for many years after their identification, the wide range of industries with a known or suspected increased risk of bladder cancer, and our ignorance of the carcinogenic potential of many materials used in current manufacturing should be a cause for continuing concern" (R. R. Hall, "Superficial Bladder Cancer," *British Medical Journal* [1994]:910–13).

6. D. T. Silverman, "Urinary Bladder," in *Cancer Risks and Rates*, NIH Pub. 96-691, A. Harras, ed. (Bethesda, Md.: National Cancer Institute, 1996, pp. 197–99). Routine screening for bladder cancer is not done. Thus earlier detection or improved diagnostic techniques are unlikely explanations for the recent increases in rates. R. A. Schulte et al. (eds.), "Bladder Cancer Screening in High Risk Groups," *Journal of Occupational Medicine* 32 (1990):787–845.

7. Environmental Protection Agency, 1992 *Toxic Chemicals Release Inventory: Public Data Release*. EPA 745-R-001 (Washington, D.C.: EPA, 1994, p. 79).

8. U.S. Department of Health and Human Services, *Seventh Annual Report on Carcinogens* (Research Triangle Park, N.C.: USDHHS, 1994, p. 389).

9. E. M. Ward et al., "Monitoring of Aromatic Amine Exposure in Workers at a Chemical Plant with a Known Bladder Cancer Excess," *Journal of the National Cancer Institute* 88 (1996):1046–52.

10. R. Ouellet-Hellstromt and J. D. Rench, "Bladder Cancer Incidence in Arylamine Workers," *Journal of Occupational and Environmental Medicine* 38 (1996):1239–47; J. D. Rench et al., *Cancer Incidence Study of Workers Handling Mono- and Di-arylamines Including Dichlorobenzidine, ortho-Toluidine, and ortho-Dianisidine* (Falls Church, Va.: SRA Technologies, 1995); "Study Finds Bladder Cancer Threat among Conn. Plant Workers," *Boston Globe*, September 21, 1995, p. 42.

11. Francis Collins, Richard Klausner, and Kenneth Olden, statement on cancer, genetics, and the environment before the Senate Committee on labor and Human Resources, March 6, 1996 (U.S. Department of Health and Human Services press release).

12. National Cancer Institute, *Understanding Gene Testing*, NIH Pub. 96-3905 (Bethesda, Md.: NCI, 1995).

13. G. Marra and C. R. Boland, "Hereditary Nonpolyposis Colorectal Cancer: The Syndrome, the Genes, and Historical Perspectives," *Journal of the National Cancer Institute* 87 (1995):1114-25; N. Papadopoulos et al., "Mutation of a *mutL* Homolog in Hereditary Colon Cancer," *Science* 263 (1994):1625–29.

14. Bert Vogelstein, "Heredity and Environment in a Common Human Cancer," lecture at Harvard Univ. Medical School, May 3, 1995). In exploring the use of the term "sporadic" by cancer researchers, historian Robert Proctor observed, "The presumption is apparently that heredity is orderly, while environmental causation is chaotic, perhaps even indecipherable. . . . Genetics offers hope for new forms of therapy, but also seems to imply resignation with regard to the possibility of prevention." See R. N. Proctor, *Cancer Wars: How Politics Shapes What We Know and Don't Know about Cancer* (New York: Basic Books, 1995, p. 245).

15. Five to 10 percent is the estimate most often cited. A recent prospective cohort study of more than 100,000 women placed this figure even lower—at about 2.5 percent. See G. A. Colditz, "Family History, Age, and Risk of Breast Cancer: Prospective Data from the Nurses' Health Study," *Journal of the American Medical Association* 2 (70) (1993):338–43.

16. D. Holzman, "Mismatch Repair Genes Matched to Several New Roles in Cancer," *Journal of the National Cancer Institute* 88 (1996):950–51.

17. "Cancer Prevention" (pamphlet) (Bethesda, Md.: U.S. Department of Health and Human Services, n.d.).

18, G. Edlin, *Human Genetics. A Modern Synthesis*, 2d ed. (Boston: Jones & Bartlett, 1990). Quotations are from pages 184–204.

19. C. E. Rosenberg, *The Cholera Years: The United States in 1832, 1849, and 1866* (Chicago: Univ. of Chicago Press, 1962, pp. 1–60).

20. Some researchers argue that "delayed childbirth" among white women explains much of the elevated incidence of breast cancer in the northeastern states. See S. R. Sturgeon, "Geographic Variation in Mortality from Breast Cancer among White Women in the United States," *Journal of the National Cancer Institute* 87 (1995):1846–53.

21. M. P. Madigan, "Proportion of Breast Cancer Cases in the United States Explained by Well-Established Risk Factors," *Journal of the National Cancer Institute* 87 (1995):1681–85.

22. D. J. Hunter et al., "Cohort Studies of Fat Intake and the Risk of Breast Cancer—A Pooled Analysis," *New England Journal of Medicine* 334 (1996): 356–61; D. J. Hunter and W. C. Willett, "Diet, Body Size, and Breast Cancer," *Epidemiology Reviews* 15 (1993):110–32; E. Giovannucci et al., "A Comparison of Prospective and Retrospective Assessments of Diet in the Study of Breast Cancer," *American Journal of Epidemiology* 137 (1993):502–11. The role of dietary fat in creating breast cancer risk remains uncertain in part because the range of fat intake among the various groups of women studied has so far been relatively narrow.

23. As two leading researchers have observed, energy intake from fat has been declining as breast cancer has increased: Hunter and Willett, "Diet, Body."

24. Drs. Devra Lee Davis, Samuel Epstein, and Janette Sherman are among the researchers calling for a more ecological approach to diet. See S. S. Epstein, "Environmental and Occupational Pollutants Are Avoidable Causes of Breast Cancer," *International*

Journal of Health Services 24 (1994):145–50; and J. Sherman, *Chemical Exposure and Disease: Diagnostic and Investigative Techniques* (Princeton, N.J: Princeton Scientific Publishing, 1994, p. 83).

25. Consumption of animal fat (or meat) is most strongly linked to colon and prostate cancers. See W. C. Willett, "Diet and Nutrition," in *Cancer Epidemiology and Prevention*, 2d ed. D. Schottenfeld and J. F. Fraumeni, Jr., eds. (Oxford: Oxford Univ. Press, 1996, pp. 438–61).

26. N. Krieger, "Exposure, Susceptibility, and Breast Cancer Risk," *Breast Cancer Research and Treatment* 13 (1989):205–23.

27. This topic is currently under exploration by Dr. Mary Wolff, who is interested in all factors, including childhood diet and level of physical activity, that contribute to the onset of puberty in girls. M. S. Wolff, "Organochlorines and Breast Cancers," presentation at the American Public Health Association, New York, November 20, 1966. See L. M. Walters et al., "Purified Methoxychlor Stimulates the Reproductive Tract in Immature Female Mice," *Reproductive Toxicology* 7 (1993):599–606; P. L. Whitten et al., "A Phytoestrogen Diet Induces the Premature Anovulatory Syndrome in Lactionally Exposed Female Rats," *Biology of Reproduction* 49 (1993):1117–21; R. J. Gellert, "Uterotropic Activity of Polychlorinated Biphenyls and Induction of Precocious Reproductive Aging in Neonatally Treated Female Rats," *Environmental Research* 16 (1978):123–30.

28. See, for example, R. Doll and R. Peto, *The Causes of Cancer: Quantitative Estimates of Avoidable Risks of Cancer in the United States Today* (Oxford: Oxford Univ. Press, 1981); and a rebuttal by S. S. Epstein and J. B. Swartz, "Fallacies of Lifestyle Cancer Theories," *Nature* 2(89) (1981):127–30.

29. Described in Proctor, *Cancer Wars*, pp. 54–74. See also M. Kaidor and K. A. L'Abbe, "Interaction between Human Carcinogens," in *Complex Mixtures and Cancer Risk*, H. Vainio et al., eds., IARC Scientific Pub. 104 (Lyon, France: International Agency for Research on Cancer, 1990, pp. 35–43).

30. The American Cancer Society does not discuss environmental factors in its recent report on cancer prevention. See American Cancer Society, *Cancer Risk Report: Prevention and Control, 1995* (Atlanta, Ga.: ACS, 1995). See also K. R. McLeroy, "An Ecological Perspective on Health Promotion Programs," *Health Education Quarterly* 15 (1988):351–77.

31. Illinois Department of Public Health, *Cancer Incidence in Illinois by County, 1985–87*, Supplemental Report (Springfield, Ill.: IDPH, 1990, pp. 7–8).

32. Rachel Carson on environmental human rights: Senate testimony hearings before the Subcommittee on Reorganization and International Organizations of the Committee on Government Operations, "Interagency Coordination in Environmental Hazards (Pesticides)," U.S. Senate, 88th Cong., 1st sess., June 4, 1962.

33. Carson, *Silent Spring* (Boston, Mass.: Houghton Mifflin, 1962, pp. 277–78).

34. R. Perera, "Uncovering New Clues to Cancer Risk," *Scientific American* (May 1996):54–62; S. Venitt, "Mechanisms of Carcinogenesis and Individual Susceptibility to Cancer," *Clinical Chemistry* 40 (1994):1421–25; G. W. Lucier, "Not Your Average Joe," (editorial), *Environmental Health Perspectives* 103 (1995):10.

35. Harvard Center for Cancer Prevention, "Harvard Report on Cancer Prevention," *Cancer Causes and Control* 7 (Suppl. 1) (1996):3–59; D. Trichopoulous et al., "What Causes Cancer?" *Scientific American* (September 1996):80–87.

36. Proctor, *Cancer Wars*.

37. 10,940 is 2 percent of 547,000, the projected figure for total cancer deaths in 1995. See American Cancer Society, *Cancer Facts and Figures—1995*, rev. (Atlanta, Ga.: ACS, 1995).

38. The environmental analysts Paul Merrell and Carol Van Strum have argued that the concept of acceptable risk is tolerable only because of the anonymity of its intended victims. See P. Merrell and C. Van Sturm, "Negligible Risk: Premeditated Murder?" *Journal of Pesticide Reform* 10 (1990):20–22. Likewise, the molecular biologist and physician John Gofman has argued, "If you pollute when you DO NOT KNOW if there is any safe dosage (threshold), you are performing improper experimentation on people without their informed consent. . . . If you pollute when you DO KNOW that there is no safe dose with respect to causing extra cases of deadly cancers, then you are committing premeditated random murder" (J. W. Gofman, memorandum to the U.S. Nuclear Regulatory Commission, May 21, 1994).

39. M. Eubanks, "Biomarkers: The Clues to Genetic Susceptibility," *Environmental Health Perspectives* 102 (1994):50–56.

40. Carson, *Silent Spring*, p. 248. See also M. J. Kane, "Promoting Political Rights to Protect the Environment," *Yale Journal of International Law* 18 (1993):389–411.

41. This principle was endorsed in 1987 by European environmental ministers in a meeting about the deterioration of the North Sea. [K. Geiser, "The Greening of Industry: Making the Transition to a Sustainable Economy," *Technology Review* (August/September 1991):65–72.] See also T. O'Riordan and J. Cameron (eds.), *Interpreting the Precautionary Principle* (London: Earthscan, 1994).

42. Devra Lee Davis, quoted in "Is There Cause for 'Environmental Optimism'?" *Environmental Science and Technology* 29 (1995):366–69.

43. This principle has been embraced by the International Joint Commission in their Eighth Biennial Report on Great Lakes Water Quality (Washington, D.C., and Ottawa, Ontario: International Joint Commission, 1996, pp. 15–17). See also discussions of proof in T. Colborn et al., *Our Stolen Future: Are We Threatening Our Fertility, Intelligence, and Survival?—A Scientific Detective Story* (New York: Dutton, 1996); and G. K. Durnil, *The Making of a Conservative Environmentalist: With Reflection on Government, Industry, Scientists, the Media, Education, Economic Growth, and the Sunsetting of Toxic Chemicals* (Bloomington: Indiana Univ. Press, 1995).

44. My ideas on this topic are inspired in part by those of biologist Mary O'Brien. See M. H. O'Brien, "Alternatives to Risk Assessment: The Example of Dioxin," *New Solutions: A Journal of Environmental and Health Policy* 3 (Winter 1993):39–42; and K. Geiser, "Protecting Reproductive Health and the Environment: Toxics Use Reduction," *Environmental Health Perspectives* 101 (Suppl. 2) (1993):221–25.

Science in Environmental Conflicts

Connie P. Ozawa

Many sociologists and other social scientists believe it is important to think criti-cally about science and to understand how scientific endeavors are, at their heart, social endeavors—that science itself is a social construct. The three pieces in this section show how policymakers utilize scientific research in their decision making about environmental issues. In this reading, Connie Ozawa, professor of urban studies and urban planning at Portland State University, examines how environ-mental science is often used as a "weapon" by individuals or groups with conflicting goals. While science has an aura of being detached and neutral, it is, in fact, per-meated with politics.

The April 20, 1993, *New York Times* reported another environmental con-troversy (Strum 1993). The Port Authority of New York and New Jersey filed an application to the U.S. Army Corps of Engineers to dredge con-tainer-ship berths in Newark Bay and dump the dredged material in the ocean. What had been an annual rite met delay when routine analysis of the silt to be displaced was found to contain dioxin. Two groups, environmentalists con-cerned about the impact of dioxin on marine life, such as endangered whales and sea turtles, and the coastal tourism industry, advocated storing the dredged material on containment islands or barges until economical and effective decontamination technologies became available. Although ultimately resolved, more than three years later the Port Authority was still waiting for permit approval, losing revenues from nearly half the harbor's prime container ship berths.

Action on the application was delayed for a number of reasons. First, because no federal standards existed for dioxin contamination in the ocean, the EPA regional office attempted to establish acceptable levels for this case. The threshold for contamination initially set by EPA was later modified, requiring more testing, resulting in further delay, and fueling debate over testing methods. Second, although the U.S. Army Corps of Engineers had conducted research on safe methods of ocean disposal, no research specifically on the dumping of dioxin had been performed and environmentalists raised doubt about the ability to prevent the spread of dioxin-contaminated silt across the ocean floor. Finally, technological advances during the 36-month delay led to continual refinement of the scientific data on which regulators, environmentalists, and the port were basing their decisions. Rather than engendering greater confidence in the numbers, such modifications had the contrary effect of increasing skepticism about the effectiveness and stability of the government standard. If the numbers were revised once, what would prevent subsequent reconsideration?

Though the heat underlying the conflict over the harbor dredging was generated by the ethical, economic, and ideological implications of alternative actions, much of the public debate focused on the technical issues of standards, criteria, and testing methodologies. This dispute illustrates what has become a common role of science in environmental conflict—science used as a weapon in the arsenal of warring public policy actors.

... In this [essay], a distinction is made between the terms "dispute" and "conflict." Dispute refers to vocalized or articulated disagreements over what ought to be done. Conflict is the underlying basis for the disagreement—the perceptions (accurate or not) of an undesirable distribution of consequent costs and benefits and/or the more subtle redistribution of political control over similar decisions in the future.

Environmental disputes arise not only from the perceptions of unfavorable, potential consequences of proposed actions, but also the sense of the legal rights and recourse awarded individuals and groups, and validated and institutionalized in national legislation in the United States. Federal legislation, starting earlier but most exemplified by the National Environmental Policy Act [NEPA] (1969), mandates that decisions affecting the environment ensure that adverse impacts are mitigated to the fullest extent possible. What is adverse, of course, is subject to interpretation, but clearly the presumption of NEPA is that impacts can be identified and evaluated prior to actual implementation of a proposed action. The Clean Air Act instructs the Environmental Protection Agency to issue air quality criteria that "accurately reflect the latest scientific knowledge useful in indicating the kind and extent of all identifiable effects on public health and welfare" (Clean Air Act 1967). Importantly, these laws and many others include citizen suit provisions that explicitly award citizens the legal right to question government actions.

Legislation like these have set the stage for decision makers and others to spotlight the scientific and technical elements in environmental disputes. Rather than having a role of equal standing with the human actors in a conflict, however, science ought best be viewed as a prop in the hands of those enacting environmental conflicts. This article is about the multiple ways science is used in environmental conflict. In addition to identifying and describing these multiple roles, I argue that a traditional image of science is essential in order for science to be used in these ways. I also describe an emerging alternative role, one that is more consistent with the social constructionist image of science. Before examining the various purposes for which this prop called science is used, let us first proceed through a brief review of how science became so central in environmental conflict.

The Authority of Science

In the economic and cultural context of the later twentieth century United States and, indeed, in much of the industrialized world, science is looked upon as a source of authority. This authority derives from a popular notion of the scientific endeavor. Science is conceived as a process that yields an objective, rational, politically neutral body of knowledge. Decisions consistent with scientific knowledge, therefore, command acceptance.

A principal feature of the popular conceptualization of the scientific enterprise is its strict methodological prescriptions. According to a philosophy of science dominant through the 1960s, known as logical positivist empiricism, the primary test of truth is the replicability of experimental findings. Hiskes and Hiskes (1986:10–11) write that logical positivist empiricism assumes that:

1. Data obtained through careful experiment and observation are objective;
2. There is one universally valid logic for science; and
3. Through rigorous application of logic to data, science gradually makes progress toward the ancient Greek ideal of theoria.

According to the logical positivist empiricist view, data are incontrovertible and unchanging. The observations of any two rational persons witnessing the same event would be identical. Data accumulated through the repetition of similar events eventually leads to the development of theory that integrates abstract concepts and generalizable principles to explain diverse phenomena. Logic is linear and one-directional. In short, this view implies that the products of work undertaken through the scientific method are absolute and without ambiguity.

... [S]cientists and their spokespersons have aggressively fought to reaffirm and protect the image of a neutral science. Proponents of unconditional financial support for scientific research by the federal government have argued that the scientific community is, and ought to be allowed to remain, self-monitoring and autonomous. The scientific community has been called a priesthood, an estate, and a republic (Lapp 1965; Polanyi 1972; Price 1965) and scientists, accordingly, have been described as objective, disinterested, uncorruptible, and impartial (Wood 1964). Uniform standards of validating fact and the self-imposed discipline of the scientific method are offered as guarantees that science is a depersonalized and selfless quest for truth.

Four Roles for Science

If scientific work is viewed as completely outside the social and political bickering and battling that occurs among individuals and groups in society, then a powerful role for science in environmental conflict would be nearly unassailable. Based on the assumption that the scientific method does indeed ensure the political neutrality of knowledge thereby produced, stakeholders in environmental conflict have crafted four important roles for science. These are the roles of science as discoverer, mechanism of accountability, shield, and tool of persuasion.

Science as Discoverer

The role for science most easily associated with an idealized conceptualization of the scientific method is the role of discoverer. In this role, a scientist working in relative isolation from contemporary social and political skirmishes incidentally uncovers a condition the researcher, from her own personal value framework, deems worthy of wider discussion or public action. When Oregon State University student Eric Forsman chanced upon a spotted owl in the Pacific northwest in 1968, he had no intention that this and subsequent encounters would fuel a debate between protecting an endangered species or an endangered livelihood twenty years later. Rowland and Molina's discovery of the correlation between CFCs and the ozone hole over Antarctica was similarly not inspired by an ambition to change public policy. However, each event contributed significantly to the debates over logging in the Northwest and reductions in the use of ozone-depleting gases, respectively.

In most cases, science plays the role of discoverer or educator only at the earliest stages of, or even prior to, conflict development. The role of science as discoverer reflects an idealized image of a scientist's quest to understand conditions in the physical world. One need not look very far, however, to recognize the critical effect that the researcher's personal value framework has on how that researcher interprets new information. Although biologists working for large timber companies would be unlikely to experience the same intimate moments of solitude with the forest as Forsman, had spotted owls crossed their paths in the late 1960s, it is doubtful that their responses would have been the same as his. Rare wildlife species simply are not a high priority for timber company employees.

Science as a Mechanism of Accountability

A second role for science might best be understood by looking back a half century to the New Deal period when Franklin D. Roosevelt established a number of independent government agencies like the Tennessee Valley Authority. These independent, specialized agencies were built on the assumption that certain types of decisions ought to (and could) be based on technical expertise, not politics. Congress reacted swiftly to try to ensure that decisions by such agencies were in fact based on non-partisan expertise and not politics by enacting the Administrative Procedures Act (APA) in 1946. The APA stipulates procedures for agency decision making, which essentially prescribes that agencies keep a record of their decision-making process and that decisions are consistent with a reading of that record. As agency decisions became more technical in nature (as with the regulation of new technologies and control of air and water pollution) and as new legislation awarded legal standing to citizen groups to challenge agency decisions, decision makers paid increasing attention to technical and scientific studies relevant to their decisions.

Since the 1970s, more than twenty new administrative agencies have been created, most having to do with environmental and health and safety regulation. Corresponding legislation have reinforced the need for decision makers to provide explicit technical documentation to support policy decisions. Statutes, such as the Occupational Safety and Health Act (OSHA), the Toxic Substances Control Act (TSCA), and Resource Conservation and Reclamation Act (RCRA) make explicit reference to the technical basis for decisions. Finally, judges have conceived their role as ensuring that agency decisions are reasonably consistent:

> [The] court has a supervisory function of review of agency decisions. This begins with enforcing the requirement of reasonable procedure, fair notice, and opportunity for the parties to present their case, and it includes examining the evidence and fact findings to see both that the evidentiary fact findings are supported by the record and that they provide a rational basis for inferences of ultimate fact. (Levanthal 1974:511)

Although decision making without the benefit of technical expertise in areas such as environmental policy would be foolhardy, a primary goal of these decision-making prescriptions was accountability. As long as agency decision makers were constrained by the technical experts' interpretations of the physical conditions and alternative actions, Congress assumed that raw politics would be constrained.

Science as a Shield

Astute decision makers quickly recognized that by framing decisions around boundaries drawn by technical studies, they could build a rationale that would protect them from the political fallout of publicly unpopular decisions. By presenting such information as definitive with respect to policy decisions, the decision maker attempts to create the illusion that science is arbitrating between multiple policy viewpoints or decision alternatives. For example, a decision maker may claim that because certain soil hydrologists have agreed that a particular tract of land proposed for development meet criteria defining a wetland, a development permit must be denied. In effect, the decision maker is claiming that the scientific findings (i.e., the determination that the land is a wetland) preclude a decision to allow development and thus absolve him of responsibility and shield him from the wrath of unhappy constituents. As one writer noted with regard to Congressional deliberations regarding policies for protecting health and the environment, "turning the job of defining adequate standards over to the 'experts' relieves Congressmen [sic] of the burden of resolving difficult controversies" (Melnick 1983:251).

The political expediency of this tactic is obvious, but the logic is questionable. In practice, the decision-maker exercises considerable discretion in formulating a response to scientific reports. He may accept the findings and rule otherwise (for other specified reasons, such as economic hardship, for example), he may seek additional advice, or he may order additional study. Throughout his term in office, Ronald Reagan avoided dealing with the acid rain issue raised by groups in the northeastern United States and Canada. Rather than heeding experts who believed the available evidence indicated a causal relationships between smokestack emissions from the industrial mid-west and rising acidity levels and ailing forests further north, Reagan preferred to listen to those scientists who cautiously avoided affirming a connection. Deciding to adopt the decision alternative suggested by a scientific finding or to wait for further confirmation is a political act.

Science as a Tool of Persuasion

Once science is recognized as a source of authority for justifying decisions, it is a small step to see its power in persuading the polity of the legitimacy of one policy or decision alternative over others (Dickson 1984; Nelkin and Pollack 1981).

Like religion and the rule of the monarchy prior to the Age of Enlightenment, science is used in twentieth-century decision making as a primary source of legitimacy to gain political support:

> By invoking the authoritative canons of scientific reasoning and method, public authorities and others having a stake in technical issues seek to demonstrate the rationality of their position and thereby gain political support and acceptance. (Brickman 1984:108)

In this role, science can be used either to support advocated positions in environmental conflicts or "to prevent policy being made around a rival scientific conclusion" (Collingridge and Reeve 1986). Opponents of a proposal might attempt to prevent a decision by either presenting alternative scientific data or analysis or by questioning the assumptions or interpretations of scientific reports that support the proposal. One well-known example of this strategic use of science in regulatory decision making is the tobacco industry's effort to stall restrictions on cigarette smoking by attempting to discredit studies linking cigarette smoking to lung cancer. In nearly any environmental conflict today, participants routinely raise questions about the assumptions, data, and models used in analyses that support opposing viewpoints.

If science and politics are separate, the boundary between science and policy is blurry at best. Science can play the role of discoverer only at very early stages of an environmental conflict, to flag a concern for action. But how that concern is framed and whether it is acted upon is a political decision. Science as a mechanism of accountability similarly serves to moderate the abuse of delegated decision-making authority, but by no means eliminates administrative discretion.

Casting science in the role of a tool of persuasion or a shield is a politically motivated act on the part of the user (policy actor) to capitalize on the authority of science derived from its image as politically neutral. However, as quickly as one set of policy advocates attempt to appropriate science to support their preferred policy or decision alternative, opposition groups move to undermine their position by discrediting the scientific basis of that position. The politics that are imbedded in science are readily uncloaked.

Social Constructionism and Scientific Uncertainty

If scientific work was indeed as free of the idiosyncrasies of the investigator as the ideal described earlier would suggest, environmental conflict would not end, but the scope of the disputes would be narrowed considerably with each additional contribution from scientists with relevant expertise. Disputes would revolve around what to do in response to a given situation, not around defining the conditions themselves. For example, if EPA knew absolutely that dioxin in concentrations below a given amount would not endanger marine life directly or indirectly, debate over the dredging of the New Jersey harbor might be narrowed to a discussion of dumping method, location, or timing.

However, much in the literature of the social studies of science suggests that scientific work is not free of political content. Irrespective of the rigidities of the scientific method, a multitude of discretionary judgments are made during the course of a scientific investigation by the researcher. Thomas Kuhn describes the progression of scientific inquiry as a temporally bound consensus among scientists. According to Kuhn, researchers perceive curves in the distribution of data points on a graph in patterns that fit pre-existing theory (Kuhn 1982). While researchers have identified discretionary judgments in laboratory research (Latour 1979), the predictive sciences relied on for illuminating conditions in environmental conflicts are fraught with even higher levels of discretion (Bacow 1980). For example, in predicting the potential impacts of the construction of a road through a forest, wildlife biologists would need to make assumptions about a seemingly endless list of items, including the geographic boundaries of the study area, the species to be studied, conditions in surrounding wooded lands, the migration patterns of animals under changed conditions, and the level of environmental devastation occurring during the road building period itself from the intrusion of heavy construction equipment. While these assumptions and others like them are to some extent constrained by conventions of practice, many cases are sufficiently unique to make such cross-references arguably uninformative.

The choice of assumptions, boundaries, and definitions of variables are replete with methodological uncertainties and indeterminacy (Klapp 1992; Wynne 1992). For example, the selection of a model to simulate meteorological conditions predicting air pollution plumes cannot be determined through any kind of scientific exercise, but is ultimately a judgment based on the researchers' assessment of the similarities between model parameters and real life conditions or the match between available data and the variables used in the model.

Wynne (1992) has further differentiated methodological uncertainty. He identifies ignorance as contributing to uncertainty in scientific analysis. Simply, scientists are unable to account for factors of which they are unaware. In contrast to the popular belief that scientific knowledge and method recognize and attempt to reduce uncertainties, Wynne (1992:115) argues:

> It is more accurate to say that scientific knowledge gives prominence to a restricted agenda of defined uncertainties—ones that are tractable—leaving invisible a range of other uncertainties, especially about the boundary conditions of applicability of the existing framework of knowledge to new situations.

Another type of uncertainty encountered in science has been described as statistical uncertainty. In theory, statistical uncertainty can be eventually reduced as more and more data are accumulated. However, in practice, decisions are made long before sufficient data are obtained.

Dealing with uncertainty requires a judgment on the part of researchers in the course of their work. How one selects methodologies, models, measuring devices, indeed even one's choice of scientific theory, is seldom rigidly defined by current practice. These discretionary elements are influenced by social and political factors such as the individual's institutional affiliations, source of research funds, and disciplinary training (Knorr-Cetina 1982). Viewed in this light, scientific work that carries the signature of the individual researcher and acceptance of scientific work by the scientific community more accurately represents a consensus among scientists, rather than objective fact.

An Alternative Role: Science as a Tool of Facilitation

In the highly contentious context of the 1980s, the art of utilizing scientific argumentation for furthering political objectives flourished. The authority of science was exploited by groups on multiple sides of any given debate, prolonging decisions on particular conflicts for several years in many cases and increasing expenses for government, private developers and industry, and community organizations. A sentiment was developing in the academic sphere that science in the role of arbiter in environmental conflict was a misuse of scientific work. Moreover, to sustain the image of science as authoritative with respect to decisions that were inherently political is a displacement of political power from elected politicians to the hands of an elite corps of scientific experts (Dickson 1984). Reports on Ronald Reagan's heavy-handed oversight of the selection of scientists to serve on advisory committees such as the Environmental Protection Agency's Science Advisory Board made such suspicions all the more disturbing (Ashford 1984).

The question remains, can science play a role in resolving environmental conflict? Over the past decade, an alternative role for science has been emerging as a by-product of decision-making innovations that include explicit negotiations among individuals and representatives of groups engaged in an environmental dispute. In one version of environmental mediation, the scientific and technical information necessary to understand current conditions and to identify possible options for action is one of the first topics on the agenda (Carpenter and Kennedy 1988; Crowfoot and Wondolleck 1990; Susskind and Cruikshank 1987). Almost from the start, the negotiating group discusses what kind of technical knowledge is pertinent. The more particular discretionary judgments encountered in scientific and technical investigations are openly discussed and subject to agreement (Ozawa and Susskind 1985). These judgments include decisions about the kind of information needed, data collection techniques, analytical models and methodologies, how to deal with statistical and methodological uncertainty and, sometimes, the disciplinary training and institutional affiliation of the researcher. Finally, the interests and concerns of various groups with a stake in the decision are explicitly acknowledged and a period of time is set aside in the negotiations to address them. The outright recognition of competing interests serves as a signal to stakeholding groups (and the public) that such issues will be addressed in the decision-making process. With such assurance, stakeholders contending to dominate the decision process are less inclined to posture behind admittedly disputable technical argumentation, as they do in more adversarial procedures in which winning on the technical points likely means protecting their interests, and are more willing to focus on collectively accumulating and making sense of relevant data and analyses. Because the discretionary nature of assumptions is acknowledged, sensitivity analysis or the substitution of variables or values for specific variables is easily accommodated by the negotiating group, again, defusing potential disputes over technical aspects of the decision.

An early example of this approach was a 1986 rulemaking procedure conducted by the U.S. Environmental Protection Agency (EPA) (Ozawa 1991). In response to a lawsuit filed against the agency for failure to regulate carcinogenic polycyclic organic matter (POMs) under Section 112 of the Clean Air Act, the agency invited representatives from key stakeholding groups, including wood stove manufacturers, national and local environmental organizations, and various state agencies from four states to develop emission standards for wood-burning stoves, the third largest source of POMs. Operating under a strict deadline, the group successfully crafted a proposal that was supported by all participants.

This agreement was achieved through a carefully structured procedure. At the first meeting, the group agreed to defer discussion of specific political concerns until after a solid technical basis for the rules was jointly constructed. The group labored long hours to develop this foundation of technical knowledge. Data and existing studies were collected from all known sources and closely scrutinized by technical experts from the EPA, the industry, and the environmental organizations, independently and together as a group. The discretionary nature of research assumptions and the inevitable statistical and methodological uncertainties were uncovered and debated. For example, it was widely accepted that wood stoves equipped with a catalyst emit fewer particulates than non-catalyst models, but no data existed to indicate how quickly catalysts degrade. The relative performance of catalyst-equipped stoves over the long term was thus highly uncertain. The testing performance of all models, for that matter, was disputable since emission rates vary according to basic factors such as how users stack wood and the age, type, and wetness of wood.

Discussion of assumptions regarding testing procedures, degradation rates, and countless other factors accentuated the fact that the rules were, at their core, political, not technical, products. However, this realization did not lead participants to ignore the science, as some might fear, but rather encouraged them to look more soberly at what scientific evidence existed to guide their deliberations. . . .

The role of science in this regulatory negotiation comes closest to the traditional role of discoverer, described in an earlier section. Scientific knowledge was shared not simply to prove the superiority of one policy alternative over another, but to educate all participants about the status and quality of available information. Science in this regulatory negotiation went beyond the role of discoverer, however. By working together to construct a joint understanding of the technical aspects of the standard-setting task, groups with competing political interests were also learning to listen to another and to appreciate one another's talents, skills, and knowledge base. Discussing mundane issues such as the way most people stack wood or the dominant type of wood burned in particular regions of the country provided a relatively calm atmosphere conducive to dialogue. Importantly, those with specialized expertise were explicitly asked and reminded that their role was to educate, not intimidate, the group on technical issues. The meetings also provided an opportunity for informal discussions and the formation of coalitions that facilitated the exchange of interest-related information and development of a fuller understanding of and mutual respect for all legitimate claims. In this case, science provided an opportunity for participants to develop a constructive pattern of interaction.

Conclusion

This example suggests that an alternative role for science in environmental conflict may be crafted. However, the decision-making process must be deliberately structured to ensure the following conditions. First, access to scientific expertise and analysis must be open to all stakeholding parties. Second, the agenda for negotiations must clearly set aside a period for addressing explicitly political concerns in order to discourage participants from stubbornly posturing behind technical positions that they believe will afford them political gains. Finally, experts invited to participate in the decision-making process must commit to share scientific information in order to educate, not intimidate, the stakeholders. If these conditions are met, a discussion of relevant technical information can provide an opportunity for parties to gain a fuller understanding of both the technical and political dimensions of the dispute. Science can be used as a tool of facilitation.

. . . [N]egotiating the scientific basis for environmental decisions may represent a way to maintain dialogue and develop a constructive understanding of the multiple perspectives of a given environmental conflict.

Note

Ozawa, Connie P. 1996. "Science in Environmental Conflicts." *Sociological Perspectives* 39, no. 2: 219–31.

References

Ashford, Nicholas. 1984. "Advisory Committees in OSHA and EPA: Their Use in Regulatory Decision-making." *Science, Technology, and Social Values* 9: 72–82.

Bacow, Lawrence. 1980. "The Technical and Judgmental Dimensions of Impact Assessment." *Environmental Impact Assessment Review* 1: 109–124.

Brickman, Ronald. 1984. "Science and the Politics of Toxic Chemical Regulation: U.S. and European Contrasts." *Science, Technology, and Human Values* 9: 107–111.

Carpenter, Susan, and W. J. D. Kennedy. 1988. *Managing Public Disputes*. San Francisco, CA: Jossey-Bass.

Clean Air Act. 1967. U.S. Code 42.

Collingridge, David, and Colin Reeve. 1986. *Science Speaks to Power*. New York: St. Martin's Press.

Crowfoot, James E., and Julia M. Wondolleck. 1990. *Environmental Disputes: Community Involvement in Conflict Resolution*. Washington, DC: Island Press.

Dickson, David. 1984. *The New Politics of Science*. New York: Pantheon Books.

Hiskes, Anne L., and Richard P. Hiskes. 1986. *Science, Technology, and Policy Decisions*. Boulder, CO: Westview.

Klapp, Merrie G. 1992. *Bargaining with Uncertainty: Decisionmaking in Public Health, Technological Safety, and Environmental Quality*. New York: Auburn House.

Knorr-Cetina, Karin D. 1982. "Scientific Communities of Transepistemic Arenas of Research? A Critique for Quasi-Economic Models of Science." *Social Studies of Science* 12: 101–130.

Kuhn, Thomas. 1982. "Normal Measurement and Reasonable Judgment." Pp. 75–93 in *Science in Context*, edited by Barry Barnes and David Edge. Cambridge, MA: MIT Press.

Lapp, Ralph E. 1965. *The New Priesthood: The Scientific Elite and the Uses of Power*. New York: Harper & Row.

Latour, Bruno. 1979. *Life in the Laboratory*. Beverly Hills, CA: Sage.

Levanthal, Harold. 1974. "Environmental Decisionmaking and the Role of the Courts." *University of Pennsylvania Law Review* 122: 509–555.

Melnick, R. Shep. 1983. *Regulation and the Courts: The Case of the Clean Air Act*. Washington, DC: The Brookings Institution.

Nelkin, Dorothy, and Michael Pollack. 1981. *The Atom Besieged*. Cambridge, MA: MIT Press.

Ozawa, Connie P. 1991. *Recasting Science: Consensual Procedures in Public Policy. Making*. Boulder, CO: Westview.

Ozawa, Connie P., and Lawrence E. Susskind. 1985. "Mediating Science-Intensive Policy Disputes." *Journal of Policy Analysis and Management* 5: 23–39.

Polanyi, Michael. 1972. "The Republic of Science: Its Political and Economic Theory." *Minerva* 1: 54–73.

Price, Don K. 1965. *The Scientific Estate*. Cambridge, MA: The Belknap Press of Harvard University.

Strum, Charles. 1993. "Dredging Stays Mired in a Debate Over Dioxin." *The New York Times*, Section B, p. 1.

Susskind, Lawrence, and Jeffrey Cruikshank. 1987. *Breaking the Impasse: Consensual Approaches to Resolving Public Disputes*. New York: Basic Books.

Wood, Robert C. 1964. "Scientists and Politics: The Rise of an Apolitical Elite." Pp. 41–72 in *Scientists and National Policy-Making*, edited by Robert Gilpin and Christopher Wright. New York: Columbia University Press.

Wynne, Brian. 1992. "Uncertainty and Environmental Learning: Reconceiving Science and Policy in the Preventive Paradigm." *Global Environmental Change* 2: 111–127.

PART

Social Movements

22

Risk and Recruitment
Patterns of Social Mobilization in a Government Town
Thomas E. Shriver

The pieces in this section address a central theme in sociology: social change—the transformation of society and culture over time. One way that change occurs is through collective action, especially through the actions of social movement organizations. Environmental movements have affected many environment-related issues in recent decades. Often, people become involved in environmental organizations when they perceive something in their community to be unhealthy or dangerous. However, as Thomas Shriver explains, involvement in collective action can be risky. Shriver discusses two examples of social movement activity in Tennessee— one that is "low risk" and one that is "high risk." In a clear demonstration of the importance of the impact that systemic forces such as culture, government, and politics can have on an individual or group's behaviors, this piece shows that sometimes people who otherwise would be interested in taking action fail to do so because they believe that the immediate costs to themselves are too great. Though this essay is about the failure of a community to successfully take political action, we can pull a crucial message from its findings—the need for social scientists, activists, and advocates to work together to develop workable, community-relevant tools for positive social change.

Since the 1980s there has been an increased awareness of technological disasters and their impact on communities. This recognition has been spurred by a changing political climate and by high-profile technological disasters, such as those at Three Mile Island and Love Canal. In these communities, residents successfully mobilized to challenge corporate polluters. Increasingly, communities around the country have formed grassroots environmental organizations to protest such contamination and resist the threat to their health and well-being. The actual number of these grassroots environmental organizations is difficult to assess, but estimates range from 6,000 to 10,000 (Edelstein 1988; Goldman 1991).

Despite the recent emergence and growth of this grassroots wing of the environmental movement, some communities have remained quiescent in the face of environmental problems. One such community is Oak Ridge, Tennessee, home of the federal government's Oak Ridge Nuclear Reservation. In the last 15 years, the government has documented significant off site releases of potentially harmful substances used in the production of nuclear weapons. It would be reasonable to expect residents to mobilize on the basis of potential threats from exposure to materials related to weapons production. Residents have not mobilized despite their history of involvement in civil rights activism, peace activities, and environmental issues such as strip mining in East Tennessee. What factors account for the absence of environmental activism in a community with a material basis for environmental grievances and a cadre of experienced, knowledgeable activists?

Few contemporary social movement analysts focus on the *absence* of mobilization or quiescence. Rather, they emphasize the variables influencing mobilization. Such analyses nevertheless offer clues for examining quiescence. Of particular note is McAdam's (1986) conception of high- and low-risk activism. Focusing on risks to the physical being, he argues that recruitment to activism is less likely when the aggrieved perceive or expect negative outcomes of their activism. Was activism challenging weapons production activities perceived as too risky for Oak Ridge residents? What role did fear of retribution play in the absence of mobilization in Oak Ridge? Were fears for *physical* safety the only risks associated with activism? I examine two mobilization efforts in the community and describe how one experienced moderate success while the other failed from the beginning. In both cases I examine the importance of perceived risks of activism in shaping recruitment and participation patterns in grassroots social movement activity. . . .

Conceptual Framework

. . . McAdam analyzes the 1964 Freedom Summer Project to illustrate a case of high-risk activism. Those who participated in the project were subjected to physical retribution—even murder—for their participation. McAdam points out that much of the past work done on recruitment was based on low-risk activism and is therefore less applicable to cases where activism is high-risk. . . .

In this essay, I expand on the high-risk activism literature by exploring the dynamics of mobilization and its absence in Oak Ridge, Tennessee. . . . I focus on perceptions of both *social* and *physical* risks associated with participation in activism against the economic base of the community.

Methods

Data sources for this research project include interviews, document analysis, and observation. The primary data for this [essay] were in-depth interviews with 83 respondents. . . .

I interviewed a cross-section of the population that had direct experience with and knowledge of the Oak Ridge community. The potential for bias always exists when conducting social research and is particularly acute in qualitative work. However, I made every effort to be objective throughout each stage of the project. The in-depth interviews were conducted with major subgroups in the population: scientists and laborers; old-timers and newcomers; Reservation employees and housewives; African Americans and Whites; residents with health problems they attribute to weapons production and residents who deny production-related health problems; and community leaders such as the mayor, officials at the Oak Ridge National Laboratory, and the founding editor of the city newspaper.

In addition, the demographics of the respondents represent a cross-section of the population. Fifty-two percent of the respondents were female, while 48 percent were male. Eighty-four percent of the respondents were White, while the remaining respondents were African American (slightly over 13 percent) and other (slightly over two percent). Over 90 percent of the respondents lived in Oak Ridge and slightly over 80 percent were either directly or indirectly dependent on the Oak Ridge Reservation for their employment. Over 60 percent of the respondents had lived in Oak Ridge for at least 25 years.

The 83 in-depth interviews ranged from one to six hours (longer interviews were spread out over two or three days) and were audio-taped. The nearly 250 hours of taped interviews were transcribed and coded. Additional data were gathered from the Oak Ridge and Knoxville daily newspapers, formerly classified Department of Energy (DOE) documents, State of Tennessee public documents, and various local historical accounts housed in the public library's Oak Ridge Reading Room. I also attended Oak Ridge public meetings and observed behavior in public settings such as the American Museum of Science and Energy and the Oak Ridge Visitors' Bureau.

Background: The Oak Ridge Community

The city of Oak Ridge, Tennessee, was created in 1942 as part of the Manhattan Project, the government's program to build the world's first atomic bomb. The land that now contains Oak Ridge was selected by the military as one of three production sites (along with Hanford, Washington, and Los Alamos, New Mexico). The Army confiscated the land, displacing four small farming and mining communities. As native residents were moved out, barbed wire fences were constructed around the new government site. The population of Oak Ridge grew quickly and by 1945 had reached 75,000 (Johnson and Jackson 1981). Workers were not told about the nature of the military project on which they worked.

On August 6, 1945, an atomic bomb fueled by Oak Ridge enriched uranium was dropped on Hiroshima, Japan. Seventy-one thousand Japanese civilians died. Japan surrendered nine days later. Only at this point were residents of Oak Ridge informed of their role in the war effort. With the city's original mission accomplished, many residents left the area. By the end of the 1940s, the city's population had decreased to approximately 30,000. In 1949, the Army tore down the fences and Oak Ridge became an "open" city. Production activities continued for 50 years in Oak Ridge around the production of nuclear weapons.

Throughout its history, Oak Ridge has been economically dependent on the federal government. In 1960, 84 percent of the city's revenues came from federal aid. Oak Ridge leaders have consistently searched for ways to diversify the local economy by recruiting other industry, but attempts have failed. In the 1980s, the city tried to recruit new high tech industries into Oak Ridge. Officials mapped out a plan for a new Tennessee Technology Corridor to stretch from Oak Ridge to Knoxville. However, the initiative failed and today the community remains largely dependent on the government and its economic activities in Oak Ridge (Shriver 1995; Shriver et al. 2000). The production activities at the Reservation remained unregulated by the government until the 1980s, when environmental violations began to be publicized. In 1983, the DOE acknowledged that it had "lost" 2.4 million pounds of mercury into nearby streams between 1950 and 1977. Large amounts were discharged into East Fork Poplar Creek, which runs through the city of Oak Ridge (Shriver and Cable 1995). Also during the 1980s, four waste water ponds at the Reservation were found to be leaking metal plating wastes, acids, and solvents into the ground water. In 1989, the Oak Ridge Reservation was placed on the Environmental Protection Agency's (EPA) National Priority List. This set in motion a series of activities required by federal Superfund law to clean up wastes and protect human health and the environment (Cable, Shriver, and Hastings 1999).

The Cold War ended in 1989 and, as the country moved away from its emphasis on nuclear buildup, nuclear warhead production in Oak Ridge also ended. Attention shifted from production to waste disposal activities and the dismantling of nuclear warheads. The Reservation's Y-12 plant is now dismantling "secondary," the second stage of warheads that includes parts of highly enriched uranium. Y-12 is also the national clearinghouse for bomb grade uranium and has been referred to as the "Fort Knox of highly enriched uranium." Thus, the end of the Cold War did not lessen the burden on Oak Ridge's environment (Shriver 1995).

In September 1994, Y-12 operations were shut down after federal safety inspectors found numerous counts of noncompliance with safety rules. Subsequent investigations by local inspectors found another 1,300 violations of written procedures. Environmental problems have not been limited to the Reservation's Y-12 plant. For example, the Reservation's X-10 plant has become the national repository for uranium-233. In December 1992, a cesium-137 leak contaminated X-10 workers. In 1993, DOE developed a multi-year plan to spend two million dollars on the cleanup of radioactive fields that had been contaminated with cesium-137 in the 1960s. Also in 1993, 40,000 pounds of uranium were discovered in the pipes of the Reservation's K-25 plant. Cleanup costs are estimated at $7.5 billion. K-25's TSCA incinerator is the only incinerator in the United States licensed to burn mixed wastes with PCBs. It is used to burn such waste from around the country as well as that which is already stored at the Oak Ridge Reservation. In the fiscal year 1994, a record 5.74 million pounds of wastes were burned, exceeding the projected goal of 4.2 million pounds (Shriver 1995). Thus, at the turn of the 21st century, new environmental threats appear in Oak Ridge, along with increasing revelations of past environmental negligence.

Despite their environmental problems, most Oak Ridge residents are proud of their community and the role they played in building the first atomic bomb, the bomb that "ended World War II." According to one resident, "As far as I'm concerned, the Manhattan Project and the first atomic bomb saved hundreds of thousands of lives, if not millions. And it's something to be proud of." For the first 40 years of its existence, Oak Ridge was referred to as "The Atomic City," a message that was displayed on greeting signs at entrances to the community. The town's motto was changed to "The Vision Lives On" several years ago. Many Oak Ridge residents disapproved of changing the city's motto because they felt the community should remain proud of its involvement with the atomic bomb. One respondent expressed his disgruntlement over the name change: "That makes me mad. That was done by some of the new people that were scared it would give them a bad name. No, we like the atomic symbol and we like the name, 'Atomic City.'" ...

Perceptions of Risk Associated with Activism: A Comparison of Two Movements

In this section I examine two attempts at activism in the Oak Ridge community, paying particular attention to residents' subjective interpretations of risks. I begin by discussing peace movement activities in the community and compare these to mobilization efforts around environmental grievances.

Peace Activism in Oak Ridge

In the early 1980s, Oak Ridge residents began organizing around peace activities. The community was a popular site for peace organizing because of its role in the production of the first atomic bomb and its legacy of building nuclear warheads. Oak Ridge residents successfully mobilized around peace issues but were careful *not* to criticize nuclear production activities in their community. . . .

While these activists rallied for peace, they remained sensitive to the community's economic dependence on the government installation. They refused to criticize the Oak Ridge Reservation directly. In a letter to the local newspaper, an activist addressed Y-12 workers, saying: "If you aren't led to stop this kind of work, you might be led to pray as you work that [the nuclear weapons] will never be used." Another activist discussed the community's dependence on weapons production: "We kept raising the issue of, 'what if Oak Ridge has all their eggs in one basket as an industry.' We are strictly a military industry. Isn't this just as dangerous as any other kind of one-industry town? And what happens if peace breaks out? But people just didn't hear us."

Oak Ridge activists formed The Oak Ridge Peace-Making Alliance (ORPAX) in early 1982. Among its founding members were a Unitarian minister, a First Christian minister, and a Roman Catholic nun. Nearly all of ORPAX's members were Oak Ridge natives. Some of the scientists who had come to Oak Ridge for the Manhattan Project joined the organization. ORPAX was concerned with the morality of the arms race. They wanted peaceful production activities to come from Oak Ridge but were also concerned with maintaining the community's economic stability and reputation.

Explaining her participation in ORPAX, one former member said:

> I felt that Oak Ridge, of all places, needed this kind of activism. My son was ten and I don't think adults realize the concern kids have about a cold war reality and arms buildup. I think adults learn how to repress it, but kids come to consciousness in a world with all these weapons aimed at other people and don't know what to do about it. It was really on [my son's] mind. Like at school, they would say, What are you going to do for vacation this summer?" And [my son] would reply, "We're going to the Grand Canyon unless nuclear war breaks out." So I got involved.

The group's primary goal was to foster an alternative consciousness about the morality of nuclear war and the use of nuclear energy. ORPAX established some connections with national peace groups but did not formally affiliate with any because, as one activist explained, "we wanted to maintain that position of autonomy." Members did get involved in some lobbying campaigns and sent a delegation to the state representative to register their concerns about the morality of nuclear war. ORPAX was fully aware of the community's economic dependence on the Reservation and were careful not to be too critical of its activities. An ORPAX activist explained: "Our main concern was with peace issues. We were taking the position that nuclear energy could be used either positively or negatively." The group was careful not to bring in local environmental issues because they saw this as too threatening to the community. The activists were concerned with the reputation of Oak Ridge and with their own social standing in the community.

The community's reaction to ORPAX was mixed. Some residents likened ORPAX members to the pacifists of the 1930s who, they believed, slowed the nation's response to Hitler's Germany. Others residents supported the group's efforts without becoming members. An activist recalls: "I remember going to choose glasses and a woman sitting there had picked up that I was the doctor's wife that was getting in the paper and I remember her saying very snootily, 'Well, I certainly don't agree with your politics.'"

ORPAX planned to hold annual peace observances on the anniversary of the bombing of Hiroshima, Japan. These observances were to consist of prayer services and open discussions about peace issues. In 1983, ORPAX reluctantly sponsored a larger demonstration involving outsiders to commemorate Hiroshima Day. One of the principal organizers of the demonstration recalls that ORPAX members were willing to commemorate the day in some way but did not want to be seen as a part of the "peacenik crowd." More importantly, they did not want the demonstration to include any criticism of their community. Another activist summed up the ORPAX position: "Oak Ridgers saw it as an insider/outsider issue. If you are an Oak Ridger and you come in and talk then fine. But if you are from outside and come in and try to talk about moral purposes then they are going to shut you off." Ground rules were established for the demonstration to protect the community's image. An activist remembered that "there were to be no signs with things like 'merchants of death.' We really wanted to make it a remembrance of the people who died. We weren't there to condemn."

A few outside activists had suggested explicitly connecting Y-12 to Hiroshima but they met staunch resistance. A prime mover in this effort reflected: "I wanted to include a handout on what the plants did but the Oak Ridgers opposed me. I said, 'Listen, the uranium came from Oak Ridge.' They were that much in denial that they wanted that kind of separation of issues. I insisted. I said we didn't have to be judgmental or re-fight World War II but we have just got to say this. People off the streets need to understand Oak Ridge's role."

The Oak Ridgers prevailed and all agreed not to incriminate the Y-12 plant or any part of the Reservation's activities. However, the demonstration did not proceed as planned. An activist explained, "Some outside groups came in from over towards Nashville and they did everything we agreed not to do and people associated it with ORPAX." Another activist added, "They agreed but then they came and they had built a coffin and were marching with a sign saying 'Bury Y-12 before it buries you!' And this was the picture on the news."

After the demonstration, ORPAX was unwilling to sponsor public activities in Oak Ridge. As one former member put it, "ministers were having to do damage control the whole week after." Residents felt that the reputation of the community had been threatened and they were less and less willing to come to the meetings. Oak Ridgers' support for ORPAX declined steadily after the incident until the organization officially disbanded in 1985.

Peace activism in Oak Ridge raises two important issues. First, Oak Ridge remains something of a "closed" community with residents insulating themselves from outsiders. They are particularly intolerant of nonresidents who attempt to criticize the community. Second, Oak Ridgers are willing to organize, even around something as controversial as nuclear weapons production, as long as long as the perceived risks are relatively low. In the case of peace activism, Oak Ridge residents held strong beliefs regarding the importance of peace and the potential dangers of nuclear weapons. Despite their concerns activism was halted after Reservation activities (and thus the community) were implicated. As long as there was a safe distance between peace activism and Reservation activities residents were willing to participate. But when Reservation activities were pulled in and the perceived risks became too high citizen involvement ended. Peace activism in Oak Ridge illustrates the importance of perceived risk associated with activism in determining involvement/continuation in organizing activities.

Environmental Activism in Oak Ridge

The importance of risks associated with activism in determining involvement is even more pronounced in the organizing efforts around local environmental problems. Unlike peace activism, which experienced some degree of success, efforts to organize around local environmental issues failed from the beginning. There has been resistance from individual activists, but there has not been collective mobilization despite various attempts to mobilize the community. The following discussion focuses on the efforts of Save Our Cumberland Mountains (SOCM), a popular regional environmental organization.

SOCM formed in 1972 to work on strip-mining issues in East Tennessee. For many East Tennessee residents, SOCM represented their first exposure to environmental activism. The environmental and social costs of strip mining combined with a democratic leadership style helped SOCM grow rapidly. SOCM has between 100 and 120 members from Oak Ridge; one of the Oak Ridge members designed the SOCM logo. Over the years, Oak Ridge residents have been among SOCM's most supportive members on most of the environmentally based issues. For example, when a landfill was proposed for a local community,

Oak Ridge residents helped lead a protest march on the Cumberland Trail. Oak Ridge members worked to pass a resolution stopping garbage trucks from passing through their community on their way to nearby cities. They have also been among the most generous financial supporters of SOCM's work on environmental issues. Oak Ridge residents are willing to organize around local environmental problems as long as they can attribute the cause of the problem to a source outside the local community. However, when production activities at the Reservation are implicated residents are unwilling to participate.

Despite their support of SOCM, Oak Ridge residents refused to work on environmental issues that targeted the nuclear Reservation. According to one SOCM organizer: "We had always had a base of support among Oak Ridgers in our work against strip mining, but guess what? We did not have that same base of support when we started working on Oak Ridge issues." Another respondent added, "We have never gotten a call from anybody in Oak Ridge who said we are really concerned with what is coming out of this complex, although, SOCM members in Oak Ridge are very supportive of other issues that could be categorized as more environmental type issues."

According to an Oak Ridge resident and former organizer, "Oak Ridgers made it clear that 'no, they did not want to work on local issues.'" Another resident and SOCM member discussed a conversation with one of the other Oak Ridge members about local issues and recalled that the person responded in a shaky voice, "I don't want to hear this. I don't need to know this. I am stuck here. This is where I have to live. We're not moving and this is the end of the conversation."

This shift in interpretive framework is evident in the way Oak Ridgers attribute blame and respond to claims of illness related to the Reservation. Attribution refers to the way individuals interpret events, define causality, and make sense of their own experiences (Cable 1988; Gilbert and Malone 1995). Several residents in Oak Ridge have made claims of environmental illness and have attributed the causes of such illnesses to the production activities at the plants. Some of these individuals have gone public with their concerns and received media attention. Many Oak Ridgers responded either by dismissing the claims or by shifting blame and attributing the causes of illness to internal factors. One Oak Ridge resident dismissed the claims by stating: "If you ate fish [from the local creek contaminated with mercury] everyday for the rest of your life, you probably wouldn't get enough mercury to harm you."

Other Oak Ridge residents simply felt that the claims of illness related to Reservation activities were exaggerated or, as one resident stated, "much ado about nothing." Other residents focused the blame internally, on those making the illness claims. Another resident stated: "I'm a bit inclined to say, 'well, is it because they'd had too many cigarettes? Is it because they've had too much pepper? Is it because they put too much bug killer on their lawns?'"

However, Oak Ridge residents consistently expressed concern for the environment as long as Reservation activities were not involved. According to another Oak Ridge activist who refused to work on local issues: "Obviously the environment needs to be protected. We have some very troubling spots in the country. On the other hand, it would appear that some of the claims are extreme [referring to those activists targeting Reservation activities]." Thus, despite general concern for environmental issues, Oak Ridge residents refuse to incriminate local production activities at the Reservation.

SOCM first tried to organize Oak Ridge residents around their local environmental problems in the early 1980s. At that time, DOE's official position as a government agency was that they were not subject to existing environmental laws because of national security issues. The State of Tennessee fully supported DOE's position. SOCM argued that DOE should disclose relevant information to the public and be held accountable for the environmental problems in Oak Ridge. As one Oak Ridge SOCM member put it, "They shouldn't be poisoning us in secret."

Despite their varied efforts over the past fifteen years, SOCM members have failed to organize Oak Ridge members on Reservation-related issues. There have been individual Oak Ridge SOCM members interested in working on such issues, but they have not been able to build a support among the other members. The general position of Oak Ridge residents has been that SOCM should maintain its focus on strip-mining and other related environmental problems and that it has no business in their community because its members are uninformed on hazardous and nuclear wastes. According to a local activist and SOCM member: "A lot of people really thought that the strip mining that those people did up in the hills was really bad stuff and somebody should make them quit. Somebody ought to make them do it in a way that didn't hurt communities. But those same people did not want us to work on the [environmental] issues in Oak Ridge."

SOCM tried to organize in Scarboro, Oak Ridge's predominantly African American neighborhood. . . . Surrounded by 300-foot ridges and the city dump, Scarboro sits a quarter of a mile from the Reservation's Y-12 plant, the most heavily polluting plant.

SOCM focused on the Scarboro 'neighborhood during the last half of the 1980s after initial attempts to organize the larger community failed. They hoped to recruit new members from Scarboro and establish an Oak Ridge chapter of SOCM. But SOCM's organizing efforts failed once again. According to a SOCM organizer, "I was trying to get people from the Scarboro community active [in SOCM] and they were successfully talking me and everybody white that I brought into the meetings [into joining] the NAACP. But none of them were joining SOCM." Organizing attempts in Scarboro ended in frustration. A SOCM organizer summed up her efforts saying, "The fact is I never even got one ten dollar membership out of anybody."

Oak Ridge residents were among SOCM's most loyal and generous members. Yet, SOCM's attempts to organize around local issues failed miserably in

Oak Ridge. The explanation points to a number of variables related to the risks associated with participation in activism against local reservation activities. I examine the importance of economic dependence and job loss, fear of government retribution, and the potential loss of social standing in the community. I pay particular attention to the subjective interpretations of Oak Ridge residents in determining whether to get involved in local environmental conflicts.

Fear of Losing Jobs and Business

Part of residents' reluctance to get involved in local issues stemmed from their dependence on Reservation activities. . . . This economic dependence was not limited to the workers at the plants, but affected all community residents. As a respondent pointed out: "If Oak Ridge [Reservation] wasn't here, then the restaurants wouldn't be here—nothing would be here. I think that has a paralyzing effect on people. It's their livelihood."

While economic dependence affected the entire community, findings indicate a class difference in how this dependence was interpreted. In other words, the subjective meaning of dependence varied by class. The middle- to upper-middle-class scientific community developed unique ways to manage information regarding environmental problems. One Oak Ridge resident who was trying to organize the community described the middle upper class scientific community in Oak Ridge: "They are mostly intellectuals. Because they are scientists and highly educated, they work out contorted ways in their heads to manage the information rather than give up their home or their community." Another SOCM member explained:

> It's truly a company town. I think they [middle- to upper-class scientists] would be utterly insulted at that because they can't accept that they are as much controlled by a company as a poor blue collar worker in a mill town. But I believe they are. I think it's worse, the way they intellectualize it and distort it. They don't say bluntly like some blue collar worker, "are you kidding? It's my job." They can't be straightforward about it. They have to complicate it.

There were exceptions to this interpretive framework among the middle-class, non-scientific community residents. A local business owner had this to say about the economic risks associated with environmental activism related to Reservation activities. "As a result of that [speaking out at a public meeting], they put my name and my company name in the paper." His business was indirectly related to DOE production activities and he feared financial repercussions from his speaking about environmental concerns.

The interpretive frameworks were different among blue-collar workers in Oak Ridge. They were primarily concerned with job loss and expressed a much clearer understanding of the economic dependence and the risks involved in challenging the Reservation.... An Oak Ridge citizen who was trying to organize residents recalls this response from workers when trying to organize around environmental problems in the community: "There was this question of, 'well, are you going to say anything about Oak Ridge issues?' And these were people who are blue collar workers and were saying to me, 'Hey, we know they do surveillance on people. We know that they figure out who you are talking to and we don't want to lose our jobs.'"

A Reservation employee and activist spoke about his concerns about job loss: "I've thought about it [activism] at work. With all of the layoffs it would be easy to put me on that list to get rid of. I worry about that happening. Nothing would surprise me." Another respondent stated her concerns over discussing activism at work, stating, "I keep quiet about it at work. A few people talk at work, but I try not to routinely discuss it at work." Another stated his concerns bluntly: "DOE [Department of Energy] controls everything in Oak Ridge and everything around Oak Ridge. DOE is definitely the source of power."

Working-class residents in Oak Ridge perceived a direct tradeoff between jobs and a clean environment, what Kais and Grossman (1982) refer to as job blackmail. They were willing to support environmental organizing outside the community, but not when Reservation activities were involved. According to one resident, "The history is that the community is essentially being blackmailed by all the jobs it's gotten from DOE. The tradeoff is, 'you take all this hazardous stuff.'" So residents face a dilemma: They must balance their concerns for the environment with the risks associated with speaking out and losing their jobs—their livelihoods. While middle- and upper-class residents developed interpretive frameworks to manage information and protect their reputations, working-class residents were quick to spell out the dependence. The perceived risks are not limited to economic dependence. Residents also fear government retribution.

Fear of Governmental Retribution

Given the unique history and operations at the Oak Ridge Reservation, the risks associated with organizing around local environmental problems extended beyond the threat of job loss. Reservation employees who discuss environmental problems associated with their work faced the fear of breaking the law, being unpatriotic, or even being considered a traitor. The concerns over such reactions were expressed entirely by working-class Oak Ridge residents. One respondent put it bluntly, "You drive by the Y-12 opening of the plant and they put everybody on video camera." Another respondent who worked closely with a number of former Reservation employees shared this story: "I have talked to regular workers who, when they went to get [security] clearance, had to sign a paper saying everything they are doing falls under the Treason Act. When you sign that paper, you agree that you will be tried under espionage laws and convicted of treason and spend the rest of your life in jail—or be shot." Another respondent shared stories told to him by retired workers. According to the respondent, "When scientists were looking at classified documents, guards would actually stand four feet away with guns cocked." A respondent discussed her mother's fear of discussing anything pertaining to her work at the Reservation: "My mother was so convinced that, if she mentioned anything that was remotely about her job, she would have all kinds of bad things happen to her."

The risks associated with challenging Reservation activities effectively silenced the workers and residents, especially those who worked in blue-collar jobs. Blatant, gun-to-the-head tactics from the early war years have been abandoned, but their legacy remains. Working-class residents who are tied to the Reservation feared they might discuss the wrong thing. As a result, most played it safe by simply refusing to discuss Reservation issues at all. An Oak Ridge activist trying to organize Reservation workers explained: "You will find that when you work with people in Oak Ridge, 'Our brains are not compartmentalized into classified versus non classified.' So, instead of saying, 'I can talk to you up until this certain point,' they would just as soon not talk to you at all."

The fear of governmental retribution brings out the subjective nature of the interpretation of risks associated with activism. As Wiltfang and McAdam (1991) point out, people have subjective expectations based on their own definition of the situation. Regardless of the validity of these fears, residents reacted based on their own subjective interpretations. Thus, various stories regarding governmental retribution were passed on from one generation to the next and became "real" for many residents. Thus, residents' subjective interpretations were important predictors of involvement in such social movement activity.

Fear of Loss of Social Standing

In addition to economic dependence and the fear of government sanctioning, respondents described a third component of risk associated with activism: loss of social standing in the community. Such fears were discussed by both working- and middle- to upper-class residents, but it was particularly acute among the latter. Middle- and upper-class residents expressed more concern over losing respect and social standing in the community. Residents were cognizant of their social standing and most were unwilling to risk their positions in the community by speaking out against local environmental problems. One SOCM organizer put it bluntly: "Many of these people are retired and you say, 'what do they have to lose?' They have social standing to lose!"

Several Oak Ridge residents who spoke out against production-related activities were ostracized from the community. . . . A respondent explained that her reputation changed after she became active in environmental issues related to the Reservation: "I've had this reputation of being a trouble maker, as being an environmentalist." . . .

Others discussed being socially isolated and being treated differently after speaking out against local environmental problems. One respondent stated, "It can get you socially ostracized. There are lots of social repercussions. People treat you like a nut." Another said, "I swear I halfway expect them to take out a cross and hold it in front of them when I come into a room."

Another respondent discussed her treatment by neighbors and friends after she spoke out publicly about local environmental problems. "I was sort of treated like a leper by a lot of people who I thought were my friends. My circle of friends diminished greatly after that." But the respondent understood the residents' reactions: "I can't entirely blame the people that treated me like that, because I think they felt that, by associating with me openly, that they themselves could be tainted and affected—and in fact they could have been."

While middle- and upper-class residents were likely to discuss their social standing within the larger community, working-class residents were more likely to discuss their fears in relation to family members. Thus, an important component of the risks associated with social standing among working-class residents involves perceptions of family members. A respondent discussed the importance of family and of the difficulties faced when speaking out against local environmental problems in Oak Ridge: "For most of us ordinary people, what we have that is important is our family. And so it's a real struggle in families to figure out how they can continue to like and not think badly against this person who they think is talking against their job. That's even a real common phrase you hear a lot, 'They are against my job.'"

Part of the pressure to conform to community standards and not speak out on local environmental issues stemmed from the feeling among friends and neighbors that the community owes DOE. A working-class resident summed up the community's general position: "There is this assumption that we owe DOE because there have been all of these jobs. They don't consider the fact that people worked at ORR. They put in a day's work for a day's pay. In addition to that a lot of people were made sick, probably died. People are willing to stand by that assumption—which Oak Ridge owes DOE."

Organizing around peace activities was successful as long as such efforts did not incriminate the Oak Ridge community. In contrast, attempts to organize around local environmental grievances failed from the beginning. Residents perceived such activism as too risky since the target of their actions would be the Oak Ridge Nuclear Reservation. They feared job and business loss, governmental retribution, and potential loss of social standing in the community.

Conclusions

The case of Oak Ridge, Tennessee, illustrates the importance of the perceptions of both social and physical risks associated with activism. Oak Ridge residents have a history of involvement in civil rights activism, peace activities, and environmental organizing on issues outside their community. But they remained involved in such activities only as long as the nuclear reservation was not targeted. In the case of peace activism, Oak Ridge residents successfully mobilized until reservation activities were implicated. At this point the costs of participating became too great and residents pulled out of the movement.

The importance of perceived risks and threats associated with activism were even more pronounced in the case of mobilization efforts around local environmental problems. Oak Ridge residents were among SOCM's strongest supporters on regional environmental problems related to such issues as strip mining. But those same residents were unwilling to participate when the target became the Oak Ridge Nuclear Reservation. These findings indicate that residents may be willing to live with an unclean environment when the perceived social and physical risks associated with fighting the polluting industry are too high.

The findings indicate that fear of job loss, fear of governmental retribution, and the potential loss of social standing were salient issues that impeded collective mobilization against the local polluting industry. Oak Ridge represents a somewhat unique case, since the polluter is not the typical corporation, but a government facility. While many of the risks associated with activism are similar, important dimensions are added: secrecy, patriotism, and perceived threats to national security. . . .

Note

Shriver, Thomas E. 2000. "Risk and Recruitment: Patterns of Social Movement Mobilization in a Government Town." *Sociological Focus* 33, no. 3: 321–37.

References

Cable, Sherry. 1988. "Attribution Processes and Alienation: A Typology of Worker Responses to Unequal Power Relations." *Political Psychology* 9(1):1988.

Cable, Sherry, Thomas E. Shriver, and Donald W. Hastings. 1999. "The Silenced Majority: Governmental Control on the Oak Ridge Nuclear Reservation." Forthcoming in *Research in Social Problems and Public Policy.*

Edelstein, Michael R. 1988. *Contaminated Communities: The Social and Psychological Impacts of Residential Toxic Exposure.* Boulder, CO: Westview.

Gilbert, Daniel and Patrick Malone. 1995. "The Correspondence Bias." *Psychological Bulletin* 117:21–38.

Goldman, Benjamin A. 1991. *The Truth about Where You Live.* New York: Random House.

Johnson, Charles W., and Charles O. Jackson. 1981. *City Behind a Fence.* Knoxville: University of Tennessee Press.

Kais, Richard, and Richard L. Grossman. 1982. *Fear at Work: Job Blackmail, Labor and the Environment.* New York: Pilgrim Press.

McAdam, Doug. 1986. "Recruitment to High-Risk Activism: The Case of Freedom Summer." *American Journal of Sociology* 92(1): 64-90.

Shriver, Thomas E. 1995. "Social Control and Environmental Degradation by the Military: The Case of Oak Ridge, Tennessee." Doctoral dissertation. University of Tennessee, Knoxville.

Shriver, Thomas E. and Sherry Cable. 1995. "Government Reveals Mercury Releases from Oak Ridge National Laboratory." Pp. 1714–1718 in *Great Events from History II: Ecology and the Environment,* edited by Thomas E. Shriver and Sherry Cable. Pasadena, CA: Salem Press, Inc.

Shriver, Thomas E., Sherry Cable, Nacelle Norris, and Donald Hastings. 2000. "The Role of Collective Identity in Inhibiting Mobilization: Solidarity and Suppression in Oak Ridge." *Sociological Spectrum* 20(1): 41–64.

Wiltfang, Gregory L., and Doug McAdam. 1991. "The Costs and Risks of Social Movement Activism: A Study of Sanctuary Movement Activism." *Social Forces* 69(4):987–1010.

Transnational Protest and the Corporate Planet

The Case of Mitsubishi Corporation versus the Rainforest Action Network

Boris Holzer

Some analysts argue that as part of the process of globalization, transnational corporations (TNCs) have gained power and national governments have lost some of their ability to regulate those corporations. Boris Holzer argues that globalization is also allowing civil society—in this case social movement organizations—to influence corporate actions in new ways. Social movement organizations may turn what were once considered local matters into transnational issues. One way they can do this is by publicizing corporate activities and thereby influencing public opinion. In this reading, Boris Holzer, a sociologist at Ludwig-Maximillian's University in Germany, examines how one social movement group, the Rainforest Action Network, experienced measured success in its creative challenges to certain practices of the Mitsubishi corporation.

M y argument in this [essay] will be that the impact of globalization has given rise to a new form of confrontation between corporate power and civil society. The integration of the world economy has increased the power of global markets vis-à-vis state governments.[1] In this process the main agents of economic globalization, transnational corporations (TNCs), appear to have gained enormous power over global economic affairs. This has led to "the transfer of some powers in relation to civil society from territorial states to non-territorial TNCs." Although corporations certainly do not "rule the world" and are unlikely to do so in the future, their influence has increased considerably. Nation-states have not only lost part of their capacity to control those companies but they are also increasingly unwilling to impose regulations for the fear of losing investments and tax revenues. Yet this does not mean that corporations now enjoy almost unfettered power. Operating on the same transnational level, social movements have developed into a major challenge to corporate power.

Against this backdrop, this paper will examine the conflict between the Mitsubishi Corporation and environmental groups over Mitsubishi's involvement in rainforest logging and timber trade. By way of this case study I seek to illustrate some characteristics of the relationship between TNCs and the forces of an emerging "global civil society." In particular, I want to examine the function of what has been called a "transnational social movement sector" for the observation and regulation of globalized business practice. The regulation of large business enterprises has recently come to be seen as problematic regarding the capacities of TNCs to carry out their activities across the globe, capitalizing on the differences between national legal systems and on their own economic power. While state governments are seen as losing power, protest groups have been partially successful in challenging business practice and enforcing their standards on some corporations. The conflict between Mitsubishi and environmental groups provides an interesting example of the possibilities and dilemmas of resistance against corporate power and environmental degradation.

The two main areas of conflict between TNCs and social movements are human rights and environmental issues. Both fields feature a wealth of social movement organizations, with Amnesty International and Greenpeace being only the most prominent ones. Against the backdrop of an impending "corporate planet," the forces of civil society are thought to provide a necessary counter-balance. They are deemed capable of representing and defending the public interest against the narrow economic interest of TNCs. Civil society actors become more important at a time when the capacity of state governments to impose regulations on industry appears to be increasingly limited by the "golden straightjacket" of a global neoliberal economic regime. State governments are anxious not to chase away the volatile sources of investment and are thus unlikely to challenge corporations. Social movements in contrast are free to pursue more confrontational strategies since they are "outside and beyond the representative institutions of the political system of nation-states."

Global environmental problems make these new constraints on state politics apparent. The boundary-crossing, often global nature of environmental problems such as global warming or pollution makes it difficult to mould solutions on the established institutional mechanisms of (inter-) state politics. Although the recognition of an environmental crisis has led to a wide variety of intergovernmental activities, meetings and organizations there is an understanding that the reaction of the international political system tends to be slow and insufficient. This view is also held by most pressure groups, which seek to influence global environmental policy. It has been suggested that an emerging "global civil society" could play a major role in tackling environmental problems on a transnational level. The expansion of existing groups as well as more extensive networking among them has given rise to a "transnational social movement sector." It comprises a plethora of environmental groups operating on a transnational level or within a transnational network, including organizations such as Greenpeace and Friends of the Earth as well as grassroots movements and lobbying groups.

Amongst other things, those groups aim to control the operations of TNCs, some of which are conceived to be among the most environmentally destructive forces on the planet. In so doing they seek to mobilize support in many different locations—and more often than not they challenge companies in places far away from the actual sites of destruction. Thus, they transform local environmental conflicts into transnational issues, which may appear on the political agenda of countries far away. With regard to their impact on policy formation these groups should not be underestimated. In the field of global environmental problems transnational social movements have been able to establish themselves as "primary definers." Environmental organizations have achieved a position of credibility and expertise in the field of environmental knowledge which enables them to oppose controversial business and government projects with considerable success. Although they are not always able to achieve their immediate objectives, transnational protest groups have grown into a serious challenge for the TNCs concerned. The case of Mitsubishi will illustrate some of the mechanisms behind that challenge.

The conflict between the Mitsubishi Group and rainforest protection movements is an excellent example of the characteristics of transnational protest action. It pits one of the largest transnational corporations against a small but determined environmental group drawing on global contacts and considerable expertise in public relations. The dispute between the two sides revolves around Mitsubishi's involvement in rainforest destruction on a global scale, which includes logging operations and timber imports from the Philippines, Malaysia, Papua New Guinea, Bolivia, Indonesia, Brazil, Chile, Canada, Siberia and the United States. Among these, the operations in Sarawak (Malaysia), where Mitsubishi has a 90,000-hectare concession, have had a particularly devastating effect on an already endangered local ecosystem.

Not only for its involvement in rainforest logging, Mitsubishi has long been the target of criticism. Many of the global and diverse operations of the companies of the Mitsubishi Group bear some environmental impact, and thus it has been argued that "Mitsubishi Group may well be the single most environmentally destructive corporate force on Earth." The Mitsubishi Corporation alone, the group's general trading company (*sogo shosha*), was ranked as the seventh largest corporation on the planet in the *Fortune Magazine*'s Global 500 index of the year 2000. Mitsubishi Corporation, which does not produce consumer goods itself, is at the core of a group of 160 linked but legally independent corporations, which form a common business network or *keiretsu*. Other companies of the Mitsubishi Group like the Bank of Tokyo, Asahi Glass, Mitsubishi Electric and Mitsubishi Motors are major transnationals, too, and rank among the largest companies in their respective business sectors. In sum, the more than 160 corporations that make up the Mitsubishi Group employ over half a million people, and hold shares in approximately 1400 Japanese companies, with another 700 companies being dependent on them. Accordingly, the business activities of the group are manifold—and so is its potential for environmentally harmful activities.

Yet Mitsubishi is one among a range of TNCs, which seek to portray themselves as "responsible corporate citizens." Individual companies of the group have introduced various environmental programmes, and Mitsubishi Corporation claims "corporate responsibility to society" to be one of its primary corporate principles. In 1992, largely owing to increasing pressure by environmentalists, Mitsubishi Corporation created an Environmental Affairs Department, which has conducted a number of research projects on reforestation. Despite these efforts the operations of Mitsubishi companies remain a thorn in the side of environmental groups. They regard Mitsubishi's environmental programmes as a form of public relations, rather than as a form of corporate transformation. Pointing out the gap between advertising campaigns and reality, they can only see another case of "corporate greenwash." For instance, a perceived mismatch between advertising and business practice in Mexico recently earned Mitsubishi and one of its subsidiaries the doubtful honour of receiving the "Greenwash Award" of the Corporate Watch group.

Given the diversity of its business operations and the high visibility of its brand name, Mitsubishi is prone to be a primary target of environmentalist campaigning. It is easy to see that the various parts of Mitsubishi have some responsibility for rainforest destruction, even if the management of Mitsubishi Corporation denies direct participation in unsustainable logging. Indeed, Mitsubishi often conducts logging operations through affiliates of which it does not necessarily have a majority interest. For instance, the logging operations in Sarawak (Malaysia) were largely carried out by Daiya Malaysia, a company in which Mitsubishi had held a minority stake until it sold its shares in the mid-1990s. Yet, as the activists of Rainforest Action Network (RAN) have argued, Mitsubishi does not only contribute to forest destruction through indirect participation and co-operation but also through the activities of many members of the Mitsubishi Group. For instance, Mitsubishi's Bank of Tokyo provides financing for destructive projects, Mitsubishi Paper Mills uses tropical timber for pulp manufacturing and other Mitsubishi companies engage in road building projects through rainforest areas. Furthermore, the accumulation of "indirect" participation and the acquisition of timber from other companies make Mitsubishi one of the largest importers of timber in the world. A real change of its environmental policy might thus be an important step towards rainforest protection.

The long-standing efforts of various rainforest protection groups to achieve this were taken up and intensified by the U.S.-based Rainforest Action Network (RAN) in 1990. The group called for a boycott of Mitsubishi Corporation and various other Mitsubishi companies, including Mitsubishi Motors, Kirin Beer and Nikon Cameras. The boycott campaign was supported by protest actions at the headquarters of Mitsubishi companies, at auto shows and dealerships, and by disruptive banners at events sponsored by Mitsubishi. Furthermore, RAN launched a national advertising campaign in the United States including a full-page ad in the *New York Times*. A door-to-door canvas campaign in 1993 yielded over 400,000 petition signatures against Mitsubishi's activities. All these campaigns focused on the American public, seeking to achieve international press coverage as a side effect. In order to challenge Mitsubishi more directly, RAN asked people to send telegrams and faxes to the headquarters of Mitsubishi Corporation. The activists also had some success in lobbying local governments. For instance, Mitsubishi lost a US$137 million contract to carry out construction work at San Francisco's airport after heavy lobbying by RAN (Dow Jones News Database 11 February 1998). Over the years the Boycott Mitsubishi Campaign developed into one of the best-known consumer boycotts in the United States.

Furthermore, the Internet provided a global platform for RAN's protest actions. Both parties feature well-developed web sites. The hyperlink structure of the World Wide Web provided an opportunity for RAN to challenge Mitsubishi's claims more directly. Thus RAN designed part of their campaign web site, located at www.ran.org, as a direct rebuttal of Mitsubishi's material. Visitors of the web site were able to check RAN's claims against those of Mitsubishi by simply clicking on a hyperlink reference. Furthermore, RAN's web site directed readers to the "comment form" on the Mitsubishi site, thus encouraging them to send electronic messages to them. This finally led to Mitsubishi's disabling the "comment" option—obviously because of the inflow of critical messages. During the whole of the campaign, the medium of the Internet proved to be a useful device for the activists. In contrast to traditional media, where activists either get little attention or have to buy expensive advertising space, the WWW enables them to articulate their views in appropriate length and to confront their opponents on a "level playing field."

The continuous and multi-faceted pressure on Mitsubishi had some impact. On 27 February 1996, Minoru Makihara, President of Mitsubishi Corporation, met with RAN activists to discuss the issue. Yet the actual policy of Mitsubishi remained unaltered. Only two American subsidiaries—Mitsubishi Motor Sales America (MMSA) and Mitsubishi Electric America (MEA)—went a step further in co-operating with the environmentalists. They asked Bill Shireman, president of the Global Futures Foundation (Sacramento, CA), to act as a mediator between the opponents. In several meetings they worked out an agreement that comprised a renewed commitment by MMSA and MEA to environmental initiatives in exchange for terminating the boycott campaign. As part of a top-to-bottom environmental review, both companies pledged to phase out the use of tree-based paper and packaging products by the year 2002; to provide funding for a "Forest Community Support Program"; and to set up a system of ecological accounting. Furthermore, they promised to use their influence on other Mitsubishi companies to start similar programmes. RAN in turn agreed to end the boycott against the two companies but intended to continue it against other parts of Mitsubishi as long as necessary.

This agreement was welcomed by business and movement representatives alike. Consumer activist Ralph Nader called it a "testament to the efficacy of consumer boycotts," and Greenpeace International saw it as a "breakthrough in the field of corporate responsibility." The business side announced it as "the most significant program of environmental initiatives in quite some time." But the enthusiasm among environmentalists was not unanimous. The main concern was that the isolated agreement with two American companies might actually decrease the pressure on Mitsubishi International and other parts of the Group. And indeed, one of the first results of the agreement was a full-page ad in the *New York Times*, signed by both RAN and Mitsubishi Electric, which featured the Mitsubishi logo and did not mention the continuation of the boycott against other members of the Mitsubishi Group. On a superficial reading this must have given the impression that RAN and Mitsubishi as a whole had achieved an agreement. Furthermore, the American subsidiaries are probably two companies with only minor involvement in timber activities. Their positive approach is basically in line with their long-standing efforts to improve their environmental records. Thus, the agreement did not necessarily hit the most destructive parts of the Mitsubishi Group. Still it stands as an example of how conflicts between environmental groups can be resolved with winners on both sides—even if one side may win a bit more than the other.

From the perspective of environmental sociology and transnational studies, the case of Mitsubishi vs. RAN provides some interesting insights. It offers enough parallels to other disputes between business and consumer organizations or protest groups to warrant a more general interpretation. Similar cases include the long-standing boycott of Nestlé for their marketing of infant formula milk products in the Third World, consumer boycotts against Shell for their engagement in Nigeria and their controversial disposal plans for a North Sea oil rig (Tsoukas 1999), and several boycott campaigns against the fast food giants Burger King and McDonald's for the environmental problems associated with the production and consumption of their products. Against this background, the case of Mitsubishi demonstrates the development of environmental protest and resistance towards transnational coordination and towards more sophisticated forms of protest action. In the following I will interpret the Mitsubishi case with regard to three aspects: the phenomenon of environmental protest "out of place" based on a transnational environmental "etiquette," the concomitant power shift in the relationship between civil society and the nation-state, and the specific opportunities and difficulties protest groups are facing when they challenge TNCs.

The first point concerns the nature of the conflict between Mitsubishi and RAN as a form of transnational, boundary-crossing protest, which is gaining importance in the field of environmental problems. It is a case of *protest "out of place."* Rather than at the actual sites of the criticized business practice the decisive aspects and actions of the conflict took place in distant places. The direct protest actions of RAN and related groups focused on American Mitsubishi companies such as the Union Bank of California and Mitsubishi Motor Sales of America. They were the primary targets of boycott campaigns and protest actions, although their involvement in the criticized logging practices was minimal. Thus, the conflict provides an example of a transnational, de-localized struggle over a local "resource regime."

In this respect, protest groups can capitalize on the fact that national regulations differ considerably and are often considered insufficient to deal with environmental problems of global relevance. The globalization of communication systems has exacerbated this problem. Activities in one locale are now scrutinized by a transnational public, which necessarily comprises various value systems. TNCs have come to realize that this may lead to problems for the implementation of business decisions if we believe the words of Phil Watts of Shell International: "Conflicting demands—the fact that what is acceptable and expected in one country is not acceptable in another, or at the international level—pose major dilemmas for business." Compared with the institutional framework provided by nation-states, transnational corporations experience the normative structure of world society as fragmented and contradictory. TNCs are well equipped to deal with the demands of economic decisions, based on consistent legal regulations within nation-states. Yet they seem to have considerable difficulty meeting different, and sometimes even contradictory, demands across different cultures and value systems. This is partly because they are used to relying on what could be called an "expert culture," i.e. the possibility to establish the validity of claims in a given field of knowledge unambiguously. In a limited domain such as economic rationality this may be possible to some extent. However, if the boundaries between, for instance, economic and ecological considerations are ignored or politicized this is bound to fail. TNCs will increasingly have to deal with the problem that "politics and morality are gaining priority over expert reasoning."

As a second point it is important to note that the increasing significance of politics refers to a very specific kind of politics. The politics of transnational protest is essentially informal politics between "non-state authorities." The relevance of transnational environmental protest for local business decisions bears implications for the role of nation-states in the regulation of business practice. It seems that the conditions of possibility for the "acceptance" of business practice are undergoing a decisive change. In a system of positive law the acceptability of decisions should generally be a matter of unambiguous judgement, no matter where and when. For instance, the criticized logging operations of Mitsubishi and its partners were in accordance with local legal regulations. Yet the dispute developed quite independently of any legal considerations. Since their claims appeal to morality rather than legality, environmental groups are able to challenge or ignore legal issues if these are thought to contradict the demands of environmental protection.

Within a nation-state context, the principles of sovereignty and democratic decision-making provide for a certain congruence of *legality* and *legitimacy*. It is accepted that the democratic legislative procedure confers not only legality but also legitimacy to positive law. As for instance Immanuel Kant argued, the principle of self-government of the people in a democratic state ensures the legitimacy of legislative acts: no one would willingly do himself an injustice, and since democratic government means that the people govern themselves, democracy makes the illegitimate use of legislative power impossible. Yet on a transnational level the legitimacy of decisions is more difficult to achieve. The problem arises from the fact that legal norms, even if they have been agreed upon in a democratic procedure, are principally confined to a national context. Groups in other places which may consider themselves affected may beg to differ as to how legitimate a certain norm is for them. Thus transnational corporations are confronted with a situation where the *legitimacy* of their operations may be challenged no matter what their *legal* status is. At least concerning the acceptance of business operations, possible reactions "out of place" have to be taken into account and their handling cannot be delegated to state authorities.

The conflict between Mitsubishi and RAN exemplifies how global civil society actors can bypass state regulation and exert pressure on corporations. Despite unambiguous local legal regulations, a form of "non-state authority" was able to challenge the legitimacy of the Mitsubishi's operations. As for the efficiency of state regulation, the activists themselves questioned state governments' abilities to deal with TNCs. Following a popular line of argument in contemporary thinking about globalization, RAN's president Randy Hayes asserted: "Nation states are losing power and transnationals are gaining." Thus, social movement organizations had to step in and put pressure on the TNCs. State agencies, then, are rendered marginal in the course of conflict resolution. In the context of the Mitsubishi case the perceived "public opinion" of a transnational media audience and the co-ordinated consumer activism proved more important than legal aspects. In the course of the settlement of this dispute the protest group even assumed a role commonly associated with state governments: It was agreed that RAN would monitor the progress of the Mitsubishi companies in implementing measures of environmental protection—based on a formal agreement between the two sides. RAN thus was able to challenge the legitimacy of Mitsubishi's activities whilst almost assuming the role of a legislative body as well. They became the definers and observers of an environmental "etiquette" beyond positive law. The fact that companies have to take into account obligations and restrictions beyond legal regulation is nothing new. There have always been factors of "common decency" which are thought to be essential features of business. Yet the fact that a non-elected, civil society organization is to supervise the adherence to certain standards is remarkable in a state system governed by positive law.

As a third aspect, the history and settlement of the conflict raises the question of how large and powerful companies like Mitsubishi can be made "publicly accountable" by relatively small social movement organizations. The fact that TNCs are vulnerable to environmental protest appears to be quite remarkable regarding their economic power and accumulated expertise. The case of Mitsubishi demonstrates that one key feature of anti-corporate transnational protest is to attack a global company's most valuable asset: its brand and public image. Activists have not failed to notice that the endeavors of TNCs to build a "corporate image" make them susceptible to public pressure: "We know now how to be an effective thorn in the sides of a transnational. A company like Mitsubishi has its vulnerabilities: their public image, their logo."

Because of their (almost) global presence, TNCs are likely to be easy targets on the grounds of their public image. They often assume top positions in individual markets and seek to foster their public image and reputation through extensive as well as expensive advertising campaigns. Therefore, they usually are exemplary cases of what has been called "public exposure." The idea of *public exposure* refers to the impact of business decisions on a wider public, i.e. to the fact that enterprises touch on public interests and, vice versa, they are themselves afflicted by measures taken in the name of public interest. Especially in areas bearing a potential environmental impact such as the chemical industry, the public has grown wary of side effects and long-term consequences of decisions. Therefore, the traditional concept of business decisions as essentially *private* decisions made by or on behalf of the owners of a company, which liberals such as Milton Friedman adamantly advocate, no longer holds. Rather, these decisions are increasingly becoming *public* in nature due to their alleged impact on other people. The larger the company, the more likely it is to have such an impact. "The price of successful economic growth for a company is that it gains increased public visibility. It is thus more subject to public scrutiny and public criticism than a small company" (Willetts 1998: 225). Ultimately, all companies depend on the public acceptance of their actions and thus have to position themselves in relation to public perceptions, standards and etiquette.

It has been argued that the ensuing scrutiny and distrust of business practice is transforming large corporations into "quasi-public institutions." To some extent this public side of business decisions is reflected in the legal regulation of business practice. Yet, as the Mitsubishi-RAN and other cases demonstrate, such regulation is never exhaustive. Definitions of what is acceptable and what is not are always contested and vary from place to place. Thus TNCs cannot always benefit from the legitimacy which legal regulation should bestow on their decisions. If they are to keep and improve their public image, they have to pay attention to the interests of many different groups or "stakeholders." Due to past experience, TNCs in sensitive areas are willing to do so. Yet their efforts may always be deemed insufficient by critical observers. In this case TNCs have no ultimate authority to appeal to. They will have to deal with the criticism as actors in the public sphere, either defending themselves or seeking a compromise with their opponents.

To sum up, the conflict between Mitsubishi and RAN shows several factors which enable a determined environmental group to put leverage on a major TNC—and even on one of the largest corporations on the planet. First, it is the ability of transnational protest groups to mobilize the support of activists and consumers in distant places which increases their relative power. Second, they can successfully base their claims on the mismatch between the regulatory reach of nation-states and the boundary-crossing nature of environmental problems. Third, the "public exposure" of TNCs creates new liabilities for them and makes them more susceptible to publicly voiced criticism.

Yet the success of consumer activism does not necessarily provide an optimum solution. The fact that RAN achieved an agreement with two American subsidiaries did not satisfy all activists concerned with the campaign. Although RAN pledged to continue the campaign against other parts of the Mitsubishi Group, this may prove to be a difficult undertaking. For the very same reason that RAN was able to exert pressure on Mitsubishi in the first place, it might now face problems. The main asset of RAN was the possibility to blame the *whole* of Mitsubishi for the logging operations, thus inflicting damage on the Group's public image. This fact could now turn against the activists since it will be difficult to convey the message that the agreement applies only to isolated parts of the Mitsubishi Group. Therefore, the success of RAN may have its price regarding the overall aims of the campaign. In particular, it is not entirely clear how the local sites of Mitsubishi's criticized practices will benefit from the agreements.[2] This shows that transnational protest does not necessarily result in desirable outcomes at a local level.

From the perspective of TNCs, the challenge of transnational protest remains a difficult one. The specific problem for large enterprises and TNCs lies in the fact that they have to deal with different standards of acceptance under different socio-legal regimes. What is *legal* in one locale may well be regarded as *illegitimate* somewhere else. The way in which the Mitsubishi companies dealt with this problem could be a model for future conflicts. On the one hand, they succumbed to the pressure and two Mitsubishi companies made a remarkable commitment to environmental principles, which could persuade other companies to do the same. On the other hand, the sheer diversity of the operations of a TNC like Mitsubishi makes it difficult to envisage this as the first step to a comprehensive corporate transformation.

The Mitsubishi case and similar confrontations between protest groups and corporations have shown that the social movement "Davids" can be successful in challenging the TNC "Goliaths." Yet this metaphor could lead one to overestimate the effects of specific, single-issue campaigns. Social movements might have to recognize that their opponents actually bear more resemblance to the many-headed Hydra than to Goliath. The latter could be felled by throwing a single stone. The fight against Hydra, however, was a more protracted task. On the part of TNCs, succumbing to public pressure in one area may serve to divert attention from other, less obvious misdemeanours. There is no need to assume that corporations will use this as a deliberate, carefully worked-out strategy. But an attentive civil society will have to keep an eye on such tendencies.

Notes

Holzer, Boris. 2001. "Transnational Protest and the Corporate Planet: The Case of Mitsubishi Corporation vs. the Rainforest Action Network." *Asian Journal of Social Science* 29, no. 1: 73–86.

1. This is not to say that we are witnessing the "end of the nation-state." Nation-states will remain important since only they can provide the political and legal framework for global markets. In implementing those policies, however, they have to abide by the new rules of the game. A balanced discussion of this controversial topic is given by Peter Dicken (1998: Chapters 3 and 8).

2. In fact, local activists might not even know of the agreement at all. This was apparently the case in Malaysia, one of the areas of the disputed logging operations, where an activist of the local Friends of the Earth group (Sahabat Alam Malaysia, Penang) told me in June 1999 that they had been unaware of the precise content of the agreement.

References

Anderson, Alice
   1993 "Source-Media Relations: The Production of the Environmental Agenda," in Anders Hansen (ed.), *The Mass Media and Environmental Issues*. Leicester/London/New York: Leicester University Press, pp. 51–68.
Beck, Ulrich
   1996 "World Risk Society as Cosmopolitan Society? Ecological Questions in a Framework of Manufactured Uncertainties." *Theory, Culture & Society*, 13: 1–32.
Bowie, Norman
   1991 "New Directions in Corporate Social Responsibility." *Business Horizons*, 34: 56–65.
Brunsson, Nils
   1989 *The Organization of Hypocrisy. Talk, Decisions and Actions in Organizations*. Chichester: John Wiley & Sons.
Chin, Christine B. N., and James H. Mittelman
   1997 "Conceptualising Resistance to Globalisation." *New Political Economy*, 2: 25–37.
Corporate Watch
   1999 "Greenwash Award of the Month: Mitsubishi." Available from: http://www.corpwatch.org/trac/greenwash/mitsubishi.html [as of 27 April 1999].
Dicken, Peter
   1998 *Global Shift. Transforming the World Economy*. London: Paul Chapman Publishing [3rd ed.].
Dyllick, Thomas
   1989 "Politische Legitimität, moralische Autorität und wirtschaftliche Effizienz als externe Lenkungssysteme der Unternehmung," in Karl Sandner (ed.), *Politische Prozesse in Unternehmen*. Berlin: Springer, pp. 205–230.
Eyerman, R., and A. Jamison
   1989 "Environmental Knowledge as an Organisational Weapon: The Case of Greenpeace." *Social Science Information*, 28: 99–119.
Freeman, R. E.
   1984 *Strategic Management: A Stakeholder Approach*. Marshfield, MA: Putnam.
Friedman, Milton.
   1970 "The Social Responsibility of Business Is to Increase Its Profits." *New York Times Magazine*, 13 September 1970, pp. 32–33 and 122–126.
Friedman, Thomas
   1999 *The Lexus and the Olive Tree*. London: HarperCollins. Gerber, Jurg
   1990 "Enforced Self-Regulation in the Infant Formula Industry: A Radical Extension of an 'Impractical' Proposal." *Social Justice*, 17: 98–112.
Greer, Jed, and Kenny Bruno
   1996 *Greenwash: The Reality Behind Corporate Environmentalism*. Penang/New York: Third World Network/Apex Press.
Haas, Peter, Robert O. Keohane, and Marc A. Levy (eds.)
   1993 *Institutions for the Earth. Sources of Effective International Environmental Protection*. Cambridge/London: MIT Press.

Habermas, Jürgen
  1989 "Volkssouveränität als Verfahren. Ein normativer Begriff von Öffentlichkeit." *Merkur*, 43: 465–477.
Hayes, Randy
  1997 "Randy Hayes—The Rainforest Action Network's Founder Targets Big Timber (Interview)." *E/The Environmental Magazine*, July–August 1997.
Karliner, Joshua
  1997 *The Corporate Planet. Ecology and Politics in the Age of Globalization.* San Francisco: Sierra Club Books.
Korten, David C.
  1995 *When Corporations Rule the World.* West Hartford, Conn.: Kumarian Press.
Lipschutz, Ronnie D.
  1992 "Restructuring World Politics: the Emergence of Global Civil Society." *Millennium*, 21: 389–420.
  1996 *Global Civil Society and Global Environmental Governance: The Politics of Nature from Place to Planet.* Albany, NY: State University of New York Press.
Memorandum
  1997 "Memorandum of Understanding": MMSA/MEA/RAN (mimeo).
Mitsubishi Public Affairs Committee
  1989 *A Brief History of Mitsubishi.* Tokyo: Mitsubishi Shoji
Ohmae, Kenichi
  1990 *The Borderless World.* New York: HarperCollins
  1995 *The End of the Nation State.* London: HarperCollins
Press Release
  1998 "Nader, Greenpeace Respond to Agreement between Rainforest Action Network, Mitsubishi Motors and Mitsubishi Electric." Washington, DC: Fenton Communications.
Rainforest Action Network
  1998 "Boycott Mitsubishi: Questions and Answers." Available from: http://www.ran.org/ran_campaigns/mitsubishi/questions.html [as of 24 April 1999].
Rheingold, Howard
  1999 "Online Activism: Rainforest Action Network Takes on Mitsubishi." Available from: http://www.ecotopia.be/ecotop/pubs/home/mitsubishi.html [as of 24 April 1999].
Ritzer, George
  1996 *The McDonaldization of Society.* Thousand Oaks: Pine Forge Press.
Shireman, Bill
  1998 "Communication Key to Bringing CEOs and Eco-Radicals Together." Global Futures Foundation. Available from: http://www.globalff.org [as of 24 April 1999]
Sklair, Leslie
  1994 "Global Sociology and Global Environmental Change," in Michael Redclift and Ted Benton (eds.), *Social Theory and the Global Environment.* London/New York: Routledge, pp. 205–227.
  1995 *Sociology of the Global System.* London: Prentice Hall/HarvesterWheatsheaf [2nd ed.].

1998   "Social Movements and Global Capitalism," in Fredric Jameson and Masao Miyoshi (eds.), *The Cultures of Globalization*. Durham/London: Duke University Press, pp. 291–311.

Smith, Jackie
1997   "Characteristics of the Modern Transnational Social Movement Sector," in Jackie G. Smith, Charles Chatfield, and Ron Pagnucco (eds.), *Transnational Social Movements and Global Politics: Solidarity beyond the State*. Syracuse, N.Y.: Syracuse University Press, pp. 42–58.

Sternberg, Elaine
1994   *Just Business. Business Ethics in Action*. London: Little Brown and Company.

Stesser, Stan
1991   "A Reporter at Large: Logging the Rainforest." *The New Yorker*, 27 May 1991, p. 48.

Strange, Susan
1996   *The Retreat of the State. The Diffusion of Power in the World Economy*. Cambridge: Cambridge University Press.

Tsoukas, Haridimos
1999   "David and Goliath in the Risk Society: Making Sense of the Conflict between Shell and Greenpeace in the North Sea." *Organization*, 6: 499–528.

Ulrich, Peter
1977   *Die Großunternehmung als quasi-öffentliche Institution: Eine politische Theorie der Unternehmung*. Stuttgart: Poeschel.

Wapner, Paul
1996   *Environmental Activism and World Civic Politics*. Albany, NY: State University of New York Press.

Watts, Philip
1998   "The International Petroleum Industry: Economic Actor or Social Activist?" in John V. Mitchell (ed.), *Companies in a World of Conflict: NGOs, Sanctions and Corporate Responsibility*. London: Royal Institute of International Affairs/ Earthscan, pp. 23–31.

Willetts, Peter
1998   "Political Globalization and the Impact of NGOs upon Transnational Companies," in John V. Mitchell (ed.), *Companies in a World of Conflict: NGOs, Sanctions and Corporate Responsibility*. London: Royal Institute of International Affairs/Earthscan, pp. 195–226.

Yearley, Steven
1995   "The Transnational Politics of the Environment," in James Anderson, Chris Brook, and Allan Cochrane (eds.), *A Global World? Re-ordering Political Space*. Oxford: Oxford University Press/The Open University, pp. 209–247.
1996   *Sociology, Environmentalism, Globalization. Reinventing the Globe*. London: Sage.

Coalition Building between Native American and Environmental Organizations in Opposition to Development

The Case of the New Los Padres Dam Project

Mik Moore

Social change occurs constantly in a number of ways. Inventions, discoveries, and the sharing of information between different communities all lead to intended and unintended social changes. Sometimes groups of people set out to intentionally bring about a predetermined social change—these groups are called "social movement organizations." In this essay, Mik Moore draws on field observations, document reviews, and interviews to explore a coalition that formed between Native American and environmental groups to resist the New Los Padres Dam project, planned for the Carmel River, California. Moore demonstrates how change occurs when different cultures share information and how social movements can influence major policy decisions. First, he shows how traditional Native American philosophy (the indigenist vision) influenced the belief systems, rhetoric, and practices of the environmental groups. Second, he explains how both Native American and environmental activists worked together to successfully educate the public and protest the proposed New Los Padres Dam.

In recent decades, Native American and environmental organizations have found considerable common ground in the shared goal of preserving wild and relatively wild areas from environmentally destructive development. In his study of native and environmentalist struggles against multinational corporations, Al Gedicks (1993) states:

> The integral connections between native survival and environmental protection have become apparent to even the most conservative environmental organizations. Now the assertion of native land rights takes place in the context of an environmental movement that is prepared to appreciate the knowledge native people have about their own environment and to accept native leadership in environmental battles. (p. 203)

Gedicks documents a number of coalitions between indigenous people's organizations and environmental groups, such as the struggle by Lake Superior Chippewa Indians and Wisconsin environmentalists to prevent Kennecott Copper Corporation from constructing a massive open-pit copper mine in Northern Wisconsin (Gedicks, 1993). More recently, Native American organizations and environmental groups are working closely together to protect Ward Valley, California, from becoming the site of a low level nuclear waste dump. Ward Valley is considered a sacred place by Native Americans, and the Colorado River Native Nations Alliance has been working together with non-native support groups and environmental justice organizations, such as the Indigenous Environmental Network, the Southwest Network for Economic and Environmental Justice, and the California Communities Against Toxics coalition, to prevent the opening of the radioactive dump there (Woodward, 1998). The coming together of Native American and environmental movements is also illustrated by the fact that the Sierra Club's (1996) magazine recently devoted an entire issue to Native Americans and environmental movements. . . .

In trying to describe and understand social movements, sociologists have drawn on the resource mobilization perspective, which focuses on the ability of organizations to make use of resources such as money, the availability of office space and communication equipment, access to professional organizing, administrative and legal expertise, political and media connections, and other assets as the most significant factor in explaining the success of a social movement (Gamson, 1975; McCarthy & Zald, 1977; Tilly, 1979). Recently, the resource mobilization perspective has been criticized by authors drawing on European new social movement theory (see Habermas, 1981; Touraine, 1981, 1988) for failing to give enough attention to the role of ideas, beliefs, and counter-discursive language and behavior (Ingalsbee, 1996; McClurg-Mueller, 1992). Carol McClurg-Mueller (1992) has argued that the resource mobilization approach focused on institutional change at the expense of personal change and that to understand social movements, it is also necessary to understand the beliefs and meanings that are created and interpreted by individuals within the micromobilization context. In his study of Earth First! activism, Timothy Ingalsbee (1996) has illustrated that the character of this environmental movement cannot be captured by simply attending to the mobilization of monetary and material resources; symbolic resources, meaning "socially constructed cognitive frameworks" (p. 264), or beliefs and ideas that provide an explanatory vision that guides action and that are represented and communicated through counter-discursive symbolic means such as costume and theatrical direct actions, are also important. "The social-psychological aspects of movements are among the most sociologically interesting and qualitatively new elements of contemporary activism, particularly in radical movements such as Earth First!" (Ingalsbee, 1996, p. 264). Where do these symbolic meaning resources and ideational materials (Tarrow, 1992), which are so important for creating commitment and motivation among social movement activists, derive from, and how are they communicated and actualized within the micromobilization context, the face-to-face interactions between movement activists?

One source of symbolic resources for environmental movements today is the *indigenist* vision (Churchill, 1993, p. 441), the traditional belief systems and practices of the native peoples of America. In his article, "I am Indigenist," Native American writer and activist Ward Churchill (1993) defines himself as indigenist in outlook, meaning that he

> draws upon the traditions—the bodies of knowledge and corresponding codes of values—evolved over many thousands of years by native peoples the world over. This is the basis upon which I not only advance critiques of, but conceptualize alternatives to the present social, political, economic and philosophical status quo. (p. 403)

Although it may be contended that there are numerous and very different cultures within the indigenous peoples of America, Churchill responds that there is also much internal variety within Western civilization, and that the differences between indigenous peoples have been exaggerated and exploited by colonizers as part of a strategy of domination. What characterizes the indigenist vision compared to the outlook associated with the dominant society? Churchill refers to the work of Mexican anthropologist, Guillermo Bonfil Batalla (1981), who writes:

> Fundamentally, the difference can be summed up in terms of [humanity's] relationship with the natural world. For the West ... the concept of nature is that of an enemy to be overcome, with man as boss on a cosmic scale.... The converse is true in Indian civilization, where [humans are] part of an indivisible cosmos and fully aware of [their] harmonious relationship with the universal order of nature. (as cited in Churchill, 1993, p. 409)

Native American beliefs about the relationship to the natural world focus on connection and communication. The natural world is not seen as separate from the human world but as animate and related to humans. Winona La Duke talks about how the Chippewa have a philosophical value system—mino bimaatisiiwin—that guides relations with others, with plants, with animals, and with the ecosystem as a whole, based on principles of reciprocity. She has written that "'the resources' of the ecosystem, whether corn, rocks, or deer, are viewed as 'animate,' and as such gifts from the Creator. Thus one could not take life without a reciprocal offering" (cited in Gedicks, 1993, pp. x–xi). Because of this relationship to the natural world, it becomes, according to Arthur Versluis (1993), "a theatre of religious revelation.... Birds, animals, plants, stars, all can have a spiritual significance. Naturally the same is true of the landscape itself" (p. 67).

In the context of actions to protect the environment, the spiritual significance of the land is thus often a prime motivating factor of Native American resistance movements. In the Ward Valley case mentioned earlier, the area is seen as sacred because of its proximity to Spirit Mountain, the birthplace of the ancestors of the tribes in the Colorado River Native Nations Alliance, because the entire valley is a spirit path (along which spirits travel), and because it is home to the desert tortoise, which is revered as a brother. These beliefs may be seen as fanciful, irrational, childish, or crazy by members of the dominant society, but they are extremely important, deeply held, and passionately embraced by followers of the indigenist vision, beliefs for which in many cases they are prepared to die. With the increase of coalitions between environmental organizations and Native American groups to resist particular development or industrial projects that threaten wild lands, these beliefs and the practices through which they are communicated are taking on an increased significance as symbolic resources that provide psychological empowerment to movement activists, both native and non-native.

The opposition movement that was organized to a proposed large dam, the New Los Padres Dam, on the Carmel River in Monterey County, California, is examined here as an example of this contemporary alliance building between Native American and environmental organizations in resistance to development projects and of the role of the indigenist vision and Native American leadership in such coalition movements.

Dam Construction and Environmental Racism

Dam construction projects have increasingly come to be seen both as major environmental despoilers and as threats to indigenous peoples and cultures and, therefore, have become the target of powerful opposition coalitions. At the same time, the building of large dams has been presented by their promoters and admirers as a miracle of modem engineering and also as a metaphor for the triumph of human society over the unpredictable forces of nature. Giant dams in themselves may be viewed as products of what Leslie Sklair (1994) calls *cultural-ideological transnational practices* in which a central assumption is that the domination and control of nature is an essential, even a spiritual, duty. This, of course, is more or less the complete antithesis of the indigenist vision described above. Patrick McCully (1996), who is campaigns director for the California-based International Rivers Network, part of an emerging international anti-dam movement, has identified several recurrent ideological themes in the arguments of dam-building corporations and government bureaucracies. These include the ideas that undammed rivers are wasted, that wild or turbulent rivers should be tamed, that rivers have no value unless they are controlled, and that dams are comparable to temples or other places of worship. Dam promoters see their projects as improving on an imperfect nature. Camille Dagenais, former head of the Canadian dam-building firm SNC, once stated, "In my view, nature is awful and what we do is cure it" (cited in McCully, 1996, p. 47). As McCully points out, "When a dam is given such a powerful symbolic role, its economic and technical rationale and potential negative impacts fade into insignificance in the decision-making process" (p. 237). However, the potential negative impacts of dams are considerable. Dams, far from being modern marvels, have a huge human and environmental cost. McCully estimates that 30 to 60 million people worldwide have been forced from their lands as a result of dam construction (McCully, 1996). There has also been a devastating impact on wildlife; for example, in the United States, dam construction is the main reason two-fifths of freshwater fish are endangered or extinct (McCully, 1996)....

This leads to a significant insight about the human cost of large dam construction. As Donald Worster (1993) puts it, "the domination of nature leads inescapably to the domination of some people by others" (p. 169). The giant dams that supposedly control wild and wasted rivers and put them to work for human society are usually devices for expropriating common resources from some users and turning them over to others. Patrick McCully (1996) writes:

The domination of rivers is one of the clearest indications of the link between the control of nature and the control of people. Large dams are not built and operated by all the society but by an elite with bureaucratic, political or economic power. The dams give this elite the ability to direct water for their own benefit, depriving the previous users of some or all of their access to riverine resources. (p. 24)

Certainly, in the case of the New Los Padres Dam Project, a key issue raised by its opponents was the question of who would truly benefit from the project—the community as a whole or a small group of developers and other business interests, at the expense of the further loss of common riverine resources in particular places that were of outstanding significance for the descendants of the indigenous population of the area, the Esselen....

... [T]he New Los Padres Dam was first proposed in 1989 as a solution to long-term water problems on the Monterey Peninsula in California. As soon as they became aware of it, the indigenous peoples of the area, the Esselen, began a passionate and outraged resistance and together with predominantly White environmental organizations, formed an effective coalition opposing the dam project. This coalition was ultimately successful in at least temporarily derailing the large dam proposal; the story of the alliance, and in particular, the role of the indigenist vision and Native American leadership in the alliance, is the focus of the research reported here.

The New Los Padres Dam Project

California, including the Monterey Bay area, experienced one of the most prolonged droughts in recent history during the years 1987 through 1992. During this period, mandatory water rationing was introduced on the Monterey Peninsula. In addition, because of overpumping of groundwater from the Carmel River basin by the Cal-Am Water Company, the primary water supplier in the area, the Carmel River dried up in the summer and fall, even during wet years, resulting in damage to the riparian environment and to the steelhead fish run. In response to this situation, the Monterey Peninsula Water Management District (the water district), which is the elected local regulatory agency legislated to manage water resources and promote water conservation, defined its mission as developing a long-term water-supply project that would provide an adequate water supply in drought years and that would restore year-round strearnflow to the Carmel River to repair the environmental damage that had occurred. A number of water-supply alternatives were looked at, including a small seawater desalination plant. On June 8, 1993, voters rejected the proposed desalination plant, and the Water District focused its energies on planning and promoting a new dam on the Carmel River....

In 1994, a large number of state and federal agencies expressed support for the dam. In February, the Environmental Protection Agency declared that it was the least environmentally damaging practical alternative. In May, the six cities in the water district passed unanimous resolutions supporting permits for the project. In November, the California Department of Fish and Game testified in favor of the project at hearings at the State Water Resources Control Board, largely because they believed that it would benefit the steelhead fish run in the Carmel River. In December, the California Coastal Commission voted unanimously to certify that the project complies with the Coastal Zone Management Act. In 1995, with permit approval from the Army Corps of Engineers, approval from the State Water Resources Control Board, and completion of Section 106 of the National Historic Preservation Act process, all the New Los Padres Dam required was approval by the voters of Measure C on the November 8, 1995 ballot, which would authorize funding for the dam. Throughout 1995, opposition efforts were therefore focused on building up resistance to the dam with the goal of achieving a rejection of Measure C by the voters in November.

The New Los Padres Dam Project was the stated mission of the Monterey Peninsula Water Management District and was also obviously strongly supported by the Cal-Am Water Company. In addition, prodevelopment and growth interests, both in the area and from outside, contributed funding to the campaign to convince voters to approve the dam in the November election. The biggest spender in the period leading up to the election was a real estate interest group called Issues Mobilization Political Action Committee, based in Los Angeles, whereas local support came from a large construction company (which would probably be involved in building roads and other support structures around the dam), local real estate interests, and the hotel industry. By October 21, the last reporting period to the Monterey County Elections Department prior to the election, supporters had outspent opponents by 2 to 1, collecting $107,215, whereas the opposition alliance had raised $41,255, the largest amount coming from the local chapter of the Sierra Club (Wolf, 1995). Who were the groups and individuals who organized the opposition to this large dam project, what was the nature of their alliance, and what kind of ideational materials did they draw on to mobilize resistance to the New Los Padres Dam Project? These questions were addressed in the research reported here.

Methods

The data employed here were collected in an 18-month period of fieldwork on the New Los Padres Dam controversy, which began in the fall of 1994. Three main sources of data were used: (a) archival and documentary materials, (b) in-depth interviews with 12 people involved in the dam dispute and/or the Native American cultural renewal movement, and (c) observation and participant observation of relevant community events and actions....

The research data were ... gathered in a number of settings, using a number of methods, and although there are important stories to tell about gaining access and building trust, the focus of this article is more what John Van Maanen (1983) would describe as a realist tale in which the field-worker "simply vanishes behind a steady descriptive narrative" (p. 46). This is so because I think the struggle against the New Los Padres Dam is an important story to tell because of what it says about the role of Native American beliefs and leadership in environmental battles today....

Opposition to the Dam

At the time of the arrival of the Spanish Empire in the Monterey Bay area in 1769, the Esselen inhabited the mountainous area, now known as the Ventana Wilderness, on the fringes of which the dam was to be constructed. Esselen village and burial sites exist on the land that would be inundated by the New Los Padres Dam reservoir, and radiocarbon dating at sites elsewhere in the Ventana Wilderness indicates they were occupied at least 4,630 years ago (Breschini & Haversat, 1993). Genealogical research by the archaeologists ... confirmed that descendants of Esselen who were at the Carmel Mission still reside in the region, and one of their spokespersons, Tom Little Bear Nason, told me that they numbered 80 people (T. L. B. Nason, personal communication, February 23, 1995). "We are small, but we're coming back," he said a number of times at public meetings. The Esselen were represented by two organizations, both of which were recognized by the water district in their efforts to fulfill requirements for the 404 Permit and Section 106 of the National Historic Preservation Act. The Esselen tribe of Monterey County had close associations with the immediate neighborhood of the dam because some of their most prominent members, the Nason family, owned a 500-acre homestead of mostly mountain land a few miles upstream of the dam site. When they first began to communicate their opposition to the dam to the Water District and the Army Corps of Engineers, their tribal chairperson was Fred Nason, who died in 1993, at which time Tom Little Bear Nason took on that position. Another grouping, the Esselen nation, affiliated with the United Tribal Families of the Central Coast of California, was chaired by Loretta Escobar-Wyer. Both organizations protested against the construction of the dam and both for the same basic reason, that it would inundate places sacred and in other ways significant to the Esselen, but there were some differences of approach and emphasis.

Neither organization was federally recognized as an Indian Tribe, and lack of federal recognition was an important issue for both groupings and for other tribes on the central coast of California, which predated the dam project. Historically, the lack of recognition is connected to the fact that for a long time, many central coast tribes were believed by official and academic opinion to have become extinct. . . . The influence of the American Indian Movement (AIM) in the area was to help initiate a cultural revitalization process among central coast tribes, which included a push for federal recognition and deliberate actions to renew and rediscover traditional tribal cultures. Not only the Esselen, but also other central coast tribes such as the Ohlone, Rumsen, Salinan, and Mutsen are all active in this cultural revitalization movement, and there is considerable coordination of activities and cooperation between tribal leaders (for example, on March 25, 1995, tribal leaders from these groups gathered together to discuss several issues local Native Americans were facing, including the New Los Padres Dam). The opposition to the dam was thus a part of this broader cultural renewal effort that involved the other local tribes and their applications for federal recognition, which had been bogged down in federal bureaucracy since 1978. One of the differences between the Esselen tribe of Monterey County and the Esselen nation was that the Esselen tribe, chaired by Nason, seemed to place less value on federal recognition, "We don't want to have a Tribal Council. Its all about protecting the land. . . . We don't need a piece of paper" (T. L. B. Nason, personal communication, December 29, 1995). . . .

The Esselen groups as activists in the anti-dam movement were also joined by many non-native supporters who participated in the Esselen organized anti-dam actions. Such non-native supporters are also significant in other Native American environmental actions, such as the occupation of Ward Valley. In the course of this research, I interviewed some of the non-native supporters about their reasons for their involvement with Native American activities. Their reasoning illustrated the significance of Native American symbolic resources as motivating ideas and meaningful cognitive frameworks for activists. One activist, Elizabeth Williams (personal communication, October 15, 1995), described that when she heard Native American leaders speak, "I thought, this is how I always felt, oh yeah, this is home." . . .

The status of non-native supporters of Native American organizations and actions is controversial. For some, they represent the frivolous appropriation of Native American culture. Nason mentioned how AIM leaders were saying, "They took our land. Now they want our religion," and went on to say, "I've argued against that. The prophecies tell us that the Red, Black, Yellow, and White nations will come together" (T. L. B. Nason, personal communication, December 29, 1995). In central coast of California cultural renewal activities and in the struggle against the New Los Padres Dam project, a strategy of deliberately reaching out to non-natives is being followed. Anne Marie Sayers (personal communication, November 3, 1995), prominent Costanoan/Mutsun tribal leader, stated at a Native American organized anti-dam gathering, "We are allowing people to become aware of our existence. We educate people of our existence. We are all Native Americans, now let us unite and become one earth and one people."

Esselen groups, with their non-native supporters and environmental and neighbourhood organizations, formed an effective coalition that defeated the proposed New Los Padres Dam. However, the arguments of the predominantly environmentalist groups and the Native American groups differed in important respects. Whereas the Esselen were opposed to building a dam in this particular place because of the threat to their sacred sites, the environmental groups were opposed to any dam, on the grounds that increasing the water supply would lead to more development and growth on the Monterey Peninsula. What were some of the pivotal issues in the struggle against the New Los Padres Dam Project? One of the most fundamental disputes was the issue of growth.

Key Issues for the New Los Padres Dam Opposition Movement

Environmental Dam or Developers' Dam?

The water district . . . concluded that there was no alternative solution to the peninsula's water problems other than the construction of the dam. The features of the dam that they chose to emphasize in public meetings and communications to the media were that it would provide a secure water supply in times of drought and that it would be environmentally beneficial. When I met with Henrietta Stern, senior planner for the water district, she emphasized to me a number of times that it was an environmental dam. "The key is restoring stream flow in the Carmel River," she told me (H. Stern, personal communication, January 18, 1995). In an article published in a local newspaper, Stern (1994) made this argument about the benefits of the dam:

> How will the dam affect the environment? The NLP project would provide year-round stream flow to the Carmel River Lagoon [at the mouth of the river] in 75 percent of water years. It would benefit about 24 miles of riverbank vegetation and wildlife, steelhead habitat, recreational and aesthetic resources, and the Carmel River Lagoon in nearly all years. (p. 6)

However, the environmental benefits of the dam were simply not believed by the Sierra Club and Citizens for Alternative Waters Solutions (CAWS) activists, and they sought to publicly discredit them. In an interview with Gruber, chair of the Sierra Club's New Los Padres Dam subcommittee, he told me: "We're calling it the *developers' dam*. It'll lead to 14,000 new homes being built. In the drought years, they'll suck up the extra water that is supposed to restore the river. Every development around here is called an environmental development" (D. Gruber, personal communication, October 12, 1995). This last comment referred to the fact that in the Monterey Bay area, controlled-growth advocates and elected representatives had been so successful in restricting development projects that such projects needed to be packaged in ways that made them more palatable to the public. Presenting them as environmental was one strategy; a similar maneuver, used elsewhere, was presenting them as promoting the performing arts (see Whitt, 1987). Richard Gendron (1996) has explored this approach in neighboring Santa Cruz County, which he described as "the developer's canny tactic of using an arts-based growth strategy as a means of . . . driving a wedge into a progressive coalition that has successfully opposed every large-scale development in the previous 20 years" (p. 551). The Sierra Club and CAWS activists were not deceived by such canny tactics and focused on the growth potential of the dam. For them, it was unquestionably not an environmental dam. . . .

[. . .] Sierra Club and CAWS campaigners emphasized that the water district's own Environmental Impact Report made clear the dam's potential for growth. In that report, it states, "If the long-term water supply project is not built, growth that is now planned for the peninsula would be constrained . . . it is clear that expansion of the water supply system would remove one obstacle to district growth" (Monterey Peninsula Water Management District, 1994a, chap. 19, p. 1). The additional water supply would make possible "roughly a 20 percent increase in housing, population and water demand" (Monterey Peninsula Water Management District, 1994a, chap. 19, p. 5). . . . On the basis of this analysis, the Sierra Club strategy in mobilizing opposition to the project was to emphasize the growth potential of the dam. The success of this strategy is illustrated by the fact that the latest dam proposal for the Carmel River (following the rejection of the New Los Padres Dam by voters) is being presented by the water district as the no-growth dam. . . .

The Sacred Sites of the Esselen

For the Esselen, the pivotal issue pertaining to the New Los Padres Dam was the destruction of their sacred places by the reservoir and dam construction. Sierra Club and CAWS activists certainly shared this concern, but for them it was one of a number of issues. For the Esselen, it was unquestionably the predominant one. In their arguments to the water district, the Army Corps of Engineers, other official agencies, and the general public, they did not oppose building a new dam per se; they opposed building it at this site, because this site was in an area that had been occupied by the Esselen for thousands of years and included numerous sites, locations, and objects of deep significance for the Esselen. Because of this, Esselen leaders argued that it was "local Esselen Native Americans who are and will always be the ones most affected by the proposed New Los Padres Reservoir project development" (Nason, 1993, cited in Monterey Peninsula Water Management District, 1994b, p. 161).

Although the term *environmental racism* was never used by the Esselen, their comments on the dam project clearly indicated that they viewed it as another example of the disrespect and disregard for Native American rights that they had so long experienced in their contacts with the dominant White society. Escobar-Wyer, chairwoman of the Esselen nation, writing to the Army Corps of Engineers, eloquently set the dam project in the context of the history of dominant society/Native American relations and argued that it created an opportunity to reverse the pattern of mistreatment, build trust between Native Americans and the dominant society, and "gain a true balance of respect" (cited in Monterey Peninsula Water Management District, 1994b, p. 162). She urged the Army Corps of Engineers to deny the 404 permit, because "in doing so trust may gain a foothold and a difficult journey of bridging two worlds can begin" (L. Escobar-Wyer, 1994, cited in Monterey Peninsula Water Management District, 1994b, p. 162). However, in the matter of the sacred sites, the Esselen nation's chairwoman believed that the water district and other agencies had ignored, disregarded, and in other ways shown lack of respect for the Esselen's testimony about the significance of the threatened places. Commenting on the Final Environmental Impact Report (EIR), she stated that it "does nothing to guarantee Indian people that they will be dealt with fairly and enjoy rights under the U.S. Constitution" and that "the lack of response and professionalism by the lead authors of this EIR demonstrates a classic example of a process of continued American Apartheidie, tokenistic and disenfranchisement policies as practiced by Federal agencies" (L. Escobar-Wyer, 1994, cited in Monterey Peninsula Water Management District, 1994c, letter no. 13).

The process of bridging two worlds ran into problems because of fundamental differences between the indigenist vision and the dominant worldview of western society. Learning from nature and from a spirit world that inhabited the natural environment are precious and deeply meaningful elements of the indigenist viewpoint that inspired and motivated Esselen-led actions to defeat the dam project. Escobar-Wyer, in an interview with the press, asked, "Why can't people relate to the spirituality of this massive structure that is Earth? It is where we learn the essentials of life. It is sacred to us" (Cone, 1994, p. A16). Because Native Americans on the central coast of California are in a process of renewing their traditional culture, the traditionally inhabited places, plant-gathering and hunting areas, ceremonial sites, and sacred places have taken on a supreme importance because they constitute a link to that culture and also to ancestral spirits who can communicate that culture to descendants today in ways not easily grasped by the Western mind. When I first arrived to talk to Nason about the threatened sacred sites, he told me that he would give me information. Then, he looked at me and said, "You'll also get information from the ancestors. They will come through you. You will feel it" (T. L. B. Nason, personal communication, March 4, 1995). After I had spent a day conversing with Little Bear and participating in a sweat-lodge ceremony, I set off to drive home, but no matter what I did, I could not get my car to start, which was unusual for this extremely reliable vehicle. I camped out for the night and called on Little Bear in the morning and told him the story. "The ancestors wouldn't let you go," he laughed. "They wanted you to sleep on the ground." It was clear to me that for Little Bear and for other Native Americans I encountered in the course of this research, there was an indigenist vision that emphasized connectedness with a communicative and animate natural environment (and thus, for Little Bear, I was learning not just from what he said, but by being on the land). The leader of the Humaya dancers, while dancing at one of the Native American–organized actions opposing the dam project, paused while dancing to inform the audience, "We learned our language through the birds. There's many secret things that we know. . . . Look to the trees, look to the Heavens, your culture's there, it has always been there." Because the culture was viewed as being discoverable in the rocks, trees, bluffs, birds, flowers, and animals of the Carmel River, it followed that Esselen activists would vehemently oppose a dam project that would irrevocably cut off access to these cultural resources. Anne McGowan (1993), the attorney representing the Esselen tribe of Monterey County, wrote to the water district to point out: "The Esselen are opposed to mitigation measures, because they believe the NLP dam should not be built; the dam will destroy irreplaceable burial sites, sacred ceremonial areas and hunting and gathering areas critical to the revival of Esselen Native American culture" (cited in Monterey Peninsula Water Management District, 1994b, p. 125). . . .

For the Esselen, there were many specific sites that had particular significance; one of the most important was the birthing rock, a prominent, freestanding natural rock monolith in the flood plain of the Carmel River, a place where traditionally Esselen women had gone to give birth and where ceremonies and dances were still held. This was located in what the dam engineers designated as the borrow area, from which rock would be quarried and crushed to provide construction material for the dam. Polomo Brennan told me how at a meeting at the dam site, one of the engineers, in earshot of Esselen representatives, had pointed to the birthing rock as a good source of construction material (J. Polomo Brennan, personal communication, March 4, 1995). He told this story to illustrate the insensitivity of the dam builders to Native American sentiments and went on to add:

> I'll make my last stand at the birthing rock. That's not going to happen in my lifetime. That's utter sacrilege. The fish altar [another sacred site] would also go. They say it'll be covered with water and so it'll be protected, but that's like saying if the Sistine Chapel were flooded, you could visit it using scuba gear. (J. Polomo Brennan, personal communication, March 4, 1995)

Close to the birthing rock was another place that the Esselen held to be especially sacred, which they called the baby burial area. In addition to a large number of specific sites, the entire course of the Carmel River was viewed as sacred by the Esselen, a spirit trail, along which the spirits of the dead travel on their way to the Western gate, the door to the land of the dead. Nason spoke of how there were many spirits along the river and as a result, "people have powerful visions and dreams here" (T. L. B. Nason, personal communication, March 4, 1995). The construction of the dam would block the path of the spirits and destroy a spiritual connection of supreme significance for the Esselen.

The Esselen believed that the water district did not have a full understanding of or respect for their sacred places. They noted that the Final Environmental Impact Report had devoted more pages to animal and plant life than to Native American cultural resources. It seemed to them that the steelhead fish were given more attention than the beliefs of the Esselen. . . . Although the water district argued that the Esselen's loss of cultural resources could be mitigated by standard procedures (for example, bedrock mortars could be removed to another site; an Esselen Cultural Center could be established), Esselen activists believed otherwise. "It's not mitigatable," Nason told me. Polomo Brennan (1994) wrote to the Army Corps of Engineers,

> The desecration, through total destruction, of two sacred sites below and under the dam is impossible to mitigate. . . . The cumulative impact of destruction, desecration, inundation, blinding, burial, translocation and other mitigation measures is the catastrophic loss of resources of immense spiritual importance to the Esselen people. (cited in Monterey Peninsula Water Management District, 1994b, p. 149)

The Opposition Movement to the New Los Padres Dam: Strategies and Alliances

Opposition to the New Los Padres Dam project gathered momentum as more and more official agencies gave the project their approval during 1994 and 1995, and the only significant hurdle remaining was the November 8 election in which voters would decide on a bond measure to provide funding for the dam. The various opposition groups and individuals adopted a number of strategies, working both together and independently to mobilize public opinion against the project.

David Brower

One of the earliest oppositional events was a public meeting in November 1994, organized by the environmental activist, Mapstead. Mapstead was known personally to Nason and, a year later, was working with him to lead a sacred hike and ride up the Carmel River. Mapstead along with many other non-native activists participated in a medicine circle preceding the hike that included Native American ceremonies. The public meeting about the dam was addressed by David Brower. Brower has played a highly visible role in the anti-dam movement in the United States and in the environmental movement in general. McCully calls him "probably the second most influential figure of the twentieth-century environmental movement" (McCully, 1996, p. 283). Brower, as executive director of the Sierra Club, was a central figure in the opposition to Bureau of Reclamation proposals for dams on rivers in the Colorado Basin, beginning with resistance to the Echo Park Dam, planned for the Green River in the 1950s (Gottlieb, 1993, p. 41; McCully, 1996, pp. 283–285; Reisner, 1993, pp. 284–285). Indeed, Gottlieb (1993) calls the Echo Park Dam fight a turning point for environmentalism. At the meeting, which was widely advertised and well-attended, Brower set opposition to the New Los Padres Dam in the context of a broader philosophy questioning the value of economic growth. "How do we handle growth? What does it cost? How can we control it? We need to do a cost-benefit analysis of growth. Growth isn't all that great if it costs more than we can afford." When Brower stated emphatically in a sonorous voice, "No more growth," the audience broke into sustained and enthusiastic applause. "We don't have to grow any more. We get more growth and then we'll need more water," he continued. Brower's talk raised fundamental questions that went far beyond the dam issue to tackle the sustainability of human society's current economic relationship with the environment. "The earth cannot sustain what we've been doing," stated Brower. Bron Taylor (1995) has identified "the rejection of economic growth and industrialization as desirable social goals" to be a common denominator of ecological resistance movements (p. 340), and this was certainly a strong theme in Brower's thinking and, to a major extent, in the Sierra Club's campaign against the dam, which constantly attacked the growth potential of the project. Questioning the sustainability of industrial society because of its devastating impact on the planet's ecosystems was also a theme of the indigenist vision, although expressed in more animate terms. "Mother Earth is sick and she needs healing" was a statement I commonly heard at Native American actions or, as Costanoan/Mutsun tribal leader Anne Marie Sayers put it at one of these events, "Everything comes from the Mother, and she's tired and weak."

Sierra Club and CAWS Strategies

The compatibility of Native American and environmental thinking enabled extensive cooperation and sharing of personnel between the groups in opposition to the dam to take place. Gruber, who chaired the Sierra Club's subcommittee, told me about how he worked together with Nason during 1995 to conduct tours of the proposed dam site for members of the Sierra Club and other interested members of the general public. Usually, such tours involved a medicine circle at the birthing rock during which Nason or other Esselen representatives would talk about the meaning of the place for the Esselen. Gruber told me, "Native American thinking and ecology are similar. Their philosophy matches ecological thinking" (D. Gruber, personal communication, February 16, 1996). The similarity for Gruber was based partly on the need to go beyond purely rational, scientific arguments to directly experience the land and view the sacred sites. "You can argue back and forth at the rational level," he told me, "but it comes down to emotions; you know it's not right at an emotional level" (D. Gruber, personal communication, October 12, 1995). . . .

Only a tiny minority of the Monterey Peninsula population ever made it to the dam site. To reach the majority, the Sierra Club spent $5,000 on a publicity campaign including advertising, mailing to the general public, organizing forums, and addressing public meetings. The purpose of this campaign, according to Gruber, was to "present ideas and quality-of-life issues. It was a matter of what do you want for life around here?' (D. Gruber, personal communication, February 16, 1996). The information about the dam that they tried to communicate was that it was not about solving environmental issues, but rather about growth. Many people who were Sierra Club members were also members of Citizens For Alternative Water Solutions (CAWS), and both organizations presented ideas to the public that challenged the water district's claims for the dam and proposed different ways to deal with the water problems on the Peninsula. CAWS was also linked to the Esselen through Polomo Brennan, who was both a campaigner and spokesperson for the citizens' action group as well as the Esselen's tribal engineer. One of the strategies of CAWS, as their name implies, was to propose various methods through which the water district could augment the water supply without building the dam. On a public radio discussion program aired just prior to the November election, Polomo Brennan pointed out (personal communication, November 2, 1995), "If we looked at retrofitting toilets, if we looked at dredging, . . . and if we looked at desalination, we would create something of the order of 15,000 acre feet of available water for whatever purposes is needed, for drought reserve, certainly." At the same time, CAWS argued that the dam would provide water for unplanned environmentally damaging growth, increase consumers' water bills by 30 percent over 4 years, degrade the quality of life, irreparably harm several world-class vineyards, and permanently end Esselen culture and religion through flooding. The Sierra Club campaign made similar kinds of arguments, not only focusing on the negative impact of the dam on the environment and quality of life on the Monterey Peninsula, but also proposing alternative solutions to the water issues. Sierra Club activist Mitteldorf (1995), writing in the local chapter's newsletter, *The Ventana*, proposed specific methods through which the water supply could be increased, including many ideas also promoted by CAWS, such as dredging the existing reservoir, building a small desalination plant to operate in drought years, and requiring dual-plumbing systems that would enable dwellings to make use of rainwater that currently runs off through storm drains. Another activist, Larigner, linked the local dam issue to the broader problems of a high water-consumption society in a relatively arid area (almost no rain falls in the Monterey Peninsula between April and October), which have been addressed by such water theorists as Reisner (1993) and Bowden (1977). . . . This approach, similar to Brower's, moved beyond discussing specific problems with the New Los Padres Dam to a critique of an unsustainable society dependent on continual growth, achieved only at the expense of unacceptable environmental depredation. The notion that Mother Nature would call this society to account has obvious parallels with what I have been describing as the indigenist vision.

In the publicity campaign organized by the Sierra Club, such philosophies informed activists but were not in the forefront of the ideas presented to the general public. This campaign, which ran through 1994 and 1995 up to the November election, involved the organization of forums and public meetings (such as a debate on the dam in August 1995 at Monterey's Navy Postgraduate School), mailing information to Sierra Club members and to the general public, display ads in the local print media, including a series of what Gruber called mosquito ads, which ran in *The Monterey County Herald* (the biggest local circulation daily) for 30 days before the ballot. These ads were short and to the point, such as "Would John Muir vote for the developers' dam?" "Quarter billion $ developers' dam = 100 dollar water bills: We can't afford them!" and "Developers' dam, environmental disaster." Both the Sierra Club and CAWS focused on the excessive development and growth issues. In reviewing the campaign, Gruber told me, "Growth is equivalent to a change in the quality of life, and people didn't like it" (D. Gruber, personal communication, February 16, 1996). CAWS ran an ad in *The Monterey County Herald* on the Saturday before the vote, which emphasized the negative impact of growth. Beside a photo of congested traffic, copy declared, "Traffic is bad enough already. The growth allowed by the dam will make it even worse. . . . GROWTH. It's what the dam is all about" (*Monterey County Herald*, 1995, p. A9).

Esselen Strategies and Actions

For both Esselen organizations, the first line of resistance to the dam project was to communicate to the water district, the Army Corps of Engineers, and other official agencies their opposition to the dam project. This was done through meetings, some which took place in Sacramento, and through formal letters, many of which, for the Esselen tribe, came from their attorney, McGowan. Settlement meetings with the water district were held as early as 1992, hearings were conducted with the Army Corps of Engineers and with the State Water Resources Control Board in Sacramento, and letters were written during 1993 and 1994 in response to the draft and final Environmental Impact reports and the report prepared by Archaeological Consulting. Commenting on these lengthy bureaucratic procedures, Nason told me: "They want us to go Western. Join the White world and go to meetings and sit on committees. I've done that. I've got burnt out on that. I'm going native now. I don't want to get sucked in" (T. L. B. Nason, personal communication, March 4,1995). He also pointed out to me that there was a huge inequality of resources between the water district and the Esselen tribe; the water district had spent $10 million researching and promoting the project, whereas the Esselen did not have access to that level of financial resources (T. L. B. Nason, personal communication, February 23, 1995). McGowan also pointed out to me the relative poverty of the Esselen. To attend hearings in Sacramento was not easy for them, as it involved taking time off work, finding low-cost accommodation and food in the city, and crowding into a van to save on transportation costs (A. McGowan, personal communication, February 21, 1995). As a result, Esselen leaders placed a great value on their traditional ways, including various ceremonies in which they prayed to spirits and ancestors for help. "It's all we have, the spirits, the ancestors. We have no other help," Nason told me (T. L. B. Nason, personal communication, March 4, 1995). Going native was seen as an effective tactic in dealing with a powerful, well-funded bureaucracy, partly because the Esselen were not wealthy but also because cultural renewal was so important to them, and cultural renewal involved the use of traditional methods. . . . All Esselen actions that I attended involved prayer and traditional ceremony, and these customs were the behavioral expression of a philosophy that emphasized interconnectedness with an animate and communicative universe. Their practice reaffirmed the importance and validity of the indigenist vision within a dominant society that questioned the relevance and even the sanity of Native American views. . . .

During the weekend before the crucial November 8th election, the Esselen tribe organized a string of public actions including a prayer ceremony at the mouth of the Carmel River attended by local spiritual leaders, both native and non-native, a reception and Native American feast, a gathering attended by traditional Native American dancers and addressed by many local Native American leaders, a sacred ride and hike along the lower 15 miles of the Carmel River, a pilgrimage to the birthing rock and the baby burial area, and a prayer circle and storytelling around a campfire, followed by sweat-lodge ceremonies. These events were attended by many of the activists in the struggle against the New Los Padres Dam, including Polomo Brennan and other CAWS members, Mapstead, and Sierra Club members; they were also attended by activists in the Native American cultural revitalization process from the Ohlone, Western Shoshone, Rumsen, Pomo, Mutsun, and Salinan tribes.

At the feast held on November 3rd, a Sierra Club member showed slides of the area of the Carmel River threatened by the dam, to which Nason provided a commentary. These were not massively attended events; for example, there were 60 or 70 people present at the feast and about 50 hikers and horse riders on the sacred hike, but they included many of the most active leaders and exemplified the cooperation between Native American and environmental groups that had characterized opposition to the New Los Padres Dam. They were also important as an opportunity for Native Americans to share their traditional beliefs, which they saw as the solution not only to the issue of the dam, but also to the much larger scale problem of the sustainability of human society on an environmentally degrading planet. . . .

Discussion and Conclusions

In the November 8, 1995, election the measure approving funding for the dam was voted down by a margin of 57 percent to 43 percent. Gruber's assessment of the meaning of the election result was that it was a "referendum on growth" and that it indicated that people on the Monterey Peninsula valued "a relatively small and uncrowded civilization, with wild areas close by" (Gruber, 1995b, p. 5). The cost of the dam and its potential for increasing the monthly water bill was significant, and he also believed that concern for the Esselen sacred sites swayed voters against the dam. "People were concerned about racism, about running over the Indians once again" (D. Gruber, personal communication, February 16, 1996). However, whereas the Native American issues were certainly not the only or even the major issues that decided the vote, for many of the activists and leaders of the opposition movement, Native American beliefs and methods were important, particularly as symbolic resources that could be drawn on in constructing psychologically empowering cognitive frameworks that provide a counter to the dominant rationality of development and growth represented by the water district. . . . Gruber, as has been mentioned, thought it important to take people to the dam site. "If you can take a few key people up there, they can reconvey the meaning of the place" (D. Gruber, personal communication, February 16, 1996). This was not a strategy for reaching the mass of the electorate but for affecting the thinking and actions of key activists. Organized trips to the dam site involved Esselen-led prayer circles at the birthing rock in which Esselen representatives offered prayers to the four directions and purified participants with the smoke of burning sage. Such traditional ceremonies functioned as symbolic actions in the way Ingalsbee (1996) has described, operating as a micro-mobilization context defined by McClurg-Mueller (1992) as a "context in which face-to-face interaction is the social setting from which meanings, critical to the interpretation of collective identities, grievances, and opportunities are created, interpreted and transformed" (p. 2). Prayer circles at the birthing rock helped construct its meaning as a highly valuable sacred place imbued with significance and memory and symbolically moved the birthing rock from a context (the dominant technocratic world-view) in which it could be merely seen as a source of crushed rock in the borrow area of the New Los Padres Dam. Native American actions frequently used symbolic means that conveyed the indigenist vision to participants. For example, on the November 4, 1995, sacred hike up the Carmel River, Nason asked participants (including Mapstead, Polomo Brennan—the CAWS activist—Sierra Club members, and other non-native supporters of the Esselen) to "make an offering of a piece of yourself, such as some hair from your head, to Mother Earth, especially at places that have been desecrated." These symbolic methods play a role in changing both personal and collective consciousness by providing an alternative cognitive framework to dominant discourses, in this case, the indigenist vision with its emphasis on relationship with, rather than domination of, nature. A more detailed understanding of the way Native American symbolic practices construct and transform meanings for participants, native and non-native, in environmental actions and their significance in mobilization efforts would certainly benefit from further empirical research. . . .

[. . .] The contention here is that Esselen actions that employed the indigenist vision played a very significant role, which is hard to quantify, in mobilizing activists against the New Los Padres Dam. Native American beliefs may not have been the most crucial factor for voters in the election, but they were certainly influential for key activists. . . .

[. . .] John Bellamy Foster (1994), in his economic history of environmental degradation, points out:

> We must begin by recognizing that the crisis of earth is not a crisis of nature but a crisis of society. The chief causes of environmental destruction . . . are social and historical, rooted in the productive relations, technological imperatives, and historically conditioned demographic trends that characterize the dominant system. (p. 12)

It follows that an end to environmental destruction comes with a transformation of society. Touraine (1981), often credited as an originator of new social movement theory, has asked "which social movement in post-industrial society will occupy the central role held by the workers' movement in industrial society?" (p. 95) and has suggested that the ecological movement "might easily provide the mould in which the main struggles that will later stir through history are to be formed" (p. 24). As Churchill (1993) points out, "indigenism stands in diametrical opposition to the totality of what might be termed 'Eurocentric business as usual'" (p. 407), and so provides a potent vision of a sustainable society based on a different value system. The role of this vision in contemporary coalitions between Native American and environmental groups opposing various kinds of development projects could usefully be explored further, and it is hoped that this research constitutes a starting point. Such coalitions, because they simultaneously address issues of environmental depredation and racism, may well have the potential for fundamental social transformation.

Note

Moore, Mik. 1998. "Coalition Building between Native American and Environmental Organizations in Opposition to Development: The Case of the New Los Padres Dam Project." *Organization and Environment* 11, no. 3: 287–313.

References

Adler, P. A., & Adler, P. (1987). *Membership roles in field research.* Newbury Park, CA: Sage.

Berkman, R. L., & Viscusi, W. K. (1973). *Damming the West.* New York: Grossman.

Bowden, C. (1977). *Killing the hidden waters.* Austin: University of Texas Press.

Breschini, G. S., & Haversat, T. (1993). *An overview of the Esselen Indians of central Monterey County, California.* Salinas, CA: Coyote.

Bullard, R. D. (1990). *Dumping in Dixie: Race, class and environmental quality.* Boulder, CO: Westview.

Bullard, R. D. (1993). *Confronting environmental racism.* Boston: South End.

Churchill, W. (1993). *Struggle for the land: Indigenous resistance to genocide, ecocide and expropriation in contemporary North America.* Monroe, Maine: Common Courage.

Cicourel, A. (1964). *Method and measurement in sociology.* London: Macmillan.

Cone, T. (1994, July 14). Dam threatens smallest tribe's sacred places. *San Jose Mercury News*, pp. A 1, A 16, A 17.

Foster, J. B. (1994). *The vulnerable planet: A short economic history of the environment.* New York: Monthly Review Press.

Gamson, W. A. (1975). *The strategy of social protest.* Homewood, IL: Dorsey.

Gedicks, A. (1993). *The new resource wars: Native and environmental struggles against multinational corporations.* Boston: South End.

Gendron, R. (1996). Arts and craft: Implementing an arts-based development strategy in a "controlled growth" county. *Sociological Perspectives*, 39(4), 539–555.

Gottlieb, R. (1993). *Forcing the spring.* Covelo, CA: Island.

Gruber, D. (1995a, November). Dam yardstick. *The Ventana*, 34(8), p, 28.

Gruber, D. (1995b, December/January). We win one! *The Ventana*, 34(9), p. 5.

Habermas, J. (1981). New social movements. *Telos*, 49, 33–37.

Holstein, J. A., & Gubrium, J. R. (1994). Phenomenology, ethnomethodology and interpretive practice. In N. K. Denzin & Y. S. Lincoln (Eds.), *Handbook of qualitative social research* (pp. 262–268). Thousand Oaks, CA: Sage.

Ingalsbee, T. (1996). Earth First! activism: Ecological postmodern praxis in radical environmentalist identities. *Sociological Perspectives*, 39(2), 263–276.

Kroeber, A. L. (1925). *Handbook of the Indians of California: Bulletin 78.* Washington: Bureau of American Ethnology.

La Duke, W. (1996). Like tributaries to a river. *Sierra*, 81(6), 38–45.

Langner, M. (1995a, November). The dam. *The Ventana*, 34(8), pp. 4–5, 20.

Langner, M. (1995b, August/September). Voters to decide on "the dam." *The Ventana*, 34(6), p. 7, 22.

May, T. (1997). *Social research: Issues, methods and process.* Philadelphia: Open University Press.

McCarthy, J. D., & Zald, M. N. (1977). Resource mobilization and social movements: A partial theory. *American Journal of Sociology*, 82(6), 1212–1241.

McClurg-Mueller, C. (1992). Building social movement theory. In A. Morris & C. McClurg Mueller (Eds.), *Frontiers in social movement theory.* New Haven: Yale University Press.

McCully, P. (1996). *Silenced rivers: The ecology and politics of large dams.* London: Zed.

McCutcheon, S. (1991). *Electric rivers: The story of the James Bay Project.* Montreal: Black Rose.

Merton, R., & Kendall, P. (1946). The focused interview. *American Journal of Sociology*, 51, 541–557.

Mitteldorf, A. (1995, October). Dam it! Is that the solution? *The Ventana*, 34(7), p. 7.

Monterey County Herald. (1995, November 4). p. A9.

Monterey Peninsula Water Management District. (1994a). *Monterey Peninsula water supply project: Final EIRIEIS.* Monterey, CA: Author.

Monterey Peninsula Water Management District. (1994b). 404 *Permit Application.* Monterey, CA: Author.

Monterey Peninsula Water Management District. (1994c). *Comments on the Final EIRIEIS.* Monterey, CA: Author.

Pinderhughes, R. (1996). The impact of race on environmental quality: An empirical and theoretical discussion. *Sociological Perspectives*, 39(2), 231–248.

Redclift, M., & Woodgate, G. (1994). Sociology and the environment. In M. Redclift & T. Benton (Eds.), *Social theory and the global environment.* London: Routledge.

Reisner, M. (1993). *Cadillac desert: The American West and its disappearing water.* New York: Penguin.

Renzetti, C. M., & Lee, R. M. (1993). *Researching sensitive topics.* Newbury Park, CA: Sage.

Sierra Club. (1996). *Sierra*, 81(6).

Sklair, L. (1994). Global society and global environmental change. In M. Redclift & T. Benton (Eds.), *Social theory and the global environment* (pp. 205–221). London: Routledge.

Stern, H. (1994, April 21). MPWMD planner provides look at the new dam. *The Carmel Pine Cone*, p. 6.

Tarrow, S. (1992). Mentalities, political cultures, and collective action frames: Constructing meanings through action. In A. D. Morris & C. McClurg-Mueller (Eds.), *Frontiers in social movement theory* (pp. 175–188). New Haven: Yale University Press.

Taylor, B. (Ed.). (1995). *Ecological resistance movements: The global emergence of radical and popular environmentalism.* Albany, NY: SUNY.

Tilly, C. (1979). *From mobilization to revolution.* Reading, MA: Addison-Wesley.

Touraine, A. (1981). *The voice and the eye: An analysis of social movements.* New York: Cambridge University Press.

Touraine, A. (1988). *Return of the actor: Social theory in postindustrial society.* Minneapolis: University of Minnesota Press.

United Church of Christ Commission for Racial Justice. (1987). *Toxic wastes and race in the United States: A national report on the racial and socio-economic characteristics of communities with hazardous waste sites.* New York: United Church of Christ.

Van Maanen, J. (1988). *Tales of the field: On writing ethnography.* Chicago: University of Chicago Press.

Versluis, A. (1993). *The elements of Native American traditions.* Rockport, MA: Element.

Verhovek, S. H. (1992, January 12). Power struggle. *New York Times Magazine*, 20–23.

Whitt, J. A. (1987). Mozart in the metropolis: The arts coalition and the urban growth machine. *Urban Affairs Quarterly*, 23, 15–36.

Wolf, P. (1995, November 2). Dam supporters outspend opponents 2–1 in campaign. *The Carmel Pine Cone*, p. 5.

Woodward, T. (1998, January). The struggle to save Ward Valley. *RESIST Newsletter*, 6–7.

Worster, D. (1983). Water and the flow of power. *The Ecologist*, 13(5), 168–174.

Hunting and the Politics of Identity in Ontario

Thomas Dunk

This selection by Thomas Dunk illustrates how social movements utilize collective action frames to present competing views on a public problem, in this case an environmental problem. Dunk uses the Ontario government's 1999 decision to shorten the spring bear hunt to show how hunting and outdoor sporting groups drew on concepts about culture and heritage to promote the idea that traditional white and masculine identities are endangered by the Ontario government's decision. The piece demonstrates how culture can be used by different interest groups as a form of political capital in claims-making regarding the use of natural resources. Finally, Dunk argues that this is why "identity politics," as opposed to a politics based on systemic political change, limits rather than fosters social, economic, and environmental justice.

On March 4, 1999, the government of Ontario decreed Regulation 88/99 under the *Fish and Wildlife Conservation Act*, thereby terminating the spring bear hunt. This action apparently was the result of lobbying by environmental groups, most notably the International Fund for Animal Welfare and the Schad Foundation. . . . The government had announced its intention to cancel the spring bear hunt approximately six weeks earlier, on January 15. This set off a storm of controversy among hunters and among the tourist outfitters who serviced the spring hunters. . . . The Ontario Federation of Anglers and Hunters (hereafter OFAH) and the Northern Ontario Tourist Outfitters (hereafter NOTO) along with several individuals launched a legal and public relations campaign to bring back the hunt and to have the right to hunt recognized under the terms of the Canadian Charter of Rights and Freedoms, particularly under section 2(b) which declares that everyone has the right to "freedom of thought, belief, opinion, and expression. . . ."[1] The protest against the cancellation of the hunt acquired international dimensions when heavy metal rocker, Ted Nugent, an avid hunter and hunting lodge owner, launched from his Michigan home a campaign to boycott hunting trips to Ontario because of the new hunting regulation and because of Canadian gun control laws.

. . . The environmentalist and animal-rightist opposition to the hunt revolved around the supposed orphaning of bear cubs.[2] Even though existing legislation restricted the killing of lactating sows, these individuals and organizations argued that it was impossible for hunters to always correctly identify the sex or condition of the bears and that as many as 274 cubs were orphaned every year. . . . In his announcement, the Minister indicated that the spring bear hunt was cancelled on ethical grounds. The government also announced a compensation package of 20 million dollars for the tourist outfitters who would suffer economically because the hunt had been cancelled, and an extension of the fall bear-hunting season.

The attempt to placate hunters and tourist outfitters with an extended autumn bear-hunting season and financial compensation were for naught. On April 12, 1999, the OFAH and NOTO, along with a number of individuals initiated an application for a judicial review, a move which if successful would have suspended the regulation canceling the bear hunt until they could appear before a panel of three judges at the Ontario Divisional Court. The legal challenge was based on two arguments.[3] One was procedural. The appellants alleged that the Minister had not followed a process that he is legally bound to respect. This would have involved, they claimed, an environmental assessment and the provision of incontrovertible proof that the action he took was necessary to achieve an end related to conservation. The OFAH and the NOTO lawyers asserted that the *Fish and Wildlife Conservation Act* had conservation as its sole aim and purpose and, therefore, the Minister could not make changes to regulations covered by the Act that were not consistent with the principles of conservation. . . . The second set of arguments was based on constitutional grounds, that the cancellation of the spring bear hunt represented a restriction of the right to hunt, a right which is covered under the terms of the Canadian Charter of Rights and Freedoms.

The appellants lost their challenge on the procedural grounds. The judge ruled that the Minister of Natural Resources had the legal authority to change regulations whether or not they could be justified in terms of conservation. . . . However, the judge did agree that there may be a triable issue under the Charter arguments. In other words, he agreed that hunting may represent a form of expression that is protected by the Charter of Rights and Freedoms. At this writing the OFAH is pursuing their Charter challenge and the campaign to have the spring bear hunt re-instituted carries on.[4]

The controversy surrounding the cancellation of the spring bear hunt in Ontario serves as an entry into a discussion of the way in which the identity politics of whiteness and masculinity influence debates about environmental controversies. The focus in this case is hunting, specifically the arguments put forward by the OFAH and NOTO regarding hunting as a meaningful activity in the culture of their members who are overwhelmingly white men. My interest pertains to the way in which concepts such as culture, tradition, and heritage are employed by groups that are composed largely of white men, a segment of the population in nation states such as Canada that have comprised the dominant norm against which subordinate and/or minority others have had to struggle.

The arguments employed by the OFAH and the NOTO exemplify a number of interesting developments in the ongoing struggles about identity, power, and relationships with nature. They indicate that culture is now recognized by these organizations as a form of political capital. The state, the legal profession and the lay public are picking up on the anthropological idea that everyone has culture—not only those odd, quaint immigrant groups, religious minorities, or remnant indigenous peoples. . . . This is not to prejudge the nature of the intention that lies behind the arguments of white men's organizations regarding their culture. The extent to which the arguments about the relationship between hunting and culture utilized by the OFAH and the NOTO are politically motivated as opposed to sincerely felt by white hunters is impossible to know. Rather than engage in a futile debate about the honesty of white hunter's claims about the place of hunting in their culture and identity, it is more fruitful to think through how this case reveals some of the limitations of identity politics as a means of achieving social, economic, and environmental justice. . . . The debates about hunting, culture, and rights reveal that, in the absence of a deeper engagement with the entire system that simultaneously involves the destruction of nature and the creation of social dislocations, alienation, and inequality, we end up in an endless cycle of claims and counter-claims about the linkages between the uses of nature, cultural traditions, and rights. . . .

Hunting as Meaningful Expression

The OFAH and the NOTO assertion that the right to hunt is guaranteed by the Canadian Charter of Rights and Freedoms is based upon an argument about the role and meaning of hunting in the culture of their members. In his argument before Superior Court of Justice Judge J. Stach, Timothy S. B. Danson, the OFAH's lawyer, cited various affidavits that had been submitted to the court. The first of his examples came from an affidavit prepared by C. Davison

Ankney, a professor in the Department of Zoology at the University of Western Ontario:[5]

> Hunting represents a fundamental expression of a hunter's identity and self-worth and search for natural freedom and personal adventure. As such the act of hunting and the hunting experience occasion moments of profound self-fulfillment. . . . Hunting is their means of spiritual renewal and reconnection with the natural world. . . . Hunting is self-fulfillment and as a fundamental expression of the hunter's sense of identity and self-worth creates for the hunter a comprehensive worldview and way of life which is intimately connected to wildlife, nature, the life cycle of both, survival, self-sufficiency, the wilderness experience, respect for nature and its power, and ensuring a sustainable annual harvest of wildlife. . . .

. . . In summarizing this first part of his argument about how the cancellation of the spring bear hunt violated his clients' right to freedom of expression, the lawyer argued that ". . . it is a profound, profound statement of expression, a way of life, how they identify themselves in the world with their family, with their friends, with their community. . . ."

One of the appellants in the case was Elsie Meshake of Aroland First Nation. Mr. Danson drew the judge's attention to the written submission and then added: "I think this is interesting because what you have here is an aboriginal perspective which is the same as the perspective of the other affiants. . . . [S]he talks about the significance of hunting and what it means to First Nations. And I think it is fair to say on the basis of this evidence, as we go down the road, that for someone of the government to tell not only my clients but [the] aboriginal community that hunting is just about killing. It is an egregious misrepresentation of what is really happening."[6]

In his submission, Mr. Danson also referred to the importance of tolerance: "the diversity in forms of individual self-fulfillment and human flourishing which ought to be cultivated in an essentially tolerant, indeed welcoming, environment not only for the sake of those who convey meaning, but also for the sake of those to whom it is conveyed."[7]

The judge suggested that aboriginal people's hunting rights were covered by section 35 of the *Constitution Act*, 1982 rather than section 2(b), and that it did not help the OFAR and the NOTO argument to invoke an analogy with the aboriginal experience.[8] Again, my concern is not with the effectiveness of the arguments from a legal point of view. My interest is in the insistence that the aboriginal and the white experience is the same.

In the words of Mr. Danson:

> No but I think it is helpful because they are still—because . . .—what I would say the white man is saying about hunting and what the aboriginal

community says about hunting in terms of [how] spiritually important it is, the meaning, being at one with nature, identifying yourself, your personal self-worth and identification, expressing that into the community that you're in with your friends, it's a way of life, it's a heritage, it's a custom. What's remarkable is that the perspective of my clients and the aboriginal perspective are the same. So that—I mean whether we have a s. 35 of *The Constitution Act* for Aboriginal People, in terms of the expression and the meaning that is being conveyed, it's the same.[9]

In his reply to the lawyers representing the Ontario government and the Schad Foundation, Gordon Acton, counsel for the NOTO, argued that in addition to the fact that his clients suffered irreparable economic harm because of the cancellation of the spring bear hunt, they also suffered another kind of injury: "Our harm of the joint applicants is not just the commercial harm. . . . There is a lack of ability to prove ourselves that we don't have to go to the local grocery store to survive. That we can sustain ourselves. It's a statement. It's a statement about who we are. That we are individually sustainable without resorting to the modern world by reaching back into a traditional activity and putting food on the table for our families. . . ."[10]

Thus, far from hunting being about the domination of nature and the taking of life, hunting is said to be about relating to and being one with nature, with one's self and one's human community. Indeed, the lawyers for the Ontario government and the Schad Foundation referred to hunting as recreational killing. This description was criticized by the lawyers for the OFAH and the NOTO. In his decision, the judge supported these arguments, referring to the characterization of hunting as recreational killing as having a "particularly pejorative connotation" that "does not lend itself to reasoned debate" and "is based, moreover, upon logically weak syllogistic reasoning."[11] Thus, even the judge seems to agree that to understand hunting requires that we adopt an anthropological mode and see it as a ritual activity which includes killing but is not necessarily primarily about killing.

These arguments represent an explicit attempt to claim a legal and moral position equivalent to that of aboriginal people in terms of the relationship to nature and in terms of cultural traditions and the legal and political rights that flow from them. Of particular significance is the emphasis on the spiritual nature of the hunting experience and the importance of hunting to white hunters' identity. This is a form of argumentation that has been used with some success by aboriginal people in Canada. . . . The importance of tolerance for minority groups and unpopular activities is a well-established element of Canadian political discourse. Thus when the OFAH and the NOTO argue that their members and clients who are overwhelmingly white and male also have a culture that involves deep spirituality, close relationships to the land, and deserves to be treated with tolerance (even if it is a minority culture that is perceived by others to involve distasteful forms of behavior), they are consciously trying to locate themselves within an established discursive field.

Hunters, Hunting, and Power

One of the discursive strategies employed by the OFAH and the NOTO is to characterize the struggle over the spring bear hunt as a struggle between wealthy, urban-based environmentalists and their young and naive supporters—Robert Schad plays this role in the spring bear hunt drama—and rural, local, small-business people and workers whose livelihoods and cultures will be ruined by further restrictions on hunting. Given what we know about the social bases of environmental concern and the environmental movement, this strategy builds on the already-existing antagonism to environmentalists and animal rights activists in rural areas.[12] . . .

In the US, . . . the "Bambi" wars—that is, the conflicts between residents of small communities who hunt and newcomers who are often opposed to hunting—are perceived to be one element in a struggle between "Tocquevillian" or "Jeffersonian" rural petty commodity producers/workers and "Keynesian" public-sector workers and members of the professional middle and upper class who increasingly inhabit rural communities and commute to their jobs and careers in cities. The locals—apparently known as "woodchucks" in some areas— perceive the local woods, whether or not they are owned in a *de jure* sense, as *de facto* property and the wildlife within them as important subsistence resources. The ethical and property concerns of their urban-dependent but rural-dwelling neighbors find little support in this cultural milieu. Game animals are likely to be perceived as food sources rather than cute or cuddly animal friends and regulations which involve adherence to seasons and licenses are likely to be seen as an annoyance to be ignored rather than a justified attempt to manage game animals.[13]

In contemporary Ontario . . . neither the spring bear hunt nor the right to hunt as a constitutional and heritage issue can be understood primarily in terms of class struggle, at least not as an expression of a struggle between rural small producers and workers and the urbanized professional classes. This is not to say that class and region are irrelevant. Indeed, I will discuss some of the ways they are important to our understanding of disputes about hunting later, but an adequate analysis of the bear hunt dispute and the right-to-hunt movement must go beyond these two variables.

The OFAH has 83,000 members. . . . [T]he OFAH claims that education levels and household incomes of members are above average and the occupations of members "run the gamut as well. Lawyers, doctors, and other professionals are a significant component of our membership as are blue collar workers. The majority of our membership is in the heavily populated urban south."[14] In the historical account of the organization which is available from its website, the OFAH traces its origins to the Toronto Angler's Association, an obviously urban-based group.[15] From this brief sketch it is clear that even if many rural small farmers and rural workers are members of OFAH, the organization cannot be characterized as having that specific a class or regional basis, although it is clear that it sees many of its opponents as urban-based. Furthermore, poaching is one of the OFAH's principal concerns. It promotes a "report-a-poacher" program and in some rural areas billboards advertising Crime Watchers programs also carry anti-poaching statements supported by the OFAH.[16] . . . In some local contexts the situation is highly racialized in that what the OFAH interprets as poaching is perceived by many Metis and First Nations individuals as an exercise of their treaty and/or constitutional rights.

The upshot of this brief discussion of the OFAH is that it cannot be understood as merely a rural organization defending the rights of small producers and workers. The kind of spatial and class analysis that has proven very useful for understanding hunting disputes elsewhere is not as useful in this case. Again, this is not, however, to argue that class has no relevance to issues such as the defense of the right to hunt.

The NOTO can, perhaps, be thought of in more straightforwardly class terms. Tourist outfitters are in essence private entrepreneurs, part of the petty bourgeoisie or middle class. Their defense of the bear hunt is at one level a defense of a class position. This is true in terms of the economic interests they have in the spring bear hunt and hunting more generally and, if the presentation by their lawyer cited above is an accurate indication, in terms of a culture in which values of independence which are expressed through hunting are highly important features. This does not mean, however, that there are not some wide variations in the actual economic circumstances faced by tourist outfitters as a group. Moreover, not all tourist outfitters are locals. Some are actually based in southern Ontario or in the US and come to northern Ontario only for the season. NOTO claims to have 650 members who operate lodges, resorts, camps, camping and trailer parks, canoe outfitting and fly-in services. "Accommodations range from luxurious resorts to little log cabins."[17] The clientele the outfitters cater to is undoubtedly as varied as the level of accommodation offered. Local petty commodity producers and workers may avail themselves of outfitters' services. The industry works hard, however, to attract wealthy hunters from both Canadian and US urban centers. . . .

At another level it is clear that hunters and the organizations that represent them, despite their attempts to claim a subject position as a misunderstood minority, have some very powerful allies. The OFAH is, despite its dispute with the provincial government over the spring bear hunt, very close to the Ontario Conservative Party and the current Ontario government. In their quest to have hunting recognized as part of Canadian heritage, hunters groups such as the OFAH are not alone. In the fall of 2000, the province of Ontario sponsored a

"Premier's Symposium on North America's Hunting Heritage." . . . The press release framed the symposium in the following manner:

> The symposium comes at an exciting time for Ontario's hunting community as the Ontario government completes a number of major initiatives that recognize the province's hunting traditions and the position of hunters as pioneers of the conservation movement. In the works are a hunting strategy for the province and legislation that recognizes hunting and fishing as heritage resource-use activities that should be safeguarded.[18] . . .

. . . In Quebec the government is officially worried about the potential loss of this part of its culture due to a decline in the number of hunters and fishers. It, along with the Federal Government, is examining ways to encourage young people to take up hunting and fishing.[19] The Ontario government for its part amended its *Fish and Wildlife Conservation Act* to include penalties for interfering with anglers and hunters and created a Hunter Apprenticeship Safety Program so that children as young as 12 may hunt, albeit with a mentor who must be at least 18 years old.[20]

Outside of Canada there is also evidence that hunters groups are now using the discourse of culture, heritage, rights, and the need for tolerance as a defense of their activities. . . . In addition to hunting heritage legislation in various US states, there is the tragic example of the Maine housewife who in 1988 was killed by a hunter while she was standing in her backyard. The hunter, despite years of experience, had apparently mistaken her for a white-tailed deer. A grand jury failed to indict the hunter. He was a local person, while the victim was a relatively recent migrant to the region. She was the wife of a professional and they had apparently moved to Maine for several reasons including the natural beauty and simpler life they expected to find. The defendant's lawyers were able to convince the jury that even though the victim was on her own property, her behavior in essence violated the local cultural norms that prevailed during hunting season. These included the fact that although she was on her own property and only a few hundred feet from her house she was not dressed in hunters' orange. Apparently in the local culture of this region of Maine, there is a sense that during hunting season the behavior of the entire population must change to conform to the fact that hunters are about. This case condenses a number of conflicts: male hunter versus female victim, a local against an outsider, and the mobile well-educated middle-class against the less mobile locals. The struggle over the ownership and meaning of terms such as victim illustrates how white hunters and their "culture" can become symbols of local concern because of the effects of large-scale processes that appear to undermine local residents' influence and control over their "own" space.[21] Thus, although in Canada and the US the hunting lobby might now adopt the language of a "minority" that needs the state to protect it in the name of human rights and tolerance, it has friends in powerful places and in some local contexts remains a hegemonic force.

White hunters do not represent a repressed minority, although it is true that hunting is a minority activity. Nor can the dispute over hunting be reduced to an expression of a class struggle between rural small producers and workers and an urban-based professional middle class charmed by an image of forests full of "Bambis." There is undoubtedly some regional conflict involved. According to a Canadian national survey conducted in 1996, 37.3 percent of those who participated in hunting lived in rural areas while 62.7 percent lived in urban areas. By comparison only 17.9 of those who participated in wildlife viewing were rural residents, as opposed to 82.1 percent who were urban dwellers.[22] Clearly, in percentage terms, hunting is far more popular in rural areas than it is among the urban population. The percentage of hunters in Ontario is far higher in the north than in the south. One third of those who hold hunting cards in the province live in northern Ontario, even though the region contains only about one-tenth of the population of the province.[23] On the other hand, there are many hunters from the urbanized southern part of the province and the OFAH draws members from all over. There are rural-versus-urban and north-versus-south dimensions to the bear hunting controversy but the issue involves far more than these variables.

White Hunters and Tradition

. . . There is not space here to provide a detailed history. Nonetheless, there are some observations that can be made that at least throw doubt on the claim that current sport hunting is the continuation of a long tradition among white males in contemporary Canada.

We should begin by distinguishing different kinds of hunting. Firstly, there is the tradition of elite hunting, the historic roots of which lie in the restrictive hunting legislation and game preserves created by the European aristocracy. This mostly northern European tradition is not particularly relevant to the hunting practices in contemporary Canada, although the tradition of "crown lands" can be said in a sense to have its origins in the medieval system of royal control of forests and other resources. . . . There is nonetheless still a tradition of wealthy hunters who keep cottages or lodges in hunting territories throughout the US and Canada. . . .

Secondly, there is tourist industry hunting. This overlaps with elite hunting in that to some extent the international class of wealthy outdoorsmen is one of its key markets. But we need to distinguish between an industry that is devoted to the provision of hunting and fishing expeditions, often in search of "trophy" animals, to paying tourists, and both the wealthy and non-wealthy hunters whose livelihoods are not dependent on hunting. There is, of course, a range of touristic hunting—from relatively low-brow fly-in camps that appeal to a working-class market, to those that offer remote and exotic settings and are aimed at a very wealthy elite....

Thirdly, there is the local middle-and working-class hunting. This consists mostly of day or weekend trips out to local bush roads and clearings that have been produced by the forest, mining, and transportation industries. To some extent, this kind of hunting can be traced to local rural "traditions" but in another sense neither this nor the second kind of hunting I have described can really be said to be very old, at least not among the white population in northern Ontario.

Ethnologists such as Bertrand Hell have insisted that contemporary hunting practices in Europe be understood as meaningful cultural systems in the present world rather than historical survivals of an earlier time.[24] The same is true of contemporary white hunting in Canada. Hunting by white men in areas such as northern Ontario is a product of industrialization and the postwar Fordist compromise that brought relative affluence to unionized, male workers, in the transportation, mining and forest industries. At an individual level, it requires a significant investment in licenses, equipment, vehicles, campers, and supplies.[25] It is also dependent upon the industrial resource extraction industry that has created the network of bush roads, pipelines, and rail beds that hunters cruise in their vehicles or walk in their search for game. What Fine says about hunting among Michigan autoworkers, can be applied to areas such as northern Ontario: "Even though workers hunted both before and after World War II, the war represented a watershed for the popularity and extent of this outdoor activity for auto men."[26] Given the relatively recent development of the industrial working-class in northern Ontario, especially in the transportation, mining and forestry industries, the hunting "tradition" among many working-class individuals cannot go back very far before World War II, and in its current form is highly dependent on the high incomes that characterized some kinds of working-class employment beginning in the 1950s. Moreover, the white population in the region is descended from a mix of European immigrants but with very large Italian, Eastern European, and Finnish elements. Many of the white communities were strongly divided along ethnic lines until the years after World War II when declining immigration from Europe, assimilation, and postwar consumer society eroded the social, economic, and cultural significance of European ethnicity. It was only in the postwar era that local populations began to identify themselves more clearly as white in opposition to the First Nations and other aboriginal populations.[27] Whatever the hunting traditions of these immigrants and their offspring were would presumably have to be located in the European countries of origin if they are to legitimately be said to go back more than three generations.

One also should not discount the influence of the marketing strategies of equipment manufacturers. Hunters represent an important market for arms manufacturers and the producers of four-wheel and two-wheel drive light trucks, campers, tents, outboard motors, all-terrain vehicles, snowmobiles, camping equipment and clothing. In the 1950s and 1960s, advertisers explicitly drew connections between fatherhood, masculinity, and the technologically proficient outdoorsman who knew how to use the latest gadgetry.[28]

The point is not that hunting was only recently invented as an activity of white men. However, contemporary white male hunting culture must be understood within a broader range of social and economic changes that have allowed the sport to have its current form and social base. . . .

New Age Hunters?

Lisa Fine begins her fascinating article on hunting and masculinity among the autoworkers of Lansing, Michigan with a scene from Ben Hamper's book, *Rivethead,* a biographical account of growing up in an autoworker's household and work and life in and around an auto plant. . . . The passage Fine uses is one in which Hamper is confronted by a co-worker who is an avid hunter about a piece Hamper published in a local magazine. The article depicted hunting and hunters in a less than flattering manner. The co-worker is described by Hamper as six feet two, 245 pounds, a former Marine and member of the NRA. He confronts Hamper, refers to him as a "dumb cunt," and informs him that "faggots" are at the top of the list of things he enjoys killing. As Fine points out, the episode illustrates how deeply the worker is invested in his identity as a hunter, and given the sexist and homophobic language he uses, how important hunting is to his masculine identity.[29]

This startling portrayal of the working-class redneck hunter fits well with a stereotypical image that is common in Canadian and US culture. As with many stereotypes, this one contains a grain of truth, or at least one can find individual examples that would seem to confirm it. As I have been arguing, though, "reality" is far more complicated than such commonplace images suggest. Hunters' organizations are now very conscious of the importance of projecting an image that is palatable to the wider public. The redneck working-class hunter has little place in the imagery that organizations such as the OFAH are promoting.

Of course, the ideals regarding masculinity in the larger culture are not static. In the current conjuncture, an older, hegemonic masculinity is increasingly under question and, in turn, is being transformed even as it is defended and reasserted. The new-age men's movement has been particularly active in trying to present an image of men as victims of an oppressive culture, a culture that demands of them forms of behavior which involve alienation from their essential selves. This discourse involves a double movement. On the one hand, it celebrates the "softer" side of men—their spirituality, their need for emotional expression, and their "wounds." On the other hand, it defends the male "need" to express a "hard" edge by getting in touch with the "darkness" inside themselves, and redis-covering the "warrior within." The claim that men are victims too and that they have a need and a right to express their essential nature is part of a strategy that allows white men to reclaim some of their lost prestige, status and power in a world where the interests, rights, and cultural values of formerly subordinate seg-ments of the population have much more influence than they once did.[30] . . .

Of course, modern men's groups which are trying to deal with male alien-ation from their supposed true inner selves search about for representative cul-tures where, they imagine, men are still "real" men because they have remained in touch with their essential masculinity—hence, the popularity in the Canadian and US men's movement of the signs and symbols of more "primitive" cultures. An emphasis on drumming and on rituals of self-discovery that are based, how-ever loosely and inaccurately, on the cultural practices of indigenous tribal soci-eties is common. . . .

Conclusions

It would be wrong to argue that the idea that hunting is part of a white male cultural tradition is a mere fabrication. Hunting has been a part of European cultures for thousands of years, although in most regions it has been a very minor activity, and it certainly has not been central to either physical or cultural survival since the agricultural revolution came to Europe. The current claims being made in Ontario about the nature of hunting, hunters' relationship to wildlife, their role in conservation, and hunting's importance to a whole way of life must be seen as the product of a set of practices, ideas and discursive strate-gies that are very much a creation of the 20th century. In this sense hunting today is hardly a traditional activity.

Of course, the larger issue is that debates about what is a "real" as opposed to a "false" customary practice may foreground the fact that this kind of identity politics is a quagmire which ultimately cannot generate social or environmental justice. The dispute about the spring bear hunt and hunting as heritage in Ontario provides another example of the limitations of identity politics when applied to environmental issues. Some subordinate groups have gained moral and political mileage through emphasizing their unique relationship to the nat-ural world. The example discussed here illustrates that this strategy can be adopted by dominant as well as subordinate groups. While the court is still lit-erally out regarding the success of white hunters' legal arguments about their right to hunt, it is clear that they can muster powerful allies to support their claims to historical and cultural legitimacy. Meanwhile the larger economic and ecological issues that impinge upon the entire society, especially on regions that historically have had close involvement in the direct appropriation of the natural world, go largely unexamined. The need for a political and ecological discourse that transcends divisive identity issues and creates a space for an inclusive discussion of meaningful social, economic and ecological sustainability and equality remains. Killing more bears will not solve the problems that beset the rural and northern regions of Ontario.

Notes

Dunk, Thomas. 2002. "Ecology and Identity Politics: Hunting and the Politics of Identity in Ontario." *CNS* 13, no. 1: 36–66.

1. Department of Justice Canada, *The Constitution Act, 1982*, Part I, Section 2, "Fundamental Freedoms."

2. The accuracy of this claim is hotly disputed by the OFAH.

3. This summary is based on my reading of the court transcript of the hearing. *Ontario Federation of Anglers and Hunters (OFAH) v. the Queen*, Ontario Court (General Division), Court File No. 99-097, April 22–23, 1999, Kenora, Ontario, Proceedings.

4. *OFAH v. the Queen*. Reasons. It should be noted that NOTO have now withdrawn from the legal challenge, a move that was very controversial within the organization. There is a widespread perception that the organization was either "bought off" with compensation payments, or, more generously to the NOTO membership, that the government threatened to hold up compensation payments to the affected tourist outfitters if the organization continued with the legal challenge.

5. Ankney is a specialist in waterfowl, although he has also participated in the debates about the biological bases of supposed differences in intelligence levels between men and women. He has defended the idea that there are biological differences in brain size and that these do correlate with differences in intelligence between men and women. See C. Davison Ankney, "Sex Differences in Relative Brain Size: The Mismeasure of Woman, Too?" *Intelligence*, 16, 1992.

6. *OFAH v. the Queen*, Proceedings, p. 127.

7. Ibid., p. 135.

8. Section 35 of *The Constitution Act, 1982* spells out the rights of the aboriginal people of Canada. Subsection (1) of section 35 states "The existing aboriginal and treaty rights of the aboriginal peoples of Canada are hereby recognized and affirmed." These aboriginal and treaty rights include the right to hunt and fish.

9. *OFAH v. the Queen*. Proceedings, pp. 154–55.

10. Ibid., p. 316.

11. *OFAH v. the Queen*. Reasons, p. 24.

12. On local working-class ideas about and images of environmentalists and nature see Thomas Dunk, "Talking About Trees: Environment and Society in Forest Workers' Culture." *The Canadian Review of Sociology and Anthropology*, 31,1, February 1994; and Thomas Dunk, "Is It Only Forest Fires That Are Natural? Boundaries of Nature and Culture in White Working Class Culture," in L. Anders Sandberg and Sverker Sörlin, eds., *Sustainability the Challenge: People, Power, and the Environment* (Montreal: Black Rose Books, 1998).

13. For an example of this kind of analysis, see Edward C. Hansen, "The Great Bambi War: Tocquevillians versus Keynesians in an Upstate New York County," in Jane Schneider

and Rayna Rapp, eds., *Articulating Hidden Histories: Exploring the Influence of Eric R. Wolf* (Berkeley: University of California Press, 1995).

14. Personal communication with Mark Holmes, Communications Officer for the OFAH.

15. At www.ofah.org/AboutUs/index.cfm (as of March 29, 2001).

16. Along the north shore of Lake Superior on Highway 17, for example.

17. The quote is from the NOTO web page (http://www.noto.net).

18. The press release was posted on the symposium's website: http://www.mnr.gov. on.ca/MNR/ps2000/media.html.

19. Martin Jolicoeur, "La chasse aux chasseurs," *L'Actualité*, 25, 20, December 15, 2000, pp. 54–58.

20. See *Hunting Regulations 2001/2002* (Ontario Ministry of Natural Resources); p. 11 for details on the program. On June 4, 2001, CBC radio in Thunder Bay announced that the Ontario government is thinking of copying a Minnesota strategy of selling lifetime hunting licences. Apparently, the Minnesota program, which has been more popular than expected, was intended to encourage more people to take up hunting. It is an age-based system. The younger the individual is when their lifetime hunting license is purchased, the cheaper it is. According to the radio report, adults can buy hunting licenses for very young children. For more information on the close relationship between the OFAH and the government of Ontario visit the following website: http://www.bmts.com.

21. Mari Boor Tonn, Valerie A. Endress, and John N. Diamond, "Hunting and Heritage on Trial: A Dramatistic Debate over Tragedy, Tradition, and Territory," *Quarterly Journal of Speech*, 79, 1993. In this struggle over space, even terms like "backyard" became highly charged symbols. At least one newspaper editor, sympathetic to the hunter, would not allow the term "backyard" to be used to describe where the victim was standing when she was shot, even though she was on her own property behind her house. The sense of hunter's control over space during the hunting season is apparently quite deep: "To many, Wood's [the victim] mere presence outside during hunting season was sufficient indication that she was *not* innocent of the danger surrounding her" (p. 172).

22. Environment Canada, *The Importance of Nature to Canadians: Survey Highlights* (Minister of Public Works and Government Services Canada, 1999), p. 9, Table 1.

23. The information upon which this statement is based was provided by the Ontario Ministry of Natural Resources. In the fall of 2000, 141,372 of the 422,963 hunting cards held in the province belonged to people with household postal codes that began with the letter P. This corresponds closely to the part of Ontario that is considered by the provincial government to be the north.

24. Bertrand Hell, *Entre chien et loup: Faits et dits de chasse dans la France de l'Est* (Paris: Maison de la Sciences de l'homme, 1985).

25. In my experience, the high cost of wild game and fish is the subject of many jokes. In most of the larger communities, it is cheaper to buy beef, pork, chicken, and fish at a grocery store than it is to replace this meat with hunted or fished produce.

26. Lisa M. Fine, "Rights of Men, Rites of Passage: Hunting and Masculinity and REO Motors of Lansing, Michigan, 1945–1975," *Journal of Social History*, 33, 4, Summer 2000, p. 805.

27. See Thomas Dunk, *It's a Working Man's Town: Male Working-Class Culture in Northwestern Ontario* (Montreal: McGill-Queens University Press, 1991) and David

Stymeist, *Ethnics and Indians: Social Relations in a Northwestern Ontario Town* (Toronto: Peter Martin Associates, 1975). In northeastern Ontario the white population remains divided along linguistic lines—French versus English.

28. Robert Rutherdale, "Fatherhood, Masculinity, and the Good Life during Canada's Baby Boom, 1945–1965," *Journal of Family History*, 24, 3, July 1999.

29. Fine, op. cit., p. 805. The scene she discusses is to be found in Ben Hamper, *Rivethead: Tales from the Assembly Line* (New York: Warner Books, 1991), pp. 135–139.

30. The most widely known of the new-age men's movement books is probably Robert Bly's *Iron John: A Book about Men* (Reading, MA: Addison-Wesley, 1990). For an insightful and critical analysis of the men's movement see Fred Pfeil, *White Guys: Studies in Postmodern Domination and Difference* (New York: Verso, 1995), and Andrew Ross, (New York: Verso, 1994).

PART

Thinking about Change and Working for Change

A Special Moment
in History
The Future of Population
Bill McKibben

*Population growth is a subject that has occupied a sub-group of sociologists—
demographers—for decades. Many demographers have argued that population
growth should be slowed for either economic or environmental reasons. However,
other sociologists have tended to minimize the impact of population growth on the
environment, preferring to focus instead on questions such as the role of capitalism,
overconsumption by people in wealthy nations, and/or social and environmental
inequalities. Journalist Bill McKibben links population and consumption, arguing
that both need to be addressed.*

Beware of people preaching that we live in special times. People have preached
that message before, and those who listened sold their furniture and climbed
up on rooftops to await ascension, or built boats to float out the coming
flood, or laced up their Nikes and poisoned themselves in some California subdivision. These prophets are the ones with visions of the seven-headed beast,
with a taste for the hair shirt and the scourge, with twirling eyes. . . .

And yet, for all that, we may live in a special time. We may live in the
strangest, most thoroughly different moment since human beings took up farming, 10,000 years ago, and time more or less commenced. Since then time has

flowed in one direction—toward more, which we have taken to be progress. At first the momentum was gradual, almost imperceptible, checked by wars and the Dark Ages and plagues and taboos; but in recent centuries it has accelerated, the curve of every graph steepening like the Himalayas rising from the Asian steppe. We have climbed quite high. Of course, fifty years ago one could have said the same thing, and fifty years before that, and fifty years before that. But in each case it would have been premature. We've increased the population fourfold in that 150 years; the amount of food we grow has gone up faster still; the size of our economy has quite simply exploded.

But now—now may be the special time. So special that in the Western world we might each of us consider, among many other things, having only one child—that is, reproducing at a rate as low as that at which human beings have ever voluntarily reproduced. Is this really necessary? Are we finally running up against some limits?

To try to answer this question, we need to ask another: *How many of us will there be in the near future?* Here is a piece of news that may alter the way we see the planet—an indication that we live at a special moment. At least at first blush the news is hopeful. *New demographic evidence shows that it is at least possible that a child born today will live long enough to see the peak of human population.*

Around the world people are choosing to have fewer and fewer children—not just in China, where the government forces it on them, but in almost every nation outside the poorest parts of Africa. Population growth rates are lower than they have been at any time since the Second World War. In the past three decades the average woman in the developing world, excluding China, has gone from bearing six children to bearing four.... If this keeps up, the population of the world will not quite double again; United Nations analysts offer as their mid-range projection that it will top out at 10 to 11 billion, up from just under six billion at the moment. The world is still growing, at nearly a record pace; we add a New York City every month, almost a Mexico every year, almost an India every decade. But the rate of growth is slowing; it is no longer "exponential," "unstoppable," "inexorable," "unchecked," "cancerous." If current trends hold, the world's population will all but stop growing before the twenty-first century is out.

And that will be none too soon. There is no way we could keep going as we have been. The *increase* in human population in the 1990s has exceeded the *total* population in 1600. The population has grown more since 1950 than it did during the previous four million years. The reasons for our recent rapid growth are pretty clear. Although the Industrial Revolution speeded historical growth rates considerably, it was really the public-health revolution, and its spread to the Third World at the end of the Second World War, that set us galloping. Vaccines and antibiotics came all at once, and right behind came population....

If it is relatively easy to explain why populations grew so fast after the Second World War, it is much harder to explain why the growth is now slowing.

Experts confidently supply answers, some of them contradictory: "Development is the best contraceptive"—or education, or the empowerment of women, or hard times that force families to postpone having children. For each example there is a counterexample. Ninety-seven percent of women in the Arab sheikhdom of Oman know about contraception, and yet they average more than six children apiece. Turks have used contraception at about the same rate as the Japanese, but their birth rate is twice as high. And so on. It is not AIDS that will slow population growth, except in a few African countries. It is not horrors like the civil war in Rwanda, which claimed half a million lives—a loss the planet can make up for in two days. All that matters is how often individual men and women decide that they want to reproduce.

Will the drop continue? It had better. UN mid-range projections assume that women in the developing world will soon average two children apiece—the rate at which population growth stabilizes. If fertility remained at current levels, the population would reach the absurd figure of 296 billion in just 150 years. Even if it dropped to 2.5 children per woman and then stopped falling, the population would still reach 28 billion.

But let's trust that this time the demographers have got it right. Let's trust that we have rounded the turn and we're in the home stretch. Let's trust that the planet's population really will double only one more time. Even so, this is a case of good news, bad news. The good news is that we won't grow forever. The bad news is that there are six billion of us already, a number world strains to support. One more near-doubling—four or five billion more people—will nearly double that strain. Will these be the five billion straws that break the camel's back?

Big Questions

We've answered the question *How many of us will there be?* But to figure out how near we are to any limits, we need to ask something else: *How big are we?* This is not so simple. Not only do we vary greatly in how much food and energy and water and minerals we consume, but each of us varies over time. William Catton, who was a sociologist at Washington State University before his retirement, once tried to calculate the amount of energy human beings use each day. In hunter-gatherer times it was about 2,500 calories, all of it food. That is the daily energy intake of a common dolphin. A modern human being uses 31,000 calories a day, most of it in the form of fossil fuel. That is the intake of a pilot whale. And the average American uses six times that, as much as a sperm whale. We have become, in other words, different from the people we used to be. Not kinder or unkinder, not deeper or stupider; our natures seem to have changed little since Homer. We've just gotten bigger. We appear to be the same species, with stomachs of the same size, but we aren't. It's as if each of us were trailing a big Macy's-parade balloon around, feeding it constantly.

So it doesn't do much good to stare idly out the window of your 737 as you fly from New York to Los Angeles and see that there's plenty of empty space down there. Sure enough, you could crowd lots more people into the nation or onto the planet. The entire world population could fit into Texas, and each person could have an area equal to the floor space of a typical U.S. home. If people were willing to stand, everyone on earth could fit comfortably into half of Rhode Island. Holland is crowded and is doing just fine.

But this ignores the balloons above our heads, our hungry shadow selves, our sperm-whale appetites. . . .

Those balloons above our heads can shrink or grow, depending on how we choose to live. All over the earth people who were once tiny are suddenly growing like Alice when she ate the cake. In China per capita income has doubled since the early 1980s. People there, though still Lilliputian in comparison with us, are twice their former size. They eat much higher on the food chain, understandably, than they used to: China slaughters more pigs than any other nation, and it takes four pounds of grain to produce one pound of pork. When, a decade ago, the United Nations examined sustainable development, it issued a report saying that the economies of the developing countries needed to be five to ten times as large to move poor people to an acceptable standard of living—with all that this would mean in terms of demands on oil wells and forests.

That sounds almost impossible. For the moment, though, let's not pass judgment. We're still just doing math. There are going to be lots of us. We're going to be big. But lots of us in relation to what? Big in relation to what? It could be that compared with the world we inhabit, we're still scarce and small. Or not. So now we need to consider a third question: *How big is the earth?*

Any state wildlife biologist can tell you how many deer a given area can support, how much browse there is for the deer to eat before they begin to suppress the reproduction of trees, before they begin to starve in the winter. He can calculate how many wolves a given area can support too, in part by counting the number of deer. And so on, up and down the food chain. It's not an exact science, but it comes pretty close—at least compared with figuring out the carrying capacity of the earth for human beings, which is an art so dark that anyone with any sense stays away from it.

Consider the difficulties. Hunter-gatherers used 2,500 calories of energy a day, whereas modern Americans use seventy-five times that. Human beings, unlike deer, can import what they need from thousands of miles away. And human beings, unlike deer, can figure out new ways to do old things. If, like deer, we needed to browse on conifers to survive, we could crossbreed lush new strains, chop down competing trees, irrigate forests, spray a thousand chemicals, freeze or dry the tender buds at the peak of harvest, genetically engineer new strains and advertise the merits of maple buds until everyone was ready to switch. The variables are so great that professional demographers rarely even bother trying to figure out carrying capacity. . . .

But the difficulty hasn't stopped other thinkers. . . . The most famous, of course, was the Reverend Thomas Malthus. Writing in 1798, he proposed that the growth of population, being "geometric," would soon outstrip the supply of food. Though he changed his mind and rewrote his famous essay, it's the original version that people have remembered—and lambasted ever since. Few other writers have found critics in as many corners. . . .

Each new generation of Malthusians has made new predictions that the end was near, and has been proved wrong. The late 1960s saw an upsurge of Malthusian panic. In 1967 William and Paul Paddock published a book called *Famine 1975!*, which contained a triage list: "Egypt: Can't-be-saved. . . . Tunisia: Should Receive Food. . . . India: Can't-be-saved." Almost simultaneously Paul Ehrlich wrote, in his best-selling *The Population Bomb* (1968), "The battle to feed all of humanity is over. In the 1970s, the world will undergo famines— hundreds of millions of people will starve to death." It all seemed so certain, so firmly in keeping with a world soon to be darkened by the first oil crisis.

But that's not how it worked out. India fed herself. The United States still ships surplus grain around the world. As the astute Harvard social scientist Amartya Sen points out, "Not only is food generally much cheaper to buy today, in constant dollars, than it was in Malthus's time, but it also has become cheaper during recent decades." So far, in other words, the world has more or less supported us. Too many people starve (60 percent of children in South Asia are stunted by malnutrition), but both the total number and the percentage have dropped in recent decades, thanks mainly to the successes of the Green Revolution. Food production has tripled since the Second World War, outpacing even population growth. We may be giants, but we are clever giants.

So Malthus was wrong. Over and over again he was wrong. No other prophet has ever been proved wrong so many times. At the moment, his stock is especially low. One group of technological optimists now believes that people will continue to improve their standard of living precisely because they increase their numbers. This group's intellectual fountainhead is a brilliant Danish economist named Ester Boserup—a sort of anti-Malthus, who in 1965 argued that the gloomy cleric had it backward. The more people, Boserup said, the more progress. Take agriculture as an example: the first farmers, she pointed out, were slash-and-burn cultivators, who might farm a plot for a year or two and then move on, not returning for maybe two decades. As the population grew, however, they had to return more frequently to the same plot. That meant problems: compacted, depleted, weedy soils. But those new problems meant new solutions: hoes, manure, compost, crop rotation, irrigation. Even in this century, Boserup said, necessity-induced invention has meant that "intensive systems of agriculture replaced extensive systems," accelerating the rate of food production.

Boserup's closely argued examples have inspired a less cautious group of popularizers, who point out that standards of living have risen all over the world even as population has grown. The most important benefit, in fact, that population growth bestows on an economy is to increase the stock of useful knowledge, insisted Julian Simon, the best known of the so-called cornucopians, who died earlier this year. We might run out of copper, but who cares? The mere fact of shortage will lead someone to invent a substitute. . . .

Simon and his ilk owe their success to this: they have been right so far. The world has behaved as they predicted. India hasn't starved. Food is cheap. But Malthus never goes away. The idea that we might grow too big can be disproved only for the moment—never for good. We might always be on the threshold of a special time, when the mechanisms described by Boserup and Simon stop working. It is true that Malthus was wrong when the population doubled from 750 million to 1.5 billion. It is true that Malthus was wrong when the population doubled from 1.5 billion to three billion. It is true that Malthus was wrong when the population doubled from three billion to six billion. Will Malthus still be wrong fifty years from now?

Looking at Limits

The case that the next doubling, the one we're now experiencing, might be the difficult one can begin as readily with the Stanford biologist Peter Vitousek as with anyone else. In 1986 Vitousek decided to calculate how much of the earth's "primary productivity" went to support human beings. He added together the grain we ate, the corn we fed our cows, and the forests we cut for timber and paper; he added the losses in food as we over-grazed grassland and turned it into desert. And when he was finished adding, the number he came up with was 38.8 percent. We use 38.8 percent of everything the world's plants don't need to keep themselves alive; directly or indirectly, we consume 38.8 percent of what it is possible to eat. "That's a relatively large number," Vitousek says. "It should give pause to people who think we are far from any limits." . . .

For another antidote to the good cheer of someone like Julian Simon, sit down with the Cornell biologist David Pimentel. He believes that we're in big trouble. Odd facts stud his conversation—for example, a nice head of iceberg lettuce is 95 percent water and contains just fifty calories of energy, but it takes 400 calories of energy to grow that head of lettuce in California's Central Valley, and another 1,800 to ship it east. ("There's practically no nutrition in the damn stuff anyway," Pimentel says. "Cabbage is a lot better, and we can grow it in upstate New York.") Pimentel has devoted the past three decades to tracking the planet's capacity, and he believes that we're already too crowded, that the earth can support only two billion people over the long run at a middle-class standard of living, and that trying to support more is doing great damage. . . .

The very things that made the Green Revolution so stunning, that made the last doubling possible, now cause trouble. Irrigation ditches, for instance, water 17 percent of all arable land and help to produce a third of all crops. But when flooded soils are baked by the sun, the water evaporates and the minerals in the irrigation water are deposited on the land. A hectare (2.47 acres) can accumulate two to five tons of salt annually, and eventually plants won't grow there. Maybe 10 percent of all irrigated land is affected.

Or think about fresh water for human use. Plenty of rain falls on the earth's surface, but most of it evaporates or roars down to the ocean in spring floods. According to Sandra Postel, the director of the Global Water Policy Project, we're left with about 12,500 cubic kilometers of accessible runoff, which would be enough for current demand except that it's not very well distributed around the globe. And we're not exactly conservationists—we use nearly seven times as much water as we used in 1900. Already 20 percent of the world's population lacks access to potable water, and fights over water divide many regions. . . .

What these scientists are saying is simple: human ingenuity can turn sand into silicon chips, allowing the creation of millions of home pages on the utterly fascinating World Wide Web, but human ingenuity cannot forever turn dry sand into soil that will grow food. And there are signs that these skeptics are right that we are approaching certain physical limits.

I said earlier that food production grew even faster than population after the Second World War. Year after year the yield of wheat and corn and rice rocketed up about 3 percent annually. It's a favorite statistic of the eternal optimists. In Julian Simon's book *The Ultimate Resource* (1981) charts show just how fast the growth was, and how it continually cut the cost of food. Simon wrote, "The obvious implication of this historical trend toward cheaper food, a trend that probably extends back to the beginning of agriculture, is that real prices for food will continue to drop. . . . It is a fact that portends more drops in price and even less scarcity in the future."

A few years after Simon's book was published, however, the data curve began to change. That rocketing growth in grain production ceased; now the gains were coming in tiny increments, too small to keep pace with population growth. The world reaped its largest harvest of grain per capita in 1984; since then the amount of corn and wheat and rice per person has fallen by 6 percent. Grain stockpiles have shrunk to less than two months' supply.

No one knows quite why. The collapse of the Soviet Union contributed to the trend—cooperative farms suddenly found the fertilizer supply shut off and spare parts for the tractor hard to come by. But there were other causes, too, all around the world—the salinization of irrigated fields, the erosion of topsoil, the conversion of prime farmland into residential areas, and all the other things that environmentalists had been warning about for years. It's possible that we'll still turn production around and start it rocketing again.

Charles C. Mann, writing in *Science*, quotes experts who believe that in the future a "gigantic, multi-year, multi-billion-dollar scientific effort, a kind of agricultural 'person-on-the-moon project,'" might do the trick. The next great hope of the optimists is genetic engineering, and scientists have indeed managed to induce resistance to pests and disease in some plants. To get more yield, though, a cornstalk must be made to put out another ear, and conventional breeding may have exhausted the possibilities. There's a sense that we're running into walls.

We won't start producing *less* food. Wheat is not like oil, whose flow from the spigot will simply slow to a trickle one day. But we may be getting to the point where gains will be small and hard to come by. The spectacular increases may be behind us. One researcher told Mann, "Producing higher yields will no longer be like unveiling a new model of a car. We won't be pulling off the sheet and there it is, a two-fold yield increase." Instead the process will be "incremental, torturous, and slow." And there are five billion more of us to come.

So far we're still fed; gas is cheap at the pump; the supermarket grows ever larger. We've been warned again and again about approaching limits, and we've never quite reached them. So maybe—how tempting to believe it! they don't really exist. . . .

But we can calculate risks, figure the odds that each side may be right. Joel Cohen made the most thorough attempt to do so in *How Many People Can the Earth Support?* Cohen collected and examined every estimate of carrying capacity made in recent decades, from that of a Harvard oceanographer who thought in 1976 that we might have food enough for 40 billion people to that of a Brown University researcher who calculated in 1991 that we might be able to sustain 5.9 billion (our present population), but only if we were principally vegetarians. One study proposed that if photosynthesis was the limiting factor, the earth might support a trillion people; an Australian economist proved, in calculations a decade apart, that we could manage populations of 28 billion and 157 billion. None of the studies is wise enough to examine every variable, to reach by itself the "right" number. When Cohen compared the dozens of studies, however, he uncovered something pretty interesting: the median low value for the planet's carrying capacity was 7.7 billion people, and the median high value was 12 billion. That, of course, is just the range that the UN predicts we will inhabit by the middle of the next century. . . .

Earth 2

Throughout the 10,000 years of recorded human history the planet—the physical planet—has been a stable place. In every single year of those 10,000 there have been earthquakes, volcanoes, hurricanes, cyclones, typhoons, floods, forest fires, sandstorms, hailstorms, plagues, crop failures, heat waves, cold spells, blizzards, and droughts. But these have never shaken the basic predictability of the planet as a whole. Some of the earth's land areas—the Mediterranean rim, for instance—have been deforested beyond recovery, but so far these shifts have always been local.

Among other things, this stability has made possible the insurance industry—has underwritten the underwriters. Insurers can analyze the risk in any venture because they know the ground rules. If you want to build a house on the coast of Florida, they can calculate with reasonable accuracy the chance that it will be hit by a hurricane and the speed of the winds circling that hurricane's eye. If they couldn't, they would have no way to set your premium—they'd just be gambling. They're always gambling a little, of course: they don't know if that hurricane is coming next year or next century. But the earth's physical stability is the house edge in this casino. As Julian Simon pointed out, "A prediction based on past data can be sound if it is sensible to assume that the past and the future belong to the same statistical universe."

So what does it mean that alone among the earth's great pools of money and power, insurance companies are beginning to take the idea of global climate change quite seriously? What does it mean that the payout for weather-related damage climbed from $16 billion during the entire 1980s to $48 billion in the years 1990–1994? What does it mean that top European insurance executives have begun consulting with Greenpeace about global warming? What does it mean that the insurance giant Swiss Re, which paid out $291.5 million in the wake of Hurricane Andrew, ran an ad in the Financial Times showing its corporate logo bent sideways by a storm?

These things mean, I think, that the possibility that we live on a new earth cannot be discounted entirely as a fever dream. Above, I showed attempts to calculate carrying capacity for the world as we have always known it, the world we were born into. But what if, all of a sudden, we live on some other planet? On Earth2?

In 1955 Princeton University held an international symposium on "Man's Role in Changing the Face of the Earth." By this time anthropogenic carbon, sulfur, and nitrogen were pouring into the atmosphere, deforestation was already widespread, and the population was nearing three billion. Still, by comparison with the present, we remained a puny race. Cars were as yet novelties in many places. Tropical forests were still intact, as were much of the ancient woods of the West Coast, Canada, and Siberia. The world's economy was a quarter its present size. By most calculations we have used more natural resources since 1955 than in all of human history to that time.

Another symposium was organized in 1987 by Clark University, in Massachusetts. This time even the title made clear what was happening—not "Man and Nature," not "Man's Role in Changing the Face of the Earth," but "The Earth as Transformed by Human Actions." Attendees were no longer talking about local changes or what would take place in the future. "In our judgment," they said, "the biosphere has accumulated, or is on its way to accumulating, such a magnitude and variety of changes that it may be said to have been transformed."

Many of these changes come from a direction that Malthus didn't consider. He and most of his successors were transfixed by *sources*—by figuring out whether and how we could find enough trees or corn or oil. We're good at finding more stuff; as the price rises, we look harder. The lights never did go out, despite many predictions to the contrary on the first Earth Day. We found more oil, and we still have lots and lots of coal. Meanwhile, we're driving big cars again, and why not? As of this writing, the price of gas has dropped below a dollar a gallon across much of the nation. Who can believe in limits while driving a Suburban? But perhaps, like an audience watching a magician wave his wand, we've been distracted from the real story.

That real story was told in the most recent attempt to calculate our size— a special section in *Science* published last summer. The authors spoke bluntly in the lead article. Forget man "transforming" nature, we live, they concluded, on "a human-dominated planet," where "no ecosystem on Earth's surface is free of pervasive human influence." It's not that we're running out of stuff. What we're running out of is what the scientists call "sinks"—places to put the by-products of our large appetites. Not garbage dumps (we could go on using Pampers till the end of time and still have empty space left to toss them away) but the atmospheric equivalent of garbage dumps.

It wasn't hard to figure out that there were limits on how much coal smoke we could pour into the air of a single city. It took a while longer to figure out that building ever higher smokestacks merely lofted the haze farther afield, raining down acid on whatever mountain range lay to the east. Even that, however, we are slowly fixing, with scrubbers and different mixtures of fuel. We can't so easily repair the new kinds of pollution. These do not come from something going wrong—some engine without a catalytic converter, some waste-water pipe without a filter, some smokestack without a scrubber. New kinds of pollution come instead from things going as they're supposed to go—but at such a high volume that they overwhelm the planet. They come from normal human life but there are so many of us living those normal lives that something abnormal is happening. And that something is so different from the old forms of pollution that it confuses the issue even to use the word.

Consider nitrogen, for instance. Almost 80 percent of the atmosphere is nitrogen gas. But before plants can absorb it, it must become "fixed"—bonded with carbon, hydrogen, or oxygen. Nature does this trick with certain kinds of algae and soil bacteria, and with lightning. Before human beings began to alter the nitrogen cycle, these mechanisms provided 90–150 million metric tons of nitrogen a year. Now human activity adds 130–150 million more tons. Nitrogen isn't pollution, it's essential. And we are using more of it all the time. Half the industrial nitrogen fertilizer used in human history has been applied since 1984. As a result, coastal waters and estuaries bloom with toxic algae while oxygen concentrations dwindle, killing fish; as a result, nitrous oxide traps solar heat. And once the gas is in the air, it stays there for a century or more.

Or consider methane, which comes out of the back of a cow or the top of a termite mound or the bottom of a rice paddy. As a result of our determination to raise more cattle, cut down more tropical forest (thereby causing termite populations to explode), and grow more rice, methane concentrations in the atmosphere are more than twice as high as they have been for most of the past 160,000 years. And methane traps heat—very efficiently.

Or consider carbon dioxide. In fact, concentrate on carbon dioxide. If we had to pick one problem to obsess about over the next fifty years, we'd do well to make it CO_2—which is not pollution either. Carbon *monoxide* is pollution: it kills you if you breathe enough of it. But carbon *dioxide*, carbon with two oxygen atoms, can't do a blessed thing to you. If you're reading this indoors, you're breathing more CO_2 than you'll ever get outside. For generations, in fact, engineers said that an engine burned clean if it produced only water vapor and carbon dioxide.

Here's the catch: that engine produces *a lot* of CO_2. A gallon of gas weighs about eight pounds. When it's burned in a car, about five and a half pounds of carbon, in the form of carbon dioxide, come spewing out the back. It doesn't matter if the car is a 1958 Chevy or a 1998 Saab. And no filter can reduce that flow—it's an inevitable by-product of fossil-fuel combustion, which is why CO_2 has been piling up in the atmosphere ever since the Industrial Revolution. Before we started burning oil and coal and gas, the atmosphere contained about 280 parts CO_2 per million. Now the figure is about 360. Unless we do everything we can think of to eliminate fossil fuels from our diet, the air will test out at more than 500 parts per million fifty or sixty years from now, whether it's sampled in the South Bronx or at the South Pole.

This matters because, as we all know by now, the molecular structure of this clean, natural, common element that we are adding to every cubic foot of the atmosphere surrounding us traps heat that would otherwise radiate back out to space. Far more than even methane and nitrous oxide, CO_2 causes global warming—the greenhouse effect—and climate change. Far more than any other single factor, it is turning the earth we were born on into a new planet.

Remember, this is not pollution as we have known it. In the spring of last year the Environmental Protection Agency issued its "Ten-Year Air Quality and Emissions Trends" report. Carbon monoxide was down by 37 percent since 1986, lead was down by 78 percent, and particulate matter had dropped by nearly a quarter. If you lived in the San Fernando Valley, you saw the mountains more often than you had a decade before. The air was *cleaner*, but it was also *different*—richer with CO_2. And its new composition may change almost everything....

... For ten years, with heavy funding from governments around the world, scientists launched satellites, monitored weather balloons, studied clouds. Their work culminated in a long-awaited report from the UN's Intergovernmental Panel on Climate Change, released in the fall of 1995. The panel's 2,000 scientists, from every corner of the globe, summed up their findings in this dry but historic bit of understatement: "The balance of evidence suggests that there is a discernible human influence on global climate." That is to say, we are heating up the planet—substantially. If we don't reduce emissions of carbon dioxide and other gases, the panel warned, temperatures will probably rise 3.6° Fahrenheit by 2100, and perhaps as much as 6.3°.

You may think you've already heard a lot about global warming. But most of our sense of the problem is behind the curve. Here's the current news: the changes are already well under way. When politicians and businessmen talk about "future risks," their rhetoric is outdated. This is not a problem for the distant future, or even for the near future. The planet has already heated up by a degree or more. We are perhaps a quarter of the way into the greenhouse era, and the effects are already being felt. From a new heaven, filled with nitrogen, methane, and carbon, a new earth is being born. If some alien astronomer is watching us, she's doubtless puzzled. This is the most obvious effect of our numbers and our appetites, and the key to understanding why the size of our population suddenly poses such a risk. ...

If you want to scare yourself with guesses about what might happen in the near future, there's no shortage of possibilities. Scientists have already observed large-scale shifts in the duration of the El Niño ocean warming, for instance. The Arctic tundra has warmed so much that in some places it now gives off more carbon dioxide than it absorbs—a switch that could trigger a potent feedback loop, making warming ever worse. And researchers studying glacial cores from the Greenland Ice Sheet recently concluded that local climate shifts have occurred with incredible rapidity in the past—18° in one three-year stretch. Other scientists worry that such a shift might be enough to flood the oceans with fresh water and reroute or shut off currents like the Gulf Stream and the North Atlantic, which keep Europe far warmer than it would otherwise be. (See "The Great Climate Flip-flop," by William H. Calvin, January Atlantic.) In the words of Wallace Broecker, of Columbia University, a pioneer in the field, "Climate is an angry beast, and we are poking it with sticks." ...

The arguments put forth by cornucopians like Julian Simon—that human intelligence will get us out of any scrape, that human beings are "the ultimate resource," that Malthusian models "simply do not comprehend key elements of people"—all rest on the same premise: that human beings change the world mainly for the better.

If we live at a special time, the single most special thing about it may be that we are now apparently degrading the most basic functions of the planet. It's not that we've never altered our surroundings before. Like the beavers at work in my back yard, we have rearranged things wherever we've lived. We've leveled the spots where we built our homes, cleared forests for our fields, often fouled nearby waters with our waste. That's just life. But this is different. In the past ten or twenty or thirty years our impact has grown so much that we're changing even those places we don't inhabit—changing the way the weather works, changing the plants and animals that live at the poles or deep in the jungle. This is total. Of all the remarkable and unexpected things we've ever done as a species, this may be the biggest. Our new storms and new oceans and new glaciers and new springtimes these are the eighth and ninth and tenth and eleventh wonders of the modern world, and we have lots more where those came from.

We have gotten very large and very powerful, and for the foreseeable future we're stuck with the results. The glaciers won't grow back again anytime soon; the oceans won't drop. We've already done deep and systemic damage. To use a human analogy, we've already said the angry and unforgivable words that will haunt our marriage till its end. And yet we can't simply walk out the door. There's no place to go. We have to salvage what we can of our relationship with the earth, to keep things from getting any worse than they have to be.

If we can bring our various emissions quickly and sharply under control, we *can* limit the damage, reduce dramatically the chance of horrible surprises, preserve more of the biology we were born into. But do not underestimate the task. The UN's Intergovernmental Panel on Climate Change projects that an immediate 60 percent reduction in fossil-fuel use is necessary just to stabilize climate at the current level of disruption. Nature may still meet us halfway, but halfway is a long way from where we are now. What's more, we can't delay. If we wait a few decades to get started, we may as well not even begin. It's not like poverty, a concern that's always there for civilizations to address. This is a timed test, like the SAT: two or three decades, and we lay our pencils down. It's *the* test for our generations, and population is a part of the answer.

Changing "Unchangeable" Needs

When we think about overpopulation, we usually think first of the developing world, because that's where 90 percent of new human beings will be added during this final doubling. . . .

We fool ourselves when we think of Third World population growth as producing an imbalance, as Amartya Sen points out. The white world simply went through its population boom a century earlier (when Dickens was writing similar descriptions of London). If UN calculations are correct and Asians and Africans will make up just under 80 percent of humanity by 2050, they will simply have returned, in Sen's words, "to being proportionately almost exactly as numerous as they were before the European industrial revolution."

And of course Asians and Africans, and Latin Americans, are much "smaller" human beings: the balloons that float above their heads are tiny in comparison with ours. Everyone has heard the statistics time and again, usually as part of an attempt to induce guilt. But hear them one more time, with an open mind, and try to think strategically about how we will stave off the dangers to this planet. Pretend it's not a moral problem, just a mathematical one.

- An American uses seventy times as much energy as a Bangladeshi, fifty times as much as a Malagasi, twenty times as much as a Costa Rican.
- Since we live longer, the effect of each of us is further multiplied. In a year an American uses 300 times as much energy as a Malian; over a lifetime he will use 500 times as much.
- Even if all such effects as the clearing of forests and the burning of grasslands are factored in and attributed to poor people, those who live in the poor world are typically responsible for the annual release of a tenth of a ton of carbon each, whereas the average is 3.5 tons for residents of the "consumer" nations of Western Europe, North America, and Japan. The richest tenth of Americans annually emit eleven tons of carbon apiece.
- During the next decade India and China will each add to the planet about ten times as many people as the United States will but the stress on the natural world caused by new Americans may exceed that from new Indians and Chinese combined. The 57.5 million Northerners added to our population during this decade will add more greenhouse gases to the atmosphere than the roughly 900 million added Southerners.

These statistics are not eternal. Though inequality between North and South has steadily increased, the economies of the poor nations are now growing faster than those of the West. Sometime early in the next century China will pass the United States as the nation releasing the most carbon dioxide into the atmosphere, though of course it will be nowhere near the West on a per capita basis.

For the moment, then (and it is the moment that counts), we can call the United States the most populous nation on earth, and the one with the highest rate of growth. Though the U.S. population increases by only about three million people a year, through births and immigration together, each of those three million new Americans will consume on average forty or fifty times as much as a person born in the Third World. My daughter, four at this writing, has already used more stuff and added more waste to the environment than most of the world's residents do in a lifetime. In my thirty-seven years I have probably outdone small Indian villages.

Population growth in Rwanda, in Sudan, in El Salvador, in the slums of Lagos, in the highland hamlets of Chile, can devastate those places. Growing too fast may mean that they run short of cropland to feed themselves, of firewood to cook their food, of school desks and hospital beds. But population growth in those places doesn't devastate the planet. In contrast, we easily absorb the modest annual increases in our population. America seems only a little more crowded with each passing decade in terms of our daily lives. You can still find a parking spot. But the earth simply can't absorb what we are adding to its air and water.

So if it is we in the rich world, at least as much as they in the poor world, who need to bring this alteration of the earth under control, the question becomes how. Many people who are sure that controlling population is the answer overseas are equally sure that the answer is different here. If those people are politicians and engineers, they're probably in favor of our living more efficiently—of designing new cars that go much farther on a gallon of gas, or that don't use gas at all. If they're vegetarians, they probably support living more simply—riding bikes or buses instead of driving cars.

Both groups are utterly correct. I've spent much of my career writing about the need for cleverer technologies and humbler aspirations. Environmental damage can be expressed as the product of Population – Affluence – Technology. Surely the easiest solution would be to live more simply and more efficiently, and not worry too much about the number of people.

But I've come to believe that those changes in technology and in lifestyle are not going to occur easily and speedily. They'll be begun but not finished in the few decades that really matter. Remember that the pollution we're talking about is not precisely pollution but rather the inevitable result when things go the way we think they should: new filters on exhaust pipes won't do anything about that CO_2. We're stuck with making real changes in how we live. We're stuck with dramatically reducing the amount of fossil fuel we use. And since modern Westerners are practically machines for burning fossil fuel, since virtually everything we do involves burning coal and gas and oil, since we're wedded to petroleum, it's going to be a messy breakup.

So we need to show, before returning again to population, why simplicity and efficiency will not by themselves save the day. . . . Speaking to the United Nations early last summer, [Bill Clinton] said plainly, "We humans are changing the global climate. . . . No nation can escape this danger. None can evade its responsibility to confront it, and we must all do our part."

But when it comes time to do our part, we don't. After all, Clinton warned of the dangers of climate change in 1993, on his first Earth Day in office. In fact, he solemnly promised to make sure that America produced no more greenhouse gases in 2000 than it had in 1990. But he didn't keep his word. The United States will spew an amazing 15 percent more carbon dioxide in 2000 than it did in 1990. . . .

What's important to understand is why we broke our word. We did so because Clinton understood that if we were to keep it, we would need to raise the price of fossil fuel. If gasoline cost $2.50 a gallon, we'd drive smaller cars, we'd drive electric cars, we'd take buses—and we'd elect a new president. We can hardly blame Clinton, or any other politician. His real goal has been to speed the pace of economic growth, which has been the key to his popularity. If all the world's leaders could be gathered in a single room, the one thing that every last social- ist, Republican, Tory, monarchist, and trade unionist could agree on would be the truth of Clinton's original campaign admonition: "It's the economy, stupid."

The U.S. State Department had to send a report to the United Nations explaining why we would not be able to keep our Earth Day promise to reduce greenhouse-gas emissions; the first two reasons cited were "lower-than-expected fuel prices" and "strong economic growth." . . .

America's unease with real reductions in fossil-fuel use was clear at last year's mammoth global-warming summit in Kyoto. With utility executives and Republican congressmen stalking the halls, the U.S. delegation headed off every attempt by other nations to strengthen the accord. . . .

Changing the ways in which we live has to be a fundamental part of deal- ing with the new environmental crises, if only because it is impossible to imag- ine a world of 10 billion people consuming at our level. But as we calculate what must happen over the next few decades to stanch the flow of CO_2, we shouldn't expect that a conversion to simpler ways of life will by itself do the trick. One would think offhand that compared with changing the number of children we bear, changing consumption patterns would be a breeze. Fertility, after all, seems biological—hard-wired into us in deep Darwinian ways. But I would guess that it is easier to change fertility than lifestyle.

Perhaps our salvation lies in the other part of the equation—in the new technologies and efficiencies that could make even our wasteful lives benign, and table the issue of our population. We are, for instance, converting our econ- omy from its old industrial base to a new model based on service and informa- tion. Surely that should save some energy, should reduce the clouds of carbon dioxide. Writing software seems no more likely to damage the atmosphere than writing poetry.

Forget for a moment the hardware requirements of that new economy— for instance, the production of a six-inch silicon wafer may require nearly 3,000 gallons of water. But do keep in mind that a hospital or an insurance company or a basketball team requires a substantial physical base. Even the highest-tech office is built with steel and cement, pipes and wires. People working in services will buy all sorts of things—more software, sure, but also more sport utility vehi- cles. As the Department of Energy economist Arthur Rypinski says, "The infor- mation age has arrived, but even so people still get hot in the summer and cold in the winter. And even in the information age it tends to get dark at night."

Yes, when it gets dark, you could turn on a compact fluorescent bulb, sav- ing three-fourths of the energy of a regular incandescent. Indeed, the average

American household, pushed and prodded by utilities and environmentalists, has installed one compact fluorescent bulb in recent years; unfortunately, over the same period it has also added seven regular bulbs. Millions of halogen torchere lamps have been sold in recent years, mainly because they cost $15.99 at Kmart. They also suck up electricity: those halogen lamps alone have wiped out all the gains achieved by compact fluorescent bulbs. Since 1983 our energy use per capita has been increasing by almost 1 percent annually, despite all the technological advances of those years.

As with our homes, so with our industries. Mobil Oil regularly buys ads in leading newspapers to tell "its side" of the environmental story. As the company pointed out recently, from 1979 to 1993 "energy consumption per unit of gross domestic product" dropped 19 percent across the Western nations. This sounds good—it's better than 1 percent a year. But of course the GDP grew more than 2 percent annually. So total energy use, and total clouds of CO_2, continued to increase.

It's not just that we use more energy. There are also more of us all the time, even in the United States. If the population is growing by about 1 percent a year, then we have to keep increasing our technological efficiency by that much each year—and hold steady our standard of living—just to run in place. The President's Council on Sustainable Development, in a little-read report issued in the winter of 1996, concluded that "efficiency in the use of all resources would have to increase by more than fifty percent over the next four or five decades just to keep pace with population growth." Three million new Americans annually means many more cars, houses, refrigerators. Even if everyone consumes only what he consumed the year before, each year's tally of births and immigrants will swell American consumption by 1 percent.

We demand that engineers and scientists swim against that tide. And the tide will turn into a wave if the rest of the world tries to live as we do. It's true that the average resident of Shanghai or Bombay will not consume as lavishly as the typical San Diegan or Bostonian anytime soon, but he will make big gains, pumping that much more carbon dioxide into the atmosphere and requiring that we cut our own production even more sharply if we are to stabilize the world's climate.

The United Nations issued its omnibus report on sustainable development in 1987. An international panel chaired by Gro Harlem Brundtland, the Prime Minister of Norway, concluded that the economies of the developing countries needed to grow five to ten times as large as they were, in order to meet the needs of the poor world. And that growth won't be mainly in software. As Arthur Rypinski points out, "Where the economy is growing really rapidly, energy use is too." In Thailand, in Tijuana, in Taiwan, every 10 percent increase in economic output requires 10 percent more fuel. "In the Far East," Rypinski says, "the transition is from walking and bullocks to cars. People start out with electric lights and move on to lots of other stuff. Refrigerators are one of those things that are really popular everywhere. Practically no one, with the possible exception of people in the high Arctic, doesn't want a refrigerator. As people get wealthier, they tend to like space heating and cooling, depending on the climate."

In other words, in doing the math about how we're going to get out of this fix, we'd better factor in some unstoppable momentum from people on the rest of the planet who want the very basics of what we call a decent life. Even if we airlift solar collectors into China and India, as we should, those nations will still burn more and more coal and oil. "What you can do with energy conservation in those situations is sort of at the margin," Rypinski says. "They're not interested in fifteen-thousand-dollar clean cars versus five-thousand-dollar dirty cars. It was hard enough to get Americans to invest in efficiency; there's no feasible amount of largesse we can provide to the rest of the world to bring it about."

The numbers are so daunting that they're almost unimaginable. Say, just for argument's sake, that we decided to cut world fossil-fuel use by 60 percent—the amount that the UN panel says would stabilize world climate. And then say that we shared the remaining fossil fuel equally. Each human being would get to produce 1.69 metric tons of carbon dioxide annually—which would allow you to drive an average American car nine miles a day. By the time the population increased to 8.5 billion, in about 2025, you'd be down to six miles a day. If you carpooled, you'd have about three pounds of CO_2 left in your daily ration— enough to run a highly efficient refrigerator. Forget your computer, your TV, your stereo, your stove, your dishwasher, your water heater, your microwave, your water pump, your clock. Forget your light bulbs, compact fluorescent or not.

I'm not trying to say that conservation, efficiency, and new technology won't help. They will—but the help will be slow and expensive. The tremendous momentum of growth will work against it. Say that someone invented a new furnace tomorrow that used half as much oil as old furnaces. How many years would it be before a substantial number of American homes had the new device? And what if it cost more? And if oil stays cheaper per gallon than bottled water? Changing basic fuels to hydrogen, say—would be even more expensive. It's not like running out of white wine and switching to red. . . .

There are no silver bullets to take care of a problem like this. Electric cars won't by themselves save us, though they would help. We simply won't live efficiently enough soon enough to solve the problem. Vegetarianism won't cure our ills, though it would help. We simply won't live simply enough soon enough to solve the problem.

Reducing the birth rate won't end all our troubles either. That, too, is no silver bullet. But it would help. There's no more practical decision than how many children to have. (And no more mystical decision, either.)

The bottom-line argument goes like this: The next fifty years are a special time. They will decide how strong and healthy the planet will be for centuries to come. Between now and 2050 we'll see the zenith, or very nearly, of human population. With luck we'll never see any greater production of carbon dioxide or toxic chemicals. We'll never see more species extinction or soil erosion. Greenpeace recently announced a campaign to phase out fossil fuels entirely by mid-century, which sounds utterly quixotic but could, if everything went just right, happen.

So it's the task of those of us alive right now to deal with this special phase, to squeeze us through these next fifty years. That's not fair—any more than it was fair that earlier generations had to deal with the Second World War or the Civil War or the Revolution or the Depression or slavery. It's just reality. We need in these fifty years to be working simultaneously on all parts of the equation on our ways of life, on our technologies, and on our population.

As Gregg Easterbrook pointed out in his book *A Moment on the Earth* (1995), if the planet does manage to reduce its fertility, "the period in which human numbers threaten the biosphere on a general scale will turn out to have been much, much more brief" than periods of natural threats like the Ice Ages. True enough. But the period in question happens to be our time. That's what makes this moment special, and what makes this moment hard.

Note

McKibben, Bill. 1998. "A Special Moment in History: The Future of Population." *Atlantic Monthly* 281, no. 5.

Cuba

A Successful Case Study of Sustainable Agriculture

Peter M. Rosset

This reading by Peter Rosset, executive director of the Institute for Food and Development Policy/Food First, describes how Cuba developed an alternative agricultural system out of necessity. After the fall of the Soviet Union, Cuba moved from a highly mechanized system of agricultural production based on high inputs of fossil fuels and chemical products to a system that is "low input rather than high input" and that relies more on organic fertilizers and biopesticides. Cuba serves as a case study of a way to produce food that is less environmentally and socially damaging than the corporate model.

Our global food system is in the midst of a multifaceted crisis, with ecological, economic, and social dimensions. To overcome that crisis, political and social changes are needed to allow the widespread development of alternatives.

The current food system is productive—there should be no doubt about that—as per capita food produced in the world has increased by 15 percent over the past thirty-five years. But as that production is in ever fewer hands, and costs ever more in economic and ecological terms, it becomes harder and harder to address the basic problems of hunger and food access in the short term, let alone in a sustainable fashion. In the last twenty years the number of hungry people in the world—excluding China—has risen by 60 million (by contrast, in China the number of hungry people has fallen dramatically).

Ecologically, there are impacts of industrial-style farming on groundwater through pesticide and fertilizer runoff, on biodiversity through the spread of monoculture and a narrowing genetic base, and on the very capacity of agro-ecosystems to be productive into the future.

Economically, production costs rise as farmers are forced to use ever more expensive machines and farm chemicals, while crop prices continue a several decade-long downward trend, causing a cost-price squeeze which has led to the loss of untold tens of millions of farmers worldwide to bankruptcies. Socially, we have the concentration of farmland in fewer and fewer hands as low crop prices make farming on a small scale unprofitable (despite higher per acre total productivity of small farms), and agribusiness corporations extend their control over more and more basic commodities.

Clearly the dominant corporate food system is not capable of adequately addressing the needs of people or of the environment. Yet there are substantial obstacles to the widespread adoption of alternatives. The greatest obstacles are presented by political-corporate power and vested interests, yet at times the psychological barrier to believing that the alternatives can work seems almost as difficult to overcome. Thee oft-repeated challenge is: "Could organic farming (or agroecology, local production, small farms, farming without pesticides) ever really feed the entire population of a country?" Recent Cuban history—the overcoming of a food crisis through self-reliance, small farms, and agroecological technology—shows us that the alternatives can indeed feed a nation, and thus provides a crucial case study for the ongoing debate.

A Brief History

Economic development in Cuba was molded by two external forces between the 1959 revolution and the 1989–90 collapse of trading relations with the Soviet bloc. One was the U.S. trade embargo, part of an effort to isolate the island economically and politically. The other was Cuba's entry into the Soviet bloc's international trade alliance with relatively favorable terms of trade. The United States embargo essentially forced Cuba to turn to the Soviet bloc, while the terms of trade offered by the latter opened the possibility of more rapid development on the island than in the rest of Latin America and the Caribbean.

Thus Cuba was able to achieve a more complete and rapid modernization than most other developing countries. In the 1980s, it ranked number one in the region in the contribution of industry to its economy and it had a more mechanized agricultural sector than any other Latin American country. Nevertheless, some of the same contradictions that modernization produced in other Third World countries were apparent in Cuba, with Cuba's development model proving ultimately to be of the dependent type. Agriculture was defined by extensive monocrop production of export crops and a heavy dependence on imported agrichemicals, hybrid seeds, machinery, and petroleum. While industrialization was substantial by regional standards, Cuban industry depended on many imported inputs.

The Cuban economy as a whole was thus characterized by the contradiction between its relative modernity and its function in the Soviet bloc's division of labor as a supplier of raw agricultural commodities and minerals, and a net importer of both manufactured goods and foodstuffs. In contrast to the situation faced by most Third World countries, this international division of labor actually brought significant benefits to the Cuban people. Prior to the collapse of the socialist bloc, Cuba had achieved high marks for per capita GNP, nutrition, life expectancy, and women in higher education, and was ranked first in Latin America for the availability of doctors, low infant mortality, housing, secondary school enrollment, and attendance by the population at cultural events.

The Cuban achievements were made possible by a combination of the government's commitment to social equity and the fact that Cuba received far more favorable terms of trade for its exports than did the hemisphere's other developing nations. During the 1980s, Cuba received an average price for its sugar exports to the Soviet Union that was 5.4 times higher than the world price. Cuba also was able to obtain Soviet petroleum in return, part of which was re-exported to earn convertible currency. Because of the favorable terms of trade for sugar, its production far outweighed that of food crops. About three times as much land was devoted to sugar in 1989 as was used for food crops, contributing to a pattern of food dependency, with as much as 57 percent of the total calories in the Cuban diet coming from imports.

The revolutionary government had inherited an agricultural production system strongly focused on export crops grown on highly concentrated land. The first agrarian reform of 1959 converted most of the large cattle ranches and sugarcane plantations into state farms. Under the second agrarian reform in 1962, the state took control of 63 percent of all cultivated land.

Even before the revolution, individual peasant producers were a small part of the agricultural scene. The rural economy was dominated by export plantations, and the population as a whole was highly urbanized. That pattern intensified in subsequent years, and by the late 1980s fully 69 percent of the island's population lived in urban areas. As late as 1994, some 80 percent of the nation's agricultural land consisted of large state farms, which roughly correspond to the expropriated plantation holdings of the pre-revolutionary era. Only 20 percent of the agricultural land was in the hands of small farmers, split almost equally among individual holders and cooperatives, yet this 20 percent produced more than 40 percent of domestic food production. The state farm sector and a substantial portion of the cooperatives were highly modernized, with large areas of monocrops worked under heavy mechanization, fertilizer and pesticide use, and large-scale irrigation. This style of farming, originally copied from the advanced capitalist countries by the Soviet Union, was highly dependent on imports of machinery, petroleum, and chemicals. When trade collapsed with the socialist bloc, the degree to which Cuba relied on monocrop agriculture proved to be a major weakness of the revolution.

Onset of the Crisis

When trade relations with the Soviet bloc crumbled in late 1989 and 1990, the situation turned desperate. In 1991, the government declared the "Special Period in Peacetime," which basically put the country on a wartime economy style austerity program. There was an immediate 53 percent reduction in oil imports that not only affected fuel availability for the economy, but also reduced to zero the foreign exchange that Cuba had formerly obtained via the re-export of petroleum. Imports of wheat and other grains for human consumption dropped by more than 50 percent, while other foodstuffs declined even more. Cuban agriculture was faced with a drop of more than 80 percent in the availability of fertilizers and pesticides, and more than 50 percent in fuel and other energy sources produced by petroleum.

Suddenly, a country with an agricultural sector technologically similar to California's found itself almost without chemical inputs, with sharply reduced access to fuel and irrigation, and with a collapse in food imports. In the early 1990s average daily caloric and protein intake by the Cuban population may have been as much as 30 percent below levels in the 1980s. Fortunately, Cuba was not totally unprepared to face the critical situation that arose after 1989.

It had, over the years, emphasized the development of human resources, and therefore had a cadre of scientists and researchers who could come forward with innovative ideas to confront the crisis. While Cuba has only 2 percent of the population of Latin America, it has almost 11 percent of the scientists.

Alternative Technologies

In response to the crisis, the Cuban government launched a national effort to convert the nation's agricultural sector from high input agriculture to low input, self-reliant farming practices on an unprecedented scale. Because of the drastically reduced availability of chemical inputs, the state hurried to replace them with locally produced, and in most cases biological, substitutes. This has meant biopesticides (microbial products) and natural enemies to combat insect pests, resistant plant varieties, crop rotations and microbial antagonists to combat plant pathogens, and better rotations, and cover cropping to suppress weeds. Synthetic fertilizers have been replaced by biofertilizers, earthworms, compost, other organic fertilizers, natural rock phosphate, animal and green manures, and the integration of grazing animals. In place of tractors, for which fuel, tires, and spare parts were largely unavailable, there has been a sweeping return to animal traction.

Small Farmers Respond to the Crisis

When the collapse of trade and subsequent scarcity of inputs occurred in 1989–90, yields fell drastically throughout the country. The first problem was that of producing without synthetic chemical inputs or tractors. Gradually the national ox herd was built up to provide animal traction as a substitute for tractors, and the production of biopesticides and biofertilizers was rapidly stepped up. Finally, a series of methods like vermicomposting (earthworm composting) of residues and green manuring became widespread. But the impact of these technological changes across subsectors of Cuban agriculture was highly variable. The drop-off of yields in the state sector industrial-style farms that average thousands of hectares has been resistant to recovery, with production seriously stagnating well below pre-crisis levels for export crops. Yet the small farm or peasant sector (20 percent of farmed land) responded rapidly by quickly boosting production above previous levels. How can we explain the difference between the state- and small-farm sectors?

It really was not all that difficult for the small farm sector to effectively produce with fewer inputs. After all, today's small farmers are the descendants of generations of small farmers, with long family and community traditions of low-input production. They basically did two things: remembered the old techniques—intercropping and manuring—that their parents and grandparents had used before the advent of modern chemicals, and simultaneously incorporated new biopesticides and biofertilizers into their production practices.

State Farms Incompatible
with the Alternative Technologies

The problems of the state sector, on the other hand, were a combination of low worker productivity, a problem predating the Special Period, and the complete inability of these immense and technified management units to adapt to low-input technology. With regard to the productivity problem, planners became aware several years ago that the organization of work on state farms was profoundly alienating in terms of the relationship between the agricultural worker and the land. Large farms of thousands of hectares had their work forces organized into teams that would prepare the soil in one area, move on to plant another, weed still another, and later harvest an altogether different area. Almost never would the same person both plant and harvest the same area. Thus no one ever had to confront the consequences of doing something badly or, conversely, enjoy the fruits of his or her own labor.

In an effort to create a more intimate relationship between farm workers and the land, and to tie financial incentives to productivity, the government began several years ago to experiment with a program called "linking people with the land." This system made small work teams directly responsible for all aspects of production in a given parcel of land, allowing remuneration to be directly linked to productivity. The new system was tried before the Special Period on a number of state farms, and rapidly led to enormous increases in production. Nevertheless it was not widely implemented at the time.

In terms of technology, scale effects are very different for conventional chemical management and for low external input alternatives. Under conventional systems, a single technician can manage several thousand hectares on a "recipe" basis by simply writing out instructions for a particular fertilizer formula or pesticide to be applied with machinery on the entire area. This is not so for agroecological farming. Whoever manages the farm must be intimately familiar with the ecological heterogeneity of each individual patch of soil. The farmer must know, for example, where organic matter needs to be added, and where pest and natural enemy refuges and entry points are. This partially explains the inability of the state sector to raise yields with alternative inputs. Like the productivity issue, it can only be effectively addressed through a re-linking of people with the land.

By mid-1993, the state was faced with a complex reality. Imported inputs were largely unavailable, but nevertheless the small farmer sector had effectively adapted to low input production (although a secondary problem was acute in this sector, namely diversion of produce to the black market). The state sector, on the other hand, was proving itself to be an ineffective "white elephant" in the new historical conjuncture, incapable of adjusting. The earlier success of the experimental "linking" program, however, and the success of the peasant sector, suggested a way out. In September 1993, Cuba began radically reorganizing its production in order to create the small-scale management units that are essential for effective organic-style farming. This reorganization has centered on the privatization and cooperativization of the unwieldy state sector.

The process of linking people with the land thus culminated in 1993, when the Cuban government issued a decree terminating the existence of state farms, turning them into Basic Units of Cooperative-Production (UBPCs), a form of worker-owned enterprise or cooperative. The 80 percent of all farmland that was once held by the state, including sugarcane plantations, has now essentially been turned over to the workers.

The UBPCs allow collectives of workers to lease state farmlands rent free, in perpetuity. Members elect management teams that determine the division of jobs, what crops will be planted on which parcels, and how much credit will be taken out to pay for the purchase of inputs. Property rights remain in the hands of the state, and the UBPCs must still meet production quotas for their key crops, but the collectives are owners of what they produce. Perhaps most importantly, what they produce in excess of their quotas can now be freely sold on the newly reopened farmers markets. This last reform, made in 1994, offered a price incentive to farmers both to sell their produce through legal channels rather than the black market, and also to make effective use of the new technologies.

The pace of consolidation of the UBPCs has varied greatly in their first years of life. Today one can find a range from those where the only change is that the old manager is now an employee of the workers, to those that truly function as collectives, to some in which the workers are parceling the farms into small plots worked by groups of friends. In almost all cases, the effective size of the management unit has been drastically reduced. It is still too early to predict the final variety of structures that the UBPCs will evolve toward. But it is clear that the process of turning previously alienated farm workers into farmers will take some time—it simply cannot be accomplished overnight—and many UBPCs are struggling. Incentives are a nagging problem. Most UBPCs are stuck with state production contracts for export crops like sugar and citrus. These still have fixed, low prices paid by state marketing agencies, in contrast to the much higher prices that can be earned for food crops. Typical UBPCs, not surprisingly then, have low yields in their export crops, but also have lucrative side businesses selling food produced on spare land or between the rows of their citrus or sugarcane.

Food Shortage Overcome

By mid-1995 the food shortage had been overcome, and the vast majority of the population no longer faced drastic reductions of their basic food supply. In the 1996–97 growing season, Cuba recorded its highest-ever production levels for ten of the thirteen basic food items in the Cuban diet. The production increases came primarily from small farms, and in the case of eggs and pork, from booming backyard production. The proliferation of urban farmers who produce fresh produce has also been extremely important to the Cuban food supply. The earlier food shortages and the rise in food prices suddenly turned urban agriculture into a very profitable activity for Cubans, and once the government threw its full support behind a nascent urban gardening movement, it exploded to near epic proportions. Formerly vacant lots and backyards in all Cuban cities now sport food crops and farm animals, and fresh produce is sold from private stands throughout urban areas at prices substantially below those prevailing in the farmers markets. There can be no doubt that urban farming, relying almost exclusively on organic techniques, has played a key role in assuring the food security of Cuban families over the past two to three years.

An Alternative Paradigm?

To what extent can we see the outlines of an alternative food system paradigm in this Cuban experience? Or is Cuba just such a unique case in every way that we cannot generalize its experiences into lessons for other countries? The first thing to point out is that contemporary Cuba turned conventional wisdom completely on its head. We are told that small countries cannot feed themselves, that they need imports to cover the deficiency of their local agriculture. Yet Cuba has taken enormous strides toward self-reliance since it lost its key trade relations. We hear that a country can't feed its people without synthetic farm chemicals, yet Cuba is virtually doing so. We are told that we need the efficiency of large-scale corporate or state farms in order to produce enough food, yet we find small farmers and gardeners in the vanguard of Cuba's recovery from a food crisis. In fact, in the absence of subsidized machines and imported chemicals, small farms are more efficient than very large production units. We hear time and again that international food aid is the answer to food shortages—yet Cuba has found an alternative in local production.

Abstracting from that experience, the elements of an alternative paradigm might therefore be:

1. *Agroecological Technology Instead of Chemicals:* Cuba has used intercropping, locally produced biopesticides, compost, and other alternatives to synthetic pesticides and fertilizers.

2. *Fair Prices for Farmers:* Cuban farmers stepped up production in response to higher crop prices. Farmers everywhere lack incentive to produce when prices are kept artificially low, as they often are. Yet when given an incentive, they produce, regardless of the conditions under which that production must take place.

3. *Redistribution of Land:* Small farmers and gardeners have been the most productive of Cuban producers under low-input conditions. Indeed, smaller farms worldwide produce much more per unit area than do large farms. In Cuba redistribution was relatively easy to accomplish because the major part of the land reform had already occurred, in the sense that there were no landlords to resist further change.

4. *Greater Emphasis on Local Production:* People should not have to depend on the vagaries of prices in the world economy, long distance transportation, and super power "goodwill" for their next meal. Locally and regionally produced food offers greater security, as well as synergistic linkages to promote local economic development. Furthermore such production is more ecologically sound, as the energy spent on international transport is wasteful and environmentally unsustainable. By promoting urban farming, cities and their surrounding areas can be made virtually self-sufficient in perishable foods, be beautified, and have

greater employment opportunities. Cuba gives us a hint of the under-exploited potential of urban farming.

Cuba in its Special Period has clearly been in a unique situation with respect to not being able to use power machinery in the fields, forcing it to seek alternatives such as animal traction. It is unlikely that Cuba or any other country at its stage of development would choose to abandon machine agriculture to this extent unless compelled to do so. Yet there are important lessons here for countries struggling to develop. Relatively small-scale farming, even using animals for traction, can be very productive per unit of land, given technical support. And it is next to impossible to have ecologically sound farming at an extremely large scale. Although it is undeniable that for countries wishing to develop industry and at the same time grow most of their own food, some mechanization of agriculture will be needed, it is crucial to recognize—and the Cuban example can help us to understand this—that modest-sized family farms and cooperatives that use reasonably sized equipment can follow ecologically sound practices and have increased labor productivity.

The Cuban experience illustrates that we can feed a nation's population well with a small- or medium-sized farm model based on appropriate ecological technology, and in doing so we can become more self-reliant in food production. Farmers must receive higher returns for their produce, and when they do they will be encouraged to produce. Capital-intensive chemical inputs—most of which are unnecessary—can be largely dispensed with. The important lessons from Cuba that we can apply elsewhere, then, are agroecology, fair prices, land reform, and local production, including urban agriculture.

Note

Rosset, Peter. 2000. "Cuba: A Successful Case Study of Sustainable Agriculture." In *Hungry for Profit: The Agribusiness Threat to Farmers, Food and the Environment*, ed. Fred Magdoff, John Bellamy Foster, and Frederick H. Buttel. New York: Monthly Review Press.

Cleaning the Closet
Toward a New Fashion Ethic
Juliet Schor

The manufacture of cloth and clothing is a major source of social injustice and environmental degradation. Far too often, clothing is made in sweatshops under abominable working conditions. This has been publicized by the antisweatshop movement. What some people do not realize, however, is that clothing production has significant environmental consequences. Schor develops three principles for "a new kind of clothing consumer": (1) emphasizing quality over quantity, (2) a return to small-scale enterprises, and (3) a movement toward socially just and environmentally responsible production practices.

I love clothes—shopping for them, buying them, wearing them. I like good-quality fabrics, such as wool or linen. I cultivate long-term relationships with favorite items, such as sweaters and scarves. I delight in a beautifully tailored suit, everything perfectly in place. And I love to find a bargain.

I confess these sartorial passions with some trepidation. Love of clothes is hardly a well-regarded trait by my friends in the environmental, simplicity, feminist, labor, and social justice movements. And for some good reasons. Much of what we now wear comes from foreign sweatshops. Textile production, with its toxic dyes, often poisons the environment. Fashion is a sexist business, which objectifies and degrades women. Young people adopt a must-have imperative for the latest trendy label. Adults have problems too: The typical compulsive shopper, deep in credit-card debt, has been supporting a shopping habit focused mainly on clothes, shoes, and accessories. Even a cursory look at the making, marketing, wearing, and discarding of clothes reveals that the entire business has become deeply problematic. But my friends have other objections that I find less compelling. Many believe that clothes are trivial—not worth spending time or effort on. Some feel they are irreversibly tainted by the excessive importance society has placed on them, or the power of the greedy behemoths that dominate the industry.

One school of thought—call it "minimalist"—takes a purely utilitarian stance. Clothes should be functional and comfortable, but beyond that, attention to them is misplaced. The minimalist credo goes like this: Buy as few clothes as possible, or better yet, avoid new altogether, because there are so many used garments around. Make sure your garments don't call too much attention to themselves. Shun labels and "designers." Purchase only products whose labor conditions and environmental effects can be verified.

This ethic has gained its share of adherents in recent years. A growing number of young people critique their generation's slavish devotion to Abercrombie, North Face, and Calvin, preferring the thrift-shop aesthetic. Simplifiers advocate secondhand stores, clothes swapping parties, and yard sales. The market for organic clothing, despite its generally inferior design and high prices, is expanding. No Logo has developed its own cachet.

Clothing minimalism is certainly a morally satisfying position. But most people do not and will not find minimalism appealing, and not because they are shallow or fashion addicted. Rather, minimalism fails because it does not recognize the centrality of clothing to human culture, relationships, aesthetic desires, and identity. Ultimately, minimalism lacks a positive vision of the role of clothing and appearance in human societies.

But what could that positive vision be? First, it will affirm the cultural importance of clothing, rather than trivialize it. It will embrace the consumer who buys conscientiously and sustainably, but who also has a prized and beautiful wardrobe hanging in her closet. It will recognize that apparel production, which after all has historically been the vanguard industry of economic development, should provide secure employment for millions of women and men in poor countries, a creative outlet for designers and consumers, and a technological staging ground for cutting-edge environmental practices. A "clean-clothes" movement has begun in Europe. Can we transform it into a "clean and beautiful clothes" movement here in the United States? If so, it holds the potential to become a model for a wider revolution in consumer practices. For if we can work it out with a commodity as socially and economically complex as clothing, we can do it with anything.

Clothes by the Pound

For an introduction to the insanity of the industry, a good place to start is a used-clothing outlet. I chose the Garment District, a hip, department-style warehouse in East Cambridge, Massachusetts. Inside, it's chockfull of every retro and contempo style one could imagine. Outside, huge eighteen-wheeler trucks deposit giant, tightly wrapped bales of clothing, gleaned from charities, merchandisers, and consumers. These clothes sell for a dollar a pound, and seventy-five cents on Friday. That's a price not too much above beans or rice.

Over the fourteen years that the Garment District has been in business, the wholesale price of used clothing has dropped precipitously, by 80 percent in the last five years alone, according to one source. Renee Weeper, director of retail services at Goodwill International, reports that prices in the salvage market dropped to two to three cents per pound by late 1999 and are now in the seven to eight cent range. In this supply-and-demand oriented "aftermarket," the price decline has been caused by an enormous increase in the quantity of discarded clothing. Throughout the 1990s, donations to Goodwill increased by 10 percent or more each year.

And what of the clothing that is not resold to consumers? The Garment District sells its surpluses to "shoddy mills," which grind up the clothes for car-seat stuffing and other "post-consumer" uses. Or they send it into the global used-clothing market, where it is sold by brokers or given away by charitable foundations. Ironically, the influx of cheap and free clothing in Africa, under the guise of "humanitarian aid," has undermined local producers and created more poverty. And that's not the only irony—the excess clothing that ends up in Africa, the Caribbean, or Asia, probably also started out there.

The Point of Production: Sweated Labor and the Poisoned Landscape

Textiles have become the vanguard industry in the emergence of a new global sweatshop, where women—who comprise 70 percent of the labor force—work for starvation wages, making the T-shirts, jeans, dresses, caps, and athletic shoes eagerly purchased by U.S. consumers. The brutal exploitation of labor and natural resources is at the heart of why clothes have become so cheap.

Consider the case of Bangladesh—which by late 2001 was the fourth largest apparel exporter to the United States. The country, with a per capita income under $1,500 per year, a 71 percent female illiteracy rate, and 56 percent of its children under age five suffering from malnourishment, is one of the world's poorest. While proponents of corporate globalization claim the process is lifting people out of poverty, a recent study by the National Labor Committee reveals otherwise. Wages among Bangladesh's 1.6 million apparel workers range from eight cents per hour for helpers to a high of eighteen cents for sewers. Workers are forced to work long hours and are often cheated of their overtime. When demand is high, they work twenty-hour shifts and are allowed only a few hours of sleep under their sewing machines in the dead of night. The workers, most of whom are between sixteen and twenty-five years old, report constant headaches, vomiting, and other illnesses. Even the "highest" wage rates meet less than half the basic survival requirements, with the result being that malnutrition, sickness, and premature aging are common. Ironically, apparel workers cannot afford to buy clothing for themselves—a group of Bangladeshi women factory workers who recently toured the United States report getting only one new garment every two years. The university caps they sew sell for more than seventeen dollars here; their share is a mere 11.6 cents per cap.

These conditions are not atypical. Disney exploits its Haitian workers who make Mickey Mouse shirts for twenty-eight cents an hour. Wal-Mart, which controls 15 percent of the U.S. market and is the world's largest clothing retailer, has Chinese factories that pay as little as thirteen cents per hour, with the norm below twenty-five cents. High-priced designers also exploit cheap labor—Ralph Lauren and Ellen Tracy pay fourteen to twenty cents, Liz Claiborne twenty-eight cents. Nike, despite years of pressure by activists, continues to exploit its Asian workforce. At the Wellco and Yuen Uren factories where its shoes are made, the company was paying only sixteen to nineteen cents per hour, requiring up to eighty-four hours per week including forced overtime, and employing child labor. A recent estimate for a Nike jacket found that the workers received an astounding one half of a percent of its sale price; a study of European jeans found a mere 1 percent went to workers. By contrast, "brand profit" accounts for about 25 percent of the price.

But low wages are only part of the horror of the global sweatshop. Many factories and worker dormitories lack fire exits and are overcrowded and unhealthy. Workers sewing Tommy, Gap, and Ralph Lauren clothes have been found locked inside the factories. They are routinely harassed—sexually and physically—by their supervisors. The Bangladeshi workers report that beatings are common and that they are forbidden to speak inside the factory. Permission to go to the bathroom is severely curtailed, many are forced to work while ill, and companies typically fire those who become pregnant. Unions are bitterly resisted, with terminations, physical harm, and intimidation by employers. The retailers who contract with local factories have tried to build a wall between themselves and their subcontractors, but this is little more than a callous ruse.

Manufacturers are also exploiting the natural environment. While clothing is not typically thought of as a "dirty" product, like an SUV, plastics, or meat, a closer look reveals that this clean image is undeserved. From raw material production through dyeing and finishing, to transport and disposal, the apparel, footwear, and accessories industries are responsible for significant environmental degradation. Consider cotton, which makes up about half of global textile production. Cotton cultivation is fertilizer-, herbicide-, and pesticide-intensive, endangering both the natural environment and agricultural workers. The crop comprises only 3 percent of global acreage, but accounts for 25 percent of world insecticide use. In some cases, the crop is sprayed up to ten times per season with dangerous chemicals, including, among others, Lorsban, Bladex, Kelthane, Dibrom, Metaphos, and Parathion. The toxicity of these chemicals ranges from moderate to high and has been shown to cause a variety of human health problems, such as brain and fetal damage, cancer, kidney and liver damage, as well as harm to birds, fish, bees, and other animals. Not surprisingly, farm workers suffer from more chemical-related illnesses than any other occupational group. Chemical run-off into the nation's drinking water has also been extensive— Aldicarb, an acutely toxic pesticide, has been found in the drinking water of sixteen states. Conventional cultivation also depletes the soil and requires large quantities of irrigation water.

Additional hazards arise from chemical-based dyeing and finishing of cloth. The most common chemical dye, used in textiles and leathers, is the so-called azo-dye, which is now believed to be carcinogenic and has been banned in Germany. Formaldehyde, pentachloraphenol, and heavy metals remain in use despite their toxicity. A little-known aspect of these toxins is their human health impact on both workers and consumers. One German study found that 30 percent of children in that country suffer from textile-related allergies, most of which are triggered by dyes. An estimated 70 percent of textile effluents and 20 percent of dyestuffs are still dumped into water supplies by factories. In South India, where the (highly toxic) tanning industry grew rapidly in the 1990s, local water supplies have been devastatingly polluted by large quantities of poisonous wastes. The various stages of textile production (from spinning, weaving, and knitting to dyeing and finishing) also require enormous energy and water use. For example, one hundred liters of water are needed to process one kilogram (2.2 pounds) of textiles.

Environmental effects can also be more indirect. The consumer rage for cheap cashmere has led to unsustainable expansion of herds in Mongolia and, subsequently, to overgrazing, desertification, and ecological collapse. The growth of new fabrics, such as the wood-based tencel, is contributing to deforestation in Southeast Asia.

Environmental impact does not end at the point of production. The globalization of the industry has led to increased pollution through long-distance transport. And eventually, the products enter the waste stream. Clothing, footwear, and accessories are a staple of municipal landfills.

Superfluity, Novelty, and Exclusivity: Hallmarks of the Clothing Industry

At the core of the disposal problem lie two developments: Clothes are cheap and Americans are buying them in record numbers. Since 1991, the price of apparel and footwear has fallen, especially women's clothing, with the drop especially pronounced after 1999. (This was most likely due to declining wages in Asia, caused by the Asian financial crisis.) It is no surprise that as clothes got artificially cheaper, Americans began accumulating more of them. Indeed, when prices are low, the pressure on manufacturers and retailers to sell more becomes intense. In 2000 alone the United States imported 12.65 billion pieces of apparel, narrowly defined (i.e., not including hats, scarves, etc.). It produced another 5.3 billion domestically. That's roughly 47.7 pieces per person per year. (Women and girls rates are higher; men and boys lower.) From Bangladesh alone we imported 1.168 *billion* square meters of cloth. That's a lot of caps.

Paradoxically, the system of low prices and high volume is anchored at the top by outrageously priced merchandise. At the high end, thousand-dollar handbags, dresses running to the many thousands, even undergarments costing a hundred dollars are the rule. A look at the nation's distribution of wealth provides one clue to why high-priced clothing is flying off the shelves: The top 10 percent of the population now own a record 71 percent of the nation's total net worth, and 78 percent of all financial wealth. (The top one percent alone own 38 and 47 percent of net worth and financial wealth.) The existence of such an upscale apparel market is a troubling symptom of a world in which some people have far too much money and far too little moral or social accountability in terms of what they do with it.

But the high-priced venues serve another purpose as well. Designer merchandise becomes available at discount stores at a fraction of its top retail price. This affordable exclusivity is part of what keeps middle-class consumers enmeshed in the system. Clothes cascade through a chain of retail outlets, prices falling at each stage. The system has led many consumers to purchase almost mindlessly when confronted with irresistible "bargain basement" prices of highly regarded designers and to spend much more on clothes than they intend or even realize. Eventually even the desirable designer merchandise ends up being sold for rock-bottom prices—on the web one can find surplus clothing sites selling clothes at a fraction of their retail prices. I found $5,000 designer dresses going for $1,000, women's coats that retail at $129 available for $22 each; men's down jackets for $12. I found Hilfiger, DKNY, Victoria's Secret. Brand-new "high-quality mixed clothing" can be had for twenty-two cents a pound.

The core features of contemporary fashion—fast-moving style, novelty, and exclusivity—also contribute to spending. A seasonal fashion cycle based on climactic needs has been replaced by a shorter timeline, in which the "new" may only last for two months, or even weeks, as in the extreme cases of athletic shoes. The exclusivity that is relentlessly pushed by marketers also contributes to high levels of spending—the product is valued *because* it is expensive. As it becomes more affordable, its value declines. Similarly, when the consumer aspires to be a fashion pioneer, she seeks rarity. The impacts of these core features of the fashion industry are profound. Many middle- and lower-middle class youth are working long hours to buy clothes. For poor youth, with limited access to money and jobs, the designer imperative has been linked to dropping out of school (because of an inadequate wardrobe), stealing, dealing, even violence. Failing to keep up with the dizzying pace of fashion innovation undermines self-esteem and social status.

But it is not only fashion-orientation that accounts for the enormous volume of clothing that is sold in this country. Shopping for clothes, footwear, and apparel have become habits, even addictions, especially for women. Just something to do because we do it. People shop on their lunch hours, on the weekend, through catalogs, or in the mall. They spend vacations at outlet malls. Americans typically have something in mind they want to buy, and, for women that something is often clothing. In my interviews of professional women who subsequently downshifted, a common refrain was the enormous superfluity of their closets. It's clear we need to get our relationship to clothes under control.

Why Clothes Do Matter

To create sustainable, humane, and satisfying apparel, footwear, and accessories industries, we need to understand the functions of these products. Their utilitarian features are obvious. We need garments to cover our bodies, hiking shoes to climb a mountain, a watch to tell time. But this is just the beginning of what clothing really *does*. Throughout history, clothing has been at the center of how human beings interact. Not always with humane purposes, of course. Clothing and footwear have long identified rank and social position. Before the nineteenth century, European governments passed sumptuary laws that regulated dress, particularly to control those of low status. Intense conflict was waged over whether one could wear a wig, choose a certain color, or sport a particular fashion style. Not surprisingly, clothing was equally central to struggles against those very inequities. Working people often asserted their social rights by choosing dress that elites deemed them unworthy of wearing. Clothing has been key to both the repression of social groups and their struggles for human dignity and justice. (Closer to home, consider how incomplete any account of the political challenges of the 1960s would be without attention to blue jeans, tie-dyed shirts, and long hair.)

Clothing has also been at the heart of gender conflicts. As early-twentieth-century American women attempted to break free of patriarchal strictures, they rejected corsets and confining dresses. In the twenties, they defied convention with cigarettes and short skirts. In the 1970s, the women's movement rejected the fashion system and created its own sartorial sensibility. Clothing has historically also been an important site of intergenerational bonding and learning between women, along with hair care and other beauty rituals. I am quite sure that I got my love of clothes from my mother. Being a good shopper, especially in the complex world of women's clothing, requires finely honed skills, which are passed on from generation to generation.

What we wear is important to the way we experience our sexuality. Our age. Or ethnicity. It allows us to show respect for others (by dressing specially for a social occasion) or to signal community (through shared garments or styles). Finally, clothing can be part of the aesthetic of everyday life. There is genuine pleasure to be gained from a well-made, well-fitting garment. Or from a piece of clothing that embodies beautiful design, craftspersonship, or artistry. Throughout history, human beings have exercised their creativity through clothing, footwear, and accessories.

In sum, dressing and adorning are a vital part of the human experience. This is why any attempt to push them into a minimalist, utilitarian box will fail. Clothes embody far more than our physical bodies; they are also a measure of our basic values and culture. So, while we may not all take great pleasure in what we wear, we should all recognize that clothes do matter. They are about as far from trivial as any consumer good can be. Which means that a new fashion ethic will be about affirming social and human values, the commitments of daily life, and our hopes and aspirations for a different kind of world.

Principles for a New Kind of Clothing Consumer

1. Quality Over Quantity. Moving from Cheap and Plentiful to Rarer and More Valuable

In the past, clothing cost more, and its use was far more ecologically responsible. Only the rich bought more clothes than they wore, in contrast to current habits. Expensive clothing was worn sparingly when newly acquired. In some places, as a garment wore out, it cascaded through a social hierarchy of uses, from esteemed social occasions to the everyday public, and eventually to the most mundane private and domestic uses. Clothing was cleaned far less frequently than today, thereby extending its useful life. Women had the skills to make clothes, and even as ready-made garments became more common, to restructure and upgrade them. Style could be attained through "refashioning" garments, rather than discarding them and buying new ones. Such refashioning could also involve new ownership. Historian Nell McKendrick has identified the origins of the eighteenth-century consumer revolution in Britain with the trickle-down from elites to servants, as maids took their mistresses' cast-off dresses and turned them into newly stylish outfits. Thus, basic principles of ecology and frugality were maintained—take only what you need, use it until it is no longer useable, repair rather than replace, refashion to provide variety.

The history of clothing practices provides guidance for fashioning a new aesthetic whose central principles are to emphasize quality over quantity, longevity over novelty, and versatility over specialization. For example, if we reject the need to keep up with fashion and can be satisfied with a smaller wardrobe, we can spend more per garment, as consumers do in Western Europe. The impact on the earth is less, and it contributes to longevity, because better clothes last longer by not skimping on tailoring or quality and quantity of yardage. Consumers are better off because high-quality clothing is more comfortable and looks better.

Ultimately we could begin to think of clothing purchases as long-term commitments, in which we take responsibility for seeing each garment through its natural life. That doesn't mean we couldn't ever divest ourselves, but that if we grew tired of a useful garment we'd find it a new home with a loving owner, kind of like with pets. Of course, to facilitate such a change, consumers would need to reject the reigning imperative of variety in clothes, especially as it pertains to the workplace and for social occasions. Just because you wore that dress to last year's holiday party doesn't mean you can't show up in it again.

With such an aesthetic, consumers would demand a shift toward more timeless design, away from fast-moving trends. Clothes could become more versatile in terms of what they can be used for, their ability to fit differently shaped bodies and to be altered. Consider the Indian sari, a simple, rectangular piece of cloth that is fitted around the body. It accommodates weight gain and loss, pregnancy, growth, and shrinkage. Couldn't designers come up with analogous concepts appropriate to Western tastes? Pants with waistbands that are flattering but also can be adjusted through double-button systems or through tailoring. Basic pieces that can be complemented by layering and accessories. Expensive, classic clothes already have some of these qualities—extra fabric for letting out and the capability to remain flattering after they have been altered.

Striving for longevity through versatility facilitates what we might call an ecological or true materialism. The cultural critic Raymond Williams has noted that we are not truly materialist because we fail to invest deep or sacred meanings in material goods. Instead, our materialism connotes an unbounded desire to acquire, followed by a throwaway mentality. True materialism could become part of a new ecological consciousness. Paying more per piece could also support a new structure of labor costs. Workers would work less, produce fewer but higher-quality items, and be paid more per hour. Such a change would help make ecologically clean technologies economically feasible.

Finally, paying more for clothes does not mean adopting the premise of social exclusivity. In luxury retailing, much of the appeal of the product is its prohibitive price or the fact that only elites have the social conditioning necessary to pull off wearing it. An alternative aesthetic would value democracy and egalitarianism through the fashioning of garments that are high-quality but affordable.

2. Small and Beautiful: Creative Clothing for Local Customers

The aesthetic aspect of clothing is and will continue to be important. But the values represented by the fashion industry are unacceptable. Despite decades of feminist criticism, the industry continues to objectify women—and increasingly men—through demeaning, violent, and gratuitously sexualized images and practices. In the late nineties we got "heroin chic," glamorizing drug abuse and poverty. Now it's teen and "tween" styles, with bare midriffs, tightly fitting T-shirts, and sexually explicit sayings emblazoned on the garments. Furthermore, the industry is comprised of megacorporations employing a small number of mostly male designers. They, in turn, produce a monolithic fashion landscape—massive numbers of copycat garments. Suddenly all that's available are square-toed shoes, or short-handled handbags, or hip huggers.

An alternative vision starts from the recognition that many young people, especially young women, yearn to be fashion designers, producing garments that are artistic, interesting, funky, visionary, and useable. And consumers are increasingly desirous of that type of individualized clothing. The industry could return to its roots in small-scale enterprises, run by the designers themselves. The British cultural analyst Angela McRobble has envisioned such a shift, calling for small apparel firms located in neighborhoods, operating almost like corner stores. They would cater to a local clientele whose tastes and needs they come to know. These face-to-face relationships between female designers and the immigrant women who labor in domestic-apparel production also have the potential to reduce the exploitation that currently characterizes the industry. Instead of driving to a mall with its cookie-cutter stores, one might walk to a converted factory housing three or four designers with workshops-cum-show rooms. The consumer could also become active in the creative process, helping to fashion an interesting or unique look for him or herself. If she didn't see what she wanted, it could be made to order, so that fit, color, and style were just right. Such a system would yield substantial savings in the areas of transport, branding, advertising, and marketing as well as a dramatic reduction in overproduction. Those savings could be used to pay decent wages, install environmentally sustainable production technologies, fund better quality materials, and support designers.

Such a vision could be realized through a combination of activist pressure, consumer mobilization, and government policies. The federal government could offer special subsidies for training and education for designers and enterprise loans to small business owners. Local governments could support apparel manufacturers through tax incentives and marketing initiatives.

3. Clean Clothes: Guaranteeing Social Justice and Environmental Responsibility

Relocalization is an important part of a movement toward a just and sustainable apparel industry. But it must go hand in hand with improvements in wages and working conditions in factories and small production units abroad. Such reform is essential to relocalizing on a global scale, because it will be the foundation for creating purchasing power in India, China, Bangladesh, and other southern countries. For now, the north must continue importing in order to provide employment for impoverished foreign workers. But as wages rise abroad, these workers can produce for their own domestic markets.

One of the most important social movements of the past decade has been the coalition of labor, student, and religious activists opposing the exploitation of garment workers around the globe. The Gap, Nike, Kmart, and others have been exposed and embarrassed by their labor practices. Students have demanded that their college's insignia clothing not be produced by sweated labor, and more than ninety institutions have complied. Most American consumers now believe that the workers who make their clothing should be paid decently, and surveys indicate they are willing to pay somewhat more to achieve that goal.

To date, however, the industry response has been inadequate. While some progress has been made, far more energy has gone into winning the PR battle than has been devoted to substantive reform. Companies remain opposed to free association in unions, which is the only true long-term solution to abuse. Nevertheless, the principle of what Europeans call "clean clothes" is making headway. In Europe, major clothing retailers have committed themselves to codes of conduct that ensure reasonable working conditions, free association, and other labor rights. For example, the British chain Marks and Spencer has joined the Ethical Trading Initiative, which is a government-sponsored initiative bringing together nongovernmental organizations (NGOs), unions, and businesses. Next, another British chain, works with Oxfam on ethical trading.

Indeed, the successes of the European clean-clothes movement are worth looking at, particularly for extending beyond labor rights into environmental impacts. In 1996, the Dutch company C&A Instituted rigorous controls over its suppliers—monitoring more than one thousand production units annually—to guarantee labor conditions and environmental impacts. It uses the Eco-Tex label for environmental certification, and many of its own brands sport it. Marks and Spencer has begun an organic cotton design project with the Royal College of Art.

The German company Otto Versand, the largest mail-order business in the world, has perhaps gone farthest in terms of environmental sustainability. It has reduced paper use in its catalogs and packaging; its mail-order facility uses wind and solar power; and it is moving to incorporate sustainability throughout its product lines. Otto subsidizes the production of organic cotton in Turkey and India, and last year offered 250,000 organic cotton products. The company has reduced the use of harmful chemicals in textiles and has certified that 65 percent of its clothing passes a strict "skin-test" for dangerous substances. In the late 1990s, Otto worked with Century Textiles (India's largest textile exporter), to phase out azo-dyes. The company has also introduced its Future Collection, which is oriented to production ecology through conservation of energy and water resources. To encourage consumers to adopt a long-term perspective, they offer a three-year replacement guarantee for all their clothes.

To be sure, the shift to just and ecologically sustainable clothing is not simple. The price of organic clothing is currently high, putting it out of reach for many consumers. But activist pressure can help solve this problem, as the European successes are showing. And the U.S. market is already increasing. Nike and The Gap have begun to use some organic cotton. If one or two major U.S. companies commit to a substantial program of organic cotton use, demand will grow and prices will fall. And even a high-priced company such as Patagonia has made some accommodations for affordability—all its clothes carry a no-questions-asked indefinite replacement guarantee and the company operates a number of discount outlets.

The successes of the European campaigns suggest that comparable progress is possible on this side of the Atlantic as well. For example, Eileen Fisher, a high-end women's retailer, has signed on to SA 8000, an international social and environmental standard. U.S. manufacturers and retailers are sensitive to the need to maintain their public image. If we can educate consumers and mobilize activists, we can "clean" the American closet. Doing so would be a substantial step toward a sustainable, but also fashionable, planet.

Note

Schor, Juliet. 2002. "Cleaning the Closet: Toward a New Fashion Ethic." In *Sustainable Planet: Solutions for the Twenty-first Century*, ed. Juliet Schor and Betsy Taylor. Boston: Beacon.

29

On the Trail of
Courageous Behavior

Myron Peretz Glazer and
Penina Migdal Glazer

Sociological studies often reveal inequalities and hidden agendas or practices by powerful actors, such as corporations. Myron and Penina Glazer explore how sociological studies can also uncover forces that challenge harmful or destructive practices. Examining the courageous behavior of American whistleblowers and environmental activists in the United States, Israel, and the former Czechoslovakia, they ask what enables ordinary people to engage in courageous behavior, such as long-term environmental activism. They identify four factors. First, activists typically have strong social networks, or social capital; second, they believe they can make a difference; third, they are willing to accumulate evidence and expertise; and fourth, they must overcome fear and intimidation.

We ask whether sociology, the study of human behavior in all its forms, can help save all of us from our most destructive proclivities. In an era of mass murder and unleashed genocide, of the production of unsafe and injurious products, of unnecessary waste and pollution caused by individual and corporate greed, can sociology assist us in unraveling the sources of evil, of behavior that intentionally harms others? Can our discipline highlight beliefs and actions that engender moral responsibility for the common good? Can sociology help isolate the social, cultural, and emotional sources of courageous behavior, behavior that embodies taking risks for the well-being of others? Can sociology, through its grounded empirical research, illuminate principles that help sustain opposition to coercive and unaccountable power?

To attempt to answer these questions, and thereby underscore the significance of sociological investigations for maintaining a viable and democratic social order, we draw upon our research on organizational whistleblowers and on community crusaders for a safe environment. Beginning in 1982 we spent six years studying sixty-four American whistleblowers in government and industry. As sociologists we were particularly interested in the values that propelled them to protest unethical and illegal activities. We emphasized the social networks which sustained them and which provided publicity, legitimacy, and legal defense. We scrutinized the organizational retaliation mounted against them as they faced firings, isolation, black lists, and character assassination. Despite their pain and loss, we also highlighted the cultural significance of their victories, their impact on corrupted institutions, and their often successful efforts to rebuild their shattered careers and lives.

After the publication of *The Whistleblowers: Exposing Corruption in Government and Industry*, we decided to continue our study of the social and cultural foundations of courageous behavior. We interviewed scores of grassroots activists in the United States, Israel, and the former Czechoslovakia who would not remain silent when their air was polluted, their wells contaminated, and their children sickened by radiation or other hazardous substances. In continuity with the whistleblowers study, we approached a wide spectrum of environmental crusaders to understand their backgrounds, values, allies, and adversaries (Glazer and Glazer 1998). We wanted to test and advance our theory that environmental crusaders, like the organizational whistleblowers, are on the front lines in exposing and demanding remediation for society's most serious problems; that they often serve as bellwethers heralding crises just over the next horizon; that, in addition, they act as moral exemplars who embody the society's highest ideals of concern for one's neighbors and community. Whistleblowers and crusaders for a safe environment constantly re-assert the boundaries beyond which others, no matter how powerful, shall not pass without encountering serious and sustained resistance. In heralding all these interrelated contributions, sociologists provide a breakwater to the flood of historical and contemporary disasters which threaten a society's moral balance.

The Courage of Ordinary People

Many scholars who study acts of courage focus primarily on military and police situations or on other heroic events that occur in the effort to save the lives of others. Such actions are dramatic, requiring, for instance, the split-second decision to charge a machine gun nest raining lethal fire on one's comrades, or to rush into a burning boiler room to rescue injured sailors before rising waters engulf them (Walton 1986; Rachman 1990). But for grassroots environmental activists and organizational whistleblowers, courage takes another form. The situations they face are usually not immediately life-threatening; rather, their battles demand a longer-term investment of time and energy. They must be ready to withstand withering criticism of their credibility, competence, and integrity. They must face attacks that may sully their reputations, isolate them from one-time friends, neighbors, and coworkers, and even cause rifts within their own families. Under extreme circumstances, grassroots activists and whistleblowers face physical threats, police harassment, and imprisonment.

Knowing these risks, we found that ordinary people in our studies took a courageous stand when they proclaimed the dangers of serious occupational and community situations. We probed into the background and situational factors that led them to commit themselves to building a sustained, collective campaign to demand accountability and remediation. We asked how gender, social class, occupation, religion, and other sociological variables influenced them. Thus, while their decisions did not reflect the single act of heroism associated with military bravery, they had to muster the social and cultural resources to spend years intensively involved in their causes, resulting in stress and the disruption of the rhythm of their lives. Friends and colleagues often experienced burnout, providing a constant reminder of the difficulties of the long-term crusade to remedy organizational irresponsibility or to ensure a safe environment.

This burnout is precisely what occurred to Cheryl Washburn, one of two courageous women who were fighting the effects of a hazardous waste site in Maine. Over time Cathy Hinds continued to demand remediation, but Cheryl Washburn began to withdraw from total involvement. She explained her inability to sustain such intense activity in words that speak for many others:

> It takes a certain person to be able to keep at it all the time. Cathy is one of those people. She's strong, and in some ways strong-willed. For me, it came down to wanting to live a normal life. I didn't want to be faced with this thing every day, or I'd be a basket case. I didn't want to go to all the meetings, to keep butting my head against a wall. As Cathy slowly started doing more, I slowly backed off. Cathy has a way with words—she knows how to put things, and I was glad to let her take over. I'm proud of what we did—and I'm proud of her for keeping it up. (Garland 1988, p. 100)

A similar situation but with a different twist confronted Frank Camps, a senior design engineer for the Ford Motor Company, who protested the construction of the Ford Pinto. Camps believed it was an unsafe car. He helped us record the impact of bureaucracy and its control of financial and career rewards in thwarting collective action. Camps' overriding sense of professional responsibility and his fear of legal liability led him to take bold action despite the risks. After going all the way up the corporation hierarchy and urging redesign without success, he ultimately sued the company to limit his own potential liability. He counted on other engineers to help expose a potentially dangerous situation to consumers, but he soon stood virtually alone. We found comparable reactions in other settings both in industry and government.

> I can recall, right after I filed the suit, other engineers said, "Go get 'em, we wish we could do it, there goes a man with brass balls." While I had tacit support, I was looking for an honest man to stand with me. I found that these guys were suddenly given promotions, nice increases in salary. Next thing I knew, I did not have the support any more. (Glazer and Glazer 1989, p. 19)

The philosopher Douglas N. Walton argues that the courageous behavior of a Cathy Hinds or a Frank Camps is rare in contemporary industrial society:

> Perhaps another reason that courage today seems an absurd or outdated virtue is the growing lack of cohesiveness in social structures and group purposes. In vast modem industrial societies, the individual feels anonymous and often loses identity with the community as a group. This phenomenon in North America has often been remarked upon. Twenty-six bystanders watched as Kitty Genovese was brutally murdered on the street. Not one even calling the police. The current expression is "nobody wants to get involved." A kind of moral anomie is described by Camus in L'Etranger—an individual fails to feel even the smallest sympathy or emotion at the death of another. The attitude seems to be a moral aimlessness, a lack of purpose beyond one's own egoistic interests. To one in this frame of mind, *courage-taking personal risks to try and save another or to help the group* or community in time of trouble *seems simply* an irrational risk— no gain at all. (Walton 1986; p. 18, emphasis added)

Invoking our approach that focuses on sustained, long-term confrontations with irresponsible organizational power, we have encountered scores of instances where people have determined to help their communities. In contrast to Walton's assertion, grassroots activists and employee whistleblowers have displayed impressive strength and resilience in caring for others and in their willingness to put themselves at risk. . . .

Why were whistleblowers and grassroots environmental activists able to engage in such behavior when so many of their fellow citizens apparently have not responded with comparable courage? Our research points to several distinctive dimensions. First, a courageous response to a community's problems does not occur in a vacuum. It is not based on a spontaneous decision. Rather, citizens call on a reservoir of social capital, of bonds that they have developed over many years. Those we studied saw their own fate as intertwined intimately with that of others. Whistleblowers expressed an irrevocable allegiance to professional, religious, and community-based values that emphasized responsibility for the well-being of others. They refused, for example, to enact decisions that resulted in the construction of dangerous nuclear power plants or that dumped toxic chemicals into an area's water supply. The whistleblowers assumed individual responsibility rather than acting as agents of hierarchical authority. . . .

Second, both whistleblowers and crusaders had a strong cultural commitment to the efficacy of action. Their past experience led them to believe that they could make a difference. . . . They embodied the democratic tradition that citizens could unite for a redress of grievances; that those in the media or in government would see the justice of their cause, and that their legitimate concerns would be met.

Third, courageous behavior depended on their taking determined action; on their ability and willingness to accumulate the evidence, and expertise. The protesters had to counter with confidence and competence the technological experts and policymakers who belittled their concerns. The whistleblowers usually had direct access to the incriminating data. . . .

To achieve their goals the environmental crusaders had to secure strategically placed allies who could supply firsthand information. At times these allies were whistleblowers within polluting corporations or regulatory agencies. Often, the crusaders also had to become adept at using public records or petitioning through the Freedom of Information Act to secure previously classified material. In this essential effort, public-interest groups often secured mountains of data necessary to prove, for example, that nuclear bomb releases had poisoned the atmosphere or that chemical waste had directly undermined the health of local residents. No campaign for environmental redress could hope to be successful without the crusaders forging a working alliance with others committed to securing governmental and corporate accountability. We designated this essential process as the creation of *alternative networks of power*.

Fourth, purveyors of courage had to overcome fear and intimidation. They knew that the reduction of serious environmental damage or the resolution of other organizational transgressions would not occur without years of struggle. The opposing government bureaucrats and corporate managers would extract a heavy price for challenging the status quo. The protesters' determination to persist in the face of such obstacles attests to their faith in the justice of their cause. Without their willingness to endure on behalf of their beliefs, without their determination to help their communities, serious problems would lie buried beneath the surface with the potential to do great damage to unknowing and unprotected victims.

The Reservoir of Social Capital

Grassroots activists and whistleblowers were deeply involved in the affairs of the community. They were not alienated or passive citizens unwilling and unable to show concern for others. Their sense of self was fully engaged in what happened to their families, neighborhoods, regions, and countries. They may have become protesters later in life, but they had been prepared by a previously developed and strongly held value system. These values included a sense of caring for others, a feeling of responsibility for their safety, and a slowly evolving but nonetheless strong commitment to act when others were in danger. The activists were con-veyers of what Robert Putnam and others have labeled social capital (Putnam 1993). For example, women such as Lynn Golumbic in Israel, Jara Johnova in Czechoslovakia, Penny Newman in the United States, and many others may have exercised civic responsibility in more conventional ways until an environmen-tal crisis propelled them into the public arena. But their community-based sys-tem of values had been developed and implemented earlier in PTAs, Green Circles, and other organizations, and had been engaged as they prepared to step forward into a leadership role. With their principles intact, they built grassroots organizations and thus joined a militant vanguard in defending communities from victimization by government and corporate bureaucracies. . . .

Each of the women cited above had borne children and had sought to raise them in the safest and healthiest areas. Penny Newman had spent countless hours volunteering to help improve the school system in Glen Avon, California. Jara Johnova had taken the dangerous step of signing the Charter 77 document peti-tioning the Czechoslovakian government to respect human rights, and for years she was part of a dissident group in Prague. Lynn Golumbic, who had moved to Israel from the United States so that she and her young family could participate in the effort to build a democratic Jewish state, joined the Association of Americans and Canadians in Israel (AACI), which focused on a wide range of social issues. For all of them, assuming a leadership role when they confronted environmental emergencies represented a major commitment and much greater public exposure. Yet this new action was grounded in past experiences and beliefs. Helping the community in a time of crisis was consistent with their strongly held values. While their new roles ruptured old schedules and made many demands, these roles were built on a foundation that could weather the inten-sive demands of community leadership (Kirp 1989; Hallie 1979; Swidler 1986).

This reservoir of social capital, this principled connection to others, is a core component of courageous behavior, for the risks of speaking out about a problem and assuming leadership could be costly, as Tom Bailie learned. Bailie, a farmer in eastern Washington state whose land abutted the Hanford Nuclear Reservation, epitomizes the grassroots activist whose past experiences prepared him to announce the existence of a crisis, even if such action was undertaken at his own peril. Bailie had deep roots in the area; he grew up on his parents' farm, raised a family, and ran for local political office. He understood that speaking out against the Hanford nuclear facility entailed substantial risks. He would be a marked man from then on, someone who had punctured the cultural fiction born in the Cold War, the commonly held faith that patriotism demanded the production of nuclear bombs and that everything was safe and under control. But Bailie urged his neighbors and other area residents to face the seriousness of their situation. He told stories of deformed animals and an epidemic of cancer, illness, and death. He drew upon his own personal history of illness and that of neighbors and friends. Bailie urged all who would listen to question the very source of their economic security. He encouraged them to challenge the government and corporations that provided their livelihood. This was not a scenario that was likely to make Bailie a local hero, and for years he and a few allies in the community were objects of ridicule and scorn. His was a dangerous position, but it was one that he was determined to hold, no matter what the personal cost.

> What I have to do goes beyond my marriage and my other family relationships. I'm sorry. I don't know what I have to do to complete this, but I'll be glad when it's done. We still have to find out what the nuclear gang has done to those of us that live here, and how it has affected our lives and affected our health. We've just scratched the surface. (Glazer and Glazer 1998)

Blue-collar and clerical workers in large nuclear organizations who saw practices that entailed serious violations of environmental health and safety standards were also inspired to speak out against such violations. Their protests emanated from a deep sense of community. The nuclear plant construction workers we studied were determined to remain in the communities where they had grown up and thus strongly identified with potential victims who could be relatives, friends, or neighbors. At times, the workers specifically referred to an obligation to the land that had been passed down by parents and ancestors. They often contrasted their sense of responsibility for the community with that of high-placed corporate executives who moved frequently to accommodate their careers and had neither roots nor long-term commitments to the local people with whom they work.

One welder and craftsman at a Texas plant became so deeply concerned about safety at the construction site that he joined a protest to the Nuclear Regulatory Commission despite the likelihood of job loss and blacklisting. His statement captured the spirit espoused by social capital theorists. With his jeans and cowboy boots and Texas drawl, Stan Miles personified the land in which he was raised:

> I was born in this state and this state means a lot more to me than just a place to live. If you will look at my work record, I've never gone out of the

state to work. I don't like to leave this land. I like the people here. It's changed a lot, though, since I was a boy. For instance, I was born in west Texas, real west Texas, west of the Pecos. You didn't have car trouble without the next person stopping, and if he had to drive eighty miles out of his way, he did, and you didn't have to pay him anything.

That's gone—all gone, and for the sake of a dollar bill. They took something that was priceless and ruined it for something made of paper. Because if you poison the water, you poison the land. How can the dollar bill replace that? This state means something to me. I was born here, my ancestors came here in 1821, my grandmother was a Comanche Indian, and they've been here for ten thousand years. (Glazer and Glazer 1989, p. 131)

Unlike the environmental crusaders and employees like Stan Miles, many other whistleblowers spoke out to protect potential victims who were unknown to them. As professionals, these whistleblowers invoked a particular brand of social capital. They felt a special responsibility for the safety of others, even if it were a public they might never know. Demetrius Basdekas, an engineer responsible for the safety of control systems essential to avoid nuclear accidents, could not remain uninvolved as he imagined potential victims of a nuclear plant disaster. He protested his supervisors' decision to grant a license to operate a new nuclear facility that he believed did not have proper safety provisions.

I said to myself, "Look, you are at a crossroads and you have to decide which way to go. You can either roll over and play dead or stand up and say what you think." I hit the wall, the red line. I could not go beyond that line. I was being asked to become a party to an act of fraud on the public where health and safety are concerned.

Management's response was to remove me, to assign a greenhorn to do the job. I was simply told that I was no longer responsible for this part of the work. As a result of this, I and other engineers who were performing in similar situations decided to take our case to the Congress and to the public. (Glazer and Glazer 1989, p. 78)

Basdekas and his colleagues would not rationalize compliance by relegating their decision to the orders of high authority. They would not assume the "agentic role" highlighted by Stanley Milgram in his famous studies of obedience to authority (Milgram 1974). . . . Milgram found that the most telling counterweight to the orders of authorities was the presence of peers or colleagues who would support the decision to protest and resist. For Basdekas that support was essential. For others the stimulus for rebellion also may be professional values and the bond of comradeship. . . .

On the Efficacy of Grassroots Action

Environmental crusaders and employee whistleblowers assumed personal responsibility to confront serious environmental and other problems that affected their communities. Their involvement began when they became convinced that remedial action would not be undertaken voluntarily after they brought the issue to their supervisors or local officials. These grassroots activists and organizational employees believed that gathering information, focusing attention, and organizing to combat social problems could make a difference. These strongly held cultural assumptions about their entitlement to speak out on public affairs and in the efficacy of united action were crucial ingredients in the protesters' willingness to be stigmatized as the bearers of bad news and reinforced their readiness to put themselves at risk. . . . While they sought distant goals, they believed fully that they could achieve them (Russell 1989).

Penny Newman and her peers represent a prime example of a local group who committed themselves to battling the chemical contamination in their community. They began to organize in 1978 when a nearby toxic waste dump, the Stringfellow Acid Pits, overflowed, sending thousands of gallons of poisoned water running through their streets and into their children's school. Their group, Concerned Neighbors, grew increasingly determined to "fight City Hall"—that is, the State of California and some of the largest corporations operating in their region that had dumped tons of dangerous chemicals into a dry pit, without sufficient regard for the safety of the community. Now, twenty years later, Newman summarized how she and her colleagues were transformed from victims to agents of their own victories (letter to authors from Newman 1997).

I look back at our 20 year battle over the Stringfellow Acid Pits and have a hard time assimilating everything that happened to us and of which we made happen. But as we start listing the accomplishments one can't help but be proud.

We were able to get a new, safe water supply installed throughout the community at no cost to the residents. . . . We made a lot of institutional changes to the Superfund program and how Americans see the issue of toxic dumps. We were the first community to get a technical advisor and become the model used for the Superfund bill in 1986 so that other communities could hire such assistance. We were the first community to have an office established to answer residents' questions and to be staffed by community volunteers—all paid for by the state. Our community has been the driving force behind each and every improvement at the site.

In more recent years we continued to have an impact both on our site and in public policy. We played a major role in the discussions about the reauthorization of Superfund. In State court, we filed the nation's largest toxic tort, consisting of 5,000 plaintiffs and 200 companies, the County of Riverside, and the State of California. We were able to reach a settlement for over $110 million, one of the first cases in which the plaintiffs were able to prove cause and effect with exposure to chemicals and community health. Another judgment that is certainly precedent setting occurred in Federal court where the court found the State of California 90–98% responsible for the cleanup of the site due to the fact that the state permitted the site, had an obligation to ensure it operated safely, and failed to do so resulting in our community being exposed and damaged.

We were the first community granted Intervenor status by a Federal court which allowed that we had a right to participate in all discussions about the site. Two hundred companies named as Responsible Parties appealed the ruling to the U.S. Supreme Court. We are still intervenors! Since then Superfund has been revised to provide a statutory right of affected communities to participate in such discussions.

In addition to the legal victories, Newman emphasized how their belief in community power transformed many of their members from shy and insecure observers to active participants in the political process. One member of Concerned Neighbors embodied this transformation.

A young Latina woman with 3 young boys at the elementary school was very concerned about the water at the school and attended a meeting where we focused on getting the water tested and filters added to the system. She was extremely shy. But her concern for her boys made her want to participate. She agreed to talk at a school board meeting, but was terrified. We practiced and practiced and that night she stood up and—with voice quivering—gave one of the most touching appeals for peace of mind in protecting the children at the school. Over the 6 month battle, she developed more confidence, was able to stand up and not shake, and helped to win that battle. But more importantly you could see the change in her. She learned that she could stand up and voice her opinion and make a difference. That she was important! This young woman had been in a very abusive marriage where she had been beaten time and time again. Through her participation in this battle, she gained enough strength, confidence and courage to leave that relationship and start a new life. She works with special education children and is doing great. (letter from Newman 1997)

Faith in their own efficacy was a crucial weapon on the side of the crusaders. It carried them forth in times of frustration, disappointment, and temporary defeat. . .

With Facts on Their Side

Because of their central organizational positions, the employee resisters to illegal or unethical behavior had surer access to incriminating evidence than did the environmental crusaders. Yet they, too, had to gather evidence that would convince a skeptical investigative reporter, attract the attention of a congressional committee, or convince public interest attorneys that a serious breach had occurred. . . . Hugh Kaufman, working for the EPA, visited scores of hazardous waste sites to document that chemical contamination was a plague undermining the health of American communities. His supervisors refused to acknowledge his evidence until he took his allegations and documentation to the media (Glazer and Glazer 1989, pp. 135–36). At the same time, Maude DeVictor, working in the Veterans Administration, gathered voluminous case material on the effects of Agent Orange (Glazer and Glazer 1989, pp. 231–32). She became a forerunner in a growing national campaign, to provide compensation for Vietnam veterans who had suffered exposure, incapacitation, and disease as a result of their wartime service. . . .

For Tom Bailie the unfolding of layer after layer of once-classified information slowly undermined the arguments of his detractors and confirmed his suspicion that the Hanford Nuclear Reservation had released dangerous emissions onto the local population. The same government and corporations that were charged with protecting the safety of the region's citizens turned out to be culpable of knowingly releasing radiation on civilian populations in the area. Despite their public-relations claims to the contrary, these bureaucratic organizations had shown blatant disregard for the safety of nearby communities. The testimony of Tom Bailie and other local residents was now supported by government documents and other hard data that had been obtained as a result of a decision by several organizations to file a Freedom of Information Act inquiry (FOIA). The Freedom of Information Act, which had been enacted in 1966, was crucial to furthering the activists' causes. To make full use of the FOIA, Bailie and the other activists had to rely on the technical assistance of allies in the press and in the Hanford Education Action League (HEAL), the local public interest group. Once they had government data confirming their claims of radiation exposure, their allegations could no longer be dismissed as the statements of paranoid personalities. Those who had risked their reputations by questioning the secrecy of the national security state or by challenging the economic viability of the defense-based economy now had the facts on their side. Family members might still be embarrassed and humiliated by the public controversy surrounding the crusaders, but they were also sobered by the evidence, which gave weight to the deadly consequences of the problem.

Not all societies have a Freedom of Information Act, and activists frequently had to use indirect methods to secure the necessary data (Flarn 1996).

In our research in Slovakia we found that prior to the fall of the Communists, a group of dedicated environmentalists believed that years of silence on environmental degradation had to end and that the time was ripe to challenge the Communists. . . . They knew they had to secure the documentation that would reveal the desperate state of the environment if they were to raise public awareness and open a serious dialogue with government officials. With retaliation, harassment, and even imprisonment a possible outcome, the environmentalists determined that their effort would be grounded in incontrovertible evidence. The best source of data was the government itself, and the activists spent months surreptitiously gathering information from officials who did not suspect that their contributions would be collated, matched, and eventually integrated into a report. Taken together, the accumulated evidence painted a dismal portrait of government deception, broken promises, and neglected environmental policy.

The publication of *Bratislava Nahlas* [Bratislava Aloud] provided documented assertions about the abysmal failure of the Communists to protect the country from air, water, and land contamination. The writers of the report, drawn from all professional fields, sought to avoid the label of political "dissidents." Rather, they claimed legitimacy as recorders of hard data that pointed to dangerous environmental conditions. The publication of the evidence substantiated their accusations and yet heightened the vulnerability of the activists. Would the government now confiscate the report, harass and arrest them, or subject them to violence? Was assassination a possibility? The *Bratislava Nahlas* group was prepared for the inevitable retaliation, having lived with its threat for years. The report opened with a strong declaration.

> In our city, the basic conditions of life have become problematic. Contamination of the atmosphere is threatening the health of virtually all inhabitants, particularly the aged, the sick, and children. . . . The degradation of values, waste, the damage to human health, and the mass problems which will impact upon future generations are immoral; nevertheless, they occur daily before the eyes of the citizens, no one is ever called to accountability. . . . The moral dimension of the Bratislava situation is, we believe, just as serious as the public health or economic viewpoints.

They ended their courageous introduction with a call to action that had never been heard in Communist Bratislava.

> We expect that the public discussion (which the document would like to introduce) will not only articulate the interests of the citizens of Bratislava, but will mobilize their forces and renew the relationship of the citizens with respect to their city. (*Bratislava-Nahlas* 1987)

Overcoming Fear

Whistleblowers and environmental crusaders undertook the challenge of exposing a major social problem and assuming a leadership role to resolve it. But they knew that they faced risks to their jobs, personal safety, reputations, and more. How did they withstand the fear of retaliation? What are the emotional resources that supported and sustained their decision?

Several factors enabled protesters to move against powerful interests. First, they transformed their anger into a positive determination to take action. Second, they depended on the trust and comradeship that developed in their local groups to sustain their motivation and commitment in the face of frustrations and setbacks. Finally, in their own personal cost–benefit analysis, they overcame the inevitable fear of retaliation by defining the environmental problems they confronted as more severe, more troubling, and more threatening than their anxiety about rejection and isolation. . . .

. . . According to Anne W. Garland, who studied women activists:

> Anger is often at the center of [activists'] transformations from private actors in restricted universes to public leaders in universes encompassing all the important issues of the day. The anger comes, of course, from a variety of sources. And it crosses the putative barriers of age and racial differences; differences in education, background, and lifestyle; and differences in religious and political belief. (Garland 1988, pp. xvi–xvii)

Trust also proved essential in sustaining commitment and solidarity among the activists. The accusations, the demeaning comments, even the threats of physical attack or arrest and imprisonment in Czechoslovakia, were all made bearable by the culture of solidarity that existed in the activists' group. . . . Pavel Šremer, a longtime dissident and one of the core group in the *Bratislava Nahlas* movement, emphasized how mutual trust in each other sustained them through difficult times. Each Bratislava participant had an assigned set of tasks, and the others had faith that it would be completed. To stop their work, to be paralyzed by the fear of what the government could do to them, would grant their adversaries an unacceptable victory. Faith in their cause and support for each other were strong incentives during their months of intensive, secret activity. . . .

Activists faced fear of failure, fear of criticism and ridicule, fear of attack and assault, and fear of imprisonment and exile, but these fears did not stop them, for they believed that the costs of inaction were even higher. . . .

For organizational whistleblowers fear takes on a particular form. To raise troublesome issues on the job is always risky. Yet as Frank Camps learned at the Ford Motor Company and Maude DeVictor at the Veterans Administration, going outside the organization with your allegations of misbehavior is much more dangerous. It almost always results in threats or actual retaliation. Whistleblowers experienced retaliation that went well beyond their expectations. Some were deliberately isolated from previously congenial colleagues, others faced dismissal from a valued position and the terror of blacklisting from all jobs in that particular industry. There were many witnesses to the career and family dislocation such punishing outcomes entailed.

These kinds of retaliation are not trivial. To withstand them requires the fortitude to stay the course no matter how severe the organizational reaction. The statement of Margaret Henderson (not her real name) best exemplifies the fear, hurt, self-depreciation that results from retaliation. The case also reflects the tenaciousness and moral certainty of her husband. Harry Henderson had joined the federal government after working as a business executive in the South. When he protested the waste of tens of millions of dollars of government funds and the severe impact on the environment, his supervisors ordered him to remain silent and threatened to ruin his career if he spoke outside the agency. When he would not retreat, they initiated a campaign to isolate and later to dismiss him. Here is how his wife described the impact on him:

> Nineteen years ago, I married a man who was outgoing, secure, bold, and optimistic. About his career he was self-confident, enthusiastic, and ambitious. As husband he was interested in and supportive of my activities. Later, as father, he was involved in the lives of his five children and made every effort to spend as much time with them as possible.
>
> From an outgoing person who was involved in the interests and activities of his family, he became withdrawn, spending whatever hours he could in isolation, poring over and over his documents, compulsively reading and rereading every memo dealing with his work situation. When he and I did sit down to talk to each other, he could speak of nothing but what was happening at work and what his supervisors were doing to him, with the pain and suffering he was going through evident in the slump of his shoulders and strained quality of his voice.
>
> All of this, of course, had an effect on the rest of the family. We were afraid to approach Harry with our own needs and concerns, having come to expect his rejection and withdrawal because he no longer had time for or interest in us. One of our children was referred for psychological counseling. I found it necessary to seek work outside the home, having to escape from his oppressive presence and influence. His sleeplessness disturbed our night's rest. We all observed the profound effect of his work situation upon him: the bold man become fearful and intimidated; the aggressive person become reticent and insecure; the optimist become hopeless; the relaxed and outgoing person become tense, withdrawn, and isolated; the well-rounded man become obsessive, paranoid, and neurotic. No longer

was he the loving spouse and father. He was the stranger who, although living among us, was not with us. (Glazer and Glazer 1989, p. 154)

Yet, despite all of this, Harry Henderson fought on with the assistance of the Government Accountability Project (GAP) which had taken his case because of its significance and because of their belief that Henderson would not back off. They were right. With GAP representing him, Henderson secured the interest of a congressional committee which held hearings on his allegations. Newspaper articles and television programs featured the seriousness of his testimony of government and industry collusion. Henderson had overcome his fear and disillusionment. His very obsessivenesses ultimately made him an unwavering adversary against government waste. He came to represent the employee who epitomizes devotion to serving the public rather than loyalty to the bureaucracy.

Employees in private industry are susceptible to even harsher retaliation and more paralyzing fear. Chuck Atchison worked as a quality assurance inspector in the nuclear industry, and after reporting several construction problems to the company, he was dismissed for overstepping his area of responsibility. Despite vindication from the Nuclear Regulatory Commission and later from the Department of Labor, he was out of a job. His whistleblowing action led to his blacklisting in the industry and to deteriorating economic circumstances.

Everything that wasn't nailed down [with] the mortgage was sold. We lost our Visa and MasterCard rights and our gasoline credit cards. Finally we lost the house in July 1983. My wife, Jeanne, has always been employed as a secretary-clerk-bookkeeper. That was the only thing that really kept us going. We let someone take over the payments on the house and found a trailer we could take over the payments on.

The company's reach seemed so pervasive that he even feared physical assault. The ambiguous circumstances of Karen Silkwood's automobile crash ten years earlier in 1974 reinforced his fear of possible organizational, revenge:

Silkwood hit the headlines again. I became paranoid if things happened like a car following me. I'd make several turns and the car would keep up with me. I feared that the company could hire gunmen that would kill someone for big dollars and get back across the border without anyone knowing it. (Glazer and Glazer 1989, pp. 146–47)

Atchison had now lost his job, his home, his credit rating, his sense of personal safety, and his self-esteem as a breadwinner. Forced to leave their familiar surroundings and to live in a mobile home without most of their possessions, the family no longer felt like respected members of the community. Atchison became a living symbol to other workers of the cost of resisting large corporations. The pressure and humiliation penetrated deeply into his sense of self. He dreamed of striking a major blow against his adversaries when in reality he had to settle for the small pleasure of knowing that they had suffered a few defeats in their rush to obtain a license and make good on their investments:

> My emotions went the full gamut from deep depression to hostility. Now most of that part is gone. The main emotion I still get is tickled to death if I see an article in the paper that makes them look a little bit worse as they go along.

Although the costs were very painful, Atchison was not content with these small victories. To fight his case of illegal firing, he contacted the Department of Labor and the press, engaged lawyers, worked with the Government Accountability Project. The reprisals enacted against him for whistleblowing extracted a heavy toll but simultaneously resulted in his developing a new reference group of environmentalists organized to fight against unsafe nuclear plants. He became a principal witness in the campaign of a local grassroots safe-energy group against the plant. He was nationally recognized in 1984 when the Government Accountability Project and the Christic Institute nominated him and several other whistleblowers as the first winners of the Karen Silkwood Award for exposing dangerous working conditions (Glazer and Glazer 1989, pp. 228–29). The citation that accompanied the award was yet another sign that the blacklisting had failed to silence him. Finally in 1988 the company came to an unprecedented agreement with Atchinson and other whistleblowers, and publicly recognized that they had been correct in raising safety concerns. Atchinson had been fully vindicated.

The Allure of Sociology

Sociology can be a significant contributor to the achievement of such vindication. By seeking out these environmental crusaders, employee whistleblowers, and other courageous people, by interviewing them and analyzing their cases from a personal, sociological, and cultural perspective, we provide both publicity and understanding of their accomplishments. By consulting with them as they pursue their cases, we sociologists can contribute useful expertise. By serving on national advisory committees, we assist their battles with the prestige and recognition of our academic positions. By teaching about them, we give our students moral exemplars in a world where irresponsibility and destruction too often hold sway.

By integrating the lives of the crusaders into our theories of resistance to abusive authority, we can emphasize the centrality of moral courage in the building and maintaining of a viable, accountable, and democratic society. . . .

Note

Glazer, Myron Peretz, and Penina Migdal Glazer. 1999. "On the Trail of Courageous Behavior." *Sociological Inquiry* 69, no. 2 (Spring): 276–95.

References

Bratislava-Nahlas. 1987. Unpublished report. Bratislava: Slovak Union of Landscape and Nature Protectors.

Erikson, Kai. 1966. *Wayward Puritans: A Study in the Sociology of Deviance.* New York: Wiley.

Flarn, Helena. 1996. "Anxiety and the Successful Oppositional Construction of Societal Reality: The Case of Kor." *Mobilization* 1: 103–21.

Freedman, Alix M., and Suein I. Hwang. 11 July 1997. "How Seven Individuals with Diverse Motives Halted Tobacco's Wars." *The Wall Street Journal.*

Garland, Anne Witte. 1988. *Women Activists.* New York: The Feminist Press.

Glazer, Myron Peretz, and Penina Migdal Glazer. 1989. *The Whistleblowers: Exposing Corruption in Government and Industry.* New York: Basic Books.

Glazer, Penina Migdal, and Myron Peretz Glazer. 1998. *The Environmental Crusaders: Confronting Disaster and Mobilizing Community.* University Park: The Pennsylvania State University Press.

Hallie, Philip. 1979. *Lest Innocent Blood Be Shed.* New York: Harper and Row.

Jackall, Robert. 1988. *Moral Mazes.* New York: Oxford University Press.

Kirp, David L. 1989. *Learning by Heart.* New Brunswick: Rutgers University Press.

Milgram, Stanley. 1974. *Obedience to Authority.* New York: Harper and Row.

Putnam, Robert D. 1996. "The Strange Disappearance of Civic America." *The American Prospect* 24:34–48.

———. 1993. "Bowling Alone: America's Declining Social Capital." *Journal of Democracy* 6:65–78.

Rachman, S. J. 1990. *Fear and Courage.* New York: W. H. Freeman.

Russell, Diane E. H. 1989. *Lives of Courage: Women for a New South Africa.* New York: Basic Books.

Schmid, Thomas W. 1985. "The Socratic Conception of Courage." *History of Philosophy Quarterly* 2:113–30.

Stout, David. 21 June 1997. "Ex-Tobacco Official Enjoys the Aftermath of the Deal." *The New York Times.*

Swidler, Ann. 1986. "Culture in Action: Symbols and Strategies." *American Sociological Review* 51:273–386.

Walton, Douglas N. 1986. *Courage.* Berkeley: University of California Press.

Index

natural gas. *See* petrochemical industry

natural sciences, xvii

nature: and corporate image, 266–67; and culture, 275; decontextualization of, 260–61; humanity's relationship to, 265–66, 297–98; institutions, non-profit, and corporations, 267–68; man-made changes to, and laws of ecology, 10–11; and mass media, 267; portrayal as carrier of human emotions, 261–62; vs. self-regulating market, 12; and social class, 263–65; and society, 288–89; television and perceptions of, 296. *See also* counterecological production

NDELA (N-nitrosodi-ethanotamine), 303

neighborhood revitalization. *See* Oakland's Fruitvale Transit Village

NEP (new environmental paradigm), 97

NEPA (National Environmental Policy Act), 327

networks of power, alternative, 455

neutrality of science, 329

new environmental paradigm (NEP), 97

Newman, Penny, 456, 459–60

Newton, J., 147

New York's Love Canal, xiv

NHANES III (Third National Health and Nutrition Examination Survey), 121

Niagara, New York, xiv

Nigeria, 247

Nike, 442

nitrogen, 420–21

nitrosamines, 303

N-nitrosodi-ethanotamine (NDELA), 303

noise pollution, 75

No Lead (North Lakes Environmental Action Defense Group), 115

nonprofit nature institutions, 267–68

nonrenewable resources, 149–50

North America, arrival/acceptance of tomatoes in, 207

Northern countries, 229–30

Northern Ontario Tourist Outfitters (NOTO), 394, 398, 400

North Lakes Environmental Action Defense Group (No Lead), 115

Northwest Brazil Integrated Development Program. *See Polonoroeste*

Nuclear Regulatory Commission (NRC). *See CANT v. LES*

nuclear technology, 280, 286–88, 456–58. *See also* Bailie, Tom

Oakland's Fruitvale Transit Village, 73–86; civic environmentalism, 85–86; environmental quality, community, and transportation, 74–78; overview, 73; restoring public access to waterfront, 84–85; road to recovery, 78–84

Oak Ridge Peace-Making Alliance (ORPAX), 346–48

O'Brien, Mary, 321

occupational hazards. *See* recycling

Occupational Safety and Health Administration (OSHA), 160–61

ocean disposal of contaminated silt, 326–27

O'Connor, James, 20

OFAH. *See* Ontario Federation of Anglers and Hunters

Office of Surface Mining Reclamation and Enforcement, U.S., 24–25

oil industry: *Exxon Valdez* spill, xiii; importation of and autos in Southern hemisphere, 230. *See also* Guadalupe Dunes oil spill; petrochemical industry

oil spill. *See* Guadalupe Dunes oil spill

old-growth forests, 148

O'Leary, Daniel J., 139

Ontario. *See* hunting in Ontario

Ontario Federation of Anglers and Hunters (OFAH), 394; Danson's representation of, 396–97; and poaching, 400; and social class, 399–400

Ontario Produce, 218–19

orcas, Sea World's use of, 265–66

Oregon, 147

organizational coupling, loose, 180–81

About the Editors

LESLIE KING is an assistant professor at Smith College, where she has a joint appointment with the Department of Sociology and the Environmental Science and Policy Program. Leslie teaches courses on world population, environmental sociology, and globalization. Her research examines how ideologies of gender, nationalism, and race/ethnicity are implicated in the construction and implementation of population policies. Leslie's recent articles have appeared in *Mobilization, Ethnic and Racial Studies,* and the *European Journal of Population.* She is currently working on a project that examines debates over immigration policy among environmentalists in the United States.

DEBORAH MCCARTHY is an assistant professor in the Sociology and Anthropology Department at the College of Charleston specializing in environmental sociology, social movements, and urban studies. Deborah has several years of experience teaching a wide range of sociology and environmental studies courses (including environmental sociology, transportation and ecojustice, urban studies, environmental justice, and brownfield redevelopment). She has published pieces on the political economy of the environment and on the relationships between philanthropic organizations and grassroots environmental groups. Her most recent article, published in *Sociological Inquiry* 2004, is titled "Environmental Justice Grantmaking: Elites and Activists Collaborate to Transform Philanthropy." Her current projects include a study of migrant farm workers' risk perception in South Carolina and a study of social justice and public transportation in the Charleston area.